FOREST CONSERVATION GENETICS

Principles and Practice

Dedication

 This book is dedicated to the memory of Dr Abdou-Salam Ouedraogo. Dr Ouedraogo, known to all his friends and colleagues simply as 'Abdou', was committed to the conservation of genetic diversity in forest tree species. While working for the International Plant Genetic Resources Institute (IPGRI) for most of the last decade, Abdou was responsible for establishing and managing a highly successful program of research on the Conservation of Forest Genetic Resources around the world. In his energetic pursuit of this goal, he visited and worked with researchers in many countries. Abdou was a unique individual, and one whose enthusiasm, humour and charm were such that he was unforgettable to all who met him.

Much of the information about tropical forest genetic resources in this book can be attributed in some way to Abdou's efforts in developing his research program and in encouraging others. He was involved in the initial discussions about the scope of this book and was a strong supporter of its aims. We can only hope that the book serves to further Abdou's vision of conserving forest genetic resources, especially in the tropics, for the benefit of future generations.

FOREST CONSERVATION GENETICS

Principles and Practice

A. Young, D. Boshier & T. Boyle (Editors)

National Library of Australia Cataloguing-in-Publication entry

Forest conservation genetics : principles and practice.

   Bibliography.
   Includes index.
   ISBN 0 643 06260 2

   1. Forest genetic resources conservation.
   I. Young, Andrew.
   II. Boyle, Timothy James Butler.
   III. Boshier, David.

   634.956

Published exclusively in Australia and New Zealand, and non-exclusively in other
territories of the world (excluding Europe, Africa, the Middle East and South America), by:

CSIRO PUBLISHING
PO Box 1139 (150 Oxford Street)
Collingwood VIC 3066
Australia

Tel: (03) 9662 7666 Int: +(613) 9662 7666
Fax: (03) 9662 7555 Int: +(613) 9662 7555
Email: sales@publish.csiro.au
Website: http://www.publish.csiro.au

Published exclusively in Europe, Africa, the Middle East and South America, and non-exclusively
in other territories of the world (excluding Australia and New Zealand), by CABI Publishing,
a Division of CAB International, with the ISBN 0 85199 504 7.

CABI Publishing
Wallingford
Oxon OX10 8DE
United Kingdom

Tel: 01491 832 111 Int: +44 1491 832 111
Fax: 01491 829 292 Int: +44 1491 829 292
Email: publishing@cabi.org
Website: http://www.cabi.org

Cover design by James Kelly
Cover photographs (from left to right) courtsey of: NASA — Goddard Space Flight Centre, Scientific
   Visualization Studio; Colin Hughes; Neil Mitchell; and Kathy Freemark
Typeset by Desktop Concepts
Edited by Alexa Cloud-Guest
Index by Russell Brooks

Printed in Australia

CONTENTS

Section II: Genetic Processes

Section V: Monitoring, Socioeconomics and Policy

Forests are biologically very diverse. They are important habitat for many of the world's species and provide significant ecosystem services in terms of nutrient and water cycling, prevention of erosion and sequestration of carbon, as well as performing a range of other physical and biological functions. Forests also provide timber and a range of non-timber forest products, such as fruits, that are critical for human survival. Nevertheless, deforestation continues at alarming rates, and sustainable management of forest ecosystems for the future has become a global environmental imperative. Within this, effective management of forest genetic resources is a key element in future forest conservation. Heritable genetic differences among individuals—genetic variation—are the raw material upon which selection acts, and the basis of evolutionary change. From a more directly utilitarian viewpoint, this natural diversity of size, form and fecundity is the variation upon which tree improvement programs act. Genetic diversity is essential for both the long-term stability and the short-term productivity of forest ecosystems.

Owing to the advent and development of molecular genetic markers over the last 20 years, we now know substantially more about the genetic structure of forest tree species and the spatial and temporal dynamics of genetic processes, such as mating and gene flow, than ever before. At the same time remote sensing technologies, coupled with geographical information systems, have begun to provide sophisticated methods for rapid inventory of forest cover across large areas of the planet, allowing previously unthinkable opportunities for detailed monitoring of the status of the global forest resource. Similarly, development and use of economic and sociological methods and models now allow critical analysis of the importance of forests to human well being.

The challenge for scientists, forest managers and policy makers alike is in working out how to integrate these complex and rather disparate sets of information, so as to allow improved decision making about the management of forests that will ensure the long-term security of the genetic, ecological and socioeconomic resource they represent—that is, how do we turn our knowledge of the amount, distribution and value of a resource into efficient resource management?

It was an intent to raise awareness of the value of integrating genetic, ecological, sociological and economic information into management of forest genetic resources that prompted us to develop and run a two-week professional course on 'Forest Conservation Genetics: Principles and Practice', co-organised by the Department of Forestry at The Australian National University, CSIRO Plant Industry, The Oxford Forestry Institute and The Centre for International Forestry Research. It is from the course, which has now run for four years, that this edited volume developed—we discovered in the first year that there was no comprehensive text that dealt with the diverse range of issues relevant to the discussion of forest conservation genetics.

This edited volume attempts to fill that gap. The book brings together in five sections a logical sequence of information on: basic genetic principles and current genetic marker technologies; population genetic processes; current threats to genetic diversity, both *in situ* and *ex situ* and finally; genetic monitoring, socioeconomics and policy development. This book is aimed at a wide audience, and as such we have tried to ensure that there is enough background material for those who find themselves out of their particular discipline in some chapters to still gain useful insights, but that there is enough detail to be of interest to the specialist.

We are indebted to our many colleagues who have given generously of their time and expertise to write the chapters in this book. Without their contributions we would have been unable to present the wide range of topics that is a feature of this synthetic discipline. The authors have often taken different approaches to organising and presenting information—however, wherever possible, we have emphasised the use of case studies to provide practical illustrations of concepts being discussed. To this end we have also included in this book, as a CD-ROM, the population genetics analysis software package POPGENE, developed by Drs Francis Yeh, Timothy Boyle and Rongcai Yang. This computer package provides readers with a hands-on tool for the analysis and interpretation of a wide range of genetic data. To provide further assistance to the reader, we have included a list of websites related to population genetics and conservation.

As well as to the contributing authors, this book owes a good deal to the students of the Forest Conservation Genetics course who stimulated our interest by asking the obvious questions to which we often had no answers. Finally, we would also like to thank the many colleagues who acted as reviewers for the 22 manuscripts that make up this volume, along with Alexa Cloud-Guest for editorial assistance.

Websites related to population genetics, conservation

There are numerous websites containing information related to forest conservation genetics and therefore only a few key ones are listed below. These addresses also provide links and search options that will facilitate finding other currently available sites.

Data analysis software

http://wbar.uta.edu/software/software.htm — EP-GED software

http://www.ensam.inra.fr/URLB/ — INRA software

http://alleyn.eeb.uconn.edu/gda/ — GDA Software for the analysis of discreet genetic data

http://www.nceas.ucsb.edu/papers/geneflow/software/index.html — Gene flow related software

http://www.ualberta.ca/~fyeh/index.htm — POPGENE population genetics software

General

http://www.geocities.com/CapeCanaveral/Lab/6572/pgchome.html — Population genetics club

http://www.metla.fi/info/vlib/forestgen/consrvat.htm — Forest genetics and tree breeding

http://www.consecol.org/Journal/ — Conservation Ecology, free on-line journal

http://www.ibin.org/cbd.htm — Indigenous people and biodiversity convention

http://userzweb.lightspeed.net/~jpthomas/sgen2.html — Molecular ecology in a spatial framework

http://iufro.boku.ac.at/ — IUFRO, International Union of Forestry Research Organizations

http://www.iimi.org/CIFOR/index.html — CIFOR, Centre for International Forestry Research

http://www.iimi.org/IPGRI/Index.htm — IPGRI, International Plant Genetic Resources Institute

Dr Sally Aitken
Department of Forest Sciences, University of British
Columbia, Vancouver, BC, Canada V6T 1Z4

Dr W. Thomas Adams
Department of Forest Science, Oregon State
University, Corvallis, OR 97331-5752, USA
email: adamsw@fsl.orst.edu

Dr David Boshier
Oxford Forestry Institute, Department of Plant
Sciences, Oxford University, South Parks Road,
Oxford OX1 3RB, UK
email: david.boshier@plant-sciences.oxford.ac.uk

Dr Timothy Boyle
United Nation Development Programme, One UN
Plaza, New York, NY 10017, USA
email: tim.boyle@undp.org

Dr Anthony Brown
CSIRO Plant Industry, GPO Box 1600, Canberra,
ACT 2601, Australia
email: anthony.brown@pi.csiro.au

Dr Jaroslaw Burczyk
Department of Biology and Environmental Protection,
Pedagogical University, Chodkiewicza 30, PL 85-064
Bydgoszcz 1, Poland

Dr Margaret Byrne
CALM Science Division, WA Herbarium,
Locked Bag 104, Bentley Delivery Centre, WA 6983,
Australia
email: margaretb@calm.wa.gov.au

Dr Shanna Carney
Department of Biology, Colorado State University,
Fort Collins, CO 80523, USA
email: secarney@lamar.colostate.edu

Dr Yousry El-Kassaby
PRT Management Inc., 4-1028 Fort St, Victoria, BC,
Canada V8U3K4
email: yelkassaby@prtgroup.com

Dr Thomas Enters
16 Jalan Tan Jit Seng, 11200 Penang, Malaysia
email: lctesea@pc.jaring.my

Dr Thomas Geburek
Institute of Forest Genetics, Federal Forestry
Research Centre, Hauptstraße 7,
A-1140 Vienna, Austria
email: thomas.geburek@fbva.bmlf.gv.at

Dr Jeffrey Glaubitz
CSIRO Forestry and Forest Products, PO Box E4008,
Kingston, ACT 2604, Australia
email: jeff.glaubitz@ffp.csiro.au

Dr James Hamrick
Department of Botany and Genetics, University of
Georgia, Athens, GA 30602, USA
email: hamrick@dogwood.botany.uga.edu

Dr Craig Hardner
CSIRO Plant Industry, Cunningham Laboratories,
306 St Carmody Rd, St Lucia, Qld 4067, Australia
email: craig.hardner@pi.csiro.au

Dr Peter Kanowski
Department of Forestry, Australian National
University, Canberra, ACT 0200, Australia
email: peter.kanowski@anu.edu.au

Dr Mathew Koshy
Department of Forest Sciences, University of British
Columbia, Vancouver, BC, Canada V6T 1Z4
email: koshy@interchange.ubc.ca

Dr Helmi Kuttinen
Department of Biology, University of Oulu, PO Box 3000, FIN-90401, Oulu, Finland
email: helmi.kuttinen@oulu.fi

Dr Jeffrey McNeely
IUCN—The World Conservation Union, Biodiversity Unit, rue Mauverney, 28, CH-1196 Gland, Switzerland
email: jam@hq.iucn.org

Dr Gavin Moran
CSIRO Forestry and Forest Products, PO Box E4008, Kingston, ACT 2604, Australia
email: gavin.moran@ffp.csiro.au

Dr Gerhard Müller-Starck
Section of Forest Genetics, Faculty of Forest Science, Technical University of Munich Am Hochanger 13, D-85354 Freising, Germany
email: mueller-starck@forst.uni-muenchen.de

Dr Brian Murray
School of Biological Sciences, The University of Auckland, Private Bag, 92019 Auckland, New Zealand
email: b.murray@auckland.ac.nz

Dr Gene Namkoong
Department of Forest Sciences, University of British Columbia, Vancouver, BC, Canada V6T 1Z4
email: gene@unixg.ubc.ca

Dr John Nason
Department of Biological Sciences, University of Iowa, Iowa City, IA 52242-1324, USA
email: john-nason@uiowa.edu

Wickneswari Ratnam
Department of Genetics, Faculty of Life Sciences, Universiti Kebangsaan, Malaysia, 43600 UKM Bangi, Selangor, Malaysia
email: wicki@pkrisc.cc.ukm.my

Dr Loren Rieseberg
Department of Biology, Indiana University, Bloomington, IN 47405, USA
email: lriesebe@bio.indiana.edu

Dr Outi Savolainen
Department of Biology, University of Oulu, PO Box 3000, FIN-90401, Oulu, Finland
email: outi.savolainen@oulu.fi

Dr Roland Schubert
Section of Forest Genetics, Faculty of Forest Sciences, University of Munich (LMU) Am Hochanger 13, D-85354 Freising, Germany
email: schubert@forst.uni-muenchen.de

Dr Frank Vorhies
IUCN – The World Conservation Union, Biodiversity Unit, rue Mauverney, 28, CH-1196 Gland, Switzerland
email: fwv@hq.iucn.org

Dr Diana Wolf
Department of Biology, Indiana University, Bloomington, IN 47405, USA

Dr Francis Yeh
Department of Renewable Resources, University of Alberta, Edmonton, AB, Canada T6G 2H1
email: francis.yeh@ualberta.ca

Dr Andrew Young
CSIRO Plant Industry, GPO Box 1600, Canberra, ACT 2601, Australia
email: andrew.young@pi.csiro.au

INTRODUCTION

Andrew G. Young, David H. Boshier and Timothy J. Boyle

INTRODUCTION

The topics of biodiversity and conservation are high on both political and scientific agendas of many countries and at a global level, with much recent attention focused on the world's forests. Forests are the most biologically diverse terrestrial ecosystems within which, trees are central to the habitat and environment of other plant and animal species. Knowledge of the biology of tree flora is therefore critical in providing a sound scientific basis to the conservation and management of the world's forest resources. By its very nature biodiversity, and hence its conservation, is complex and multi-faceted. With respect to forests, biodiversity may be viewed at the level of the forest community (ecology—ecosystems, habitats) and constituent populations of species (population genetics and demography—species, populations, genomes). The emphasis in conservation has, however, often been one dimensional, traditionally focused on the establishment of national parks for the protection of fauna, often megafauna, or particular ecosystems with little regard for hierarchy of organisation and the spatial and temporal dynamics of biological processes.

More recently, however, there have been several conservation initiatives that have included forest genetic resources at some level (see Chapter 22). Awareness that there is a genetic aspect to forest conservation is growing, although the level of understanding of the principles behind the issues and the implications for day-to-day conservation and management are often limited. Even though most natural resource planners would probably recognise genetic processes as essential components of ecosystem or species stability and adaptability, this is usually an implicit assumption, rather than a subject of explicit provision in decision making and planning. Often because genetic diversity is less visible, its possible importance is ignored, which may ultimately compromise a particular conservation program. Along with ecological considerations, genetics is a major factor in the design and implementation of species-based conservation strategies if both short-term and long-term goals of effective resource management and utilisation, and maintenance of evolutionary flexibility, are to be achieved.

Classically in the conservation of forest genetic resources, perhaps in contrast to crop genetic resources, much emphasis has been on *in situ* conservation. However, although State-managed or controlled national parks and forest reserves occur in almost every country, few have been established according to population genetic principles—their design and management has usually been determined by political, social and economic constraints. Their location—not atypically on slopes, sites of lower fertility and in stands of lesser economic value—bias their composition and limit their value for genetic resource conservation. For example, extensive removal of trees and forest from the best agricultural land has occurred to such an extent that some species are now thought of as characteristic of river courses, whereas previously their distribution was far more widespread. Large populations and indeed forest types have completely disappeared, leading to a restriction of species and genotypes within species, biased sampling of species' gene pools and subsequent possible changes in their ability to adapt on an evolutionary time scale.

Clearly, adequate actions for conservation of forest genetic resources must be wide ranging and complementary, both at species, community and ecosystem levels, incorporating both *in situ* and *ex situ*, on-reserve and off-reserve actions, as and where appropriate. Consequently some forest managers are now advocating and developing management strategies that give priority to conserving genetic diversity within production systems, or that recognise the importance of genetic considerations in achieving sustainable management. Management needs to be sustainable not only in ecological, economic and social, but also genetic terms. Few, if any, forest ecosystems are unmodified by human activities of some type, and management based explicitly on such principles requires knowledge of the effects, not necessarily adverse, of human interaction on associated gene pools, through processes such as deforestation, exploitation, fragmentation, habitat alteration, environmental deterioration, translocation and domestication.

In comparison with many crops, most forest tree species have had, until recently, little or no history of domestication, while their longevity produces overlapping generations, thereby reducing effective population sizes, and increasing minimal area requirements for conservation. Forest trees, with their variable geographical distributions, extreme life history characteristics (e.g. most long-lived plants, greater opportunity for accumulation of mutations, range of mating and dispersal systems), and the wide variety of stresses to which they are subject (e.g. pollution, logging, fragmentation) can be considered as paradigms of genetic conservation. Subpopulation structure, interpopulation variation and gene flow, all

effect variation and adaptation. From both conservation and utilisation viewpoints we need to know not only about the extent of genetic variation (allelic richness) within a species but also how it is distributed (allelic evenness)—what area retains sufficient genetic variation to constitute a viable population over the long term? Since the 1970s the development of isozymes and more recently DNA markers has facilitated more direct study of the levels and patterns of genetic variation in forest trees, providing answers to such questions.

Although there is in some cases a growing awareness of a genetic aspect to conservation, relatively little information on conservation genetics is reaching forest scientists, managers and policy makers so that it can be translated into practice. Often ecological, social or economic considerations may define processes of any particular conservation strategy. The challenge for conservationists, geneticists and foresters alike is, therefore, to establish the circumstances under which genetic considerations, although often unseen, may become limiting to the overall conservation goals and objectives of a particular program, reserve or species.

Despite the importance of the subject, the magnitude and imminence of the threats and a burgeoning scientific literature fuelled partly by rapid advances in genetic marker technologies, there is no comprehensive reference text that draws together the principles of forest conservation genetics. This book aims to fill this gap, introducing readers, with a wide variety of backgrounds, to the principles and practice of forest conservation genetics. This is done through peer-reviewed chapters, contributed by specialists on each particular topic, in the following sections: basic genetic principles; genetic processes; threats to *in situ* genetic conservation; domestication and *ex situ* genetic conservation; and monitoring, socioeconomics and policy. We envisage that someone without a specialised knowledge of genetics will be able to work through the logical sequence of topics and gain an understanding of the principles involved and how they may impinge on day-to-day planning and management decisions. The book is primarily aimed at undergraduate students of biology, ecology and forestry and graduate students of forest genetics, resource management policy and conservation biology. We hope that it will prove useful for those teaching courses in these fields and as such help to increase the awareness of genetic factors in conservation and sustainable management, in both temperate and tropical regions.

Acknowledgment

Work by David Boshier on this publication is an output from research partly funded by the United Kingdom Department for International Development (DFID) for the benefit of developing countries. The views expressed are not necessarily those of DFID. Projects R5729, R6516 Forestry Research Programme.

SECTION I
BASIC GENETIC PRINCIPLES

Conservation genetics is a synthetic discipline whose major goals are to understand the factors influencing the current and future dynamics of genetic processes, how these affect the distribution of genetic variation, and what this means for forest resource management within a social and economic context. To develop and implement effective genetic conservation strategies it is necessary to integrate information drawn from several areas of expertise including genetics, ecology, sociology and economics. Even within the general field of genetics itself, information comes from a range of subdisciplines including population genetics, quantitative genetics and molecular genetics. The three chapters in Section I provide an introduction to basic genetic principles and an overview of the most powerful genetic tools—biochemical and molecular genetic markers. This section is likely to be of most value to those readers with little or no background in genetics.

In Chapter 2, Brian Murray, Andrew Young and Timothy Boyle describe the structure, function and regulation of genes, and their organisation into nuclear, chloroplast and mitochondrial genomes. Processes of gene replication (mitosis) and recombination (meiosis) are described generally, along with discussion of mutation and heritable alteration at both the gene and chromosome level. These processes provide the raw material for evolutionary change—genetic variation. In Chapter 3, Francis Yeh introduces the fundamentals of population genetics, which is concerned with the quantitative analysis of genetic variation within populations. Concepts such as the Hardy-Weinberg Principle and gametic-disequilibrium are introduced, along with basic mathematical developments of the effects of processes such as random genetic drift, migration and selection on allele and genotype frequencies. Common genetic statistics used to measure genetic variation and population differentiation are also defined.

In the final chapter of this section, Chapter 4, Jeffrey Glaubitz and Gavin Moran provide a comprehensive introduction to biochemical and molecular marker technologies. In each case, ranging from allozyme markers through to restriction fragment length polymorphisms

(RFLPs), and the plethora of new polymerase chain reaction (PCR)-based markers (e.g. amplified fragment length polymorphisms, AFLPs), they describe how the techniques work, what kind of information they provide (e.g. dominant or codominant, patterns of inheritance) and what they are most effectively used for (e.g. assessment of genetic variation, paternity analysis, genetic mapping).

BASIC GENETICS

Brian G. Murray, Andrew G. Young and Timothy J. Boyle

SUMMARY

This chapter briefly outlines the structure, function and organisation of plant genes and genomes and introduces the basic processes involved in the replication and recombination of the genetic material. An understanding of these processes leads to an appreciation of how they function in the transmission of biological information. Most of the processes ensure that information is transmitted unchanged from cell to cell and from parent to progeny. However, mutations, heritable alterations at the gene and chromosome levels, provide the raw material for evolutionary change in populations.

2.1 Introduction

Genetics is concerned with the transfer of information in biological systems. This is a massively complex process and only a brief outline can be provided here. Detailed treatments are available in many excellent texts, two examples of which are Griffiths *et al.* (1996) and Brown (1999). The transmission of this information must balance the need for fidelity against that for novelty. In other words, it is important that the correct and exact information is passed from cell to cell within an organism but there also must be the opportunity for new information to arise and be incorporated into the gene pool of the species and for the existing information to be recombined (or rearranged) in the individuals of a population. So, in general, all the cells of a single plant contain copies of the same genes but if we look at populations of plants of a species we find that they contain a variety of different genotypes.

The physical basis for these contrasting systems is seen in the replication of the genetic material and the behaviour of the chromosomes at cell division. In somatic cells, the process of mitosis ensures that the two daughter cells derived from the initial parental cell contain exact copies of the parental chromosomes. Each chromosome must replicate itself to give rise to two chromatids before mitosis can start (see Chapter 2.6). These chromatids divide at anaphase of mitosis and consequently the two new cells that are formed at the end of the division are genetically identical (Fig. 2.1). In the production of sex cells or gametes there is an opportunity for new variation to be produced via the process of meiosis. Here there is a pairing of the replicated chromosomes, one derived from each of the parents of the plant, so that there is an exchange of genetic material by breakage and reunion, the process of chiasma formation, between the paired chromosomes. One member of each of the pairs of chromosomes then moves to opposite poles of the cell so that the new cells that are formed are genetically different from the original one (Fig. 2.2). Detailed treatments of mitosis and meiosis can be found in most genetics texts or specialised monographs (John 1990; Griffiths *et al.* 1996).

Even though all the cells of a single plant have the same genetic make up or genotype, it is still possible for

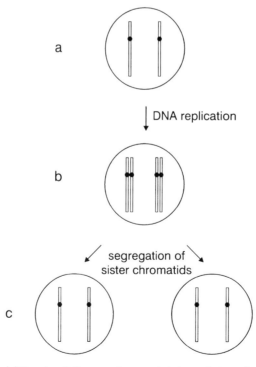

Figure 2.1 Mitotic cell division. (a) For simplicity, a nucleus containing only two chromosomes is shown. (b) Following DNA replication, the duplicated chromosomes each consist of two identical sister chromatids. (c) During mitotic anaphase sister chromatids are separated, resulting in two genetically identical cells.

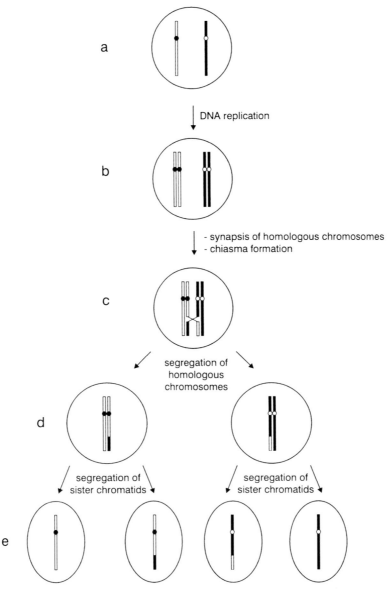

Figure 2.2   Main features of the meiotic process. (a) Meiotic cell prior to DNA replication with two homologous chromosomes. (b) Following DNA replication, the duplicated chromosomes each consist of two identical sister chromatids. (c) Chromosome pairing and chiasma formation takes place. (d) At anaphase I segregation of homologous chromosomes occurs. (e) At anaphase II segregation of sister chromatids occurs.

different parts of the same plant or cuttings derived from a single plant to look different. This is because the appearance of the plant, its phenotype, is the product of the interaction of the genotype with the environment. Thus, in many forest trees we see the production of distinct types of leaves on different parts of the same tree in response to different light intensities. Sun leaves, exposed to maximum sunlight, are thicker and more robust and have more complex arrangements of palisade parenchyma whereas shade leaves are thinner and usually have only a single palisade layer. Similarly, identical cuttings grown in different soils or temperatures may develop into distinct looking plants (Briggs & Walters 1997). This phenomenon, where the same genotype can give rise to different phenotypes, is called phenotypic plasticity: it

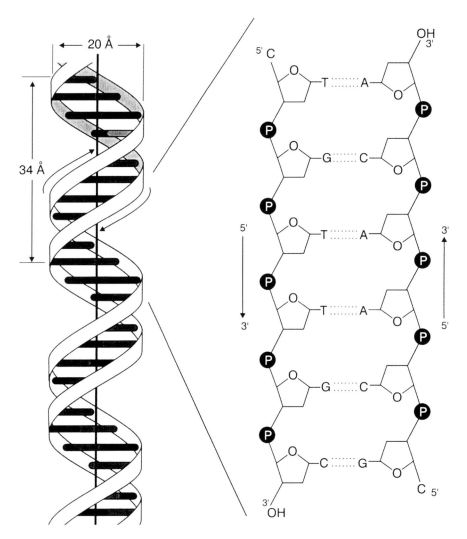

Figure 2.3   Diagram of the DNA double helix. (a) The ribbon-like strands constitute the sugar-phosphate backbones and the horizontal rungs are the nitrogenous base pairs. (b) The complementary pairing of the bases resulting in antiparallel arrangement of the two strands of the helix (from Klug and Cummings 1999).

is clearly an important adaptive process for most plants which are permanently fixed to their substrate and unable to move to a more amenable environment in the way that animals can.

2.2 Genomes and genes

The genome is the total genetic material of an organism. This definition is now broader than the traditional one that described the genome as the total of all genes carried by an individual. This reflects the realisation that the genome consists of more than genes. It is made up of deoxyribonucleic acid (DNA), which forms a double helix made up of two paired

complementary strands each consisting of a sugar phosphate backbone carrying a sequence of four possible bases, adenine (A), thymine (T), guanine (G) and cytosine (C) (Fig. 2.3). The basic unit of organisation is the deoxyribonucleotide, which consists of a single sugar linked to a phosphate group and one of the four bases. The two strands of the helix are held together by hydrogen bonds between the bases. The base pairing is not random, A pairs with T and G with C, so the two strands have complementary sequences (Fig. 2.3). This complementarity allows both strands to serve as templates in a population without the loss of information.

Genes are the units of inheritance whose behaviour is described by Mendel's Laws of segregation and independent assortment. Mendel proposed that in an individual each gene was represented twice and the two components of the gene segregate randomly from each other such that each gamete contains only one representative. These two components are called alleles and individuals may contain two of the same alleles, and are said to be homozygous, or have two different alleles and are heterozygous. Plants such as these with two alternate alleles at a single locus are called diploids (see Chapter 2.7.2). However, individual plants show only a single phenotype, even when they contain different alleles, due to the masking of one allele by the other. One allele is said to be dominant and masks the other which is recessive. The second law states that when two genes are present the segregation of one is independent of the segregation of the other. We now know that the two alleles are carried by pairs of homologous chromosomes derived from the male and female parents and the pattern of inheritance of genes can be understood when it is linked to the meiotic process. Although Mendel was correct in the interpretation of his data, not all genes show simple dominance/recessive relationships nor do all pairs of genes assort independently.

In some situations the heterozygote may show a phenotype that is intermediate between the parental phenotypes thus exhibiting incomplete dominance. Additionally, both alleles may be expressed in the heterozygote, the phenomenon of codominance. Genes may also show epistasis where one gene pair masks or modifies the expression of another. This is best observed in biosynthetic pathways where a gene fails to produce an active enzyme so that the substrate required by the next enzyme in the pathway is not available. In this case, the failure of the first gene masks or hides the action of the second gene and, therefore, is epistatic to the second and any subsequent genes in the pathway. Finally, although a diploid individual is restricted to having only two alternate alleles, different individuals may have alleles not found in others. In other words, within a population one can find large numbers of alleles called multiple alleles, which can give rise to many different phenotypes by their pairwise combination.

Genes can also be present in multiple copies as multi-gene families. Genes in multi-gene families have the same or very similar sequences although they may not all be active in every cell. Some of these inactive copies

are examples of pseudogenes that have lost their ability to function because of mutation. Multi-gene families may be clustered in tandem repeats in specific regions of the chromosome, an example is the genes that code for ribosomal RNA, which are located in the nucleolar organiser regions. Alternatively, multi-gene families may also be dispersed throughout the genome, an example of this is the *zein* family in maize that encode a series of kernel storage proteins and are present in about 100 copies in the maize genome.

Many pairs of genes do not segregate independently because they are on the same chromosome and therefore will only recombine when a chiasma forms between them during meiosis. Since each homologous pair of chromosomes usually has only a small number of chiasmata, the chances of one occurring between two genes is smaller if they are closely adjacent to each other than if they are widely spaced. The process by which genes recombine by the exchange of chromosome segments is called crossing-over. Genes that do not show independent assortment are said to show linkage. The frequency of crossing-over between genes can be used to construct genetic (recombinational) maps and these will show that the total number of linkage groups corresponds to the number of different chromosome pairs.

2.3 Gene structure

A 'typical' gene is made up of a transcription unit or open reading frame, the protein coding part of the DNA, flanked by an initiation or start codon at the beginning of the gene and a termination codon at the end (Fig. 2.4). In addition there are promoter sequences and upstream regulatory regions that control the rate of initiation of transcription at the 'start' of the gene. A typical gene has at least 1000 nucleotides but big differences can be seen between different genes. In most plants, the coding regions of many genes are discontinuous and are interrupted by non-coding regions called introns. These are removed from the mRNA by a splicing process before it is processed into a protein by the ribosomes.

2.4 Genome structure and organisation

Plants are unique in having three different genomes: the nuclear genome (nDNA), the chloroplast genome (cpDNA) and the mitochondrial genome (mtDNA). The three genomes show huge differences in size,

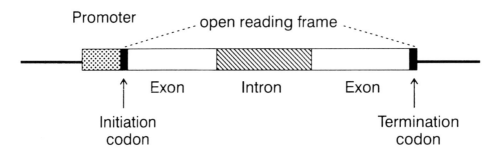

Figure 2.4 The structure of a typical gene. Information is read left to right starting with the promoter sequence.

physical organisation and in the number of genes that they contain as well as in the way that they are inherited.

2.4.1 The nuclear genome

The nDNA is the largest of the three genomes in plants and contains most of the plant's genes. The DNA in the nDNA is organised into chromosomes, linear molecules of DNA complexed with specific proteins called histones. These DNA–protein complexes show several orders of coiling that enable the chromosome to be compacted about 5000-fold from interphase to metaphase, allowing the efficient movement of the chromatids or chromosomes at mitotic or meiotic anaphase. The number of chromosomes in somatic cells (the diploid number or $2n$) is highly variable and in plants ranges from $2n = 2$ to $2n =$ about 1300. The size of the chromosomes is also variable and ranges from 1 µm to 20 µm.

The combination of these two variables means that nDNAs can vary greatly from one species to another. In the angiosperms there is a 1000-fold size variation between the largest and smallest genomes whereas in the gymnosperms, which typically have much bigger genomes than angiosperms, the range is only 14-fold (Bennett & Leitch 1998; Murray 1998). The difference in genome size between different species does not reflect differences in the number of genes that are present. It is generally agreed that all plants have about the same number of genes (but see Chapter 2.7.2 regarding polyploidy), somewhere in the range of 50 000 to 100 000, and much recent research has gone into investigating the molecular organisation of nDNAs. Some of the variation in genome size can be accounted for by the presence of multiple copies of genes (see above) but much of the variation consists of DNA that is not transcribed (i.e. it does not code for a product). A wide variety of different repetitive

elements, where the repeats may range from one or a few bases through to several thousand, have been described (Kubis *et al.* 1998; Schmidt & Heslop-Harrison 1998). Many of these repeat sequences have no known function and have been called junk or selfish DNA (Orgel & Crick 1980) but others have clear roles in the structure and function of the chromosomes. For example, short, seven-base repeats occur at the ends or telomeres of the chromosomes of most plants and are essential in replication and in maintaining chromosome integrity.

2.4.2 Organelle genomes

Both the cpDNA and mtDNA are made up of circular DNA molecules that show similarities to prokaryote genomes from which they have evolved (Margulis 1970; Martin 1999). Unlike the nDNA, which is present as a single copy, there are multiple copies of the organelle genomes within each cell. An interesting feature of the organelle genomes is that their major functions, photosynthesis and respiration, involve the products of both organelle and nuclear genes.

2.4.2.1 Chloroplast genome

Among the land plants cpDNAs are remarkably uniform in size and organisation generally falling within a size range of 120 kilobases (kb) to 180 kb. A typical cpDNA consists of one large and one small region of unique sequence DNA separated by the two copies of a large inverted repeat sequence. This repeat is present in most, though not all cpDNAs, and characteristically contains the ribosomal RNA genes required for processing of the organelle genes. Chloroplast genomes typically have 100 to 120 genes, many of which code for components of photosynthesis (Fig. 2.5) and generally lack large amounts of non-coding sequences, so that the genes are tightly

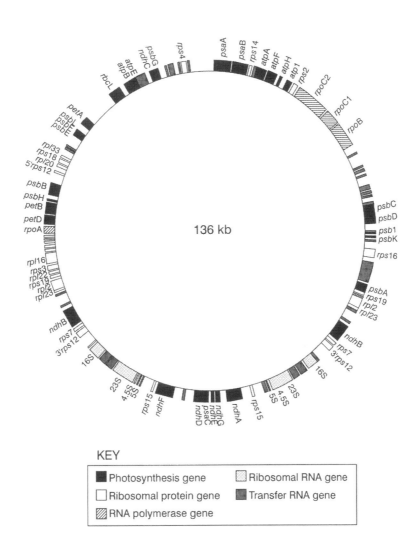

KEY

■ Photosynthesis gene	▦ Ribosomal RNA gene
□ Ribosomal protein gene	▨ Transfer RNA gene
▧ RNA polymerase gene	

Figure 2.5   The *Oryza sativa* L. (rice) chloroplast genome showing the compact nature of this genome (from Brown 1999).

packaged around the chromosome. Introns are present in many chloroplast genes.

2.4.2.2   Mitochondrial genome

In contrast to the cpDNA, the mtDNA is large and highly variable. Even the smallest mtDNAs are greater than 200 kb with most falling within the range of 300 kb to 600 kb but there are examples where the size exceeds 2500 kb. These genomes typically contain large repeat sequence(s) but, unlike the cpDNA, these repeats are highly variable although specific ones may be restricted to groups of populations or species. Recombination between these repeats can create a complex, multipartite genome structure which in its simplest form consists of a master and two smaller circles (Fig. 2.6). Plant mtDNAs typically have 40 to 50 genes, which may or may not have introns, many fewer than cpDNA. The arrangement of mtDNA is highly variable with large non-coding regions, sometimes derived from cpDNA and nDNA, separating the coding regions. Most of the mtDNA genes are involved in producing enzymes associated with glycolysis and electron transport, two important processes in cellular respiration.

2.4.2.3   Inheritance of organelle genomes

There are several unique features of organelle inheritance which are not shared with the nDNA. The nDNA is always biparentally inherited and subjected to intergenerational recombination whereas organelle genomes frequently show uniparental inheritance and can exhibit intragenerational segregation.

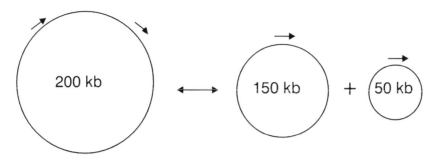

Figure 2.6   The relationship between master and slave genomes of plant mitochondrial DNA. The two smaller circles arise by homologous recombination at the large direct repeats on the master circle (indicated by arrows).

The mitochondrial genome is most often maternally inherited but in several groups of gymnosperms (e.g. Araucariaceae, Cupressaceae, Taxodiaceae) it is paternally inherited (Birky 1995), while biparental inheritance has been observed in species of Pinaceae and Taxaceae (Wagner 1992).

Chloroplast DNA is maternally inherited in most angiosperms but there is evidence for biparental inheritance in about one-third of the genera that have been investigated (Birky 1995) and there are also well-documented examples of paternal inheritance in some species (Chat *et al.* 1999). In conifers, paternal inheritance is common (Wagner 1992).

There is also some evidence of cpDNA variation within individuals of several tree species, such as *Pinus monticola* Douglas ex D. Don that points to the possibility of occasional biparental inheritance (White 1990).

2.5   Genetic code and gene expression

The information a gene contains, the genetic code, is a triplet code, where each group of three bases called a 'codon', specifies a particular amino acid. Most of the triplets code for one of the 20 amino acids found in proteins but one, the initiation codon, is found at the start of the gene and three codons, the termination codons, are found at the end of the coding sequence. As there are 4^3 (64) possible combinations of three bases, and there are only 20 amino acids, there is a good deal of redundancy, with most amino acids specified by more than one codon. Thus, changing one base in the sequence CT**A** to CT**G** has no functional effect, in both cases coding for the same amino acid, leucine. Most of this redundancy exists in the last base of the triplet sequence, a change to which is less likely

to affect the amino acid coded than an alteration to either the first or second position base, which will nearly always result in a change in amino acid.

The flow of information or the way that a gene is expressed can be summarised in basic terms as 'DNA makes RNA makes protein'. At the start of gene expression the two strands of the DNA molecule separate and the process of transcription is initiated such that a complementary copy of messenger RNA (mRNA) is synthesised. This involves the 'opening up' of the chromatin structure so that the enzymes that are responsible for transcription, DNA-dependent RNA polymerases, can access the information in the DNA. These enzymes assemble mRNA from individual ribonucleotides such that the sequence of nucleotides in the DNA dictates the sequence on the mRNA molecule. This mRNA is then exported into the cell cytoplasm where it is translated or processed by the ribosomes into protein. This involves a third type of RNA, transfer RNA (tRNA). There are about 50 different tRNAs in plants and these are capable of linking to particular amino acids in a precise fashion. Each type of tRNA has a specific region of its structure, the anticodon, which is complementary to the codon that specifies a particular amino acid. It is the function of the ribosomes, produced by the ribosomal RNA genes, to bring together the mRNA and tRNA to produce a chain of amino acids or polypeptide.

2.6   DNA replication

Replication is a complex process and involves the production of complementary copies of each of the two strands of the DNA double helix (Fig. 2.7). It is initiated at many points along the chromosome, the origins of replication, with the unwinding of the helix and the creation of two replication forks, since the

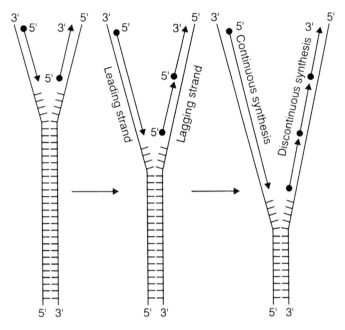

Figure 2.7   DNA synthesis showing that on the lagging strand synthesis is discontinuous resulting in the production of Okazaki fragments, but that on the leading strand it is continuous (from Klug and Cummings 1999).

synthesis of the new strands takes place in both directions from the origin of replication. A RNA primer is needed for the initiation of DNA synthesis since the enzyme involved, DNA polymerase, requires a free 3' end in order to start the production of the new DNA strand. However, as the replication fork moves away from the origin of replication, synthesis is continuous on one strand, the leading strand, and discontinuous on the other, the lagging strand. This is because the DNA polymerase only produces the complementary copy along the existing molecule in the 5' to 3' direction. On the lagging strand, small fragments called Okazaki fragments, about 1000 to 2000 nucleotides long, are produced and these are enzymatically joined following the removal of each RNA primer to produce a continuous DNA molecule (Fig. 2.7).

2.7   Where does genetic variation come from?

There are two sources of genetic variation. The process of meiosis, as we have seen (Fig. 2.2), acts to recombine gene alleles on the same chromosome through chiasma formation, and genes on different chromosomes via the random segregation of chromosome pairs at anaphase I. However, this is only rearranging the genetic variation that exists in the population or species and we need to consider how new variation can come about.

The complexities of the processes of inheritance mean that mistakes occur occasionally. If these errors are heritable they are called mutations and are the mechanism that introduces new genetic variation into the population. There are two main types of mutation: gene mutation and chromosome mutation.

2.7.1   Gene mutation

Gene mutations are changes in the nucleotide sequence of a region of the genome. For example, the replacement of one nucleotide by another or the removal of a nucleotide at almost any position in the gene will alter the triplet code downstream from the change and result in a change in the amino acids specified by the new triplets. Thus, different proteins or no protein will be produced, resulting in a change of phenotype. All genes undergo mutation, some are spontaneous, arising as errors during replication of DNA, whereas others are induced by physical or chemical mutagens. Physical mutagens include ultraviolet and ionising radiation whereas several different groups of chemical mutagens, which may act either directly or indirectly on DNA, are known. Spontaneous mutation rates can differ between different genes within the same species, for example, in

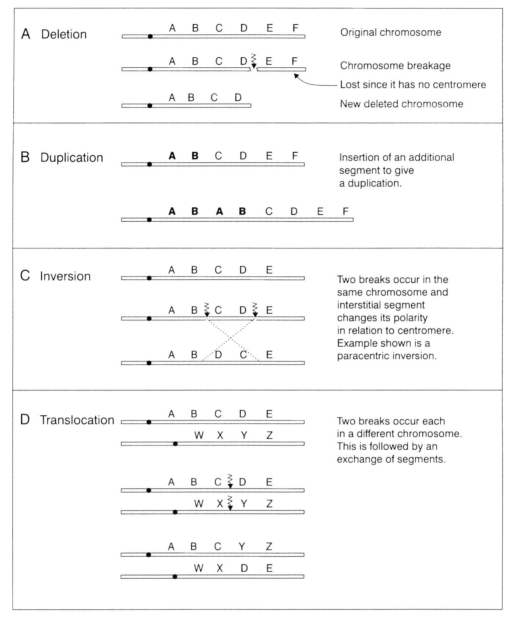

Figure 2.8 The main types of structural chromosome mutation. (a) Deletion. (b) Duplication. (c) Inversion. (d) Translocation.

maize the gene *purple* mutates at a rate of 1×10^{-5} per gamete per generation whereas for *colourless* the rate is 2×10^{-6}.

If overall mutation rates of genes in the three plant genomes are compared, there are general rate differences that exceed the variance within each genome. Mitochondrial genes typically have the lowest mutation rate, nDNA the most rapid with cpDNA being intermediate. The reasons for this are unclear but it is possible that the mechanisms of DNA repair are most efficient in mtDNA and consequently there is a lower mutation rate.

However, these are general trends only, and it is becoming more obvious that there are many exceptions. Nevertheless, these general differences in rates of genome evolution, level of diversity and mode

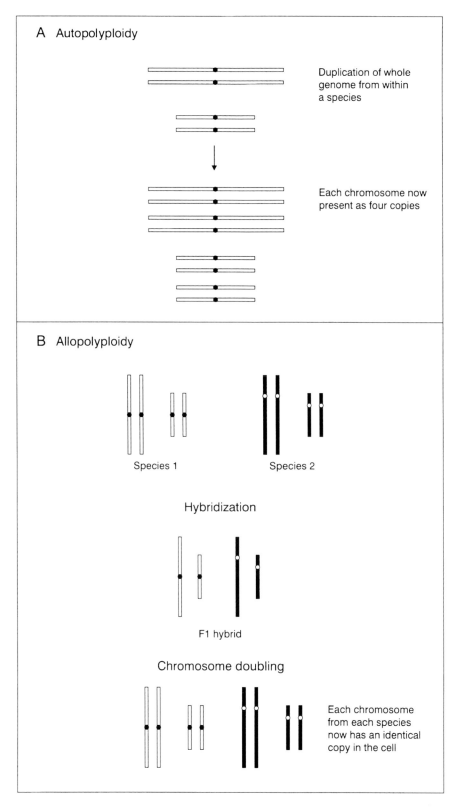

Figure 2.9   The two types of numerical chromosome mutation. (a) Autopolyploidy. (b) Allopolyploidy.

of inheritance mean that the genes within the different genomes are useful as genetic markers for examining the dynamics of genetic and evolutionary processes that operate at different speeds. For example, slowly evolving chloroplast genes are very helpful when looking for phylogenetic relationships between genera and even families, while rapidly evolving nuclear simple sequence repeats (SSRs) are useful for examining questions of paternity and parentage within populations (see Chapter 4).

2.7.2   Chromosome mutation

Chromosome mutations are of two types, those that alter the structure of the chromosome (structural mutation, Fig. 2.8) and those that alter chromosome number (numerical chromosome mutations, Fig. 2.9). They can be thought of as large-scale mutations since their effects are often visible as changes in chromosome structure. Most of the agents that cause gene mutations can also cause chromosome mutations and like gene mutations, chromosome mutations also arise spontaneously.

There are four main types of structural mutation: deletion, duplication, inversion and translocation (Fig. 2.8). These may be heterozygous, involving one member of the homologous chromosome pair, or homozygous, involving both.

(a) *Deletion*—involves the loss of a chromosome segment. The effect of deletion depends on the size of the segment that is lost and the number of genes that it contains, but in general there is a reduction in fitness. This can be due to pseudodominance where deleterious recessives, which are normally masked by their dominant allele, are expressed when the dominant allele is deleted.

(b) *Duplication*—addition of material to the chromosome. This may also reduce fitness since it results in a change in the number of copies of the genes in the duplicated segment which may give rise to aberrant development.

(c) *Inversion*—results from two breaks in the same chromosome and the reversal of the original segment when it is reincorporated into the chromosome. Because crossing-over within inversions leads to duplications and deletions in the recombinant chromatids, the gametes that contain these chromatids are usually inviable and, therefore, inversions appear to act as crossover

suppressors. In this way the linkage relationships of genes along the chromosome can be altered and large blocks of genes will all be inherited together.

(d) *Translocation*—comes about through breakage of two, usually non-homologous chromosomes which then exchange segments. This also reduces fertility and alters linkage relationships as genetically balanced gametes are only formed in translocation heterozygotes when the two mutant chromosomes and the two wild type chromosomes segregate together to opposite poles at meiotic anaphase I.

Numerical chromosome mutations, often loosely referred to as polyploidy, may involve changes in the number of single chromosomes, aneuploidy, or changes in the number of whole genomes, euploidy. Aneuploids are rare in nature, since they are usually deleterious, but are useful tools for the mapping of genes to specific chromosomes. Euploidy is very common and up to 70% of all angiosperms may be polyploid (Masterson 1994) but interestingly it is rare among the gymnosperms (Khoshoo 1959). Euploidy may involve the multiplication of the same genome, autopolyploidy, or different genomes, allopolyploidy or hybrid polyploidy (Leitch & Bennett 1997) (Fig. 2.9). Both these processes can result in gene duplication, probably the most important way that new genes are generated during evolution. One copy of the gene will be free to mutate since there is a duplicate copy that will maintain the original, probably vital, function of the locus. Autopolyploids characteristically show meiotic pairing between their component genomes and exhibit multivalent formation at metaphase I. Since they have multiple copies of the same genome they will typically show polysomic inheritance, as at each gene locus there will be more than the usual two alleles. Allopolyploids, combining different genomes, are characterised by bivalent formation at metaphase I, disomic inheritance and may show fixed heterozygosity.

2.8   Quantitative genetics

The genes that were discussed in Chapter 2.2 were characterised by having discontinuous effects on phenotype. However, many adaptively significant morphological traits show continuous variation. One source of this variation is due to polygenes, independent loci which together contribute to the same phenotype. This means that the phenotype is the result of the combined effects of many gene loci that may be involved in epistatic interactions. A quantitative approach is necessary for the analysis of this type of genetic variation. The basic principle behind quantitative genetic analysis is the partitioning of phenotypic variation (V_P) into components attributable to environmental (V_E) versus genetic (V_G) effects.

$$V_P = V_G + V_E \qquad (2.1)$$

The genetic component of variance can then be further subdivided into additive and non-additive effects. Additive genetic variance (V_A) represents that component of the total phenotypic variance that can be attributed to simple additive effects, so called because the phenotype of the progeny is quantitatively intermediate between that of the parents. In terms of single-locus variation, this corresponds to codominance. In contrast, non-additive genetic variance (V_{NA}) is the remainder of the genetic variance that is associated with non-additive genetic effects including both dominance and epistatis.

$$V_P = V_A + V_{NA} + V_E \qquad (2.2)$$

Accurate estimation of additive and non-additive components of phenotypic variation from common environment experiments requires either knowledge of mating system parameters, or the use of clonal material to estimate V_E. The most commonly measured parameter, narrow-sense heritability (h^2), is equal to $\dfrac{V_A}{V_P}$ and measures how much of the observed phenotypic variance is due to underlying additive genetic variance. This is commonly estimated for a given trait from the slope of the regression of progeny against parent values. Detailed discussion of quantitative genetic analysis is beyond the scope of this volume. In depth treatments of this topic are provided by Falconer (1989) and the papers found in Fins *et al.* (1992).

2.9   Conclusions

This chapter presents an introduction to general genetic concepts, mechanisms and basic genetic processes. More detailed treatments of the genetics of populations and the wide variety of genetic markers that have been developed to help understand the processes of genetic change in populations are presented in following chapters.

POPULATION GENETICS

Francis C. Yeh

SUMMARY

Population genetics is the quantitative study of the amount and distribution of genetic variation in populations, and the dynamics of the underlying genetic processes. This chapter provides an introduction to the analysis of gene and genotype frequencies and the measurement of genetic variation at individual and population levels. The basic mathematical models that form the framework for such analyses are developed (e.g. The Hardy-Weinberg Principle) and these are used to explore the effects of mutation, random genetic drift, sex, migration and selection on population genetic structure. The final section deals with the more complex analysis of multilocus genetic structure.

3.1 Introduction

Natural populations seldom exist in isolation; neither do the factors that shape the genetic structure of populations. Nature provides a wide range of habitats over time and space and as a result, the process of managing genetic diversity is dynamic. Population genetics is concerned with the study of genetic diversity in natural populations of a species. As noted by Frankel (1983):

> Wild species must have available a pool of genetic diversity if they are to survive environmental pressures exceeding the limits of developmental plasticity. If this is not the case, extinction would appear inevitable.

Determining how much genetic diversity exists in a species and of explaining this diversity in terms of its origin, organization and maintenance is thus of fundamental significance in the application of genetic principles to conservation biology. Genetically, organisms are structured at the following hierarchical levels:

Genes → genotypes (individuals) → populations → species.

It is essential to have a quantitative measure of this hierarchy when assessing genetic diversity. This is often based on the characterization of amount and distribution of genetic diversity in the hierarchy, that is, the population genetic structure, which is the most fundamental piece of information for species that requires genetic management (Brown 1978).

Description of population genetic structure and its dynamics has been based on the analysis of allele and genotype frequencies in sampled populations with simply inherited traits whose transmissions follow Mendelian rules. If there are k alleles at a locus, there are $k(k + 1)$ number of different possible genotypes. The estimate of allele frequencies at a locus from knowledge of genotype frequencies is forthright under the assumptions of the Hardy-Weinberg Principle (Hardy 1908; Weinberg 1908). The principle states that in a large random-mating population with non-overlapping generations, the allele and genotype frequencies will remain constant from generation to generation when there is no mutation, migration and natural selection. The Hardy-Weinberg Principle provides the foundation for all population genetic investigation. It is the single, most important theorem for the genetic analysis of natural populations.

3.1.1 Hardy-Weinberg equilibrium

To illustrate the Hardy-Weinberg Principle, let the genotype frequencies of AA, Aa, and aa at a diallelic locus in the parental population be P, Q, and R respectively, where $P + Q + R = 1$. Alleles A and a at respective frequencies p and q can be estimated in the next generation as in Table 3.1.

The constancy of allele and genotype frequencies indicates that in the absence of mutation, migration and natural selection, Mendelian inheritance by itself will preserve genetic diversity in large randomly-mating diploid populations and attain Hardy-Weinberg equilibrium (HWE). The Hardy-Weinberg proportions are attained in one generation of random mating only if allele frequencies are the same in females and males. This prerequisite is true only for populations with non-overlapping generations. In populations with more complex life histories, the Hardy-Weinberg proportions are attained gradually over several generations.

TABLE 3.1   **Illustration of the Hardy-Weinberg Principle**

	Parents		Offspring		
Mating	Frequency of mating	AA	Aa	aa
$AA \times AA$	P^2	1	0	0
$AA \times Aa$	$2PQ$	0.5	0.5	0
$AA \times aa$	$2PR$	0	1	0
$Aa \times Aa$	Q^2	0.25	0.5	0.25
$Aa \times aa$	$2QR$	0	0.5	0.5
$aa \times aa$	R^2	0	0	1
Total (next generation)		$(P + Q/2)^2 = p^2$	$2(P + Q/2)(Q + R/2) = 2pq$	$(R + Q/2)^2 = q^2$

The level of genetic diversity in a population is affected by an array of genetic, life history and ecological characteristics that collectively define the population genetic structure. It is desirable to summarize the allele data at all loci in a simple way that expresses the level of genetic diversity in each sampled population and that would permit comparison with other sampled populations. Two sets of measurable parameters have been used to characterize population genetic structure (Brown 1989). The first set is geographical parameters that describe and quantify the geographical variation patterns of populations at the single-locus level. Four genetic diversity measures are commonly used:

(a) polymorphisms (*P*) to describe what proportion of all gene loci are variable

(b) average number of alleles per locus (*A*)

(c) average heterozygosity (*h*) to describe what proportion of all gene loci are heterozygous in a typical individual in a sampled population

(d) the level of among-population differentiation (G_{ST}).

The second set is the genomic parameters that describe and quantify the non-random associations of genes or gametic disequilibria of populations. Genomic variation patterns of populations are far less investigated than geographical variation patterns in natural populations. This is because they are much harder to sample than single-locus patterns and the associated statistical and analytical complexity when studying associations of multiple alleles among many loci. Both sets of parameters describe the outcomes of selective and non-selective events in natural and domesticated populations. About 10 or more randomly chosen loci are usually studied in each sampled population to obtain an unbiased estimate of the amount and distribution of genetic diversity.

3.2 Factors that affect allele frequencies

The genetic constitution of a population depends on the frequencies of allele at different loci, and on their effects. The factors that affect allele frequencies in a random-mating population are: random genetic drift (chance fluctuation), mutation, natural selection and migration.

3.2.1 Random genetic drift

Natural populations are often finite and small in number, and random genetic drift refers to the chance fluctuations in allele frequencies in such populations. In random-mating populations, chance fluctuations can affect allele frequencies in at least two ways. The first can lead to 'one time drift' (or founder effects) that can have important evolutionary consequences. This occurs when some chance event or factor subdivides a population into fragments. Each fragment would then evolve separately. The frequency of the alleles in each fragment would depend on the sample of the original population with which it began.

More important in random-mating populations is the effect of random fluctuations in allele frequency from one generation to the next from sampling among the gametes, particularly in small populations. Special circumstances, such as the reproductive success of very few individuals, could reduce the effective number of apparently large populations to the point where random genetic drift is also important. A small number of parents that contributed most to the pollen pool due to early phenology in plants or a small number of dominant males in mate selection in animals are such examples.

3.2.1.1   *Variation in allele frequency under random drift*

To examine the influence of genetic drift, let the population number be static at *N* individuals. The production of *N* offspring would require *N* female and *N* male gametes from the parents, or 2*N* gametes. Let *p* be the frequency of allele *A* and *q* be the frequency of allele *a* in the parental generation at a diallelic locus. The probability that a sample of the parents (i.e. 2*N* gametes) contains *i* of allele *A* is the *i*th term in the binomial expansion:

$$(p + q)^{2N} = \binom{2N}{i} p^i q^{2N-i}$$
$$= \frac{2N!}{i!\,2N-i!} p^i q^{2N-1} \tag{3.1}$$

The mean or expected value of *i* is 2*Np*, the mean of the binomial distribution. Thus, mean frequency of allele *A* is $\frac{2Np}{2N} = p$ and is not changed by the sampling process. The variance of *i* is 2*Npq*. However, when *i* is expressed as a proportion of the 2*N* gametes in the sample (*i*/2*N*), its variance is:

$$\sigma^2_{(i/2N)} = \left(\frac{1}{2N}\right)^2 \times \sigma^2_i = \frac{2Npq}{(2N)^2} = \frac{pq}{2N} \qquad (3.2)$$

Thus, sampling increases the variance of allele frequency by the amount $\frac{pq}{2N}$ each generation, large only for intermediate values of p and small values of N. The occurrence of i, which is the number of A allele in the sample of $2N$ gametes produced, describes the sampling effect on allele frequency in one generation. This occurrence has $2N + 1$ possible values: 0, 1, 2, 3, ..., $2N$. The result due to sampling in the next generation depends on which of the values were produced by the first sample. Two of these values, 0 and $2N$, denote a lost allele and a fixed allele, respectively, and are of particular interest in conservation because it represent the complete loss of diversity at the locus.

Consider if the first sample produced a value of i that gave a new frequency $\frac{i}{2N}$ to the offspring generation. From the offspring generation, a new binomial distribution of possible allele frequencies would be produced because the value of p and q on which it was based had changed. This particular sample of $2N$ gametes that they contributed to the offspring had yet a third value of $\frac{i}{2N}$. Thus, in successive generations, allele frequency would drift from one value to another solely as the result of the sampling process.

There is a certain probability of fixation ($i = 2N$) or loss ($i = 0$) in each generation. After an initial phase, this probability would reach a stable value, at about $\frac{1}{2N+1}$ for a dioecious population with equal number of females and males (Li 1955). Suppose a population of number $N = 5$ has five A alleles at a diallelic locus so that $p = \frac{5}{10} = 0.5$ and ($q = 1 - p = 0.5$). In the next generation the probability that allele A will be fixed ($p = 1$; $i = 10$) is given by the binomial equation:

$$\frac{10!}{10! \times (10-10)!} \times 0.5^{10} \times 0.5^{10-10} \qquad (3.3)$$
$$= 0.5^{10} = 0.00098$$

The probability that allele A is lost ($p = 0$; $i = 0$) is given by the binomial equation:

$$\frac{10!}{0! \times (10-0)!} \times 0.5^0 \times 0.5^{10-0} \qquad (3.4)$$
$$= 0.5^{10} = 0.00098$$

Thus, the probability that the population will remain segregating for the A allele after one generation of random sampling is:

$$\begin{aligned}
&1 - \text{probability that allele } A \text{ will be fixed} - \\
&\text{probability that allele } A \text{ is lost} \qquad (3.5)\\
&= 1 - 0.00098 - 0.00098 = 0.9981
\end{aligned}$$

To demonstrate the effect of sampling on allele frequency for the above population over many generations, let p_i, the initial frequency of allele A, be 0.5 and draw a sample of $2N = 10$ gametes. Suppose the frequency of allele A, p_{t+1}, is 0.7 in the offspring (first) population. Draw a sample of $2N = 10$ gametes from the offspring population and suppose the new frequency of allele A, p_{t+2}, is now at 0.60. The sampling process would continue for 35 generations. The results from 100 such simulated populations are given in Table 3.2.

The hypothetical small population with $N = 5$ at a locus with even distribution of alleles is informative because it represents many of the important features of random genetic drift. In just five generations of random genetic drift, 18% of the populations were fixed. The remaining 82% remained segregating for a wide spread in allele frequencies. Each of the segregating population produced a spread of allele frequencies in the next generation, including some cases in which the allele has been newly fixed or newly lost. A smaller and smaller fraction of the original populations would remain unfixed with increasing time. By generation 35, all 100 populations were fixed, 44 populations for the A allele and 56 populations for the a allele. Thus, when random genetic drift has continued for a sufficient number of generations, most populations with a small number would be fixed for one allele or the other.

In 100 simulated populations of the same hypothetical population with $N = 5$, an allele that is rare in the population ($q < 0.05$) is lost in 85% of the populations in just five generations (results not presented). This is expected because the probability that a population will remain segregating for a rare allele is much smaller than the 0.9981 probability estimated for the common allele. After one generation of random sampling the probability that allele A will be fixed ($p = 1$; $i = 10$) is:

$$\frac{10!}{10! \times (10-10)!} \times 0.95^{10} \times 0.05^{10-10} \qquad (3.6)$$
$$= 0.95^{10} = 0.5987$$

TABLE 3.2 Change of allele frequency p at a diallelic locus by random genetic drift over 35 generations in 100 simulated populations of number $N = 5$ and initial allele frequency = 0.5

Allele frequency	Generation number							
	1	5	10	15	20	25	30	35
1.0	0	8	27	34	38	41	42	44
0.9	3	7	5	3	2	1	2	0
0.8	3	7	4	1	1	1	0	0
0.7	10	9	2	2	0	0	0	0
0.6	18	10	4	3	2	1	0	0
0.5	23	6	5	3	1	0	0	0
0.4	25	11	4	4	2	0	0	0
0.3	10	11	4	2	2	1	0	0
0.2	4	10	5	1	2	1	0	0
0.1	3	11	8	0	0	2	0	0
0.0	1	10	32	47	50	52	56	56
Proportion of the population fixed	1	18	59	81	88	93	98	100

and that the allele A is lost ($p = 0$; $i = 0$) is

$$\frac{10!}{0! \times (10-0)!} \times 0.95^0 \times 0.05^{10-0} \tag{3.7}$$
$$= 0.5^{10} = 9.7656^{-14}$$

Therefore, the probability that the population will remain segregating for allele A after one generation of random sampling is $1 - 0.59874 - 9.76562^{-14} = 0.4013$. Thus, the population remains segregating longer at a locus when its number is large or when the allele frequency is near 0.50. When $p = 0.50$, the average time that a population remains segregating at a locus is about 2.8N generations. In contrast, the average time that a population remains segregating at a locus is about 1.3N generations when $p = 0.10$.

The variance in frequency of allele A for generation $t + 1 (\sigma^2_{p(t+1)})$ in terms of the variance in frequency in allele A in the previous generation $\sigma^2_{p(t)}$ was derived by Wright (1942) and is of the following form:

$$\sigma^2_{p(t+1)} = \frac{(pq)_t}{2N} + \left(1 - \frac{1}{2N}\right)\sigma^2_{p(t)} \tag{3.8}$$

Equation 3.8 is of interest because the rate of change in the variance of p depends on the sample size, or effective N. This offers a method for estimating the effective population number, N_e, from the increase of variance in p between successive generations.

Therefore, rearranging Equation 3.8 and solving for effective population number:

$$N_e = \frac{(pq)_t - \sigma^2_{p(t)}}{2(\sigma^2_{p(t+1)} - \sigma^2_{p(t)})} \tag{3.9}$$

3.2.1.2 Heterozygosity under random drift

We have shown how random genetic drift could change allele frequencies. In genetic conservation, however, it is desirable to know how rapidly genetic diversity will be lost from a population of size N. The answer can be examined from the decay of heterozygosity in a finite population.

Population genetic theory predicts that the presence of small number of individuals sustained over many generations in a partially or completely isolated population will lead to depletion of genetic diversity. This reduction of heterozygosity can be rapid if population number (N) is small. The expected heterozygosity under random mating declines by a factor of $1 - \frac{1}{2N}$ per generation, providing that the sex ratio is 1:1, and there is no mutation and selection. Figure 3.1 shows that with $N = 32$, about 14% and 80% of the heterozygous genotype at a locus is lost after 10 and 100 generations of random genetic drift, respectively. When N is half of 32, at 16, about 27% and 96% of the heterozygous genotype is lost

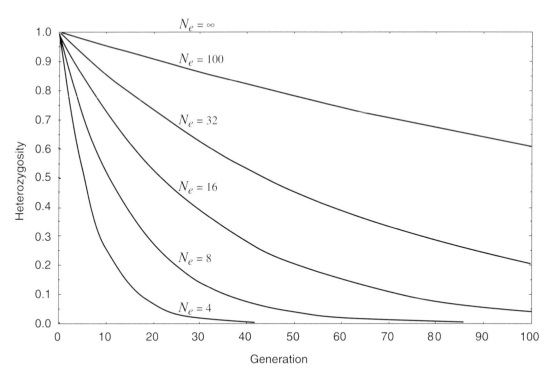

Figure 3.1   Delay of genetic diversity in a randomly mating population of effective size (N_e), without mutation or selection.

after 10 and 100 generations, respectively. If N is very small at 4, only 26% of the heterozygous genotype is left in the population after 10 generations, and by generation 40 all of the heterozygous genotype is lost.

Mating between related individuals, that is, inbreeding, further accelerates the reduction in heterozygosity (Fig. 3.2). This is because inbreeding reduces the effective population number. Under the mixed random mating and selfing model, N_e reduces to $0.5(1 + t)N$, where t is the outcrossing rate (Strobeck 1983; Pollak 1987). Thus, N_e is always smaller than the census number (N) unless the rate of outcrossing is unity. For example, with $N = 4$ and $t = 0.8$, $N_e = 3.6$. Consequently, unless a species exhibits low outcrossing, the effect of inbreeding on the decay of genetic diversity is small relative to that of random genetic drifts.

3.2.1.3   Effective population size

The effect of random genetic drift depends decisively on the effective population number (N_e) instead of the census population number (N_a). In sexual populations the number of breeding females (N_f) and males (N_m) is often unequal so that N_e is a fraction of N_a. To

illustrate this, let $N_a = N_f + N_m$. Li (1955) showed that the effective population number is given by:

$$N_e = \frac{4N_fN_m}{N_fN_m} \qquad (3.10)$$

In a plant population if N_m is only one-tenth N_f because of the small number of pollen parents, then ($N_m = 0.1N_f$) and $N_a = 1.1N_f$. From Equation 3.10, $N_e = (0.40/1.1)N_f = 0.33N_a$ indicating that the effective population number is just one-third of the census population number.

Variation in number of offspring per parent also reduces the effective population number (Li 1955) because of the relationship:

$$N_e = \frac{4N_a}{2 + \sigma_k^2} \qquad (3.11)$$

where σ_k^2 is the variance in family size. If the variation in fertility is random, $N_e = N_a$ because the family size has a Poisson distribution with mean = variance = 2. In natural populations $N_e \ll N_a$ because σ_k^2 is greater than 2, especially in plant populations where the family size is large. If the family size can be kept

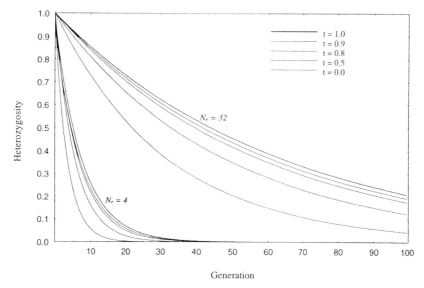

Figure 3.2   Effect of inbreeding on decay of genetic diversity in a population of effective size (N_e) of 4 and 32.

constant at 2, then $N_e = N_a$. This special case should probably be exploited in animal gene management programs for maximizing the effective population number.

3.2.2 Mutation

3.2.2.1 Mutation rates

New genetic variations arise every generation by mutation. A mutation can be defined as any change in the base sequence of the DNA in the genome. Mutations may involve the replacement of one base by another, insertion or deletion of single bases, deletion or duplication of a section of the DNA and inversion of a section of DNA.

Mutation rates per locus per generation vary between genes and between organisms. They are generally higher in multicellular organisms (1×10^{-6} to 1×10^{-4}) than in bacteria and microorganisms (1×10^{-10} to 1×10^{-8}). New mutations emerge continuously in a species despite the low mutation rates. This is because there are many individuals in a species and many genes in each individual. For a multicellular species with one million individuals (10^6) and assuming that the average mutation rate per locus is one per 100 000 gametes (1×10^{-5}), the average number of new mutations in a given generation would be $2 \times 10^6 \times 10^{-5} = 20$ per locus. Thus, mutation provides species with vast amount of new genetic variation every generation.

Fisher (1930) showed that the probability of survival of a new mutant with no selective advantage at generation 1 is $0.6321(1 - e^{-1})$ and at generation t is $e^{-(1 - l_{t-1})}$. The estimates over 40 generations are given in Table 3.3.

The probability of survival of a selectively neutral mutation is 0.6321 after one generation and it reduces to 0.1582 at generation 10. At generation 40, there is only a 4.7% chance of survival for a selectively neutral mutation. Thus, a single mutation is unlikely to survive in the population and contribute to the population genetic diversity unless it has a marked selective advantage.

TABLE 3.3 **Probability of survival of a selectively neutral mutation over 40 generations**

Generation	Probability of survival
1	0.6321
2	0.4685
3	0.3741
4	0.3121
5	0.2681
10	0.1582
20	0.0875
30	0.0608
40	0.0465

3.2.2.2 Equilibrium between forward and backward mutation

Let q_t be the frequency of allele a and p_t be the frequency of allele A in generation t at a diallelic locus. If the mutation rate at A to a is μ, and the rate from a to A is v, then in each generation q_t increases by the amount $\mu(1-q_t)$ and is diminished by the amount vq_t so that:

$$q_{t+1} = q_t + \mu(1-q_t) - vq_t = q_t + \mu - (v+\mu)q_t \quad (3.12)$$

Thus, the frequency change in allele a from generation t to generation $t+1$ is:

$$\Delta q_t = q_{t+1} - q_t = \mu - (v+\mu)q \quad (3.13)$$

When equilibrium is reached, that is, $\Delta q_t = 0$, then:

$$\hat{q}_t = \frac{\mu}{\mu+v} \quad (3.14)$$

This means that if v is one-fifth as large as μ, the frequency of allele a, q_t, will increase until it is $\frac{\mu}{2\mu+\mu} = 0.8333$.

3.2.2.3 Rate of allele frequency change under mutation

Although the allele frequency changes caused by mutation that are selectively neutral is small over a few generations, the cumulative effects of mutation over long times can become appreciable. Let q_t be the frequency of allele a and p_t the frequency of allele A in generation t at a diallelic locus. If the mutation rate at A to a is μ, and when there is no backward mutation from a to A, the increment of a due to mutation from A is $\mu(1-q_t)$. It follows that:

$$p_{t+1} = p_t - \mu p_t = p_t(1-\mu) \quad (3.15)$$

and

$$q_{t+1} = q_t + \mu(1-q_t) = \mu + q_t(1-\mu) \quad (3.16)$$

Repeated substitutions for previous generations of q as:

$$\begin{aligned} q_{t+1} &= \mu + \mu(1-\mu) + (1-\mu)q_{t-1} \\ &= \mu + \mu(1-\mu) + q_{t-1}(1-\mu)^2 \end{aligned} \quad (3.17)$$

produce a geometric series in which the last term with q_t refers to initial generation of interest:

$$q_{t+1} = \mu + (1-\mu)\mu + \mu(1-\mu)^2 + \quad (3.18)$$
$$\dots + q_t(1-\mu)^{t+1}$$

The sum of terms of this geometric series, with μ as the initial term and $(1-\mu)$ as the common ratio becomes:

$$q_{t+1} = 1 - (1-\mu)^{t+1} + q_0(1-\mu)^{t+1} \quad (3.19)$$

Thus,

$$(1-\mu)^{t+1} = \frac{1-q_{t+1}}{1-q_0} = \frac{p_{t+1}}{p_0} \quad (3.20)$$

Solving for t using log one has:

$$t\log(1-\mu) = \log\frac{p_{t+1}}{p_0} \quad (3.21)$$

and therefore

$$t = \log\frac{p_{t+1}/p_o}{1-\mu} \text{ or } \log\frac{(1-q_{t+1})/(1-q_t)}{1-\mu} \quad (3.22)$$

We may wish to ask how many generations of mutation are required to change the frequency of allele a from 0.05 (q_t) to 0.10 (q_{t+1}) when the average mutation rate per locus is one per 100 000 gametes (1×10^{-5})? Using Equation 3.22 we have

$$\begin{aligned} t &= \log\frac{(1-q_{t+1})/(1-q_t)}{1-\mu} = \log\frac{0.90/0.95}{1-10^{-5}} \\ &= \log\frac{0.94737}{0.99999} = 5406 \text{ generations} \end{aligned}$$

When there is no backward mutation from a to A, a first approximation to the rate at which the frequency of allele a changes under mutation is $E(q_t) = \mu + q_t(1-\mu)$, which is the sequence equation of an algebraic series whose general term accords with $\mu_t = (d/c) - [(d/c) - \mu_0](1-c)^t$. Substituting $\mu = d = c$:

$$q_t = 1 - (1-q_0)(1-\mu) \quad (3.23)$$

When t is large:

$$q_t = 1 - (1-q_0)e^{-\mu t} \quad (3.24)$$

When the number of generations past is the reciprocal of μ and $\mu t \approx 1$, then $q_t = 1 - e^{-1} = 0.632$.

Thus, about 40% of the population, that is q_t^2, would be changed to the aa genotype. When there is backward mutation from a to A:

$$q_{t+1} = q_t - \nu q_t + \mu(1 - q_t) \qquad (3.25)$$

and therefore

$$E(q_t) = \mu + q_t(1 - \mu - \nu) \qquad (3.26)$$

When the above sequence equation of an algebraic series whose general term accords with $E\mu_t = d + (1 - c)\mu_t$ and in which $a = \dfrac{d}{c}$, then:

$$\mu_t = \frac{d}{c} - \left(\frac{d}{c} - \mu_0\right) \times (1 - c)^t \qquad (3.27)$$

Letting $c = \mu + \nu$ and $d = \mu(\mu + \nu)$ so that:

$$q_t = \frac{\mu}{\mu + \nu} + \left(q_0 - \frac{\mu}{\mu + \nu}\right) \times (1 - \mu - \nu)^t \qquad (3.28)$$

and the approximate exponential form is:

$$q_t = \frac{\mu}{\mu + \nu} - \left(\frac{\mu}{\mu + \nu} - q_0\right) \times e^{-(\mu + \nu)t} \qquad (3.29)$$

When $t = \infty$, this expression has a limiting value less than 1, and in accordance with the equilibrium frequency in Equation 3.14:

$$\hat{q}_\infty = \frac{\mu}{\mu + \nu} \qquad (3.30)$$

What are the biological meanings of these equations? Since μ and ν are very small, the term $(1 - \mu - \nu)^t$ will approximate $1 - t(\mu + \nu)$ when t is very small, say at less than 100 generations. If the population is initially fixed for allele A, that is $q_0 = 0$, then $q_t = t\mu$. This suggests that the frequency of allele a increases linearly with time and the slope of the line is μ. However, unless the population number is exceedingly large, the linear increase in allele a is difficult to detect because μ is small. When t is very large, say at 10^6 generations, the term $(1 - \mu - \nu)^t$ is near zero so that whatever the initial frequency of a, its allele frequency eventually will reach the equilibrium frequency, at $\dfrac{\mu}{\mu + \nu}$.

3.2.3 Migration

In many species, the population is a network of subpopulations with intermittent migration between neighbouring subpopulations, that is, a metapopulation. Mating within each subpopulation can be random and the level of genetic differentiation among the subpopulations depends on the extent of migration and on the relative allele frequencies in the donor and recipient subpopulations. When migration is extensive, the entire population approximates a single random-mating unit. When migration is limited, the subpopulations may be quite different. Consequently, even if the population is distributed continuously over an area, there may be local differentiation when the population is large relative to the distance that any genotype (e.g. individual) and/or gamete (e.g. pollen) can migrate to another subpopulation and interbreed with the residents.

Assume that genotypes and/or gametes from neighbouring subpopulations migrate at the rate, m. In the next generation, a proportion, m, of genes will be descended from migrants and $1 - m$ are descendants of residents in each subpopulation. Let the initial frequency of allele A at a diallelic locus before migration in a given subpopulation be p and that the same allele in the neighbouring subpopulations has an average frequency $\bar{P}$. After one generation of migration, the frequency of A, p_1, in this subpopulation is:

$$p_1 = (1 - m)p + m\bar{P} = p - m(p - \bar{P}) \qquad (3.31)$$

The change in allele frequency A, Δp, in this subpopulation is:

$$\Delta P = p_1 - p = p - m(p - \bar{P}) - p = -m(p - \bar{P}) \qquad (3.32)$$

Thus, Δp is zero when either m or $p - \bar{P}$ is zero. Therefore, unless there is no migration, the allele frequency in this subpopulation will continue to change until it becomes the same as in the neighbouring subpopulations.

The difference in allele frequency between this subpopulation and neighbouring subpopulations after one generation is:

$$\begin{aligned} p_1 - \bar{P} &= p - m(p - \bar{P}) - \bar{P} = p - mp - \bar{P} + m\bar{P} \\ &= (1 - m)p - (1 - m)\bar{P} = (1 - m)(p - \bar{P}) \end{aligned} \qquad (3.33)$$

After two generations of migration, the frequency of A in this subpopulation is:

$$p_2 = p_1 - mp_1 + m\bar{P} = p(1 - m)^2 - m^2\bar{P} + 2m\bar{P} \qquad (3.34)$$

The difference in allele frequency between this subpopulation and the neighbouring subpopulations after two generations is:

$$\begin{aligned} p_2 - \bar{P} &= p(1 - m)^2 - m^2\bar{P} + 2m\bar{P} - \bar{P} \\ &= (1 - m)^2(p - \bar{P}) \end{aligned} \qquad (3.35)$$

TABLE 3.4

Genotype	AA	Aa	aa
Fitness	1	$1 - hs$	$1 - s$
Frequency of zygotes	p_t^2	$2p_t q_t$	q_t^2
Proportions of adults	p_t^2	$2p_t q_t (1 - hs)$	$q_t^2 (1 - s)$

After t generations of migration, the difference in allele frequency between this subpopulation and the neighbouring subpopulations is:

$$p_t - \bar{P} = (1 - m)^t (p - \bar{P}) \text{ or } (1 - m)^2 = \frac{p_t - \bar{P}}{p - \bar{P}} \quad (3.36)$$

As an example of the use of Equation 3.34, suppose a large continuous population of annual plants fragments into two smaller units as result of habitat disturbance. The allele frequency A in one unit is the same as before fragmentation, at 0.65. However, the frequency of this allele in the second unit is 0.45. What rate of migration for 20 years between these units is required to bring the allele frequency in the second unit to 0.60? In this example $t = 20$, $p_t = 0.60$, $p = 0.45$ and $\bar{P} = 0.65$. Thus:

$$(1 - m)^{20} = \frac{0.60 - 0.65}{0.45 - 0.65} = 0.3333 \quad (3.37)$$

and $1 - m = \sqrt[20]{0.3333}$

and therefore $m = 0.0535$.

This would suggest that in any generation about 5.4% of the alleles at the locus in the second unit must be newly derived from the first unit in order to increase the frequency of allele A from 0.45 to 0.60 in 20 generations.

How much migration must there be before subpopulations come to resemble each other? Nei (1973) has defined a measure that compares the genetic diversity within a subpopulation with the genetic diversity between subpopulations as:

$$G_{ST} = \frac{H_T - \hat{H}_S}{H_T} \quad (3.38)$$

in which

$\hat{H}_S = 1 - \sum_i p_{is}^2$, the average genetic diversity in a subpopulation

$H_T = 1 - \sum_i \bar{p}_i^2$, is the genetic diversity in the total population

p_{is}^2 = frequency of allele A_i in subpopulation s

$\bar{p}_i^2$ = average frequency of allele A_i in the entire population.

If we assume that the number of subpopulations is large and the mutation rate is very small, then, at equilibrium between migration and mutation:

$$G_{ST} = \frac{1}{1 + 4N_e m} \quad (3.39)$$

in which m is the proportion of breeding individuals that are migrants. If $N_e m \gg 1$, $H_T \cong \hat{H}_S$ and there is little genetic differentiation between the subpopulations. If $N_e m \ll 1$, then $H_T \gg \hat{H}_S$ and there is significant genetic differentiation between the subpopulations.

3.2.4 Natural selection

3.2.4.1 Change in genotype frequency under selection

When a genotype is at a selective disadvantage, s, it contributes proportionally less offspring to the next generation than other genotypes. This decrease in reproductive fitness is termed a loss of fitness. Let A be the favorable allele and a be the unfavourable allele at a diallelic locus in a random-mating population. Let p_t be the frequency of A and $1 - p_t$ or q_t be the frequency of a among new zygotes in generation t. This situation can be represented in Table 3.4.

We may choose different values of h to consider dominance ($h = 1$), intermediate dominance ($h = 0.5$) or recessive ($h = 0$). If there is complete dominance, the frequency of allele a in the next generation of zygotes, q_{t+1}, is:

$$q_{t+1} = \frac{q_t^2(1 - s) + p_t q_t}{p_t^2 + 2p_t q_t + (1 - s)q_t^2} = \frac{(1 - s)q_t^2 + p_t q_t}{1 - sq_t^2} \quad (3.40)$$

TABLE 3.5

Genotype	AA	Aa	aa
Fitness	$1 - s_1$	1	$1 - s_2$
Frequency of zygotes	p_t^2	$2p_t q_t$	q_t^2
Proportions of adults	$p_t^2(1 - s_1)$	$2p_t q_t$	$q_t^2(1 - s_2)$

The change in allele a after one generation, Δq, is:

$$q_{t+1} - q_t = \frac{q_t^2(1+s) + p_t q_t}{1 - sq_t^2} - q_t = -\frac{sq_t^2(1-q_t)}{1 - sq_t^2} \quad (3.41)$$

The actual frequency of the allele at the time is the primary factor affecting the rate of change of allele frequency under selection. When s is moderate, the rate of change increases dramatically for intermediate allele frequencies, and declines rapidly for low and high allele frequencies. This is the reason for relative persistency of even severely disadvantaged alleles at low frequencies.

We may estimate the number of generations required to produce a given amount of change in allele frequency. For example, if allele a is recessive with absolute disadvantage, that is, $s = 1$, its frequency in the next generation of zygotes, q_{t+1}, is:

$$q_{t+1} = \frac{q_t^2(1-1) + p_t q_t}{p_t^2 + 2p_t q_t + (1-1)q_t^2} \qquad (3.42)$$
$$= \frac{(1-q)q_t}{(1-q_t)(1+q_t)} = \frac{q_t}{1+q_t}$$

The frequency of allele a in the q_{t+2} generation of zygotes is:

$$q_{t+2} = \frac{q_{t+1}}{1 + q_{t+1}} = \frac{q_t}{1 + q_t} \Big/ 1 + \frac{q_t}{1 + q_t} = \frac{q_t}{1 + 2q_t} \quad (3.43)$$

and the frequency of allele a in the q_{t+n} generation of zygotes is:

$$q_{t+n} = \frac{q_t}{1 + nq_t} \qquad (3.44)$$

Solving for n:

$$n = \frac{q_t - q_{t+n}}{q_t q_{t+n}} \qquad (3.45)$$

What will be the frequency of allele a in the q_{t+n} generation of zygotes when the Aa genotype has the highest fitness and if there is complete dominance, that is, $h = 1$? This situation can be represented in Table 3.5.

The change in the frequency of allele a per generation due to selection is:

$$\Delta q = \frac{p_t q_t(s_1 p_t - s_2 q_t)}{1 - s_1 p_t^2 - s_2 q_t^2} \qquad (3.46)$$

and selection will keep both alleles in the population. An equilibrium is reached when $\Delta q = 0$ and the equilibrium frequency of allele a is:

$$\hat{q} = \frac{s_1}{s_1 + s_2} \qquad (3.47)$$

Thus, the equilibrium frequency depends on the relative fitness of the two homozygous genotypes in the population.

3.2.4.2   Equilibrium between selection and mutation

What is the consequence when selection against an allele is opposed by mutation? Let the mutation rate from allele A to a be μ at a diallelic locus in a population of N zygotes. Let us assume that the initial frequency of allele a is very small and that backward mutation is negligible. There are $2Np$ A alleles and hence $2Np\mu$ new a alleles arise by mutation in each generation. At equilibrium, that is, $\Delta q = 0$, the number of new mutations is equal to the number eliminated by selection:

$$2N\hat{p}\mu = 2N\hat{q}s(\hat{p}h + \hat{q}) \text{ or } \hat{p}\mu = \hat{q}s(\hat{p}h + \hat{q}) \quad (3.48)$$

If allele a is recessive ($h = 0$), $\hat{p}\mu = \hat{q}^2 s$ and since $\hat{p} \cong 1$:

$$\hat{q} = \sqrt{\frac{\mu}{s}} \qquad (3.49)$$

If $h = 0.5$, that is, semidominance, Equation 3.46 reduces to:

$$\hat{q} = \frac{2\mu}{(1+\mu)s} \approx \frac{2\mu}{s} \qquad (3.50)$$

For partial dominance, q is much less than ph so that:

$$\hat{q} \approx \frac{\mu}{hs} \qquad (3.51)$$

3.2.4.3 Equilibrium between selection and migration

What is the consequence of when selection against an allele is opposed to by migration? Wright (1940) showed that when the fitness (W) of heterozygote is intermediate between the two homozygotes at a diallelic locus, that is, $W_{AA} = 1$, $W_{Aa} = 1 - s$ and $W_{aa} = 1 - 2s$, the joint effect of selection and migration on allele frequency is:

$$\Delta q = -sq(1-q) - m(q - \bar{q})$$
$$= sq^2 - (s+m)q + m\bar{q} \qquad (3.52)$$

At equilibrium, that is, $\Delta q = 0$:

$$\hat{q} = \frac{(s+m) \pm \sqrt{[(s+m)^2 - 4sm\bar{q}]}}{2s} \qquad (3.53)$$

If m is much smaller than s, allele frequencies in the subpopulations are mainly conditioned by the direction of selection, with only a weak diluting effect of the immigrants. Thus, we expect a significant degree of subpopulation differentiation, depending upon the local conditions of selection. In contrast, when m is extensive and is much greater than s as in wind-pollinated plants, the effect of immigrants will outweigh that of selection. The equilibrium allele frequencies of subpopulations will not differ greatly from the average allele frequency of the total population. Consequently, we expect an insignificant differentiation of the subpopulations.

When s and m are the same size, then $s + m = 2s$ if selection is against allele a (i.e. s is positive) and $s + m = 0$ if selection favours allele a (i.e. s is negative). Substituting into Equation 3.53 we have:

$$\hat{q} = \sqrt{\bar{q}} \text{ for } s \text{ is positive and}$$
$$\hat{q} = 1 - \sqrt{1 - \bar{q}} \text{ for } s \text{ is negative} \qquad (3.54)$$

The allele frequencies of subpopulations can differ from each other and we expect considerable differentiation among subpopulations under differential selective forces.

3.3 Analysis of population differentiation

A diploid population is considered to be in HWE when the alleles at a given locus are randomly distributed throughout the population and there is no association between the pair of alleles that an individual receives from its parents. Deviations from HWE can arise from several factors, including population subdivision, Wahlund's Principle, migration from outside , non-random mating, sampling of siblings, sex-specific differences in allele frequency, presence of null alleles and selection. Thus, it is always important to determine whether there are significant deviations from HWE at sampled loci in the survey of genetic diversity in natural populations. When the genotype proportions in a population deviate significantly from HWE at several independent loci it may suggest non-random mating, or that the population consists of discrete *demes* and is subject to migration from outside sources. However, when genotype proportions at a locus are not in HWE for several populations it may indicate the presence of null alleles, or may be evidence for selection.

3.3.1 Wahlund Principle

Natural populations are often subdivided. This is because allele frequencies among populations may diverge, either because of natural selection or by chance if the subpopulations are small. What will be the genetic effect of population subdivision?

Let us assume there are k subpopulations of numbers $n_1, n_2, \dots n_k$ and that mating is random within the subpopulations. If the frequency of allele A at a diallelic locus is $p_1, p_2, \dots p_k$, the mean of genotype AA in the whole population is:

$$\overline{p^2} = \frac{n_1 p_1^2 + n_2 p_2^2 + \dots + n_k p_k^2}{n_1 + n_2 + \dots + n_k} \qquad (3.55)$$

The variance of p in the entire population, σ_p^2, is $\overline{p^2} - \bar{p}^2$ in which:

$$\bar{p}^2 = \frac{n_1 p_1 + n_2 p_2 + \dots + n_k p_k}{n_1 + n_2 + \dots + n_k} \qquad (3.56)$$

Thus:

$$\overline{p^2} = (\sigma_p^2 + \bar{p}^2), \ AA = \overline{p^2} = \bar{p}^2 + \sigma_p^2; \qquad (3.57)$$
$$Aa = 2pq = 2\overline{pq} - 2\sigma_p^2; \text{ and } aa = \overline{q}^2 + \sigma_p^2$$

The consequence is that frequencies of *AA* and *aa* homozygotes in the entire population are greater than the Hardy-Weinberg proportions while *Aa* heterozygotes in the entire population are less than the Hardy-Weinberg proportions. This observation of an inbreeding-like effect, that is, increase of homozygotes at the expense of heterozygotes, was first noted by Wahlund (1928) and is termed the Wahlund Principle.

3.3.2 *F*-statistics

The genetic structure of subdivided populations can be analyzed by *F*-statistics using the correlation between uniting gametes (Wright 1943, 1951). Three parameters measure the deviations of genotype frequencies in a subdivided population:

(a) F_{IS} which is the correlation between two uniting gametes to produce the individuals relative to the subdivisions

(b) F_{IT} which is the correlation between two uniting gametes to produce the individuals relative to the total population

(c) F_{ST} which is the correlation between two gametes drawn at random from each subpopulation.

Computationally, F_{IS}, F_{IT} and F_{ST} are of the following forms:

$$F_{IS} = \frac{H_S - H_I}{H_S} \qquad (3.58)$$

$$F_{IT} = \frac{H_T - H_I}{H_T} \qquad (3.59)$$

$$F_{ST} = \frac{H_T - H_S}{H_T} \qquad (3.60)$$

Where H_I is the observed average heterozygosity of individuals within subpopulations and a measure of the within-population heterozygote deficit; H_S is the expected average heterozygosity and a measure of the global heterozygote deficit; and H_T is the total heterozygosity (gene diversity) when all subpopulations are pooled together and a measure of the among-population heterozygote deficit.

Rearranging Equations 3.58, 3.59 and 3.60 will show that the three parameters are related by the following equation:

$$F_{ST} = 1 - \frac{1 - F_{IT}}{1 - F_{IS}} \qquad (3.61)$$

Although F_{ST} is always positive, estimates of F_{IS} and F_{IT} may be positive or negative, indicating heterozygote deficit and excess, respectively. If all subpopulations are in HWE, $F_{IS} = 0$ and $F_{IT} = F_{ST}$. However, even if Hardy-Weinberg proportions are attained, genetic differentiation due to allele frequency differences among the subpopulations can lead to significant F_{IT} and F_{ST} values.

F_{ST} is a measure of the extent of genetic differentiation among subpopulations and is the most widely used *F*-statistic. For loci with only two alleles, F_{ST} is the standardized variance in allele frequencies among populations and can be estimated as:

$$F_{ST} = \frac{\sigma_p^2}{\bar{p}\bar{q}} \qquad (3.62)$$

The term σ_p^2 is the weighted sum of squared deviations of the individual subpopulation allele frequencies from the average allele frequency divided by the number of subpopulations, and $\bar{p}$ and $\bar{q}$ are the weighted average allele frequencies. An F_{ST} estimate of less than 0.050 represents a low level of genetic differentiation. The range of 0.051 to 0.150 and 0.151 to 0.250 for F_{ST} is typical of moderate and large genetic differentiation, respectively, and an F_{ST} above 0.250 suggests very large genetic differentiation. However, even when the estimate of F_{ST} is 0.050 or less, it does not imply that genetic differentiation is negligible (Wright 1978).

As an example of the use of Equations 3.61 and 3.62 to compute *F*-statistics, suppose five populations were studied at a diallelic locus. Allele *A* was fixed in two populations and its frequency in the remaining three populations was 0.8, 0.7 and 0.5. The observed frequency of heterozygotes in the three polymorphic populations was 0.22, 0.27 and 0.38, respectively. Here,

$H_I = [(2)(0) + 0.22 + 0.27 + 0.38]/5 = 0.174$

$H_S = [(2)(0) + 2(0.8)(0.2) + 2(0.7)(0.3) + 2(0.5)(0.5)]/5$
$\quad = 0.248$

$H_T = 2\{[(2)(1) + 0.8 + 0.7 + 0.5]/5\} \{[(2)(0) + 0.2 + 0.3 + 0.5]/5\} = 2(0.8)(0.2) = 0.320.$

Thus,

$F_{IS} = (0.248 - 0.174)/0.248 = 0.298$ and indicates a 29.8% heterozygote deficit relative to Hardy-Weinberg expectations at the locus

$F_{IT} = (0.320 - 0.174)/0.320 = 0.456$ and indicates a 45.6% heterozygote deficit relative to Hardy-Weinberg expectations at the locus

$F_{ST} = (0.320 - 0.248)/0.320 = 0.225$ and indicates a 22.5% among-population genetic differentiation at the locus.

3.3.3 G-Statistics

Nei (1973) proposed an alternate analysis of population subdivision that does not rely on knowledge of genotype frequencies and can be estimated from allele frequencies in terms of the expected heterozygosities within and between populations. The relative magnitude of gene differentiation among the subpopulations can be measured by G_{ST}, the coefficient of gene differentiation outlined in Equation 3.38. In Nei's notation:

$$G_{ST} = \frac{D_{ST}}{H_T} \qquad (3.63)$$

in which H_T is the gene diversity in the total population and D_{ST} is the average gene diversity among subpopulations. Computationally:

$$H_T = 1 - \sum_k \left(\sum_i p_{ik}/s \right)^2 \text{ and}$$

$$D_{ST} = \sum_i \left[\sum_k p_{ik}^2 - \sum_k \left(\sum_i p_{ik}/n \right)^2 \right] \qquad (3.64)$$

in which p_{ik} is the frequency of allele k in subpopulation i, and S is the number of subpopulations.

G_{ST} is identical to F_{ST} if there are only two alleles at a locus. It is the most widely used measure of genetic differentiation because it is not affected by the reproductive system of the species, the number of alleles per locus and the pattern of evolutionary forces such as migration, mutation and selection. The value of G_{ST} varies from 0 to 1 and a value of less than 0.050 represents a low level of genetic differentiation. The range between 0.050 and 0.150 indicates moderate genetic differentiation, 0.151 and 0.250 is representative of large gene differentiation, and above 0.250 is characteristic of very large gene differentiation. The estimate of G_{ST} is highly dependent on the value of H_T. When H_T is small, for example, in species with low level of genetic diversity caused by severe habitat disturbance, G_{ST} may be large even when the absolute genetic differentiation is small.

The analysis of genetic differentiation can be extended to any level of hierarchical subdivision. For example, populations can nest within geographical regions and individuals can nest within populations within geographical regions. A large number (i.e. 10) of monomorphic and polymorphic loci which are random samples of the genome should be surveyed in order to obtain an unbiased estimate of gene differentiation among the subpopulations.

3.3.4 Genetic distance

Genetic distance is a measure of the amount of genetic divergence between populations and is useful for grouping the populations. Among the many estimates of genetic distance using differences in allele frequency among populations, Nei's standard genetic distance, D_S, (Nei 1972) has been most frequently used. It is given by:

$$D_s = -\log_e[J_{xy}/(j_x j_y)^{0.5}] \qquad (3.65)$$

where $J_{xy} = \sum x_i y_i$, $J_x = \sum x_i^2$, $J_y = \sum y_i^2$, and x_i and y_i denote the frequency of the ith in populations X and Y, respectively. If population differentiation is largely the result of isolation by distance, then genetic distance and geographical (spatial) distance are expected to be positively correlated. The value of D_s varies from 0 to 1. Based on isozyme data from several studies, Nei (1974) reported that species were characterized by genetic distances of 0.1 to 1.0, subspecies and varieties by 0.02 to 0.20 and races by 0.01 to 0.05.

In studies where populations are closely related and genetic drift (i.e. small effective population size) is the primary determinant of evolutionary differentiation, the modified Cavalli-Sforza distance, D_A, is recommended for use (Nei *et al.* 1983). It is given by:

$$D_A = \frac{1}{r} \sum_{j=1}^{r} \left(1 - \sum_{i=1}^{k} \sqrt{x_{ij} y_{ij}} \right) \qquad (3.66)$$

in which x_{ij} and y_{ij} are frequencies of allele i at locus j in populations X and Y, respectively, k is the number of alleles at locus j and r is the number of loci.

3.3.5 Multilocus genetic structure

The analysis of genetic diversity assumes random recombinations of genes at different loci. In such case, single-locus estimates of genetic parameters and their average across loci are adequate for describing the genetic diversity pattern. However, many selective and non-selective evolutionary forces could cause non-random association of alleles among the loci. Thus, it is often desirable to study the joint effects of genes at different loci on the population structure of natural populations, that is, the multilocus genetic structure.

The study of gametic disequilibrium is a key to analyzing the multilocus genetic structure. The term 'gametic disequilibrium' rather than 'linkage disequilibrium' has been used to refer to the non-random associations between independent as well as linked loci. Consider a pair of loci **A** and **B**, gametic disequilibrium between allele A_i at locus **A** and allele B_j at locus **B** is measured by

$$D_{ij} = g_{ij} - p_i q_i \qquad (3.67)$$

where g_{ij} is observed frequency of gamete $A_i B_j$, p_i is frequency of allele A_i and q_j is frequency of allele B_j.

This definition of gametic disequilibrium can be extended to the case of multiple loci. The range of gametic disequilibrium defined above is a function of allele frequencies. Thus, we cannot compare the D values for two pairs of loci when their allele frequencies are different. For this reason, Lewontin (1964) suggested dividing this measure by its maximum possible value. This normalised measure has the same range of -1 to $+1$ for all allele frequencies, but it still depends on the allele frequency (Lewontin 1988). Two common methods of analysing multilocus genetic structure in natural populations are:

(a) correlation between zygotic loci (Smouse & Neel 1977)

(b) multilocus heterozygosity (Brown *et al.* 1980; Brown & Feldman 1981).

3.3.5.1 Correlation between zygotic loci

The first step in studying the correlation between zygotic loci (Smouse & Neel 1977) is to code the genotypes. For example, the genotypes *AA*, *Aa* and *aa* at a diallelic locus are represented by the vector $Y' = (1, 1/2, 0)$. The next step was the construction of the matrix of correlations between pairs of loci (R) for each population:

$$R = \begin{bmatrix} 1 & r_{12} & . & . & r_{1k} \\ r_{21} & 1 & . & . & r_{2k} \\ . & . & 1 & . & . \\ . & . & . & 1 & . \\ r_{k1} & r_{k2} & . & . & 1 \end{bmatrix} \qquad (3.68)$$

The genetic expectation of the pairwise correlation between loci *i* and j, $E(r_{ij})$, is:

$$E(r_{ij}) = \frac{D_{ij}}{\sqrt{p_i(1-p_i)q_i(1-q_i)}} \qquad (3.69)$$

Yeh *et al.* (1985) studied isozyme differentiation from *Pinus contorta* spp. *latifolia* Dougl. in western North America using canonical discriminant functions that were constructed on the basis of correlation matrices using Equation 3.68. Two significant discriminant functions accounted for 38% of total variation in 20 polymorphic loci. Their scatter plot of the 17 populations showed that while southern populations were in one cluster, northern populations were disjointly distributed over the whole domain. Thus, there was greater genetic differentiation among the northern populations. This is consistent with the population structure of the species since northern populations often occur in small, more isolated stands and must adapt to narrow ecological niches. In contrast, southern populations are in large, dense stands with potential for extensive gene flow and must adapt to the relatively stable environments. The rich structure of genetic diversity associated with geography contrasts two earlier single-locus analysis of population structure of the same species (Yeh & Layton 1979; Dancik & Yeh 1983) which revealed only slight population differentiation. This argues for a multilocus approach to studying population genetic structure.

3.3.6 Multilocus heterozygosity

The analysis of gametic disequilibria in natural populations typically requires the computation of individual pairwise disequilibrium coefficients (Brown *et al.* 1980). The number of pairwise disequilibrium coefficients becomes exceedingly large with many loci and alleles. Thus, it is desirable to have a set of summary statistics that adequately describes the multilocus structure within and among subdivided populations. One such set of statistics is constructed from the distribution, $f(K)$, of the number of heterozygous loci in a sample of n gametes assayed for each population. There are $n(n-1)/2$ possible comparisons for the n gametes and the resulting distribution estimates the observed mean and the observed second to fourth moments about the mean. Using sampling variance and assuming that the sampling distributions of sample moments approximate normality, it is possible to construct the upper and lower 95% confidence limits of the sample

TABLE 3.6 Components of total and average variances in the number of heterozygous loci in *Cunninghamia lanceolata* (Lamb) Hook

Component	
Single-locus effect	
Mean gene diversity (MH)	4.086
Variance in gene diversity (VH)	0.307
Wahlund effect (WH)	−0.113
Total	4.280
Two-locus effect	
Mean disequilibrium (MD)	1.092
Wahlund effect (WC)	3.720
Interaction between MD and WC (AI)	0.055
Variance of disequilibria (VD)	1.116
Covariation in interaction (CI)	−0.072
Mean variance	6.277
(MH + MD + AI + VD + CI)	
Total variance	9.147

moments to test for presence of gametic disequilibria or association among loci (Yeh *et al*. 1994).

The total and mean variance in the number of heterozygous loci (second moment) can be appointed into single-locus and two-locus effects (Brown & Feldman 1981). The total variance has three single-locus and three two-locus effects. The three single-locus effects are:

(a) mean gene diversity (MH)

(b) variance in gene diversity (VH)

(c) single-locus Wahlund effect (WH).

The three two-locus effects are:

(a) mean disequilibrium (MD)

(b) two-locus Wahlund's covariance (WC)

(c) MD × WC (AI).

The mean variance has one single-locus effect in MH and four two-locus effects in MD, AI, variance of disequilibrium (VD) and covariation in the interaction between MD and WC (CI).

Yeh *et al*. (1994) studied multilocus heterozygosity on population structure in *Cunninghamia lanceolata*

(Lamb.) Hook. Fifteen of the 16 observed variances exceeded their upper 95% confidence limits, indicating the significance of two-locus gametic disequilibria in *Cunninghamia lanceolata*. The observed third and fourth moments exceeded their upper or lower 95% confidence limits in all populations except for the third moment in three of 16 populations. This signified predominance of three-locus and four-locus gametic disequilibria. The comparison of the single-locus and two-locus Wahlund effect indicates negligible WH but an appreciable amount of WC (Table 3.6). Thus, there is a greater level of genetic differentiation among populations when many loci are considered jointly. The high WC and low AI are characteristics of multilocus associations that are affiliated with a strong effect of population subdivision. The component due to variation in disequilibria accounts for 19% (1.116/5.911) of the two-locus effect. This suggests that the most common multilocus gamete types usually differ from one population to another. Such a result would arise under the founder and/or the diversifying selection hypothesis. These findings have important implications for *ex situ* conservation. For example, when gametic disequilibria are widespread and when population subdivision is a major cause of gametic disequilibria, pooling of trees from different populations should be avoided because this would disrupt the disequilibria and result in an increase in the rate of decay of additive genetic variance.

The analysis of multilocus genetic structure has generated many requests for computer programs. I have made available POPGENE without charge on the World Wide Web (URL http://www.ualberta.ca/~fyeh/). It is also included with this book on CD ROM. POPGENE is a Microsoft® Windows-based program for analysing genetic variation among and within populations. It computes comprehensive genetic statistics (e.g. allele frequency, gene diversity, genetic distance, *G*-statistics, *F*-statistics) and complex genetic statistics (e.g. gene flow, neutrality tests, linkage disequilibria, multilocus structure) for codominant and dominant markers in haploid and diploid data.

3.4 Conclusions

A single chapter cannot possibly bring together a proper balance and sufficient depth of understanding of population genetics. To those who lack experience in mathematics, a first glance at the chapter may give the impression that the subject matter is advanced and difficult. This is far from the truth. The material I present here is elementary, but fundamental to our understanding of population genetics.

3.5 Acknowledgments

I am indebted to many of my students and colleagues who reviewed the chapter. While they may not agree in total with my final selection and presentation, their suggestions have been taken in to the improvement of the final version.

GENETIC TOOLS: THE USE OF BIOCHEMICAL AND MOLECULAR MARKERS

Jeffrey C. Glaubitz and Gavin F. Moran

SUMMARY

Several types of genetic marker are available to forest conservation geneticists, including isozymes, terpenes, restriction fragment length polymorphisms (RFLPs), microsatellites, random amplified polymorphic DNAs (RAPDs) and amplification fragment length polymorphisms (AFLPs). The techniques underlying these markers are explained in terms that are accessible to those unfamiliar with molecular biology. The strengths and weaknesses of the various marker systems are elucidated via comparison. Guidelines are given for choosing the appropriate markers for particular applications in conservation genetics.

4.1 Introduction

The past 20 years have seen rapid advances in the technologies available for assessing genetic diversity at the molecular level. This chapter provides detailed discussion and illustration of the available genetic markers, their relative strengths, limitations and applicability to different research. Genetic markers have many applications in forest conservation genetics (e.g. Box 4.1). Genetic markers can be divided into three main classes—morphological markers, biochemical markers and the more recent DNA-based markers. Morphological markers that display Mendelian inheritance (e.g. chlorophyll deficiency) are rare in trees and will not be considered further in this chapter. Biochemical markers can be either at the protein level (e.g. isozymes) or at the level of organic chemicals (e.g. terpenes). A variety of DNA markers have been developed and most of these have been successfully applied to forest tree species.

4.2 Biochemical markers

4.2.1 Isozymes

The advent of isozymes as genetic markers in the early 1970s heralded a great advance for population genetic studies of forest tree species, since only rare morphological markers were available up to that time. The sudden availability of 20–30 Mendelian loci per species enabled estimates to be made of genetic diversity and mating system parameters for populations of tree species. Today isozyme markers are often still the best tool to answer many research questions in conservation genetics of forest trees.

Isozymes are different molecular forms of the same enzyme. These forms can be separated by gel electrophoresis as shown in Figure 4.1a.

Plant tissue is ground in a buffer, the homogenate is absorbed onto paper wicks and subjected to electrophoresis. Extraction buffers for seedlings can be simple (e.g. Conkle *et al.* 1982; Moran *et al.* 1989b) whereas for mature tissue such as needles, buds and leaves more complicated buffers containing chelating agents, antioxidants and other chemicals are used (e.g. Cheliak & Pitel 1984). After homogenisation, isozymes are separated in an electric current on the basis of differential charge, size and/or shape. Resolution of isozymes in tree species has been primarily by horizontal starch gel electrophoresis (Conkle *et al.* 1982; Moran & Bell 1983; Liengsiri *et al.* 1990; Adams *et al.* 1991). Cellulose acetate gels (Coates 1988) and acrylamide gel electrophoresis can also be used. Visualisation, or 'staining', of isozymes is achieved by the biochemical coupling of the pertinent enzymatic reaction to a colour-generating redox reaction, often involving chlorogenic tetrazolium salts and the cofactor nicotinamide adenine dinucleotide (NAD) or its phosphate (NADP). In starch gel electrophoresis, each gel can be cut into slices and each slice stained for a separate enzyme (Fig. 4.1a). Enzyme

---

Box 4.1 Uses of molecular markers in forest conservation genetics

Molecular or genetic markers can be used in forest conservation genetics in the following ways:

1   measuring genetic diversity and differentiation in natural, managed and breeding populations;

2   estimating rates of gene flow or migration;

3   characterising the mating system;

4   analysing paternity or parentage;

5   assessing seed orchard efficiency;

6   DNA fingerprinting or verification;

7   for quality control in breeding;

8   studying phylogeny or taxonomy; and

9   genetic linkage mapping/quantitative trait loci analysis/marker-assisted selection.

a

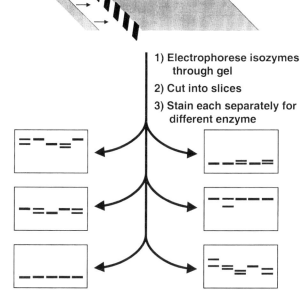

Tissue

Grind in buffer,
Absorb into wick

paper wick

Place wicks on starch gel

1) Electrophorese isozymes
   through gel
2) Cut into slices
3) Stain each separately for
   different enzyme

b

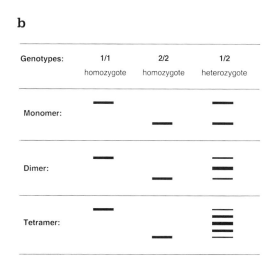

Figure 4.1 Isozymes. (a) Steps in isozyme analysis. (b) Diagrammatic representation of isozyme genotypes for enzymes of different quaternary structure.

staining recipes are detailed in Brewer and Sing (1970), Shaw and Prasad (1970), Vallejos (1983) and Wendel and Weeden (1989). When a gel is stained for an enzyme, isozyme bands appear, constituting the phenotype. The banding patterns are interpreted genetically and bands assigned to loci. Scoring of gels is manual and cannot be automated easily. Photographic records of gels are usually kept for reference.

Isozymes vary in their quaternary structure (i.e. the number of polypeptide subunits that combine to form a functional enzyme). Monomeric enzymes (e.g. shikimate dehydrogenase, aconitase) are composed of a single polypeptide—their heterozygotes appear as two bands (Fig. 4.1b).

Dimeric enzymes (e.g. malate dehydrogenase, aspartate aminotransferase, glucose phosphate isomerase) are composed of two polypeptide subunits—heterozygotes at such loci have three-banded phenotypes with an intermediate, heterodimeric band, composed of a polypeptide from each allele, between the other two homodimeric allozyme bands (Fig. 4.1b). Multimeric enzymes (e.g. glutamate dehydrogenase and menadione reductase) have more complex banding patterns. Enzymes typically have the same quaternary structure in most plant species.

Isozyme bands may be coded by a single genetic locus in which case the electrophoretic variants (alleles) are called allozymes. However, multiple loci, known as

'enzyme systems', are detected by some stains. Isozyme loci are codominant (see Chapter 1) as most or all alleles are identifiable as distinct bands on the gel. The haploid megagametophytic tissue of conifer seeds has enabled the genetic inheritance of isozyme loci to be determined easily without needing to make controlled crosses (Conkle *et al*. 1982; Mitton 1983; Moran *et al*. 1983; Adams *et al*. 1991). This tissue is genetically equivalent to the maternal contribution to the embryo of that seed. In angiosperm tree species crosses have been available in only a few species (Moran & Bell 1983) and genetic inheritance is often based on segregation in open-pollinated arrays as has been done in *Alnus* (alder) (Bousquet *et al*. 1987), *Casuarina* (casuarinas) Moran *et al*. (1989b), and *Acacia* (acacias) (Playford *et al*. 1993; McGranahan *et al*. 1997). There is usually a common number of allozyme loci for each enzyme for all species within a plant genus or even
a family.

4.2.2   Terpenes and essential oils

Terpenes in plants consist of a large and diverse group of hydrocarbons made up from the five-carbon building block isoprene [$CH_2=C(CH_3)CH=CH_2$] or similar five-carbon units (Hanover 1992). They are generally volatile and highly aromatic compounds. They are divided broadly into three groups, with monoterpenes being made up of two five-carbon units, and with sesquiterpenes and diterpenes being made up of three and four units, respectively. Methods for determination of terpene composition via gas-liquid chromatography are outlined in Squillace (1976). For conifers, monoterpenes from the resin have been the primary marker of choice. The essential oils that are commercially harvested in several broad-leaved forest tree species such as *Eucalyptus* (eucalypts) and *Melaleuca* (melaleucas) are also terpenes (Weiss 1997; Doran 2000).

Recent reviews (Hanover 1992; Baradat *et al*. 1995b) consider the genetic inheritance of terpene compounds and outline the limitations in genetic interpretation. Within a species there are many possible compounds or potential markers and these can display considerable qualitative and quantitative variation, and can be used to define chemotypes (e.g. Butcher *et al*. 1994). The mode of inheritance can vary from a single dominant gene to apparently polygenic and the expression can be significantly affected by environmental effects and stages of development. The

main drawback to the use of terpenes as genetic markers is the difficulty of assigning exact terpene genotypes to trees. The considerable statistical difficulties of analysing data consisting of proportions of different terpene compounds in conifers have been reviewed by Birks and Kanowski (1988, 1995). Unless there is commercial interest in the terpenes themselves there seems little point in using them as genetic makers given the ready availability of isozymes and DNA markers in forest trees.

4.3   DNA-based markers

Biochemical markers examine the products of genes. This analysis can be complicated by effects of gene expression, of epistasis (interactions between genes), and of the redundancy of the genetic code. The advent of restriction enzymes and the polymerase chain reaction (PCR) has allowed assessment of genetic variation directly at the DNA level. A range of powerful and rapidly developing techniques are now available—we will discuss below the most commonly used techniques.

First, however, it will be helpful to define the term 'locus' in terms of molecular genetics. In the classical sense the term refers to a particular nuclear gene. However, with the advent of DNA markers the meaning of this term has changed. As most of the DNA in the nuclear genome of trees does not code for genes (see Chapter 1), most DNA markers tag intergenic regions. Hence, in molecular genetics, a marker locus simply refers to the unique position within the genome (nuclear, chloroplast, mitochondrial) of the DNA segment detected, and no longer implies that the DNA segment detected is necessarily part of a gene. For the purposes of this chapter, a Mendelian locus occupies a unique position within the nuclear genome (as opposed to either of the organelle genomes), and therefore has two segregating copies (alleles) in each diploid organism, one derived from its maternal parent and one derived from its paternal parent.

4.3.1   Restriction fragment length polymorphisms

Restriction fragment length polymorphisms (RFLPs) are the original DNA marker, and were developed in the late 1970s (Botstein *et al*. 1980). Their development was facilitated by the discovery of restriction enzymes, which are 'molecular scissors' that cut DNA at specific sequences (recognition sequences). For example, the restriction enzyme called

*Eco*RI cuts DNA at every occurrence of the sequence 'GAATTC'. The resulting pieces of DNA produced are called 'restriction fragments'. Digestion (i.e. cutting) of the DNA from a single pine cell with *Eco*RI produces roughly five million of these restriction fragments. Hundreds of these enzymes are now commercially available for use by molecular geneticists, covering many unique recognition sequences [usually 4 or 6 base pairs (bp) long].

Genetic differences between chromosomes can result in differences in the lengths of particular restriction fragments, or 'length polymorphisms'. Substitutions occurring in the DNA can result in a sequence difference within a particular recognition sequence (restriction site), leading to either the loss or gain of a particular restriction site and a length difference in the fragment produced. Alternatively, insertions or deletions of segments of DNA between two restriction sites may occur, changing the length of a particular fragment. The RFLP process allows the detection of these length polymorphisms in particular restriction fragments via hybridisation with labelled probes (Box 4.2). These probes are single-stranded DNA molecules that match part of the sequence of certain restriction fragments and hence will bind to them via complementary hybridisation. A DNA fragment from anywhere in any of the three genomes may be used to generate a probe—however, Mendelian genotypes will only be revealed by a probe derived from single-copy nuclear DNA. Probes from other species can also be used provided that they are similar enough in sequence (i.e. have sufficient homology) to hybridise to the corresponding DNA of the species being studied.

4.3.1.1   *Restriction fragment length polymorphisms in organelle DNA*

The high copy number and small size of organelle genomes make them far more amenable to RFLP analysis than nuclear genomes, particularly in conifers. Hence, most RFLP work performed in trees has been with chloroplast DNA (cpDNA). Another factor facilitating cpDNA studies in trees is the high degree of sequence conservation in the chloroplast genome, which allows the use of heterologous probes from other genera, or even families.

In Pinaceous species, the contrasting modes of inheritance of chloroplast and mitochondrial genomes (see Chapter 1) provides a unique opportunity to trace both maternal and paternal lineages through the species' recent evolutionary history, by examining

mitochondrial DNA (mtDNA) and chloroplast DNA (cpDNA) polymorphisms, respectively. Such studies in pines reveal a much higher degree of population differentiation in mtDNA than in cpDNA, as would be predicted by the expectation that dispersal via seed (tracked by mtDNA in Pinaceous species) is much more limited than dispersal via wind-disseminated pollen (tracked by cpDNA) (Dong & Wagner 1993, 1994; Latta & Mitton 1997). A remarkable result from cpDNA work in angiosperms is the apparent ease with which organelle genomes can be transferred from one species to another in some genera via hybridisation, a phenomenon known as 'chloroplast capture' (Rieseberg & Soltis 1991). In angiosperm trees this phenomenon has been observed in both American and European oaks (Whitemore & Schaal 1991; Petit *et al.* 1993, 1997) and in *Eucalyptus* (Steane *et al.* 1998). In such situations, cpDNA has limited utility for phylogenetic analysis.

4.3.1.2   *Restriction fragment length polymorphisms in repetitive nuclear DNA*

Most of the early RFLP studies of tree nuclear genomes have been done using probes that hybridise to repetitive DNA (e.g. Rogstad *et al.* 1988; Strauss & Tsai 1988; Kvarnheden & Engstrom 1992). Detecting repetitive DNA in large genomes via RFLP analysis is far less technically demanding than detecting single copy sequences. This approach can be very useful for distinguishing individuals and clones (e.g. Rogstad *et al.* 1991; Kennington *et al.* 1996). A drawback to this approach is that the complexity of the multi-banded hybridisation patterns produced precludes straightforward genetic interpretation in Mendelian terms, which is a necessary prerequisite for most traditional population genetic analyses (see Chapter 2).

4.3.1.3   *Mendelian restriction fragment length polymorphisms*

Use of single or low copy nuclear probes in RFLP analysis provides data that can be readily interpreted in Mendelian terms. This approach is more tractable in angiosperms because of their smaller genomes. The first study of this sort in trees uncovered unidirectional introgression of genes from one *Populus* species (aspen) into another within a sympatric zone (Kiem *et al.* 1989). Most of the work with this type of marker in trees has been for the purposes of genetic linkage mapping (i.e. locating many genetic markers relative to each other on linkage groups corresponding to chromosomes) with the eventual goal of accelerating

Box 4.2 The restriction fragment length polymorphism process

Steps in the RFLP assay are illustrated in Figure 1.

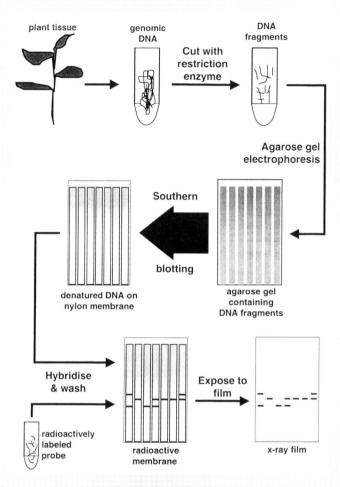

Figure 1 The restriction fragment length polymorphism (RFLP) process.

Genomic DNA is extracted from plant tissue (usually leaves) and then digested with a restriction enzyme. The resulting DNA fragments are separated by size on an agarose gel, with larger fragments migrating more slowly (from the top of the gel in the figure) than the smaller fragments. The DNA fragments can be visualised in the gel by ultraviolet illumination after staining with ethidium bromide. The DNA fragments are then covalently bound in denatured (single-stranded) form onto a nylon membrane by Southern blotting. The nylon membrane is then incubated in a solution containing the radioactively labelled, single-stranded probe (hybridisation). Excess probe other than that binding to DNA fragments of matching sequence is then removed from the membrane by repeated washes. The membrane is then exposed to X-ray film, and the radioactively tagged DNA fragments are visualised as bands. Sizes of the bands can be calculated by comparison with bands of reference DNA size markers run in at least one or more lanes of the gel (not shown). Probes are then stripped from the membrane by washing it in a very hot (95°C) detergent solution. The membrane can then be reused many times with different probes.

The DNA-level genotypes and corresponding RFLP banding patterns for a hypothetical Mendelian locus with two alleles are shown in Figure 2.

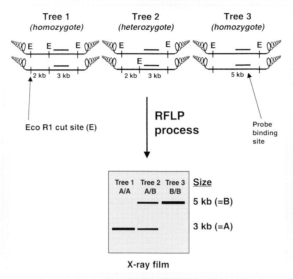

Figure 2 The three possible genotypes at an RFLP marker locus with two alleles. Allele A = 3 kb and allele B = 5 kb.

tree breeding programs via early, marker-assisted selection for desirable traits influenced by genes that are linked to some of the markers (e.g. Devey *et al.* 1994; Bradshaw *et al.* 1994; Mukai *et al.* 1995). Population or conservation genetics work in trees with single or low copy number nuclear RFLP markers has been fairly limited (Kiem *et al.* 1989; Lui & Furnier 1993; Fjellstrom & Parfitt 1994; Glaubitz 1995; Lerceteau *et al.* 1997; Parani *et al.* 1997; Butcher *et al.* 1998; Byrne *et al.* 1998). This is probably because of the advent of random amplified polymorphic DNA (RAPD) markers (see Chapter 4.3.5), which quickly became very popular as they are far less technically challenging to use in trees than are Mendelian RFLPs. However, RAPD markers have several disadvantages relative to RFLPs (see Chapter 4.4).

4.3.2 Polymerase chain reaction

All DNA markers other than RFLPs are based in some way upon PCR (Mullis & Faloona 1987) (Box 4.3). This powerful technique provides a means by which billions of copies of a particular 'target' DNA fragment can made, starting from a few copies within a complex DNA mixture (i.e. all of the DNA from numerous cells).

PCR products usually cannot exceed about 3 kb in size because of limitations of the PCR process. PCR products can be derived from anywhere in the genome (nuclear, chloroplast, mitochondrial), provided that the sequences flanking the target region to be amplified by PCR are known. Specific PCR-amplified Mendelian loci from the nuclear genome are called 'sequence tagged sites' (STSs).

4.3.3 Detection of polymorphism in products of polymerase chain reaction

PCR products from different trees (or, in the case of nuclear DNA, from different alleles at a given locus within an individual) can differ in their length or sequence. Length differences may be detected by gel electrophoresis, although small differences in length (i.e. 1–10 bp) need to be resolved on high resolution polyacrylamide gels rather than agarose gels. However, PCR products of identical length but different sequence will be indistinguishable without further manipulation.

Methods to detect differences in PCR products of identical length derived from the same locus will now be discussed.

Box 4.3 The polymerase chain reaction process

The PCR reaction is performed in cycles, with each cycle consisting of short (about 30 s) incubations at three different temperatures (Fig. 1).

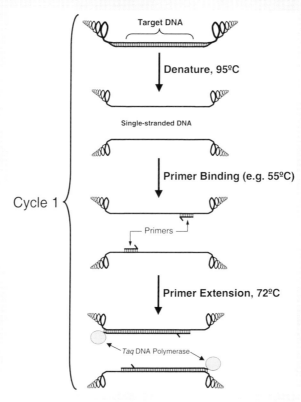

Figure 1   The polymerase chain reaction—cycle 1.

PCR reactions are carried out in thermocyclers, which are essentially fully programmable heating blocks. A cycle begins with denaturation of the sample genomic DNA at 95°C. This is followed by primer binding (or annealing), at an optimal temperature, usually about 50°C to 65°C. Primers are short, synthetic, single-stranded DNA molecules. Specific pairs of custom-made primers are used in PCR reactions, and define, by their sequences, which particular target DNA fragment will be copied, or amplified. Primers are usually 18 to 30 bases long. Primer pairs are designed to bind to opposite, complementary strands a short distance apart from each other (75–3000 bases) along the DNA. The optimal annealing temperature is generally empirically determined to allow for efficient amplification of the intended target with minimal non-specific amplification of non-target sequences, and depends mainly on the length and sequence of the primers. The segment of DNA between the two primer binding sites is amplified in the PCR reaction. The primers serve as starting points for the synthesis of new DNA strands complementary to the original template strands. The enzyme DNA polymerase adds bases (nucleotides) to the 3' ends of the growing DNA strands. This step of the reaction, called primer extension, is performed at 72°C. This rather high temperature is used because the DNA polymerase was obtained from a bacterium (*Thermus aquaticus*, or *Taq* for short) that lives in hot springs. The optimal temperature for the activity of this enzyme is 72°C. *Taq* polymerase was chosen for use in PCR because it is able to withstand repeated exposure to the high temperature (95°C) used in the denaturation step.

A single cycle in PCR is then composed of the following three steps:

1   denaturation at 95°C

2   primer annealing (e.g. at 55°C)

3   primer extension at 72°C.

Figure 2 considers only the DNA strands synthesised in the first cycle, which are of indeterminate length. These strands serve as templates for synthesis in the second cycle of single DNA strands of the expected length. Only after the third cycle are double-stranded PCR products of appropriate length produced. At this point in the reaction, two of these will have been produced for every template molecule present in the DNA sample originally added to the reaction. The total number of PCR products produced will then nearly double every cycle after the third cycle. After 27 more cycles (30 cycles in total), more than 100 billion PCR products are produced.

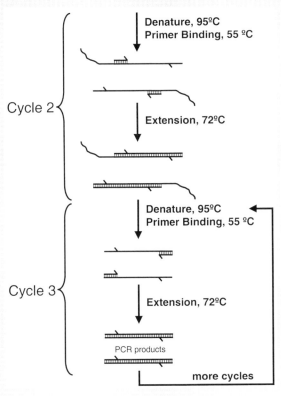

Figure 2   Cycles 2 and 3.

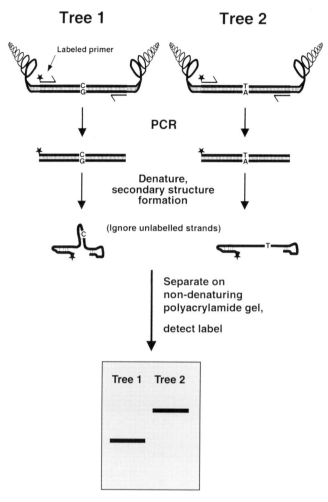

Figure 4.2 Polymerase chain reaction (PCR)–single strand conformation polymorphism. In this example a segment of the chloroplast genome from two trees is amplified by PCR. The two PCR products are identical except for the bases at the position shown, with this sequence difference causing a detectable difference in the secondary structures of the single-stranded DNA. One of the PCR primers is labelled on the 3' end with radioactivity or fluorescence (star).

4.3.3.1   DNA sequencing

The most comprehensive analysis of genetic differences between PCR products can be obtained by sequencing. Direct sequencing of PCR products is now fairly routine, particularly with the advent of cycle sequencing methodology (Murray 1989). However, this method is generally too expensive for population and conservation genetics applications, for which large sample sizes are required. Also, analysis of Mendelian loci with this method is complicated in heterozygotes by the simultaneous generation of sequences from both alleles. Interpretation can be challenging in such cases, particularly when small deletions of DNA occur in one of the alleles. For this reason, direct sequencing of PCR products can be applied more successfully to organelle DNA loci or to repetitive nuclear DNA such as ribosomal DNA where 'consensus' sequences are obtained (e.g. Savard *et al.* 1994). For phylogenetic studies, direct sequencing of PCR products is the method of choice, as the most powerful methods of phylogenetic analysis require sequence data, optimally from more than one gene (Nei 1987).

4.3.3.2   Restriction fragment length polymorphisms

Another, less comprehensive way to detect differences among PCR products is to digest them with restriction enzymes. For this purpose, several restriction enzymes with four base recognition sequences are usually used. Digestions with each enzyme are analysed separately by agarose or polyacrylamide gel electrophoresis, and genetic differences can be detected as changes in the resulting banding patterns, as in RFLPs. This method has been applied to forest trees by Tsumura *et al.* (1997) and by Harry *et al.* (1998).

4.3.3.3 Single strand conformation polymorphism and denaturing gradient gel electrophoresis

Even when many restriction enzymes are used in PCR–RFLP, sequence differences can go undetected if they do not occur within any of the restriction sites of the enzymes employed in the analysis. Alternative methods of detecting nearly any sequence difference in PCR products that are cheaper than DNA sequencing include PCR–single strand conformation polymorphism (SSCP) and PCR–denaturing gradient gel electrophoresis (DGGE).

SSCP analysis (Hayashi 1992) allows, with minimal manipulation, the detection of greater than 80–90% of the possible single-nucleotide substitutions within PCR products (or restriction fragments of PCR products) up to 400 bases long. PCR products are denatured by heating prior to their resolution on a cooled, non-denaturing polyacrylamide gel. Single-stranded DNA molecules develop 'secondary structure' under non-denaturing conditions by folding upon themselves and partially binding to themselves via internal complementary hybridisation. A great variety of secondary structures are possible and the precise shape taken is highly sensitive to sequence differences. The different shapes produced often migrate differently on the non-denaturing polyacrylamide gel (Fig. 4.2).

Several studies in forest trees have used PCR–SSCP analysis (e.g. Watano *et al.* 1995; Dumolin-Lapegue *et al.* 1996; Bodenes *et al.* 1997; Quijada *et al.* 1997).

DGGE (Myers *et al.* 1987; Sheffield *et al.* 1989) utilises polyacrylamide gels containing a gradient of increasing concentrations of the denaturing agents urea and formaldehyde. PCR products are loaded on the gel in double-stranded form and, upon electrophoresis, reach a point in the gel where they begin to denature. Complete denaturation of the products is prevented by the attachment of a 'GC clamp' to one end prior to loading (Sheffield *et al.* 1989). As the binding of G to C is much stronger than

that of A to T, the GC clamp will resist denaturation under conditions that denature sequences of 'normal' GC content. Hence, the PCR products partially denature up to the GC clamps, starting from the region of lowest GC content. The partially denatured DNA molecules cannot migrate further in the polyacrylamide gel, and form a tight band at the point where they partially denatured. PCR products of slightly different GC content will often stop migrating at different positions in the gel (Fig. 4.3). PCR–DGGE, though technically more demanding to perform than PCR–SSCP, has shown promise in its first application in forest trees (Temesgen *et al.* 1998).

4.3.4 Microsatellite markers

A powerful derivative of PCR technology are microsatellite markers (Box 4.4). These markers are also known as simple sequence repeats (SSRs). As this latter name implies, they consist of segments of DNA containing tandem repeats of simple motif sequences, usually one to five bases, that are amplified by PCR. Tandem repeats of many simple sequence motifs, in particular the dinucleotide repeats CA or GA, are abundant in most eukaryotic nuclear genomes, and are distributed throughout these genomes in dispersed locations. They therefore comprise a particular class of dispersed repetitive DNA. These microsatellite repeats are often flanked by unique sequences, occurring only once in the genome.

Microsatellites were first developed for use in the human genome (Weber & May 1989) and were later found to be abundant in plants (Morgante & Olivieri 1993). They are useful because they are highly polymorphic markers, as there are often many alleles at a microsatellite locus, each allele having a different number of tandem repeats. This occurs because there is a high mutation rate for the number of repeats at microsatellite loci, as DNA replicating enzymes make 'mistakes' in the number of repeats at a relatively frequent rate. Larger microsatellites, containing more repeats, tend to be more polymorphic (Beckman & Weber 1992). The high degree of polymorphism and codominance of microsatellites makes them extremely informative. They have a very high discrimination power—for example, at eight microsatellite marker loci, each with 10 alleles, there are more than 83 trillion theoretically possible multilocus genotypes. Hence, they are the markers of choice for 'DNA fingerprinting' applications, for example in forensic work or paternity analysis in humans. Applications in

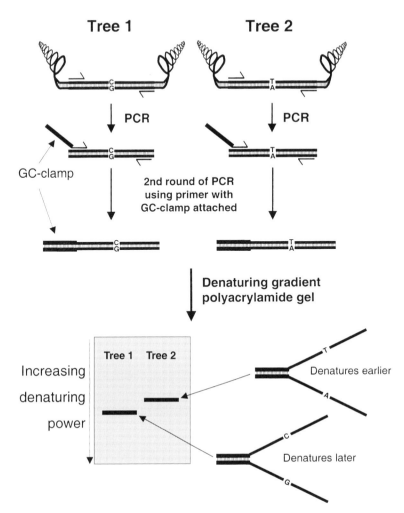

Figure 4.3 Denaturing gradient gel electrophoresis (DGGE). The same hypothetical chloroplast DNA polymorphism as the one in Figure 4.2 is here detected by DGGE. The stronger base pairing of the CG pair in the PCR product from Tree 1 causes it to resist denaturation longer than that from Tree 2. The GC clamp prevents further denaturation, and the partially denatured Y-shaped molecules stop migrating in the gel.

tree species include analysis of mating systems, paternity and patterns of gene flow. They will also prove to be of great utility for quality control in tree breeding programs and for the certification of genetically improved seed and planting stock.

The first microsatellites developed in trees were from *Pinus radiata* D. Don (radiata pine) (Smith & Devey 1994) and they have since been developed in many forest tree species, including *Quercus* spp. (oaks) (Dow *et al*. 1995; Barrett *et al*. 1997; Isagi & Suhandono 1997), *Eucalyptus* (Byrne *et al*. 1996), *Pinus strobus* L. (eastern white pine) (Echt *et al*. 1996), *Picea abies* (L.) H. Karst (Norway spruce) (Pfeiffer *et al*. 1997), and several tropical trees (Chase *et al*.

1996a; White & Powell 1997; Dawson *et al*. 1997; Steinkellner *et al*. 1997). Mononucleotide microsatellites (i.e. with repeat motifs of only 1 base) have also been uncovered in the pine chloroplast genome (Vendramin *et al*. 1996). Development of methods for the production of DNA libraries enriched for microsatellite sequences (e.g. Kijas *et al*. 1994; Edwards *et al*. 1996) has increased the efficiency of microsatellite development.

The power of microsatellites has been convincingly demonstrated in a study of the tropical tree species *Pithecellobium elegans* Ducke that detected long-distance gene flow via pollinators between isolated trees (Chase *et al*. 1996b). This study had important conservation genetic implications, suggesting that, in this species at least, spatially fragmented conservation populations of scattered trees may still be genetically viable (see Chapter 10).

4.3.5 Random amplified polymorphic DNA

The RAPD marker system (Williams *et al.* 1990) is a modification of PCR in which a single, short primer of only 10 bases long is used (rather than the 2 flanking primers of around 20 bases each used in regular PCR). To allow binding of this short primer to matching sequences, low annealing temperatures (e.g. 37°C) are used in RAPD reactions. RAPD primers are of arbitrary sequence. By chance, sequences matching a given RAPD primer will occur in many places throughout a large, eukaryotic genome. Occasionally, two such matching sequences will occur within a PCR-amplifiable distance (less than 3000 bp) and in the proper orientation (i.e. on opposite strands of the DNA) to allow PCR amplification. In large genomes, several loci are usually amplified with any given RAPD primer, resulting in 3 to 20 detectable bands on an agarose gel.

The RAPD marker method detects DNA polymorphisms that result in the loss or gain of amplification at a locus. Amplification can be lost owing to substitution within a primer binding site, deletion of a primer binding site or a large insertion between two primer binding sites. Amplification can be gained from substitutions or insertions that create new primer binding sites, or deletions between two existing primer binding sites (bringing them within PCR-amplifiable range of each other). Hence, there are typically only two possible alleles at a RAPD marker locus—that leading to amplification of a band (the 'presence' allele) and that from which a band is not produced (the 'absence' or 'null' allele). Alternative presence alleles of different size are seldom observed. Different sized bands produced from a single RAPD primer are generally assumed to be derived from different RAPD loci.

In the case of a RAPD marker locus from the nucleus, only two 'phenotypes' are observable—presence or absence of the corresponding band. Hence, at the level of genotype, presence/presence homozygotes are indistinguishable from presence/absence heterozygotes, as both lead to the observable 'phenotype' of the presence of the band. Complete absence of the band is only observed in absence/absence homozygotes. Hence, the absence allele is recessive to the presence allele, and RAPDs are therefore dominant markers.

As a given RAPD primer will generate bands from the DNA of most eucaryotic species, a set of RAPD primers can be applied to any species. The only prior knowledge needed to apply RAPDs to a new species is how to extract DNA of sufficient purity for PCR amplification. As, in theory, more than one million unique 10-base RAPD primers can be synthesised, the number of RAPD loci that can be generated is well beyond conceivable needs. Sets of RAPD primers are commercially available—in total, around 2000 unique primers can be purchased from suppliers.

The technical ease of RAPD markers and their application to any species has led to their use in many studies in forest trees, both in genetic linkage mapping (e.g. Tulsieram *et al.* 1992; Grattapaglia & Sederoff 1994; Nelson *et al.* 1994) and population genetic applications (e.g. Mosseler *et al.* 1992; Chalmers *et al.* 1994; Isabel *et al.* 1995; Nesbitt *et al.* 1995; Schierenbeck *et al.* 1997). The use of RAPD markers has led to rapid advances in the molecular genetics of trees.

RAPD marker bands may originate from any position within cellular DNA, including nuclear DNA (single, low copy or repetitive sequences), cpDNA or mtDNA. In *Pseudotsuga menziesii* (Mirb.) Franco (Douglas-fir) a high proportion (45%) of polymorphic RAPD bands displayed maternal inheritance and were thus mitochondrial in origin (Aagaard *et al.* 1995). The authors attribute this higher than expected proportion of RAPD bands of mitochondrial origin to the prevalence of inverted repeats in plant mitochondrial genomes, as well as to the possible large size of the mitochondrial genome in *Pseudotsuga menziesii* (Aagaard *et al.* 1995). Hence, to ensure appropriate analysis, the inheritance of all RAPD markers should be confirmed before they are used in population genetic studies.

4.3.6 Amplification fragment length polymorphism

A recent, powerful genetic marker that can be thought of as a combination of RFLP and PCR technology has been called amplification fragment length polymorphism (AFLP) (Vos *et al.* 1995). This rather complex method is outlined in Figure 4.4a.

Box 4.4 Microsatellite development

The process of microsatellite development is illustrated in Figure 1.

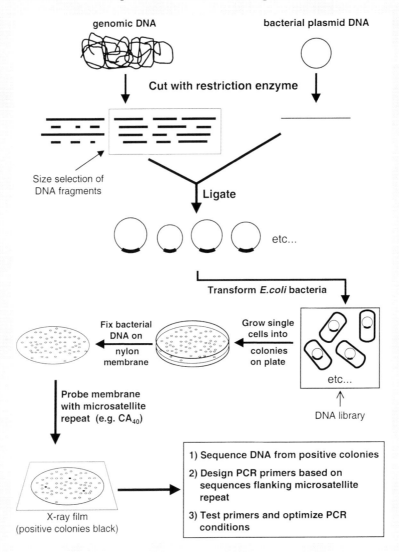

Figure 1   Development of microsatellites.

Genomic DNA is cut with a restriction enzyme and then fragments from 200 to 1000 bp in size are selected from an agarose gel and purified. The size-selected DNA is then ligated into a restriction site in bacterial plasmid DNA to make many circular, recombinant plasmids, each containing a particular DNA fragment from the foreign genome. The recombinant bacterial plasmid DNAs are introduced, or transformed, into the bacterium *Escherichia coli* wherein they proliferate and are passed on upon cell division. The resulting collection of millions of bacteria, each containing a particular foreign DNA fragment, is referred to as a DNA library (DNA libraries are also the source of the probes used in RFLP analysis). Individual bacteria from the library can be separated on nutritive media and grown into colonies. A portion of each colony is transferred to a nylon membrane in replicate fashion by carefully laying the nylon membrane on top of the plate and then peeling it off. The bacteria on the membrane are lysed with detergent and their released DNA is then fixed in position upon the membrane using sodium hydroxide. To identify the rare colonies that harbour microsatellite-containing DNA fragments, the membrane is probed with microsatellite repeat motifs that are radioactively labelled. Positive colonies on the X-ray film are identified and then picked from the original plate by aligning the X-ray film with the plate. Positive colonies are then grown in culture and their plasmid DNAs are purified. The foreign insert DNA portions of the plasmids are then sequenced. PCR primers are designed from unique sequences flanking the microsatellite repeats. Optimisation of PCR conditions (e.g. annealing temperature, magnesium concentration, primer concentrations) to achieve reliable but specific amplification comprises the final stage in microsatellite development.

A DNA sample is simultaneously digested with two restriction enzymes (e.g. *Eco*RI, a 6-base, or rare cutter, and *Mse*I, a 4-base, or frequent cutter). Two short, synthetic, double-stranded adaptor sequences, each with overhanging bases complementary to those produced by one of the restriction enzymes (Fig. 4.4b), are then ligated, or joined, to the ends of the restriction fragments.

Two PCR primers are then added to the reaction, matching sequences on each of the two adaptors, respectively, and the mixture is subjected to PCR amplification. Detection of the fragments produced after their separation on high resolution polyacrylamide gels is facilitated by the presence of a radioactive (or fluorescent) label on one of the PCR primers. Hence, the basic AFLP assay consists of these four successive steps:

1   digestion with restriction enzymes

2   ligation of adaptors

3   PCR amplification

4   electrophoresis and detection.

The AFLP reaction produces many bands from a single assay—from the complex genomes of trees, the basic AFLP assay will produce too many bands to analyse. The number of bands produced is reduced by the addition of extra bases to the 3' end of the PCR primers. These extra bases, termed selective bases, make the primer binding site extend beyond the restriction enzyme recognition sequence, further into the restriction fragment (Fig. 4.4b). They reduce the number of fragments produced from a PCR amplification step, as only those fragments with sequences internal to the restriction enzyme recognition sites that match these selective bases will

be amplified. Usually, one to three selective bases are added to either or both primers. The addition of one selective base to each of the primers will reduce the number of fragments produced by 16-fold. The addition of three selective bases to each primer, a necessity for the large genomes of most trees, should reduce the number of fragments produced by 4000-fold. Generally, 30 to 150 fragments are manageable in terms of electrophoretic separation and analysis.

AFLPs, like RAPDs, are predominantly dominant markers with only two alleles—the presence and absence of a given band. The absence of a band can result from the loss of one of the corresponding restriction sites, which will generally result in the production of a restriction fragment too large for PCR amplification. Alternatively, an evolutionary event resulting in an insertion of DNA between the relevant restriction sites can also produce a fragment that is too large for PCR. A new band can arise by the gain of a suitable restriction site, or by a deletion between two restriction sites, putting them within PCR-amplifiable range of each other.

Although AFLPs are a relatively recent development, there are already several examples of their application in forest trees (Cervera *et al.* 1996a; Akerman *et al.* 1996; Beismann *et al.* 1997; Gaiotto *et al.* 1997; Marques *et al.* 1998). It is a powerful technique of great promise, the main advantage being the large number of loci uncovered in a single assay.

4.4 Comparison of marker systems

We will compare the strengths and limitations of the most commonly used marker systems, using Table 4.1 as a point of orientation. The comparison is limited in

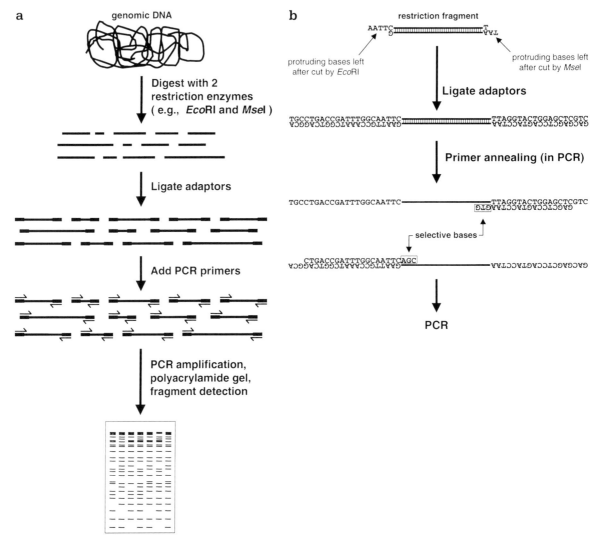

Figure 4.4 Amplification fragment length polymorphism (AFLP). (a) The basic assay. (b) Ligation of adaptors and the use of 'selective bases' in the AFLP assay. Only those *Eco*RI/*Mse*I fragments with sequences just internal to the restriction sites that match the selective bases on the 3' end of the primers will be amplified by PCR. (See text for detailed description).

scope to applications where the intent is to examine variation at Mendelian loci.

4.4.1 Number of available loci

The major limitation of isozymes is the number of available loci (Table 4.1). In some pine species over 40 isozyme loci have been resolved in seedling and megagametophyte tissues (e.g. Buchert *et al.* 1997) but this number reduces substantially in bud, needle and other tissues. However, in angiosperm genera like *Eucalyptus*, *Melaleuca* and *Casuarina*, the number of loci available is often fewer than 30. For applications relating to broad population genetic structure and mating systems these number of loci are generally adequate. However, for fine-scale studies of gene flow, paternity and spatial structure, the level of genetic discrimination is usually not high enough. In organisms with low levels of genetic diversity the number of available polymorphic loci can be a serious limitation to the use of isozymes.

Expression of isozymes varies between plant tissues. Typically the greatest number of isozyme loci can be resolved in seed tissues or in just germinated seedlings. Many of these loci are not active enough to be assayed

TABLE 4.1   Comparison of commonly used genetic markers

AFLP, amplification fragment length polymorphism; RAPDs, random amplified polymorphic DNA; RFLP, restriction fragment length polymorphisms; SSR, simple sequence repeats or microsatellites

	Isozymes	RFLP	SSR	RAPD	AFLP
Theoretical no. of loci	30–50	Unlimited	10 000	Unlimited	Unlimited
Practical no. of loci	30–50	100s	10s	1000s	1000s
Degree of polymorphism	Low to moderate	Moderate to high	Very high	Moderate to high	Moderate to high
Dominance	Codominant	Codominant	Codominant	Dominant	Dominant
Null alleles	Rare	Extremely rare	Occasional	Presence/absence	Presence/absence
Transferability of loci	Across families	Among related genera	Within subgenera	Within species	Within species
Reliability (reproducibility)	Very high	Very high	High	Low to medium	Medium to high
Amount sample required (per assay)	Milligrams of tissue	2–10 mg DNA per lane	25–50 ng DNA	5–10 ng DNA	25 ng DNA
Ease of assay	Easy	Difficult	Easy to moderately difficult	Easy	Moderately difficult
Can be automated?	Difficult	Difficult	Yes	Yes	Yes
Multiplexing (loci/ assay)	Gel slices (6)	1–3	1–9	5–20	20–100
Costs					
Equipment:	Cheap	Expensive	Expensive to v. expensive	Moderate	Expensive
Development:	Cheap	Expensive	Very expensive	Moderate	Moderate
Assay:	Cheap	Expensive	Moderate to expensive	Moderate	Moderate

in other tissues. Thus, only a limited subset of isozyme loci can be assayed across the life cycle stages for most tree species. This epigenetic variation has been a distinct drawback in population and conservation genetic studies of rare tree species. In such cases a comprehensive and direct sampling of vegetative tissue from individuals in populations is desirable, as seed availability is often limiting. Another disadvantage of isozymes is that they represent a non-random sample of the genome, as most are glycolytic enzymes.

In contrast, all DNA markers theoretically have an 'infinite' or very large number of available loci. In practice, financial and time constraints mean that, at best, thousands of loci can be uncovered using AFLPs and RAPDs, hundreds using RFLPs and only tens of loci using microsatellites. The latter low number is due to the high cost and effort involved in microsatellite development.

4.4.2   Degree of polymorphism

The drawback in microsatellites of the small number of available loci is compensated for by the very high degree of polymorphism in these markers. Isozymes, however, have relatively low levels of polymorphism. As few as five or six microsatellite loci can often answer many conservation genetic questions (e.g. paternity, pollen flow) that cannot be answered with 30 or more isozyme loci.

Although isozymes may give lower absolute levels of genetic diversity, the patterns of population genetic structure that they reveal are generally similar to those from the DNA markers (e.g. Byrne *et al.* 1998; Butcher *et al.* 1998). Hence, they are still the markers of choice for studies of genetic diversity and mating systems in natural populations in most species, except those that display low isozyme variability such as *Acacia mangium* Willd. (Moran *et al.* 1989c; Butcher *et al.* 1998). For testing hypotheses about changes in levels of genetic diversity, however, isozymes, because they have only moderate levels of variability, may not provide enough statistical power.

RFLPs generally show a level of diversity intermediate to that of isozymes and microsatellites. As RAPDs and AFLPs are based on the presence or absence of bands, there is a maximum of two alleles available at each locus—therefore, they are not particularly diverse at the single locus level. However, because so many loci can be uncovered with these markers, enough polymorphic loci can usually be found for most purposes.

4.4.3   Dominance

With RAPDs and AFLPs the advantage of the many polymorphic loci generally available is tempered by the dominant nature of these assays. This translates into a large loss of genetic information, resulting in estimates of population genetic statistics of much lower precision relative to those obtained with an equal number of loci from codominant markers such as isozymes or RFLPs (Lynch & Milligan 1994). Hence, with RAPDs and AFLPs, many more loci need to be assayed to obtain sufficient statistical power to answer a particular question.

4.4.4   Null alleles

Null alleles are not very common in any of the codominant marker systems, where alleles are distinguishable by mobility differences on the gels. In RFLPs, null alleles are extremely rare, as they are only observed when a deletion has removed all or most of the probe binding site. In isozymes they are occasionally observed, and are due to mutations causing the loss of enzyme function. Null alleles at isozyme loci seem to be more prevalent for less specific enzymes (i.e. with more loci per enzyme system) such as esterases and peroxidases.

For microsatellites, null alleles are the result of mutations that prevent PCR amplification, such as sequence changes within one of the primer binding sites. Null alleles in microsatellites appear to be

observed more frequently than in isozymes or RFLPs (Jarne & Lagoda 1996; Band & Ron 1997; Kijas *et al.* 1997; Gockel *et al.* 1997). The undiagnosed presence of null alleles causes genotyping errors, and is detected only when they are frequent enough to sometimes occur in homozygous form. However, the occasional blank lane that results can easily be misinterpreted as a failed reaction. The resulting occasional genotyping errors are more problematic in applications where correct genotyping is critical, such as mating system or paternity analyses.

4.4.5   Transferability of loci

Isozymes provide a common set of loci, functional in all plant species, that are very useful for comparative studies across taxonomic categories (e.g. Hamrick & Godt 1989). In contrast, comigrating RAPD or AFLP bands from different species cannot be assumed homologous (i.e. derived from the same locus), even in closely related species. Homology of RAPD or AFLP bands should be confirmed before they are employed in phylogenetic studies across species. Microsatellites are, in general, transferable only to related species (i.e. within subgenera). A strength of RFLPs relative to the other DNA markers is that even anonymous genomic probes usually transfer well to any other species in the genus, and even to closely related genera. More conserved probes (i.e. from conserved genes) will transfer even further—for example, chloroplast gene probes will transfer across plant families.

4.4.6   Reliability and reproducibility

One of the critical characteristics of a genetic marker system is its reliability (reproducibility). This is the greatest weakness of RAPD markers, which sometimes display low reproducibility (e.g. Ellsworth *et al.* 1993; Muralidharan & Wakeland 1993; Skroch & Nienhuis 1995), due to their sensitivity to reaction conditions and DNA template quality. RAPDs are particularly difficult to reproduce across laboratories, even when the same DNA samples are used by the different researchers (e.g. Penner *et al.* 1993; Jones *et al.* 1997). The best studies with RAPD markers are those conducted with a great deal of caution, prescreening the bands for those that are reproducible, and monitoring reproducibility throughout by partial replication.

In contrast to RAPDs, AFLPs have shown a medium to high degree of reproducibility both within and across laboratories (Akerman *et al.* 1996; Jones *et al.* 1997).

This is probably due to the use in AFLPs of longer PCR primers, allowing more stringent PCR conditions. The currently very popular RAPD markers probably will be supplanted soon by AFLPs which, as well as being more reproducible, generate far more polymorphic loci per assay. However in light of the complexity and multiplex nature of the AFLP assay it would be prudent to carefully monitor reproducibility when using these markers as well.

Isozymes and RFLPs are both very highly reproducible assays as they are not subject to the potential for artifacts inherent in simultaneous PCR amplification of multiple loci. Microsatellites, based on PCR of single loci using relatively long primers and stringent conditions, also display high reproducibility. The main challenge to microsatellite reproducibility lies in accurate sizing of alleles, which can differ by as little as one or two bases.

4.4.7 Amount of sample required

Isozyme assays can be performed on small amounts of tissue (e.g. single embryos, germinants) but fewer loci can be detected in mature vegetative tissue (e.g. leaves, buds). All PCR-based assays (RAPDs, AFLPs, microsatellites) require only small amounts of DNA (5–50 ng), whereas relatively large amounts of DNA are required (2–10 µg) for each lane of a Southern blot for RFLP analysis of Mendelian loci. This is a considerable advantage for PCR-based assays, particularly where plant tissue is limiting. Also DNA extraction will be more efficient as small-scale DNA isolations are cheaper and quicker.

4.4.8 Ease of assay

Isozymes and RAPDs are the least technically demanding to perform while RFLPs, due to the many steps involved, are the most challenging. Maximum sensitivity and cost efficiency are achieved in RFLPs when radioactive detection is used—this adds to the training required and also creates a need for a licenced laboratory facility. AFLPs are more difficult than RAPDs because there are more steps involved and because polyacrylamide gels are needed to achieve adequate resolution of the many fragments produced. Polyacrylamide gels are more challenging to use than the agarose gels used for RAPDs. For microsatellites, the PCR step is very similar to RAPDs but again polyacrylamide gels are required for adequate resolution of the small size differences between alleles. However, it is possible to run microsatellites on

automated DNA sequencers, which are essentially polyacrylamide gel setups with automated laser detection of fluorescent-labelled DNA fragments. Some of these machines also automate the gel loading process. Although such machines are very expensive, they do greatly reduce the effort needed for the assay, and most of them will also partially automate the scoring of microsatellites. AFLPs can also be run on automated sequencers, again reducing the effort required. When microsatellites or AFLPs are separated on manual polyacrylamide gels, detection is best achieved using radioactive labels, which further adds to the technical difficulty of these assays.

4.4.9 Automation

Most conservation genetics applications require large sample sizes in order to obtain sufficient statistical power. Automation of marker assays provides one means of attaining the high throughput needed for such sample sizes. Isozymes and RFLPs are predominantly manual processes, not amenable to automation. As isozymes are highly efficient markers, this is not a concern. In contrast, attainment of high throughput with RFLPs is impeded by their laborious nature. Unlike isozymes and RFLPs, all the PCR-based methods are quite amenable to automation. The PCR step can be automated by the use of pipetting robots. Size separation and scoring can be automated using DNA sequencer machines, some of which also automate the gel loading process. As these machines use fluorescent label technology, there is no need for radioactivity.

4.4.10 Multiplexing

Another important point of comparison of marker systems is the number of loci analysed per assay. The simultaneous analysis of multiple loci in a single assay is called multiplexing. In isozymes, each gel is sliced into about six horizontal slices and each slice is stained for a different 'enzyme system'. As enzyme systems are sometimes comprised of more than one Mendelian locus, up to about 10 isozyme loci can be simultaneously analysed. The lowest degree of multiplexing is achieved by RFLPs. Banding patterns produced by RFLP probes that bind to more than three or four loci are usually too complex to be interpreted genetically (i.e. as genotypes at Mendelian loci). Of the DNA markers, RAPDs and, in particular, AFLPs have the highest degree of multiplexing, uncovering many loci simultaneously. However, as these markers are both dominant, all bands are interpreted as

---

Box 4.5 Best DNA markers for various applications

The various applications for the use of DNA markers can be divided into three groups.

1   Applications that require large number of loci:

- measuring genetic diversity and differentiation

- estimating rates of gene flow or migration (between populations)

- genetic linkage mapping or quantitative trait loci localisation.

Markers of choice: AFLPs or RFLPs

2   Applications that require high discrimination power:

- characterising mating systems

- analysing paternity or parentage

- characterising patterns of gene flow or migration within populations

- assessing seed orchard efficiency

- quality control in breeding, DNA fingerprinting or cross verification.

Marker of choice: microsatellites

3   Applications that require DNA sequence information:

- phylogeny and taxonomy

Marker of choice: PCR and sequencing

---

presence/absence 'phenotypes' rather than as genotypes at Mendelian loci. Microsatellites can also be multiplexed. Microsatellites of different size ranges can be run in the same lane of a gel, and, when fluorescent detection is used, microsatellites labelled in different colours can also be run in the same lane. With three available colours and three size ranges, it is possible to run as many as nine microsatellites in a single lane of an automated sequencer, greatly improving cost efficiency. When AFLPs are run on DNA sequencers, further multiplexing can be achieved by using primers labelled with different fluorescent dyes.

4.4.11 Cost

Isozymes are much cheaper at all stages (equipment, development, assay costs) than any of the DNA markers. RAPDs are the cheapest of the DNA markers, followed by AFLPs, which are more expensive due to the need to use large polyacrylamide gels for their separation and radioactive detection. RFLPs are expensive at all stages, their development cost being increased by the need for a DNA library as a source of

probes. The major drawback of microsatellites is that they are very expensive to develop, the greatest costs, other than skilled labour, being DNA sequencing and the synthesis of many custom PCR primers. Greater sizing accuracy and a greater degree of automation are obtained if the microsatellites are run on automated DNA sequencers—however, these are very expensive. Assay costs for microsatellites are also higher on these machines. The cost of an automated DNA sequencer will be offset somewhat by reduced labour costs.

4.4.12 Which marker?

None of the markers is more competitive than the others in all aspects. However, the very low cost of isozymes along with their ease of use and the speed with which data can be obtained combine to make them very attractive markers for use in conservation genetics. Hence, an efficient strategy would be to use isozymes wherever possible as the first tool for answering the question at hand—DNA markers could then be used to supplement the information obtained from isozymes whenever the number of loci and/or the degree of polymorphism are insufficient. Another clear-cut result is that AFLPs are likely to completely replace RAPDs. The higher equipment costs of AFLPs and the somewhat more complicated assay are easily compensated for by their higher degree of multiplexing and greater reproducibility relative to RAPDs. With RFLPs there is a tradeoff between very reliable and informative data (with multiple, codominant alleles from potentially many loci) compared with the considerable expense and labour involved. Microsatellites give much more information per locus from fewer loci but have a very high development cost.

Deciding which DNA marker to use (i.e. beyond using isozymes wherever possible) depends on the situation—major factors are the financial resources, and availability of equipment and skilled personnel. It also depends on the study objectives, with different types of studies having different requirements. Groupings of applications for conservation genetics according to these requirements, and guidelines for the best DNA markers, are given in Box 4.5.

4.5 Future markers

DNA marker technologies are evolving rapidly, with the Human Genome Project acting as a major catalyst. DNA markers of the future will be more automated, more miniaturised and highly multiplexed, allowing very large numbers of loci to be accessed in parallel. A promising new marker technology developed for the human genome and exhibiting all of these features are single-nucleotide polymorphisms (SNPs), which are based upon the very exciting 'DNA chip' technology (Wang *et al.* 1998). Spin-offs of these futuristic technologies will eventually become available for application to forest trees.

4.6 Acknowledgments

The authors would like to thank Penny Butcher, Livinus Emebiri and two anonymous reviewers for helpful comments and suggestions for improvements to the manuscript. Box 4.2 figure 1 was derived from an original kindly supplied by Penny Butcher.

SECTION II
GENETIC PROCESSES

Genetic processes and their dynamic nature are fundamental to any consideration of both the conservation (*ex situ* and *in situ*) and use of forest genetic resources. Whether the objective is to preserve existing diversity, conserve evolutionary potential or maintain options for future generations while satisfying the needs of the present (World Commission on Environment and Development 1987), maintenance of genetic processes is essential. Timely identification as to which of these processes, and under what circumstances, they are limiting may save resources, whereas ignorance of them, either unconscious or wilful, may condemn costly initiatives to immediate or long-term failure. The basic genetic principles discussed in Section I showed how allelic and genotypic frequencies followed the Hardy-Weinberg equilibrium in populations of infinite size, under conditions of random mating, without migration, selection or mutation. The chapters in this section look at the processes behind each of these assumptions and examines how they interact to effect the gene pools of trees.

In Chapter 5, David Boshier considers the basic processes of reproductive biology (sexual systems, incompatibility mechanisms, flowering patterns, pollination processes) and how they combine to influence mating patterns in both temperate and tropical tree species. He identifies situations that lead to non-random mating in natural and disturbed forest systems, discussing methods available for quantifying mating patterns, and provides examples of their application to conservation issues. Levels of gene flow are critical in determining patterns and dynamics of genetic diversity. With the capacity to counteract the effects of selection and/or genetic drift through small population size, levels of gene flow are therefore crucial in any consideration of population differentiation and viability. However, with gene flow in plants resulting from a combination of pollen and seed dispersal, accurate measurement has been restricted both by the discriminatory powers of genetic markers (see Chapter 4) and the sampling regimes of studies. Typically views of gene flow in plants range from restricted, through idiosyncratic to extensive—varying among taxa, populations, individuals and seasons. In Chapter 6, James Hamrick and John Nason review current knowledge of gene flow in trees, the relative roles of pollen and seed dispersal in both localised and wider scale genetic structure, along with the implications for management and significance from both ecological and evolutionary viewpoints.

Selection is regarded as the other of the two 'strong' genetic forces that determines the level and pattern of genetic variation within species. With the changes in environmental variables caused by climate change, pollution, harvesting and disturbance, there may be significant effects on selective pressures and interactions with other genetic forces. In Chapter 8, Gene Namkoong, Mathew Koshy and Sally Aitken review the possible consequences of changing the selective pressures on forest tree populations, through encountering novel environments. Many such impacts on forests may also reduce effective population sizes, either temporarily or in the long term. Consideration of the dynamics of small populations and the consequences of reduced population size is therefore important for many issues related to the conservation of forest trees. These consequences include both genetic effects (increased inbreeding, genetic drift) and non-genetic effects (density-dependent reproductive success, density-dependent mortality). The relative importance of these in determining the fate of populations has aroused controversy in conservation biology. In Chapter 7, Outi Savolainen and Helmi Kuittinen describe the consequences (both genetic and non-genetic) of reduced population size and discuss their significance, using both theoretical and empirical evidence, in determining the fate of small populations. Within this context and those of disturbance and fragmentation, they discuss the application of dynamic metapopulation models to forest trees.

MATING SYSTEMS

David H. Boshier

SUMMARY

Patterns of mating vary with the reproductive biology and spatial structure of a species, and combine to influence levels and dynamics of genetic diversity. Knowledge of all aspects of a species' mating system is necessary to understand the distribution of genetic variation between individuals, gene flow within and between populations, what effect human influence may have on these and, therefore, how to best conserve and utilise this variation. This chapter covers the basic processes of tree reproductive biology (sexual systems, incompatibility mechanisms, flowering patterns, pollination processes), their integration to produce observed mating patterns in both temperate and tropical species and some of the methods of quantifying mating systems. Trees are shown to be predominantly outcrossed, but in some species a variety of conditions may lead to a degree of selfing or inbreeding owing to related matings. Abnormal levels of inbreeding may have immediate implications for seed collection, and *ex situ* conservation strategies, while their long-term maintenance may lead to losses of diversity with longer term adaptive consequences.

5.1 Introduction

In trees, as with other plants, mating patterns vary depending on the attributes of a particular species' reproductive biology and spatial structure, which combine to influence the levels and dynamics of genetic diversity (Bullock & Bawa 1981; Hamrick & Murawski 1990; Loveless 1992). Consequently, achievement of both short-term and long-term goals of effective resource management and maintenance of evolutionary flexibility, through the design and implementation of species-based conservation strategies, require an understanding of the basic processes of tree reproductive biology (sexual systems, incompatibility mechanisms, flowering patterns, pollination processes), and how they combine to produce observed patterns of gene flow and genetic variation. Plants have a diverse array of systems that influence their reproduction and, hence, evolution. Such systems are controlled genetically and are therefore not constant, but have the flexibility to respond to changing conditions. The maintenance of genetic variation, enabling adaptability to a range of environmental conditions over both space and time, is an important part of the reproductive process itself and hence of any conservation efforts. Generally trees have been shown to carry a heavy genetic load of deleterious recessive alleles (e.g. Williams & Savolainen 1996), such that inbreeding, in particular selfing, may lead to reduced fertility and slower growth rates in progeny (Hodgson 1976; Park & Fowler 1982; Sim 1984; Griffin & Lindgren 1985; Griffin 1991). The need to reduce the possibility or effect of inbreeding and maintain diversity in naturally outcrossing species is evident, while maintenance of breeding system flexibility will be a priority for species that naturally combine outcrossing and inbreeding. Whatever the situation, information on actual levels of outcrossing or inbreeding will be important for natural or managed ecosystems, and to the success of forestry breeding programmes. Similarly, it is essential for effective and representative seed collections, for *ex situ* conservation or for reforestation and breeding, to have knowledge of the genetic base being sampled.

The importance of the generation and use of scientifically based information for genetic conservation is well illustrated by changes in views of mating patterns in tropical trees. Early discussion of the reproductive biology of tropical trees, extrapolating from the low density of many tree species to speculate on breeding systems, suggested that most tropical rainforest trees were self-pollinated

and inbred (Corner 1954; Federov 1966). Subsequently, work based primarily on studies of self-compatibility and cross-compatibility and on observations of pollinator behaviour indicated strong barriers to selfing and led to the conclusion that tropical trees are predominantly outcrossed (Janzen 1971; Bawa 1974; Zapata & Arroyo 1978; Bawa *et al.* 1985). Despite such evidence, the view of tropical trees being predominantly self-pollinated can still be found (e.g. Willan 1985). Although it is recognised that knowledge of reproductive biology is very limited for most tropical tree species (e.g. Bawa *et al.* 1990; National Research Council 1991), the same is true for many temperate species. The reality is that detailed knowledge has been mainly restricted to tree species of economic importance [e.g. *Pinus* (pines), *Eucalyptus* (eucalypts), *Leucaena leucocephala* (Lam.) De Wit], whereas for most other species knowledge has not advanced much beyond the anecdotal or initial observation stage. Inevitably many species under threat of loss of genetic diversity are those of least current economic importance and for which knowledge is most restricted. This chapter describes the basic processes involved in tree reproduction, their influence on mating patterns, methods for quantifying such patterns and implications for the development of conservation strategies.

5.2 Sexual systems

Categorisation of a species' sexual system is the first stage in understanding its mating system. A range of sexual systems are found in trees, although they can be broadly grouped into three main types:

(a) *dioecious*: all individuals in a population are either male or female

(b) *hermaphrodite*: an individual with both male and female function—it may have either monoecious (single sex) or hermaphrodite (both sexes) flowers

(c) *monoecious*: hermaphrodite individuals in which anthers and gynoecia occur in different flowers— male and female function are separated.

Such sexual systems offer a way to reduce the proportion of self-pollination and, hence, the degree of inbreeding that may occur—dioecy representing the extreme where selfing is completely prevented. Characteristically there are large differences between forest types in the proportion of sexual systems (Table 5.1); the predominance of monoecy in temperate

TABLE 5.1   Comparison of the distribution of sexual system in trees by forest type (% of species)

Sexual system	Neotropical montane wet	Neotropical lowland dry	Nigeria lowland wet	Sarawak lowland wet	Temperate	Temperate
Hermaphrodite	68	68	47	60	7	12
Monoecious	11	10	13	14	74	75
Dioecious	21	22	40	26	19	13
References	Tanner (1982)	Bawa (1974)	Bawa & Opler (1975)	Bawa & Opler (1975)	Bawa (1974)	Bawa & Opler (1975)

forests being a reflection of the high proportion of coniferous species.

There may be gradations between the main categories (e.g. andromonoecy, where a hermaphrodite bears both male and hermaphrodite flowers, and gynodioecy, where female and hermaphrodite genets coexist; Richards 1986), with some species showing a range of sexual types (e.g. *Fraxinus excelsior* Boiss., Schultz 1892). In this respect it is important to distinguish between morphology and functionality; for example in *Swietenia* spp., on first inspection the flowers appear to be hermaphrodite. Closer observation reveals that the flowers are either male or female, with the respective ovaries or anthers being vestigial (non-functional; Lee 1967). In other cases, although flowers are clearly hermaphrodite, individual trees may be more effective as either male or female, such that the potential number of crosses in the population is severely reduced and the actual situation more closely approximates to dioecy (e.g. *Cordia collococca* Aubl., *Cordia inerma* I.M. Johnst., Opler *et al.* 1975; Lloyd 1979). Similarly, perhaps where the resources required for seed production are limiting, functionality within a tree may vary within a season or from year to year, such that a tree may act principally as a female one year and as a male in the following year, even leading to the misclassification of monoecious species as dioecious (Condon & Gilbert 1988). In other cases the sex of individual flowers may be influenced by the status of other flowers within an inflorescence or on the same tree, with resource limitation and an optimum sex allocation strategy suggested as the cause (Diggle 1993; Brunet & Charlesworth 1995). In *Calliandra calothyrsus* Meissn., where hermaphrodite flowers are held on sequentially opening inflorescences, seed set in the first flowers to open leads to effective andromonoecy with a much higher proportion of male flowers among the later opening flowers than in the absence of fertilisation (Chamberlain 1998).

5.3 Incompatibility mechanisms

Self-incompatibility is a genetically controlled mechanism which reduces the prevalence of inbreeding in a population (Sedgley & Griffin 1989), and may be characterised as prezygotic and/or postzygotic. Prezygotic self-incompatibility is particularly well developed, and studied, in the angiosperms and in most cases operates via inhibition of pollen tube growth at some stage in the stigma or pistil (Fig. 5.1), preventing fertilisation, and is under the control of one or more genes with multiple alleles (Box 5.1).

Self-incompatibility in a species can be detected through controlled crosses, with the comparison of seed set between self-pollinations and cross-pollinations (Box 5.2), although such studies do not distinguish between prezygotic and postzygotic incompatibility. Indeed, despite increasing evidence for the occurrence of the latter, often referred to as 'late acting self incompatibility' (Seavey & Bawa 1986), in

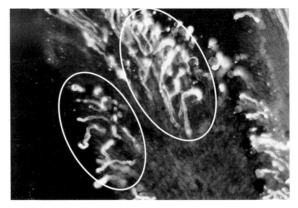

Figure 5.1   Incompatible controlled cross, showing inhibition of pollen tube growth (circled areas) in the stigma of *Cordia alliodora* (24 h after pollination). See Box 5.1 for a detailed explanation. Photograph: David Boshier.

Box 5.1 Prezygotic incompatibility mechanisms

Gametophytic self-incompatibility

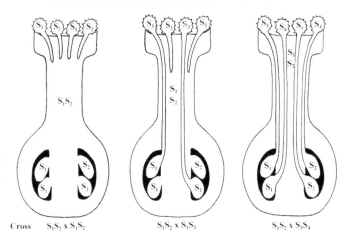

Figure 1   Gametophytic self-incompatibility. Drawing: Rosemary Wise.

Control: haploid genome of pollen grain and diploid genome of pistil tissue, inhibition generally within the stigma or style.

Consequences: prevents selfing and mating with closely related individuals (sharing both incompatibility alleles). The number of incompatible crosses within a population is determined by the number of incompatibility alleles within the population, reduction in population size may lead to loss of incompatibility alleles and a reduction in fertility.

Example: *Leucaena diversifolia* (Schltdl.) Benth. (Pan 1985).

Homomorphic sporophytic self-incompatibility

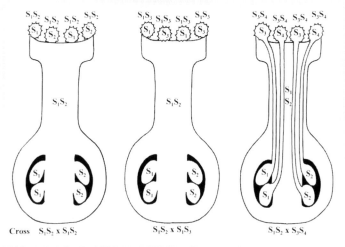

Figure 2   Homomorphic sporophytic self-incompatibility. Drawing: Rosemary Wise.

Control: diploid genome of pollen grain and diploid genome of pistil tissue, inhibition generally at stigmatic surface.

Consequences: prevents selfing and mating with related individuals (sharing one incompatibility allele). The number of incompatible crosses within a population is determined by the number of incompatibility alleles within a population, reduction in population size may lead to loss of incompatibility alleles and a reduction in fertility.

Example: *Ulmus americana* Aiton (Ager & Guries 1982).

Heteromorphic sporophytic self-incompatibility

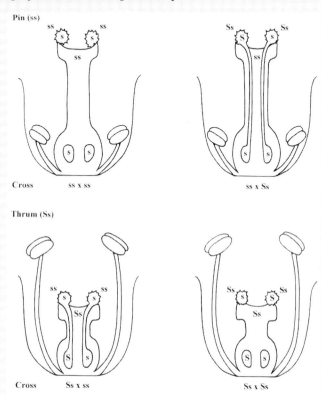

Figure 3   Heteromorphic sporophytic self-incompatibility. Drawing: Rosemary Wise.

Control: diploid genome of pollen grain and diploid genome of pistil tissue; one super gene locus, diallelic.

Consequences: prevents selfing and mating with 50% of population (reduction in effective population size); no control of mating between related individuals.

Examples: *Cordia* spp. (Kanashiro 1986; Boshier 1995).

several species (e.g. *Liquidamber styraciflua* L., Schmitt & Perry 1964; *Sterculia chicha* A. St.-Hil., Taroda & Gibbs 1982; *Ceiba* spp., Gibbs 1988; *Tabebuia* spp., Gibbs & Bianchi 1993), little attention has been given to its study, which requires detailed histological techniques to determine the exact stage of pollen tube or ovule arrest. The phenomena may represent a variety of inhibition mechanisms, none of which has received the same attention to date, with respect to evidence for genetic control, as the prezygotic mechanisms described above (Gibbs & Bianchi 1993). While prezygotic self-incompatibility mechanisms generally have not been found in gymnosperms, postzygotic mechanisms are particularly well developed and generally are manifested during the very early stages of embryo development. It is often

---

Box 5.2 Use of controlled pollinations to study incompatibility

The following index was suggested by Sorensen (1971, Relative Self Fertility); Zapata & Arroyo (1978, Index of Self Incompatibility) to measure self-incompatibility, using controlled pollinations.

$$RSF \text{ or } ISI = \frac{\text{fruit set from self-pollination}}{\text{fruit set from cross-pollination}} \qquad (1)$$

The resultant value of the RSF/ISI reflects the following possibilities:

(a) > 1 = self-compatible

(b) > 0.2 < 1 = partially self-incompatible

(c) < 0.2 = mostly self-incompatible

(d) 0 = completely self-incompatible.

RSF/ISI values should be interpreted with caution, however, as levels vary between individuals of the same species, as well as among species. Counts should include only viable seed.

---

difficult, however, to distinguish between incompatibility and embryo failure owing to inbreeding depression, although the latter might be indicated by variation in timing of embryo or fruit abortion or variation in seed set from selfing. Many *Eucalyptus* and *Banksia* species appear, from controlled crosses, to be self-compatible but still show a mixture of selfing and outcrossing, with inbreeding depression apparently reducing the viability of selfs at various stages (e.g. reduced seed set, growth rate, Eldridge *et al.* 1993). Studies, initially in Costa Rica (Bawa 1974, 1979; Bawa *et al.* 1985) and elsewhere (e.g. Yap 1980; Chan 1981; Ha *et al.* 1988), have shown most tropical trees to be self-incompatible. However, some well-known tree species, for example *Leucaena leucocephala*, are self-compatible and typically show low levels of genetic variability with a lack of adaptation to a wide range of sites and susceptibility to pest attack (Hughes 1998).

Apomixis (asexual reproduction), via vegetative propagation or agamospermy (seed development without fertilisation), is a feature of a number of tree species and has consequences for levels and patterns of genetic variation. Vegetative propagation under natural conditions is most commonly known in some temperate angiosperm tree genera (e.g. *Populus*, *Prunus*, *Ulmus*). In such cases large clones take up more space than single (non-clonal) genotypes, such that distances between individuals may be much greater and effective population sizes much smaller

than is immediately apparent. Vegetative reproduction may thus restrict the gene pool and evolutionary opportunities of a species. In conifers vegetative propagation is generally restricted to species of a prostrate, creeping habit, where branches root adventitiously (e.g. *Podocarpus alpinus* R. Br. ex Hook. f., *Juniperus communis* subsp. *nana* L., Richards 1986). Agamospermy occurs in some angiosperm species from one of several developmental processes (e.g. Richards 1986; see Sedgley & Griffin 1989 for details). In most forms of agamospermy there is no recombination (somatic recombination and/or various forms of autosegregation), so that individuals genetically identical to their parent are generated from seed. The degree to which agamospermy occurs in tree species is uncertain and so far there has been little research to confirm or refute suggestions that it may be relatively common in rainforest tree species (Kaur *et al.* 1978). Agamospermy has been reported mainly from species in the Clusiaceae (*Clusia* spp., Maguire 1976; *Garcinia* spp., Ha *et al.* 1988; Richards 1990), Dipterocarpaceae (*Shorea* and *Hopea* spp., Kaur *et al.* 1978, 1986) and Melastomataceae (Renner 1989). In the absence of cytological research it is impossible to distinguish between self-compatibility and agamospermy where pollination is a necessary stimulus to asexual seed development (pseudogamy), such that the extent to which species classified as self-compatible are agamospermic is unknown. Agamosperms might be expected to suffer reduced fitness compared with sexually reproducing species, through a lower generation of variability which may reduce their ability to adapt to changing environments. They are also likely to suffer from the accumulation of disadvantageous mutations, with poor recombination and migration of advantageous mutations, owing to reduced sexual function. However, most agamosperms, as with trees that exhibit vegetative propagation, appear to retain some level of sexual reproduction and are, more or less, facultative rather than obligate, with the capacity to produce seed in the absence of pollination [e.g. *Artocarpus heterophyllus* Lam. (jackfruit); Pushpakumara 1997].

5.4 Flowering

Knowledge of a species' flowering phenology, both within individual trees and within and between populations, is of fundamental importance in understanding its genetic structure and in ensuring an adequate genetic base in seed for conservation or reafforestation. The study of flowering patterns has been limited, however, with most studies focused on flowering at the ecosystem level (e.g. Daubenmire 1972; Frankie *et al.* 1974; Lieberman 1982; Newstrom *et al.* 1994) or individual tree level. However, it is the population level, that is the interaction of flowering between conspecifics, that is critical in influencing mating patterns and hence central to issues of conservation genetics. How individual trees flower over time, with respect to one another and to pollinators, will not only affect mating patterns, but may also alter the patterns and levels of genetic variation, through influencing seed production levels. The long lived nature of trees leads to overlapping generations which reduce effective population sizes, but the degree of this will be influenced by the age at which trees reach reproductive maturity.

Most studies of within–population flowering have been in temperate regions, particularly in coniferous seed orchards, where there has been a demand to quantify and predict the effect of floral phenology on seed yield and genetic quality (e.g. Ehrenberg & Simak 1956; Brown 1971; Koski 1980; El-Kassaby *et al.* 1988; Erickson & Adams 1989; Askew & Blush 1990; El-Kassaby & Reynolds 1990; Boes *et al.* 1991; Burczyk & Chalupka 1997). Some of these studies have shown problems such as inadequate pollen supply and parental imbalance in seed production with about 20% of the clones producing some 80% of the seed crop [e.g. *Pinus radiata* D. Don (Monterey pine), Griffin 1982; *Pinus taeda* L. (loblolly pine),

Schmidtling 1983; *Pseudotsuga menziesii* (Mirb.) Franco (Douglas-fir), El-Kassaby *et al.* 1989a). A similar imbalance was also found in a young seed orchard of the neotropical tree *Bombacopsis quinata* (Jacq.) Dugand (Sandiford 1998).

Variation in flowering time at all levels (inflorescence, tree, population) may represent an important way for a species to adapt ecologically and physiologically to its environment, with promotion of strategies such as cross-pollination and competition for pollinator services. Inevitably individuals of outcrossing species must flower in relative synchrony to effect appropriate gene exchange and seed set (Bawa 1983). Such synchrony within a species may also give a strong competitive advantage for pollinators over the period of flowering (Opler *et al.* 1976). Within the mass flowering events, characteristic of wet lowland dipterocarps in South-East Asia and thought to be triggered by the El Niño phenomenon, there is a high degree of synchrony within species and asynchrony between species, suggested as an adaptation for sequential pollination by thrips (Chan & Appanah 1980; Appanah & Chan 1981).

Flowering is also often influenced by climate such that the same species occurring in seasonally dry climates may show a different flowering pattern when in aseasonal wet climates (Newstrom *et al.* 1994). In some cases in the wetter environment flowering appears to be more prolonged (e.g. *Hamelia patens* Jacq., Frankie *et al.* 1974) with a possible increase in the degree of asynchrony [e.g. *Cordia alliodora* (Ruiz & Pav.) Oken., Boshier & Lamb 1997], while in others flowering becomes more irregular (e.g. *Shorea* spp., Yap & Chan 1990; *Andira inermis* Humb., Bonpl. & Kunth; *Ceiba pentandra* Gaertn., Newstrom *et al.* 1993). In a *Pseudotsuga menziesii* seed orchard, lower temperatures were shown to reduce the degree of asynchrony of flowering (El-Kassaby & Ritland 1986a).

The flowering phenology of tropical trees has been described as a continuum between 'mass-flowering' species, producing many new flowers each day over a week or less, and 'steady-state' species, producing small numbers of new flowers daily for many weeks (Gentry 1974; Opler *et al.* 1980). The reality is that flowering, within species or populations, represents a continuum between these two extremes: some species or trees may produce few flowers over short or long periods, while others produce many flowers over short or long periods (e.g. *Cordia alliodora*, Boshier 1992; *Bombacopsis quinata*, Sandiford 1998). Similarly all populations show a degree of flowering synchrony, rather than any simple dichotomy between synchrony and asynchrony (Augspurger 1983). In comparisons between species, mass-flowering species have shown, both qualitatively and quantitatively, greater population flowering synchrony than steady state species (Janzen 1967; Opler *et al*. 1976; Augspurger 1983). However, within some species the degree of synchrony increases as the mean length of flowering decreases (El-Kassaby *et al*. 1984, 1988; Askew 1986; Boshier 1992; Sandiford 1998). Highly asynchronous flowering within a species creates structure by limiting the number of potential mates and reducing effective population sizes. Within a population, the relative contributions of individual trees to pollen flow and seed production may change markedly from year to year. Adjacent trees may mate in one year but not in the next, owing to variation in flowering synchrony and/or the quantity of flowering. Hence, seed collected under such circumstances may contain much less genetic diversity than would otherwise be expected. Low levels of flowering synchrony may result in high numbers of related crosses, particularly if there is higher synchrony among near-related neighbours. Asynchronous flowering and yearly variation could, however, increase genetic diversity by altering the nearest flowering neighbours of a tree over time. Self-incompatible species that show highly asynchronous flowering may be more susceptible to reductions in population size with respect to reproduction, both in terms of compatible pollination and reduced diversity (e.g. *Shorea siamensis* Miq., Ghazoul *et al*. 1998).

In monoecious species there is often some degree of spatial or temporal separation between male and female inflorescences that reduces the incidence of self-pollination. In many conifers, female strobili are concentrated in the apex of the crown, while the males occur predominantly in the middle and lower parts. Where there is temporal separation there is inevitably a degree of asynchrony within the population to ensure that pollen is available when female flowers are receptive, and a wide variety of strategies appear to promote cross pollination (Bawa 1983). In *Cupania guatemalensis* Radlk., a monoecious neotropical tree, the duration of male and female phases varies widely among trees; some trees start with the female, rather than the male phase, whilst others show only a male phase in some years (Bawa 1977). Work on species of

the neotropical vines *Gurania* and *Psiguria* showed such temporal variation in sex expression, within a season and/or related to size, that the species were originally described as dioecious (Condon & Gilbert 1988). In *Asterogyne martiana* H. Wendl. ex Drude the asynchrony is at the level of the inflorescence, with each inflorescence having a distinct male and female phase (Schmid 1970). In *Juglans* the degree of temporal separation between male and female flowers varies from species to species (Gleeson 1982) and is also affected by weather, such that in *Juglans nigra* L. high temperatures reduced the degree of temporal separation between male and female flowers, increasing the chance of self-pollination (Masters 1974).

5.5 Pollination vectors

In all conifers pollen transfer is wind mediated. In the angiosperms the freeing of floral organs from a protective role to the ovary has led to a vast floral diversification and coadaptation with animal pollinators to varying degrees of specialisation. Although animal pollination predominates in angiosperm tree species, wind pollination occurs in many temperate species (e.g. *Quercus* spp., *Fraxinus* spp., *Salix* spp.), but is less common in tropical species (e.g. *Ateleia herbert-smithii* Pittier, Janzen 1989; *Artocarpus heterophyllus*, Pushpakumara 1997). Bullock and Bawa (1994) suggested, however, that wind pollination may be more common in tropical angiosperms than thought previously. Determination of pollination vectors and their effect on mating patterns, gene flow and reproduction are, however, integral to a species conservation. Indeed where pollination is highly specialised, the study of pollinators and their management may be just as important as that of other aspects of reproductive biology, as was illustrated by the failure of *Bertholletia excelsa* Humb. & Bonpl. (Brazil nut) trees to produce nuts owing to a lack of pollinators once the surrounding forest was cleared (Brune 1990). Related to this is the need to understand the biology of the pollinators. Many require alternate host species to provide food, so that maintenance of adequate pollinator populations may require active management of such host plants in the maintenance or restoration of forest (e.g. *Byrsonima crassifolia* Humb., Bonpl. & Kunth, Frankie *et al*. 1993), or in plantations. Efficient seed production in seed orchards may also require the provision of suitable habitat for pollinators [e.g. *Bixa orellana* L. (anatto) and

Passiflora spp. for *Bertholletia excelsa* nuts, Adalgisa 1985). Therefore, seed orchard design may need to take into account specific aspects of the pollination biology of the species concerned, if seed production is to be abundant. In the case of wind pollination, seed orchard design may be adjusted to account for prevailing winds (e.g. Dyson & Freeman 1968).

The pattern of a pollinators movements will also potentially effect the pattern of mating and, hence, the effectiveness of a pollinator as an agent of gene flow. Different species of insect may vary in their patterns of movement and, therefore, influence gene flow in a variety of ways. Trap lining (Janzen 1971), whereby pollinators move among many plants over a relatively large area in patterns sometimes repeated from day to day, is seen in some of the larger bees (e.g. *Xylocopa* spp.), bat, butterfly and hummingbird species (Linhart 1973; Heinrich 1975; Heithaus *et al.* 1975; Gilbert 1975; Frankie 1976; Frankie *et al.* 1976; Ackerman *et al.* 1982). Trap lining is typical of some tree species that produce relatively few flowers over extended periods and would be expected to lead to high levels of long distance gene flow, but with a predominance of particular non-random matings. On mass flowering species most pollinator movements are within trees, leading to a large amount of self-pollination, but interactions among pollinators (Frankie & Baker 1974; Rausher & Fowler 1979), interspecific and intraspecific variations in nectar production (quality, timing) appear to act as stimuli to movement between trees (Frankie & Haber 1983), leading to cross pollination. Mark and recapture experiments on *Andira inermis* showed that of 70 bee species collected on the flowers, only eight (6 of which were *Centris* spp.), were recorded as moving between conspecific trees (Frankie *et al.* 1976). Although intertree movement was low, it was sufficient to account for most, if not all, the observed fruit set. Most movement was to near-neighbouring trees, but some moved further than one kilometre.

Similarly the degree to which a species pollination syndrome is specialised will influence whether mating patterns are fairly constant and whether they may change from one environment to another. Tree species with specialist pollinators are also more likely to face threats from reductions in both pollinator and tree populations, where successful pollination may become a limiting factor (Compton *et al.* 1994; Ghazoul *et al.* 1998). In contrast, species with unspecialised pollination syndromes are only likely to suffer

problems when there is a general loss of fauna in an area. Within the same species, however, different pollinators may lead to variation in different mating patterns, with different insects acting as pollinators in some species with a prolonged flowering season.

5.6 Mating in trees— quantification and patterns

Although studies of self-compatibility and cross-compatibility may indicate strong barriers to selfing, and while flowering patterns indicate limits to mating, neither indicate actual levels of outcrossing and the extent to which mating between related individuals may occur. Crude indicators of a species' breeding system may be obtained from various tests (Box 5.3), although such measures are constrained by several factors and limited in their use (see Dafni 1992 for techniques and discussion of limitations). The use of genetic markers such as allozymes, and more recently DNA markers (e.g. restriction fragment length polymorphisms, amplification fragment length polymorphisms, microsatellites, see Chapter 4.3) offer the means of providing precise estimates. Together with field observations, these techniques permit the effect of factors such as stand size, density, spatial distribution and flowering phenology on mating system, gene flow and genetic structure to be studied more directly (Brown 1990; Wagner 1992; Boshier *et al.* 1995a, 1995b).

Given the number of tree taxa and the limited resources available, research has inevitably focused on species of commercial interest and consequently, information has been limited and biased towards coniferous taxa of the northern temperate forests (Muona 1989; National Research Council 1991; Hamrick 1994a). More recent tropical studies (reviewed by Murawski 1995) have begun to redress the balance, and research is now tackling a much wider range of taxa which may be taken as representative of others, sharing, for example, similar attributes of pollination, phenology and sexual system. Already some consistency is apparent for most tree taxa, which are characterised by effective mechanisms for outbreeding and gene flow, individual longevity and fecundity, and little or no history of domestication (Brown & Moran 1981; Yeh 1989). Tree taxa generally exhibit relatively high levels of genetic diversity at the individual tree, population and taxon levels, a few notable exceptions notwithstanding (e.g. *Pinus resinosa* Aiton, Mosseler 1992), with both

Box 5.3 Use of pollen to ovule and seed to ovule ratios and controlled pollinations to determine breeding systems

Pollen to ovule ratios

The ratio between the number of pollen grains (P) and ovules per flower (O) has been used to distinguish between different breeding systems on the basis of standard ratios established by Cruden (1977), as shown in Table 1 below.

Table 1   Standard ratios between the number of pollen grains (P) and ovules per flower (O)

Breeding system	P:O range
Obligate selfers (including cleistogamy)	2-7–39.0
Facultative selfers	31.9–396.0
Facultative outcrossers	244.7–2588.0
Obligate outcrossers	2108.0–195 525.0

Preston (1986) showed, however, that standards for evaluating breeding systems based on $P:O$ need to be set at the level of the family.

Seed to ovule and fruit to flower ratios

High seed production (e.g. > 80% of flowers setting seed) under natural conditions is often associated with self-compatibility (see Figure 1).

Figure 1   Comparison of pod set (on multi-flowered capitula) between (a) self-compatible *Leucaena leucocephala* and (b) self-incompatible *Leucaena salvadorensis*. Photographs: Colin Hughes.

The ratio between the number of seed and ovules in multi-ovulated flowers has been used as a measure of distinguishing between outcrossed and self-compatible/inbred species (Wiens 1984; Bawa & Buckley 1988). Outcrossed species tend to have lower seed to ovule ratios and flower to fruit ratios than selfed/inbred species, and in *Gliricidia sepium* Humb., Bonpl. & Kunth crosses between related half sibs showed significantly lower seed to ovule ratios than outcrossed material (Simons 1996). Where flowers are held on a multi-flowered head (e.g. mimisoid legumes) seed to ovule ratios are often similar, the largest difference being in the flower to fruit ratio, where there is the greatest chance for differences in pollen source to be expressed (Bawa & Buckley 1988). As for pollen to ovule ratios, comparisons may be possible only at the family or genus level.

Estimation of selfing rate

The frequency of selfing (S) can be estimated by comparing the level of seed set from open-pollinated flowers (P_O), with seed set from controlled selfing (P_S) and controlled cross-pollination (P_X; Charlesworth & Charlesworth 1987):

$$S = \frac{P_X - P_O}{P_X - P_S} \tag{1}$$

the average level of heterozygosity and proportion of polymorphic loci high in relation to those of non-woody plants (Hamrick *et al.* 1992). Levels of genetic diversity vary with mating system, with higher levels in most predominantly outcrossing taxa. In contrast, inbred species show lower levels of diversity, but greater interpopulation variation. Most alleles (typically 70–80%) are common across most populations (Hamrick 1994a; Loveless 1992; Moran 1992; Müller-Starck *et al.* 1992), even in rare taxa (Bawa & Ashton 1991), those occurring as small and isolated populations (Moran & Hopper 1987; Moran 1992), those which are relatively fragmented (Hamrick 1994a) and those which have been subject to substantial human modification (Harris *et al.* 1997). Most of the genetic variation is found within, rather than between, populations (among-population G_{ST} values typically > 0.1 < 0.2), although such differences are often of major significance for adaptation or production, as has been shown in provenance trials of many species (e.g. Eldridge *et al.* 1993). Species with disjunct distributions (e.g. *Acacia mangium* Willd., *Eucalyptus caesia* Benth., *Eucalyptus pulverulenta* Link, Moran *et al.* 1989c; Moran 1992) typically show very high differentiation between populations (G_{ST} > 0.3, Moran 1992) and in many widespread species a larger proportion of the variation is often among regions (sometimes corresponding to recognisable geographical regions), rather than between populations within regions (e.g. *Casuarina*

cunninghamiana Miq., *Eucalyptus delegatensis* R.T. Baker, Moran 1992; *Acacia auriculiformis* A. Cunn. ex Benth., Wickneswari & Norwati 1993; *Cedrela odorata* L., Gillies *et al.* 1997; *Acacia mangium*, Butcher *et al.* 1998). As such, many widely distributed species appear to be characterised by a higher hierarchical ordering of population structure, so patterns not evident on a small scale may become apparent at a larger scale (Loveless 1992). Such studies can play an important role in defining more efficient strategies for conservation and genetic improvement (Moran 1992).

Various genetic parameters derived from molecular marker studies may be used to provide estimates of the mating system (see Ritland 1983 for detailed descriptions). From studies of population differentiation, heterozygote excesses or deficiencies, as departures from Hardy Weinberg equilibrium, and given by Wright's fixation index (F), or F_{IS} (see Chapter 3), may be used as indicators of inbreeding within populations. Calculations of F are based on the assumption that the population is in inbreeding equilibrium and that selfing is the only form of inbreeding occurring. As the assumptions may obscure many other factors contributing to inbreeding, this method should only be used as a preliminary means of studying a species' mating system (Ritland 1983). More frequently the mating system of a species is assessed through outcrossing rate estimates, based on

---

Box 5.4 Outcrossing rate estimates and assumptions of the mixed mating model

Estimation of the mating system describes the way in which gametes from parents in a population unite. The mixed mating model separates offspring into those resulting from inbreeding (both selfing and related matings) and those produced by outcrossing to a random sample of pollen from the whole population. The mixed mating model is based on certain assumptions (Clegg 1980; Brown *et al.* 1985), namely:

(a) maternal genotypes outcross at the same rate to a homogeneous pollen pool

(b) for each maternal parent, progeny genotype classes are independent, identically distributed, multinomial random variables

(c) alleles at different loci segregate independently

(d) genetic markers are not affected by selection or mutation between the time of mating and progeny evaluation.

Although many populations deviate from the model's assumptions, multilocus estimates of outcrossing rate (t_m) are much less sensitive to departures from the model than single locus estimates (t_s) and in practice it appears that such departures generally result in relatively minor effects (Ritland & Jain 1981; Shaw & Allard 1982; Ritland 1985; Brown 1990). The nature of departures from the assumptions of the model can also indicate how mating occurs within a population. Consequently, even in a self-incompatible species, estimates of outcrossing rate can be useful in assessing other types of inbreeding (e.g. Perez de la Vega & Allard 1984).

Outcrossing rates theoretically range from $t = 0$ (complete selfing) to $t = 1.0$ (outcrossed to a random sample of the populations pollen pool). Values significantly lower than $t = 1.0$ indicate a degree of inbreeding, which may be due to selfing and/or mating between related individuals (effective selfing). In selfing all loci are effected equally, in contrast to related matings, such that a mean single locus value significantly lower than the multilocus estimate indicates effective, rather than actual, selfing. Values significantly greater than 1.0 may occur owing to disassortative mating (mating preferentially to unrelated individuals) or sampling effects, such as the inability of the model to cope with multiple heterozygous individuals (Brown *et al.* 1985).

Computer programmes that estimate outcrossing rates (t_m, t_s) as well as other mating parameters derived from other models (e.g. effective selfing, correlated paternity) are readily available (e.g. MLTR, ESR; Ritland 1990).

---

the mixed mating model (Allard & Workman 1963; Brown & Allard 1970; Ritland & Jain 1981; Box 5.4). Studies of tree species that have provided estimates of actual outcrossing rates generally indicate high levels of outcrossing (Table 5.2). In conifers, despite the lack of incompatibility mechanisms preventing selfing, the spatial separation of male and female cones on individual trees, combined with generally high levels of genetic load decreasing seed set under inbreeding, are fairly effective at reducing selfing, leading to outcrossing rates generally greater than 0.85. Lower outcrossing rates tend to be associated with small, isolated populations. In *Eucalyptus* spp. where

controlled pollinations show varying degrees of self-compatibility (Eldridge *et al.* 1993), isozyme studies show a mixed mating system, with outcrossing predominant, but estimates of multilocus outcrossing rates vary from $t_m = 0.59$ to $t_m = 0.83$ between species (Moran 1992; Eldridge *et al.* 1993). Variation in outcrossing rates between populations of the same species also appears to be a common phenomenon in mixed mating species (e.g. *Banksia cuneata* A.S. George, Schoen 1982; Coates & Sokolowski 1992). Few tropical angiosperms studied so far appear to have truly mixed mating systems (e.g. *Cavanillesia platanifolia* Humb., Bonpl. & Kunth, Murawski *et al.*

TABLE 5.2   Examples of mulilocus outcrossing rate (t_m) estimates in temperate and tropical trees

Species	Populations	Mean t_m	Range of t_m	References
Temperate				
Abies lasiocarpa	1		0.78–0.99	Neale & Adams (1985)
Juglans nigra	1	0.91		Rink *et al.* (1989)
Larix laricina	5		0.54–0.91	Knowles *et al.* (1987)
Liriodendron tulipifera L.	3		0.55–0.86	Brotschol *et al.* (1986)
Picea engelmannii	1	0.86		Shea (1987)
Pinus attenuata	1	0.92		Burczyk *et al.* (1996)
Pinus banksiana	1	0.91		Cheliak *et al.* (1985)
Pinus contorta	3		0.93–0.98	Perry & Dancik (1986)
Pinus ponderosa	2		0.81–0.96	Farris & Mitton (1984)
Pinus sylvestris L.	3		0.92–0.99	Muona & Harju (1989)
Pseudotsuga menziesii	8		0.86–0.96	Shaw & Allard (1982)
Quercus robur L.	1	0.98		Bacilieri *et al.* (1996)
Tropical				
Acacia auriculiformis	2		0.92–0.93	Moran *et al.* (1989d)
Acacia crassicarpa A. Cunn. ex Benth.	2		0.93–0.99	Moran *et al.* (1989d)
Bertholletia excelsa	1	0.85		O'Malley *et al.* (1988)
Brosimum alicastrum	1	0.88		Hamrick & Murawski (1990)
Cavanillesia platanifolia	2		0.21–0.66	Murawski & Hamrick (1992)
Cordia alliodora	1	0.98		Boshier *et al.* (1995b)
Eucalyptus grandis W. Hill ex Maiden	21	0.84		Eldridge *et al.* (1993)
Eucalyptus urophylla S.T. Blake	2		0.90–0.91	House & Bell (1994)
Pithecellobium pedicilare	1	0.95		O'Malley & Bawa (1987)
Platypodium elegans Ducke	1	0.92		Hamrick & Murawski (1990)
Pinus caribaea Morelet	4		0.85–0.92	Matheson *et al.* (1989)
Pinus kesiya Royle ex Gordon	4		0.68–0.97	Boyle *et al.* (1991)
Pinus oocarpa Schiede ex Schltdl.	5		0.81–0.96	Matheson *et al.* (1989)
Shorea congestiflora P.S. Ashton	1	0.87		Murawski *et al.* (1994a)
Shorea trapezifolia P.S. Ashton	2		0.54–0.62	Murawski *et al.* (1994a)

1990; *Ceiba pentandra*, Murawski & Hamrick 1992). Dioecy, self-incompatibility or synchronous flowering generally leads to high levels of outcrossing, associated in some cases, with small but significant levels of inbreeding that results from mating between relatives associated with local genetic structure, temporal variation, density or pollinators (O'Malley & Bawa 1987; O'Malley *et al.* 1988; Murawski *et al.* 1990; Loveless 1992; Boshier *et al.* 1995b).

Recognition of the possible influences of such demographic and genetic factors on mating patterns in plants has led to new approaches for estimating mating systems. The 'effective selfing model' (Ritland 1984, 1986; Ritland & Ganders 1985, 1987) estimates the total amount of apparent selfing (*E*) that results from a combination of one or more of: self-fertilisation, mating to relatives or mating in spatially structured populations (within local areas with associated genetic relatedness). The 'correlated selfing and paternity model' (Ritland 1988, 1989) permits estimation of the extent to which matings are correlated, such as where pollinator behaviour leads to non-random mating. Similarly seed within the same multi-seeded fruit may share the same male parent if the pollen leading to fertilisation came from one source and one pollination event, or there is limited pollen carry over (Brown *et al.* 1985). Correlation of outcrossed paternity may also occur when the number of potential male parents in a population is low or

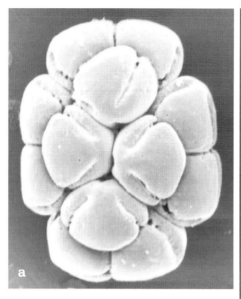
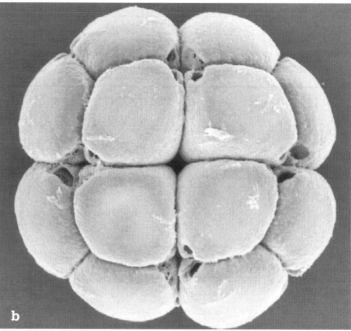

Figure 5.2   Sixteen-grained polyads in (a) *Leucaena multicapitula* and (b) *Xylia xylocarpa*. Photographs: Colin Hughes.

mating is mainly between near neighbours, or if pollen is deposited as multiple grain units on a stigma, such as the polyads found in many legume species (e.g. *Acacia melanoxylon* R. Br., Muona *et al.* 1991 and those shown in Fig. 5.2).

As such, even where populations are predominantly outcrossed, there may be variation between trees in sampling of the pollen pool. While the effective selfing and correlated paternity models allow more detailed assessment of such mating patterns, paternity analysis examines the relative contributions of different fathers to mating. Paternity analysis requires the genotype and location of all potential males to be known, therefore limiting use of the technique to small isolated populations. Most tree populations are not isolated, however, and to take account of this Adams and Birkes (1989, 1991) proposed the neighbourhood model, particularly for use in wind-pollinated conifers, where the local population is defined as a specified area around a sample mother tree. Pollen that has fertilised embryos is, therefore, assumed to come from one of three sources:

(a)   self fertilisation

(b)   cross fertilisation to males within the neighbourhood

(c)   cross fertilisation to males outside the neighbourhood.

The model also allows individual male reproductive success to be related to various factors that may influence it (e.g. distance from mother tree, fecundity, phenology, Burczyk *et al.* 1996). When most mating is among neighbours the model evaluates the relative mating success of each male within the neighbourhood with respect to a particular mother tree, while also indicating where pollen is primarily from trees outside the neighbourhood (NEIGHBOR computer program, Burczyk *et al.* 1993). For a natural stand of *Pinus attenuata* Lemmon and a *Pseudotsuga menziesii* clonal seed orchard outcrossing rates were both 0.97 (Burczyk *et al.* 1996; Burczyk & Prat 1997). The neighbourhood model showed that male reproductive success generally increased with pollen fecundity, proximity and phenology overlap with the mother tree with, in both cases, many males (about 11) from within each neighbourhood mating with each female, although about 54% of matings were by pollen from outside of the neighbourhood (11 m and 30 m radius, respectively). In contrast, Hamrick (1994a) concluded that in neotropical tree populations, two contrasting trends are responsible for the general mating patterns observed to date.

(a) Individual trees appear to receive pollen from relatively few pollen donors but the genetic composition of the pollen received varies greatly from tree to tree.

(b) Although a high proportion of fertilisation is effected by nearest neighbours, a significant proportion of pollen movement occurs over relatively long distances. As such, this is consistent with the relatively low heterogeneity observed in allele frequencies among populations separated by one to several kilometres.

Spatial and temporal structuring of populations, associated with corresponding variation in flowering, the nature of incompatibility mechanisms, and stand composition and density, may result in temporal variation in outcrossing rates for stands and individual trees (Murawski *et al.* 1990; Murawski & Hamrick 1991; Boshier *et al.* 1995a; Murawski 1995). While many populations of many tree species appear predominantly outcrossed, variation for estimates of multilocus outcrossing rates among trees may be high, with several factors leading to variation and reduction in outcrossing rates. Identifying under what circumstances and to what extent such inbreeding or variation in mating patterns occurs naturally is important in understanding how disturbances, human or other, may affect the genetic integrity of populations. The remainder of this chapter considers briefly examples of the influence of such factors on mating.

5.6.1   Spatial structure

Genetic structuring within tree populations is typical, particularly for those taxa with wind-dispersed seed, near neighbours usually having more alleles in common than more distantly separated individuals, even when occupying relatively homogeneous habitats (Linhart 1989; Hamrick 1994a; Ledig 1992; Moran 1992; Yang & Yeh 1992; Hamrick *et al.* 1993). In natural populations of *Eucalyptus* spp. such neighbourhood 'family' groups appear to lead consistently to a degree of effective selfing, with outcrossing rates averaging about 0.75 (Eldridge *et al.* 1993). In plantations, however, such family structure is broken up and seed collected there shows a corresponding increase in outcrossing (e.g. *Eucalyptus regnans* F. Muell.—0.74 natural stand, 0.91 plantation, Moran *et al.* 1989a; Fig. 5.3). Related mating owing to local structure was also apparent in two dioecious species, *Brosimum alicastrum* Sw.

(Hamrick & Murawski 1990) and *Gleditsia triacanthos* L. (Surles *et al.* 1990). In many cases, however, pollen dispersal is much more extensive than any such localised genetic structure and there is no or little inbreeding (e.g. *Banksia brownii* Baxter ex R. Br., Sampson *et al.* 1994; *Cordia alliodora*, Boshier *et al.* 1995a, 1995b). In three neotropical species (*Calophyllum longifolium* Willd., *Spondias mombin* L., *Turpinia occidentalis* G. Don) occurring naturally at low densities, mating patterns were strongly affected by the spatial distribution of reproductive trees, although still showing high levels of outcrossing. Where trees were clumped, most matings were with near-neighbours, whereas with evenly spaced trees, most matings were over several hundred metres and well beyond the nearest reproductive neighbours (Stacy *et al.* 1996).

5.6.2   Temporal structure

The degree of flowering synchrony between neighbours may also increase the tendency towards inbreeding, such that in an outcrossed population of *Cordia alliodora* some trees, surrounded by a few trees of similar genotype, with which flowering was highly synchronous, showed related mating, while other asynchronous flowerers in the same group were outcrossed (Boshier *et al.* 1995a). Similarly in a study of *Tachigali versicolor* Standl. & L.O. Williams, although overall the population was outcrossed ($t_m = 1.00$), the monocarpic (adults die after flowering) and pulsed nature of flowering (intervals of 4–6 years) led to great variation between seed crops in their genetic composition, producing a Wahlund effect in genotypic frequencies (Loveless *et al.* 1998). In a *Pseudotsuga menziesii* seed orchard the timing of flowering affected the outcrossing rate, with a greater degree of selfing occurring among the late ($t_m = 0.89$) and early ($t_m = 0.94$) flowering trees compared to the intermediate group ($t_m = 0.97$, El-Kassaby *et al.* 1988). Among five unrelated ramets in another *Pseudotsuga menziesii* seed orchard, the estimate of multilocus outcrossing rate ranged from $t_m = 0.50$ to $t_m = 1.07$, with the highest levels of inbreeding associated with ramets flowering much earlier than the rest of the orchard and/or with a high degree of overlap between pollen release and female receptivity on the same tree (Erickson 1987). Similarly the estimate of multilocus outcrossing rate was lower in *Abies lasiocarpa* Hook. and *Picea engelmannii* Engelm. trees with above-average male cone production (Shea 1987). Studies of some serotinous cone pine species

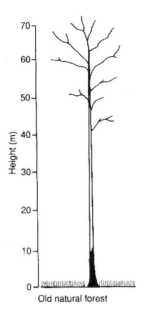

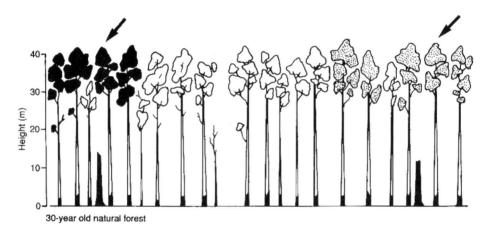

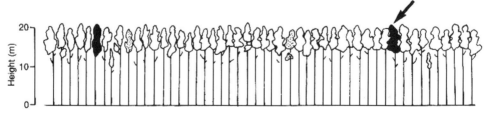

Figure 5.3   *Eucalyptus regnans*—neighbourhood inbreeding in a natural stand and increased outcrossing in a plantation. After an old natural stand is killed by fire the seedlings growing under a large tree may be mainly from that tree. At about 30 years to 40 years a group of neighbouring trees may be from the same mother tree, now a dead stump (full sibs, half sibs, selfs). Since family groups exist in the stand (one shaded black, another shaded grey), seed collected from a tree in such a group (arrow) is likely to have been pollinated by neighbouring close relatives. In a plantation the relatives are widely dispersed among unrelated trees and seed collected from any tree (arrow) is most unlikely to have been pollinated by a relative (from Eldridge *et al*. 1993). Drawing: S.M. House.

(e.g. *Pinus contorta* Douglas ex Loudon, Perry & Dancik 1986; *Pinus banksiana* Lamb., Cheliak *et al.* 1985) and two eucalypt species (*Eucalyptus delegatensis*, Moran & Brown 1980; *Eucalyptus stellulata* Sieber ex DC., Brown *et al.* 1985) where seed of successive years is held on the tree at the same time, allowing estimates of mating systems 1, 2, 3 and more years earlier, showed a consistent increase in outcrossing rate with the age of the seed. Seed viability in these genera appears to be associated with heterozygosity, as inbred material is selected against, although in some species the degree of selection may vary depending on the availability of resources (e.g. *Eucalyptus camaldulensis* Dehnh., James & Kennington 1993).

5.6.3 Density

Differences between, and annual variation in, outcrossing rates for individual trees of several neotropical tree species are consistent with changes in local flowering densities and the spatial patterns of flowering individuals (Murawski & Hamrick 1991). Species occurring at low densities appeared to combine significant levels of correlated mating with long-distance gene flow, whereas higher density species showed more random mating, generally over shorter distances. *Cavanillesia platanifolia*, a self-fertile tree species, showed year-to-year variation in outcrossing rate; the lower outcrossing level seemed to be related to lower flowering densities (1987—t_m = 0.57, 74% trees in flower; 1988—t_m = 0.35, 49% trees in flower; 1989—t_m = 0.21, 32% trees in flower; Murawski *et al.* 1990; Murawski & Hamrick 1992). Similarly increased levels of inbreeding were found in some lower density conifer stands [e.g. *Pinus ponderosa* Douglas ex Lawson & C. Lawson, Farris & Mitton 1984; *Larix laricina* (Du Roi) K. Koch, Knowles *et al.* 1987], presumably owing to the increased proportion of self pollen in the pollen cloud.

5.6.4 Pollination vectors

Depending on the particular characteristics of a pollination vector, mating patterns may vary not only from tree to tree but also within the tree canopy. Within conifers, levels of selfing typically appear to be slightly higher in lower parts of the canopy, compared with the upper (e.g. Shen *et al.* 1981; Omi & Adams 1986; Burczyk *et al.* 1991), although this may partly reflect differential self fertility between individuals. In angiosperms, however, there has been no evidence of variation at different crown levels, but there have been differences owing to sampling from limited parts of the crown (t_m = 1.09 whole crown sample, t_m = 0.75 sample from three panicles, Boshier *et al.* 1995a) or low flower production (Murawski *et al.* 1990; Murawski & Hamrick 1991). In such cases there appears to be either correlated biparental mating owing to limited sampling of the pollen pool by pollinators, or an actual increase in the rate of selfing, depending on the species and the strength of the incompatibility mechanism. In the bird-pollinated *Eucalyptus ramelliana* F. Muell. there was a high degree of relatedness between seed within the same fruit (mean 3.85 fathers per fruit), but staggered flowering over the whole tree led to a wide variety of pollen sources being sampled and a correspondingly high outcrossing rate (t_m = 0.89, Sampson 1998).

5.6.5 Conclusion

The degree of relatedness that results from mating in trees can be influenced by a variety of factors over a hierarchy of levels (e.g. seed, fruit inflorescence, tree) that results from mating in trees. In many cases observed mating patterns are likely to result from interactions between several factors, rather than any one particular factor. Any increased, abnormal levels of inbreeding may be unimportant from an evolutionary viewpoint, with selfed individuals selected against at various stages of regeneration (seed production, seedling establishment and growth). They may, however, be critical in terms of the levels of diversity that are sampled for *ex situ* conservation, tree breeding or plantation programmes. The inclusion of progeny from parents with a high level of selfing or biparental mating will bias results from open-pollinated progeny trials, as well as leading to low rankings of parents that may actually have high breeding values (Shaw & Allard 1982; Sorensen & White 1988). In the following chapters it will be seen how patterns of mating may vary as a result of a variety of interventions and how knowledge of such changes can be used to influence management practices for the sustainable use of the genetic resources.

5.7 Acknowledgments

My thanks are due to an anonymous reviewer for their helpful comments. This publication is an output from research partly funded by the United Kingdom Department for International Development (DFID) for the benefit of developing countries. The views expressed are not necessarily those of DFID. Projects R5729, R6516 Forestry Research Programme.

GENE FLOW IN FOREST TREES

James L. Hamrick and John D. Nason

SUMMARY

Pollen and seed immigration into forest tree populations promotes genetic continuity by transfering genetic variation among populations. The homogenising effects of gene flow, therefore, counter the disruptive effects of selection and genetic drift. Gene flow rates are typically estimated in two ways: indirectly from the distribution of genetic variation among populations or directly via the detection of immigrant pollen and seed. Gene flow rates estimated indirectly from spatial genetic structure measure effective rates of immigration pooled over several populations and generations. Reviews of the distribution of allozyme genetic variation demonstrate that tree species typically possess more genetic diversity than herbaceous species and that more of this variation occurs within individual populations. These data support the conclusion that tree species generally experience more gene flow than most herbaceous species. In only a few studies have the relative contributions of pollen and seed flow been distinguished. These indirect estimates indicate that pollen dispersal is responsible for much higher levels of gene migration than seed dispersal.

Genetic markers can be used to directly estimate gene flow by detecting immigrant pollen and seed genotypes. The relatively few direct measurements of gene flow into tree populations demonstrate that gene immigration via pollen is often more than 25% over distances of several hundred metres. Direct estimates of seed movement are rare but indicate that seed flow is more restricted than pollen flow over comparable distances. This conclusion is tentative, however, as seed movement has not been characterised for tree species with light, wind-borne seeds or with strong-flying animal vectors.

6.1 Introduction

Forest tree species are subject to evolutionary forces that differentially affect the distribution of genetic variation within and among their populations. Natural selection favouring adaptation to environmental gradients or patchily distributed habitat variation, for example, may produce genetic differentiation among populations or population subdivisions. Factors such as small population size and inbreeding, which reduce within-population genetic variation via genetic drift, further contribute to the development of genetic heterogeneity among populations. In contrast to these potentially disruptive forces, gene exchange among populations, or gene flow, homogenises spatial genetic variation. Although the evolution of population genetic structure results from complex interactions of selection, drift and gene flow, tree species with high rates of gene flow should have relatively more genetic variation distributed within and less among populations than species with more limited gene movement.

Plants disperse their genes during two independent life cycle stages: pollen dispersal before fertilisation and seed dispersal after fertilisation and embryo development. Since 'gene flow' is defined as the effective movement of genes between populations or population subdivisions, fertilisation of an ovule by an immigrant pollen grain and the subsequent establishment of that seedling represents a gene flow event. Seed dispersal, however, constitutes gene flow when immigrant seeds become established in pre-existing populations. When seed dispersal results in the founding of a new population, it may or may not contribute to the homogenisation of spatial genetic structure depending on the degree of genetic correlation between the founders (Whitlock & McCauley 1990; Hamrick & Nason 1996). This dual mode of gene dispersal, while not unique to plants, is an important difference between plants and most animals. Recognition of the two stages of gene dispersal is essential for understanding the distribution of genetic diversity within and among plant populations (Levin 1981; Whitlock & McCauley 1990). Since pollen and seed dispersal are independent events, plant species with effective long-range pollen dispersal but relatively localised seed dispersal [e.g. a *Quercus* (oak) species] may have less genetic variation among geographically separated populations than occurs among groups of related individuals (i.e. family

structure) separated by a few metres (Hamrick *et al.* 1993; Hamrick & Nason 1996).

Despite its important consequences for the genetic composition of plant populations, accurate estimates of gene flow rates and of pollen and seed dispersal distances are relatively rare. This absence of empirical data has allowed considerable debate concerning the magnitude and importance of gene flow for plant species. Some authors (e.g. Ehrlich & Raven 1969; Levin & Kerster 1974) have argued that gene flow is relatively unimportant in shaping the genetic structure of species whereas others (e.g. Ellstrand & Marshall 1985; Hamrick 1987) have held that gene flow has a major influence on the spatial distribution of genetic diversity. In this chapter, we review various methods of estimating gene flow into plant populations. We then present data indicating that gene flow is extensive in many forest tree species.

6.2 Methods of estimating gene flow

Two general methods are used to estimate gene flow:

(a) *indirect*—historical levels of gene flow are estimated based on the distribution of genetic diversity among populations

(b) *direct*—contemporary gene flow is estimated based on observations of the movement of pollen and seed vectors, marked pollen or seed, or the identification of immigrant alleles or genotypes.

6.2.1 Indirect measures

The two most commonly used indirect methods of estimating gene flow utilise the distribution of genetic diversity among populations. The first method partitions total genetic diversity into within-population and among-population components. Historical rates of gene flow can be estimated indirectly from the proportion of the total diversity found among populations (F_{ST}, Wright 1931, 1951; G_{ST}, Nei 1973; see Chapter 3). For a set of populations F_{ST} is used to estimate gene flow in terms of the effective number of immigrants per generation (N_em) as:

$$N_em = \frac{1-F_{ST}}{4F_{ST}} \qquad (6.1)$$

where, N_e is the effective number of individuals in the population and m is the immigration rate. With this procedure, data from hundreds of plant allozyme studies can be used to estimate N_{em}. This method does not, however, allow independent estimates of immigration (m).

A second method of calculating N_{em} is based on mean allele frequencies in populations in which the alleles occur (conditional average frequencies) (Slatkin 1980). Quantitative estimates of N_m can be obtained from the mean frequency of private alleles (i.e. Slatkin 1985; Barton & Slatkin 1986). The relationship is:

$$\log_{10}[\bar{p}(1)] = a\log_{10}(N_{em}) + b \qquad (6.2)$$

where $\bar{p}(1)$ is the mean frequency of private alleles and a and b are constants that vary with population sample sizes (Barton & Slatkin 1986).

Slatkin and Barton (1989) demonstrated via simulation that both the F_{ST}-based and $\bar{p}(1)$-based methods provide reasonably accurate estimates of N_{em} under a variety of conditions. However, the F_{ST}-based method is empirically preferable since it uses all the gene frequency data and, thus, is less sensitive to errors in the estimation of rare allele frequencies.

Recently the use of indirect methods to estimate gene flow rates has been criticised (e.g. Bossart & Pashley-Prowell 1998). First, indirect methods estimate historical levels of gene flow and may not represent current rates of gene flow. Second, these methods require several assumptions, the most important being that populations are in genetic and demographic equilibrium (Porter 1990). Finally, the F_{ST}-based method assumes that genetic structure among populations is caused by a balance of gene flow and genetic drift and that selection plays no role in the distribution of genetic variation among populations. Such assumptions are difficult to test. However, reviews of the plant allozyme literature (Hamrick & Godt 1989, 1996a; Hamrick *et al.* 1992) have shown that more than 40% of the variation in F_{ST} values among plant species is associated with life history traits of the species. Unexplained variation among species is probably due to differences in experimental procedures (e.g. number and choice of loci, number of populations sampled, sample sizes, geographical scale of the study) as well as to unknown evolutionary and ecological factors in the history of the species.

6.2.2 Direct measures
Several approaches directly estimate pollen and/or seed immigration rates into populations. Perhaps the simplest and least technologically demanding is to observe the movement of pollen or seed dispersal agents (i.e. animals) from one plant to another or from one population to another. While seemingly straightforward, this approach is actually fraught with problems. For example, because only a fraction of the pollen obtained from a given plant may be deposited on the next plant visited (pollen carryover, *sensu* Thomson & Plowright 1980), measurements of pollinator movement may significantly underestimate actual pollen dispersal distances (Schaal 1980). Estimating gene flow from the movements of seed dispersal agents is subject to similar problems. Forest trees present special difficulties for the direct observation of dispersal agents due to their height and large, often obscured, canopies. Also, this approach is not applicable to wind-pollinated species.

A technically more demanding approach is to mark pollen or seeds with dyes, paint, rare earths or radioactive tracers and to monitor their movement. Although this approach better represents actual patterns of pollen and seed dispersal, assumptions must be made concerning whether marking influences pollen or seed dispersal and whether immigrants are successfully incorporated into the recipient population. Moreover, this approach is most often used to describe pollen and seed movement within populations rather than among populations (i.e. gene flow) necessitating the questionable assumption that patterns of pollen and seed movement within populations are indicative of movement among populations. The height of forest trees also limits the feasibility of this approach.

The availability of codominant, presumably neutral, genetic markers (e.g. allozymes, microsatellites) coupled with the development of advanced statistical procedures provides the most accurate method of directly estimating gene flow rates for natural plant populations. The simplest method is to locate an individual or cluster of individuals with a unique allele within a natural or experimental population. From the observation of this allele within progeny arrays of other individuals or in seedling cohorts within the population, patterns of pollen and seed dispersal can be quantified. This approach is superior to the two procedures just described since assumptions concerning fertilisation success are unnecessary. Further, it is often relatively easy to genotype juveniles and to collect and genotype seeds from forest trees. There are two limitations of this procedure, however.

First, population-level patterns are often extrapolated from studies of only one or a few individuals carrying a unique allele. Second, although this method describes pollen and seed movement within populations, it may not be useful for determining gene flow among populations or population subdivisions because of the low frequency of the alleles involved. Thus, other direct techniques are needed to estimate interpopulation gene flow.

The most straightforward method estimates gene flow from differences in allele frequencies between adult and seed generations. If m represents the rate of gene flow into a population (i.e. the proportion of immigrant alleles), $(1 - m)$ is the proportion of the alleles that are not immigrants, q_t the allele frequency in some generation t, and $\bar{q}$ the mean frequency of the allele in populations surrounding the recipient population (i.e. the donor population), the expected relationship between gene flow and gene frequency change is:

$$q_t = q_{t-1}(1-m) + \bar{q}m \qquad (6.3)$$

As q_t, q_{t-1} and $\bar{q}$ can be estimated directly by surveying the appropriate population and generation, this equation can be reorganised to estimate m:

$$m = \frac{q_{t-1} - q_t}{q_{t-1} - \bar{q}} \qquad (6.4)$$

Since this procedure assumes that other evolutionary factors (e.g. selection) do not affect allele frequencies in the recipient population, it is most robust when an allele is absent from the recipient population and is, therefore, insensitive to the action of other evolutionary factors. In this case, the equation above reduces to:

$$m = \frac{q_t}{\bar{q}} \qquad (6.5)$$

This approach was applied to an isolated subpopulation of the neotropical leguminous tree *Tachigali versicolor* Standl. & L.O. Wms. in the forest of Barro Colorado Island, Central Panama (Loveless *et al.* 1998). A cluster of five trees isolated by about 500 m were monomorphic for one of two alleles at a single allozyme locus. When 48 progeny from each of the five adult trees were analysed, the frequency of the alternative allele was 0.021 (i.e. 4.2% of the progeny were heterozygous for the marker allele). Since the

frequency of the less common allele was 0.200 in the Barro Colorado Island population as a whole, the rate of gene flow was estimated as $m = 0.105$. Since pollen is haploid, a gene flow rate of 10.5% represents a pollen immigration rate of 21% (i.e. 21% of the progeny resulted from pollen movement beyond 500 m).

In practice, it is unusual to have a relatively common allele in the donor population that is absent from the recipient population. However, in populations with relatively few flowering adults, the number of multilocus combinations of alleles produced in their gametic arrays is a limited subset of the possible number of multilocus pollen gametes (the product of the number of alleles at each locus). Therefore, genetic analyses of several progeny arrays may identify multilocus gametes that could not have been produced by adults within the recipient population. These 'apparent' pollen flow events are equivalent to observing alleles in progeny arrays that are absent from the adult population (see above example for *Tachigali versicolor*). However, estimates of apparent gene flow need to be adjusted to account for the immigration of gametes that could have been produced by adults in the recipient population (i.e. 'cryptic' gene flow).

With conifers it is straightforward to estimate the total rate of gene flow since conifer seeds consist of haploid female gametophytic tissue and the diploid embryo. As a result, the genotype of the male gamete can be determined directly by subtracting the female gametic contribution from the embryo's genotype. For example, if the embryo has a three-locus genotype of A1 A3 B1 B1 C1 C2 and the genotype of the female gamete is A1 B1 C2, the male gamete is A3 B1 C1. If this male gamete represents an apparent gene flow event, its observed frequency in the recipient population (q_s) provides an estimate of the rate of pollen gene flow as:

$$m = \frac{q_s}{\bar{q}_m} \qquad (6.6)$$

where $\bar{q}_m$ is the expected frequency of this gamete in the donor population. Since immigrant gametes with different genotypes will usually be detected, estimates of q_s and $\bar{q}_m$ can be pooled to obtain a more robust estimate of pollen gene flow. Friedman and Adams (1985) used this approach to demonstrate that pollen flow into a *Pinus taeda* L. (loblolly pine) seed orchard averaged about 36%.

In angiosperms, as in gymnosperms, the rate of apparent pollen flow provides a minimum estimate of total pollen flow. However, formulation of an unbiased estimate of the total pollen flow rate is more complicated for angiosperms. In angiosperms, the exact male gamete often cannot be observed directly since, in heterozygous progeny bearing the same genotype as their heterozygous mother, either allele can be assigned to the paternal gamete. Devlin and Ellstrand (1990) developed a procedure that uses the rate of apparent pollen gene flow to obtain an estimate of total pollen flow adjusted for cryptic pollen flow. This procedure is widely used to estimate rates of pollen flow into populations of angiosperm trees (e.g. Schnabel & Hamrick 1995; Dow & Ashley 1996; Nason & Hamrick 1997).

6.3 Estimates of gene flow into forest tree populations

6.3.1 Indirect methods

Hamrick *et al.* (1992) reviewed allozyme data for woody plants (predominantly trees) and calculated average levels of genetic variation within and among populations of woody plants with different life history and ecological traits (Table 6.1). Their summary indicated that woody species have somewhat more genetic diversity (H_{es}) than herbaceous species, while within their populations woody plants maintain much more genetic diversity (H_{ep}) than herbaceous species. The large differences in mean within-population genetic diversity occurs because woody species have significantly less genetic differentiation among their populations (G_{ST}; Hamrick *et al.* 1992). There are numerous ecological and life history characteristics that differ between herbaceous and woody plant species that could explain the observed differences in genetic structure. In particular, small herbaceous species may have shorter life spans, form smaller breeding units, and have patchier population structures than many woody species, especially trees. Thus, genetic drift may have more influence on the genetic composition of populations of herbaceous species. In addition, for the taller forest trees, pollen and seeds are released at greater heights and, as a consequence, are dispersed longer distances (Hamrick & Godt 1996a). Due to the greater potential mobility of their pollen and seeds, trees should experience higher rates of gene flow than herbaceous species with the same pollination and seed dispersal mechanisms. This conclusion is supported by Hamrick and Godt

(1996a) who demonstrated that outcrossing, wind-pollinated woody plants have significantly lower G_{ST} values than herbaceous species with the same breeding system (0.094 versus 0.218, respectively). Furthermore, estimates of N_m calculated for forest trees are generally high enough ($N_m > 1.0$) to counteract genetic drift, whereas N_m values for many herbaceous plants fall well below 1.0.

Within woody plants, gymnosperms have higher gene flow estimates than angiosperms and species with larger geographical ranges have higher gene flow than species with more limited geographical ranges. Table 6.1 indicates also that wind-pollinated woody species tend to have higher rates of gene flow than woody species with other breeding systems while woody species with gravity-dispersed seeds tend to have less gene flow and more genetic differentiation among their populations. It is not surprising that species with the greatest potential for long-distance pollen and seed movement (e.g. outcrossing species with ingested or wind-dispersed seeds) would have more gene flow but it is less clear why the geographical range of a species should affect gene exchange among populations. Perhaps species with limited ranges tend to have smaller, more discontinuous populations while more widespread species tend to have larger, more continuous populations. Hamrick and Godt (1996b), for example, report that species in the genus *Pinus* with discontinuous ranges generally have more genetic heterogeneity among their populations (e.g. *Pinus halepensis* Mill., $G_{ST} = 0.300$, Schiller *et al.* 1983; *Pinus longaeva* D.K. Bailey, $G_{ST} = 0.164$, Hamrick *et al.* 1994) than pines with more continuous distributions (e.g. *Pinus rigida* Mill., $G_{ST} = 0.039$, Guries & Ledig 1982; *Pinus contorta* Douglas ex Loudon, $G_{ST} = 0.036$, Wheeler & Guries 1982).

Insect-pollinated tree species and those that occur at lower densities (e.g. many tropical trees) might be expected to have less gene flow than wind-pollinated species that occur at higher densities (e.g. many north temperate trees). Summaries of the allozyme literature do not support this expectation, however, as temperate and tropical tree species have similar G_{ST} values (0.092 and 0.119, respectively, Hamrick *et al.* 1992). This conclusion is supported by recent studies of the distribution of genetic diversity among geographically separated populations of two neotropical tree species, *Cordia alliodora* (Ruiz & Pav.) Cham. ($G_{ST} = 0.117$, Chase *et al.* 1995) and *Enterolobium cyclocarpum* (Jacq.) Griseb.

TABLE 6.1 Estimated levels of allozyme genetic diversity within species (H_{es}) and populations (H_{ep}), the proportion of the total diversity at polymorphic loci among populations (G_{ST}), and the number of migrants per generation (N_m) for long-lived woody plants (after Hamrick *et al.* 1992)

Categories	N^A	H_{es}[B]	H_{ep}	G_{ST}	N_m
Herbaceous plants	441	0.140b	0.099b	0.305a	0.51b
Long-lived woody species	195	0.177a	0.148a	0.084b	2.73a
Taxonomic status					
Gymnosperms	89	0.169a	0.151a	0.073b	3.17a
Angiosperms	102	0.183a	0.143a	0.102a	2.20b
Geographic range					
Endemic	20	0.078c	0.056d	0.141a	1.52b
Narrow	45	0.165b	0.143c	0.124a	1.77b
Regional	115	0.169ab	0.194b	0.065b	3.60a
Widespread	11	0.257a	0.228a	0.033b	7.33a
Breeding system					
Mixed-animal	11	0.075c	0.035b	0.122a	1.80a
Outcrossing-animal	51	0.211a	0.163a	0.099a	2.28a
Outcrossing-wind	128	0.173b	0.154a	0.077a	3.00a
Seed dispersal					
Gravity	23	0.144cd	0.141bc	0.131a	1.66b
Gravity-attached	10	0.115d	0.104bc	0.099ab	2.28ab
Attached	26	0.204ab	0.144abc	0.065ab	3.60ab
Ingested	27	0.231a	0.208a	0.051b	4.65a
Wind	103	0.160bc	0.149b	0.076b	3.04ab

[A]N, number of entries for each category.
[B]Means followed by different letters in a column are significantly different at the 5% probability level.

(guanacaste) ($G_{ST} = 0.039$, Rocha & Lobo 1996). Furthermore, direct estimates of pollen flow for several insect-pollinated trees (see later) has demonstrated that they experience high rates of pollen immigration over several hundred metres.

The question also arises concerning the relative importance of pollen versus seed dispersal to overall gene flow. The distribution of genetic diversity for biparentally inherited nuclear genes such as allozymes is influenced by both seed and pollen movement. In contrast, chloroplast (cpDNA) or mitochondrial (mtDNA) genes are typically inherited from a single parent. Chloroplast genes are usually (but not always) maternally inherited in angiosperms and paternally inherited in gymnosperms while mitochondrial genes are inherited maternally in both taxa and, as a consequence, are dispersed only by seeds. Thus, if pollen and seed flow differ significantly, the degree of interpopulation differentiation should vary between maternally, paternally and biparentally inherited genes. The available forest tree data indicate that maternally inherited genes have more interpopulation differentiation than nuclear genes. El-Mousadik and Petit (1996) list seven tree species for which the ratio of gene flow via pollen (m_p) can be compared to gene flow via seeds (m_s). The $\frac{m_p}{m_s}$ ratios ranged from 1.8 for the animal-pollinated, wind-dispersed *Eucalyptus nitens* Maiden to 500 for the wind-pollinated, largely gravity-dispersed *Quercus petraea* (Matt.) Liebl. indicating that gene flow via pollen greatly exceeds gene flow via seed for many tree species (Box 6.1).

The relatively low estimates of gene flow via seeds appear to be inconsistent with the observation that following the last glacial epoch many forest tree species expanded their ranges northward at rates of several hundred metres per year (> 25 km per generation) (Davis 1983; Huntley & Birks 1983).

Box 6.1   Differential rates of pollen versus seed mediated gene flow in
            Pinus flexilis

The most comprehensive study of geographic variation for different types of molecular and biochemical marker loci is that of Latta and Mitton (1997). They analysed random amplified polymorphic DNA (RAPD) (biparentally inherited), allozyme (biparentally inherited), cpDNA (paternally inherited) and mtDNA (maternally inherited) loci for seven populations of *Pinus flexilis* in the Rocky Mountains of central Colorado, USA. Like most pines, *Pinus flexilis* has wind-dispersed pollen. Unlike most pines, however, *Pinus flexilis* seeds are distributed by Clark's nutcracker (*Nucifraga columbiana* Wilson). Estimates of the proportion of genetic diversity found among populations (F_{ST}) and indirect estimates of gene flow (N_m) were made for all marker loci, as shown in Table 1:

Table 1   Estimates for marker loci

Marker type	F_{ST}	N_m
cpDNA	0.013	38.0
Allozymes	0.016	15.0
RAPD	0.172	1.2
mtDNA	0.679	0.3

In contrast to other types of markers, values of F_{ST} varied greatly among the nine RAPD loci, indicating that factors other than gene flow had affected their distribution.

Latta and Mitton (1997) conclude that gene flow via pollen is much greater than gene flow via seeds ($\frac{m_p}{m_s} = 160$), even though Clark's nutcrackers have been observed to move seeds several kilometres.

Hewitt (1996) addressed this apparent inconsistency by proposing that these range expansions resulted from long-distance seed dispersal events that established colonies well ahead of the main wave of expansion. Surrounding areas were then colonised from these foci. Such colonisation patterns coupled with the relative lack of effective seed dispersal into established populations could explain current patterns of cpDNA variation in species such as the eight species in the European white oak complex (probably dispersed by jays, Petit & Kremer 1993; Dumolin-Lapegue *et al.* 1997; Petit *et al.* 1997) and *Alnus glutinosa* L. (wind and water dispersed, King & Ferris 1998).

6.3.2   Direct estimates

The ability to identify individual seeds or seedlings that result from long-distance pollen or seed dispersal by multilocus genetic analyses has greatly enhanced direct estimations of gene flow rates. It is interesting that more tropical than temperate tree species have been studied (Table 6.2). Furthermore, most of these studies are of pollen flow while gene flow via seeds is largely unexamined (but see Dow & Ashley 1996; Schnabel *et al.* 1998).

Estimates of pollen movement are quite heterogeneous among studies but the general consensus is that pollen flow can be quite extensive (> 25%) over distances as great as one kilometre (Table 6.2). Some of the heterogeneity in gene flow rates among species is almost certainly due to constraints imposed by varying amounts of assayable genetic variation as well as to differences among study sites in the degree of spatial isolation among populations. Studies of sites isolated by several hundred metres are more likely to detect long-distance pollen moves than studies of sites embedded in continuous forests where, due to technical limitations, pollen movement can only be documented over shorter distances. Below, we discuss

TABLE 6.2 Direct estimates of pollen-mediated gene immigration into tree populations based on multilocus paternity analyses

Species	Family	Immigrant pollen (%)	Distance (m)	References
A. Temperate trees				
Gleditsia triacanthos	Caesalpinaceae	40–70	> 200	Schnabel & Hamrick (1995)
Pinus flexilis E. James	Pinaceae	6	> 5000	W.S.F. Schuster & J.B. Mitton (unpublished data)
Pinus sylvestris L.	Pinaceae	26	> 200	Harju & Muona (1989)
		38	> 100	Nagasaka & Szmidt (1984)
		21–36	> 100	El-Kassaby *et al.* (1989b)
Pinus taeda	Pinaceae	28–48	> 200	Friedman & Adams (1985)
Pseudotsuga menziesii (Mirb.) Franco	Pinaceae	40	> 100	Neale (1984)
		29–52	> 100	Smith & Adams (1983)
		4–25	> 500	Wheeler & Jech (1986)
Quercus macrocarpa	Fagaceae	70	> 200	Dow & Ashley (1996)
B. Tropical trees				
Calophyllum longifolium Willd.	Cluisaceae	62	> 210	Stacy *et al.* (1996)
Cordia alliodora	Boraginaceae	3	> 280	Boshier *et al.* (1995b)
Enterolobium cyclocarpum	Mimosaceae	60	> 250	Apsit & Hamrick (in review)
Ficus (three sp.)	Moraceae	> 90	> 1000	Nason & Hamrick (1997)
Pithecellobium elegans Ducke	Mimosaceae	15	> 350	Chase *et al.* (1996b)
Platypodium elegans Vogel	Papilionaceae	36–77	> 100	Hamrick & Murawski (1990)
Spondias mombin	Anacardiaceae	44	> 200	Nason & Hamrick (1997)
Tachigali versicolor	Caesalpinaceae	25	> 500	Loveless *et al.* (1998)
Turpinia occidentalis G. Don	Staphyleaceae	1–27	> 130	Stacy *et al.* (1996)

several case studies of gene flow into isolated populations of forest trees (see also Box 6.2).

6.3.2.1 Burr oak

Quercus macrocarpa Michx. (burr oak) is a wind-pollinated oak with large, heavy acorns. Dow and Ashley (1996) used four microsatellite loci to study pollen and seed dispersal into an isolated population of 62 mature *Quercus macrocarpa* in Illinois, USA. Since microsatellite loci have numerous alleles, parentage of acorns and established saplings could be determined with a high degree of certainty. Their results indicated that 71 of the 100 saplings examined had only one parent in the stand while 14 saplings had neither parent after adjustment for cryptic gene flow. Thus, it appears that over 70% of the saplings resulted from pollen flow into this stand while 14% resulted from seed dispersal. Estimates of pollen flow (62%) based on analyses of 284 acorns were consistent with the estimate of pollen flow obtained from saplings.

6.3.2.2 Honey locust

Gleditsia triacanthos L. (honey locust) is dioecious, insect-pollinated and has large flat pods that can be dispersed by animals such as deer. Schnabel and Hamrick (1995) used 18 polymorphic allozyme loci to determine rates of apparent pollen flow over two years into two honey locust populations located in northeastern Kansas, USA. Trees were isolated by 140 m to 240 m at one site and by 85 m to 100 m at the second site. Paternity exclusion analyses were used to estimate the proportion of seeds sired by immigrant pollen. Single-parent and parent-pair exclusion analyses on naturally established seedlings were also used to estimate gene flow into one site over 12 years and into a second site over 22 years. These analyses yielded high minimum estimates of pollen flow into each site (17–30%). In both sites there was significantly less gene flow in years of high fruit production. In a second analysis, Schnabel *et al.* (1998) found that a minimum

---

Box 6.2 Landscape level pollen flow in neotropical strangler figs

Ficus spp. (figs) are pollinated by small species-specific wasps (Ogaonidae, Chalcidoidea). Wasp larvae develop in fig ovaries within closed inflorescences (called syconia) of fig trees. Mature inseminated female wasps carrying pollen from a mature syconia fertilise female-phase receptive inflorescences in which both wasps and seeds will be produced. Since the female wasps carry pollen from a single pollen donor, fruits fertilised by a single female wasp contain seeds that are full sibs. Nason *et al.* (1996, 1998) analysed several full-sib progeny arrays for six species of strangler figs located in Barro Colorado Island Nature Monument in Central Panama. The fig species in this area occur at very low densities and flower year around with approximately equal numbers of trees in the male or female phase at any given time. If the maternal genotype is known, genetic analyses of several full-sib progeny allows the inference of the multilocus genotype of each pollen parent. As a result Nason *et al.* (1996) could estimate how many pollen donors were represented in the fruit crops of each tree. Surprisingly, for most species and trees the estimated number of pollen donors greatly exceeded the number of male-phase trees known to be on Barro Colorado Island. From estimates of the number of pollen donors and from the known densities of these trees, mean areas for the overall breeding population could be obtained. These areas ranged from 108 km^2 to 632 km^2, indicating that these small wasps disperse pollen over distances up to 14 km.

---

of 2.1% of the established seedlings resulted from seed immigration over at least 100 m.

6.3.2.3 Guanacaste

Enterolobium cyclocarpum is pollinated by bees and hawk-moths and has large ear-shaped pods. Large animals such as horses and cows eat the pods and disperse its seeds. *Enterolobium cyclocarpum* is a dominant tree of the seasonally dry forests of Central America. The once continuous tropical dry forest is now highly disturbed and fragmented by agriculture and ranching. As a result, dry forest tree species are usually found in small isolated fragments or along streams or escarpments. Aspit & Hamrick (in review) used allozyme loci to measure pollen immigration rates into four fragments isolated by 250 m to 500 m. Apparent gene flow rates ranged from 17% to 24% while total pollen immigration ranged from 25% to 93%. Such high rates of pollen flow should prevent the loss of genetic diversity in individual populations or in the region as a whole. However, estimates of N_m based on observed rates of pollen flow and effective population sizes (N_e usually < 10) indicate that current levels of genetic diversity among populations ($G_{ST} = 0.051$) should increase (predicted $G_{ST} \approx 0.200$) and that some low frequency alleles may be lost from individual populations.

6.3.2.4 Jobo

Spondias mombin L. (jobo), a neotropical tree often found in second growth forests, is pollinated by small insects and produces fleshy animal and water dispersed fruits. Nason and Hamrick (1997) studied several populations of this species within the Barro Colorado Island Nature Monument in Central Panama. Two research sites were located in continuous forests while other sites were on islands in Lake Gatun (part of the Panama Canal). Study sites in the continuous forest were isolated by 100 m to 200 m and had pollen immigration rates of about 45%. The five island sites had estimated pollen flow rates of 60% to 100% indicating that most effective fertilisations were accomplished by pollen that had immigrated at least 80 m to 1000 m. Reduced seed set and lower germination rates on the more isolated island sites relative to the continuous forest sites led to the conclusion that small, island populations experienced higher rates of self-fertilisation, but that inbred seeds either abort or fail to germinate; viable seeds are predominately those produced by long-distance pollen dispersal.

6.4 Conclusions

Both indirect and direct estimates of gene flow into forest tree populations indicate that gene flow can be high. Except for particularly small spatial isolates, rates of gene flow may be high enough to counter genetic drift and/or inbreeding. Also, for species that have been studied, gene flow via pollen is more extensive than gene flow via seeds. However, tree species with light, wind-dispersed seeds (e.g. *Betula*) or seeds dispersed by particularly mobile animal vectors may have longer distance seed movement.

These conclusions have several implications for the management of forest genetic resources. First, the high rates of gene flow experienced by many forest tree species indicate that seed orchards or seed production areas will experience high rates of pollen contamination (e.g. Friedman & Adams 1985; Harju & Muona 1989). Consideration, therefore, needs to be given to placing seed orchards well away from local non-selected stands. Second, it would appear from the high levels of gene flow observed that management practices that leave only a few seed trees to recolonise otherwise clear-cut areas might be successful. Studies of several tree species, however, indicate that seed set and seedling germination is reduced for such trees presumably due to pollen limitation and higher rates of selfing (Smith *et al.* 1988; Nason & Hamrick 1997). Finally, with such high rates of gene flow one would expect, except for very low frequency alleles, that most of a tree species' alleles would be found within each population. However, forest trees occurring along environmental gradients usually have significant genetic differentiation (Libby *et al.* 1969). Thus, even in the face of the high gene flow rates, natural selection is capable of producing genetic combinations that are adapted to local environmental conditions.

SMALL POPULATION PROCESSES

Outi Savolainen and Helmi Kuittinen

SUMMARY

Some forest tree populations have small geographical distributions or occur in fragmented populations, and the distribution of more species is becoming fragmented as land use increases. In fragmented populations the genetic variability within populations is reduced, and the variance between populations increases. In small populations homozygosity increases because of inbreeding. This may have deleterious consequences because of inbreeding depression. The rates of inbreeding and drift depend on the genetically efficient population size. Even populations with large census size can be genetically small if only part of the individuals participate in reproduction, or if there is clonal reproduction. A short bottleneck will have minor effects on overall variation, whereas a long-term reduction in population size will deplete variability. Similarly, once variability is depleted due to a long-term bottleneck, the recovery of variation by mutation takes a long time after the increase of population size. Because of the long time to reach equilibrium between mutation and drift, the current population sizes do not always correlate positively with the level of variability. For neutral loci such as molecular markers, the distributions of allele frequences depend mainly on the balance between mutation, migration and drift, but the level and distribution of quantitative variation is more likely to be affected as well by selection.

7.1 Introduction

Forest tree populations are often considered large and continuous, but many tree species occur naturally as very small populations. There are several well-studied *Eucalyptus* species that have small isolated populations (Moran 1992). *Washingtonia filifera* H. Wendl. (California fan) palms endemic to California and Mexico are similarly found in isolated small populations across their range. *Pinus radiata* D. Don (radiata pine) has only five natural populations, and the populations on oceanic islands are very small (Moran & Bell 1987). At the northern margin of their range in southern Finland, *Quercus robur* L. (oak) and other noble hardwoods (Mattila *et al.* 1994) are found in low numbers. *Pinus sylvestris* L. (Scots pine) occurs in scattered small populations at the southern and western margins of the distribution. Many currently widespread populations are thought to have previously gone through population size bottlenecks, for example *Pinus resinosa* Aiton (red pine) (Fowler 1965) or *Acacia mangium* Willd. (tropical mangium) (Moran *et al.* 1989c). These examples show that small population processes, such as genetic drift (see Chapter 3), can be important in shaping the genetic structure of natural populations. Many formerly large populations are experiencing reductions in population size owing to human disturbance (see Chapters 9 & 10). Similarly, the consequences of small population size also need to be considered in artificially generated or maintained populations, as when collecting seed samples, maintaining gene banks or breeding populations, or establishing seed orchards (see Chapters 14 & 15).

Large populations maintain ample genetic variability, which assures a possibility to respond to current selection pressures. Furthermore, continuous mutation provides for future evolutionary response. Small populations may lose variability and their ability to respond. The low population size can also give rise to inbreeding and inbreeding depression. For these reasons, it is important to understand genetic processes discussed in this chapter.

Small populations are those where genetic sampling, that is, genetic drift can have a significant effect. In a population of 10 individuals chance events can have more overall significance than in a population of 100 or 1000 individuals (Box 7.1). Drift and inbreeding take place in any finite population, albeit at a slower rate in large than in small populations. The relative significance of drift is determined in balance with the other evolutionary forces: selection, migration, recombination and mutation. The influence of these factors also always depends on population size. The allelic distribution within the population of effective size N_e depends on selection ($N_e s$), migration ($N_e m$), recombination ($N_e r$) and mutation ($N_e \mu$) (where s is the selection coefficient, m, migration rate, r, recombination rate and μ, mutation rate). Thus, if variation in some trait does not influence survival or reproduction (e.g. some random molecular marker), the population allele frequencies will be determined by the balance of genetic drift and migration, whereas in the same population, selection might largely influence the fate of genes regulating cold tolerance, and genetic drift might have hardly any role. Also, the same absolute number of migrants will have relatively less genetic significance in a small than a large population, because in a small population the allelic frequencies could be still governed by drift. Further, the effect of recombination is also related to population size. We will concentrate the discussion on the effects of genetic drift, but the population sizes where drift will predominate depend on the other evolutionary forces.

Genetic processes take place in an ecological context. Populations can become extinct for many reasons, some of which could be so rapid that genetics is irrelevant (Lande 1988). Small population size can have consequences that lead to rapid extinction before genetic drift has had a chance to operate. There could be a complete failure of pollinators, or a small population of a dioecious species could consist completely of one gender and fail to reproduce. Direct human disturbances can destroy the habitat and the populations causing immediate disappearance of the populations. In many cases environmental changes, demographic fluctuation and genetic effects may all interact (e.g. see Lande 1995). Global change may be so rapid that small populations are unable to respond, while larger populations could have sufficient variation. Environmental changes may lead to a demography with high variance in reproductive output, and this, in turn, could lead to inbreeding. It is, thus, important to understand also the joint effects of ecology, demography and genetics.

7.2 Genetically small populations

The vulnerability of populations to genetic drift does not depend directly on the actual size of the

Box 7.1 Genetic drift

In a finite population allele frequencies fluctuate in unpredictable ways because of random genetic drift. Figure 7.2 shows how neutral alleles have been sampled during 20 generations in 10 populations of effective size $N_e = 8$ or $N_e = 50$. Initially the allele frequencies were identical (0.5) in all populations but because of sampling, frequencies fluctuate and the populations differentiate from each other. In five generations one of the alleles has already gone to fixation in some populations, and if the process continues, all populations will finally be monomorphic and homozygous for one of the alleles. Which allele goes to fixation cannot be predicted for individual populations but with enough replicate populations the proportion of populations with each allele fixed depends on the initial allele frequencies. In Figure 7.2, both alleles have the probability of 0.5 of going to fixation. Drift is more important in the smaller population than in the larger population: the increase in variance of the allele frequencies between populations and the rate of fixation is higher in the populations of $N_e = 8$ than in those of $N_e = 50$ for the same number of generations (Table 7.1).

In small populations weakly selected variation may be effectively neutral, because chance effects may be stronger than selection.

Inbreeding coefficient, F, is the probability that two alleles are identical by descent, that is, they are copies of an allele from a common ancestor. The probability of sampling the same allele twice in each generation is $1/(2N_e)$. In generation t:

$$F_t = 1 - \left(1 - \frac{1}{2N_e}\right)^t \tag{1}$$

Expected heterozygosity (H_e) decreases at the same rate as the inbreeding coefficient increases. In the generation t, the expected heterozygosity is:

$$H_{e(t)} = \left(1 - \frac{1}{2N_e}\right)^t H_{e(0)} \tag{2}$$

where $H_{e(0)}$ is the heterozygosity in the original population. The amount of additive genetic variance within a population (V_A) decreases at the same rate as expected heterozygosity.

When the size of a population is reduced drastically the population goes through a population bottleneck. Genetic variability is reduced in the first and successive generations due to sampling effects. After the population starts to grow again, new mutations will gradually accumulate and variability is restored very slowly. The reduction of average heterozygosity depends both on the size of the bottleneck and population growth, such that in rapidly growing populations much of the variability may be retained even after a bottleneck with a very low population size. The reduction in the number of alleles is determined mostly by the size of the bottleneck, because drift efficiently eliminates rare alleles (Nei *et al.* 1975). When a new population is being established with a few individuals, random sampling of alleles from the source population causes a genetic bottleneck. This is called a founder effect.

When a population is partitioned in several small subpopulations little overall change in mean allele frequencies due to drift occurs, but the variation is redistributed compared to the initial state. Wright's F-coefficients (Wright 1951, 1965) are commonly used to quantify the change in the set of subpopulations (see Chapter 3):

(a) F_{IS}—reduction in heterozygosity of an individual due to non-random mating within subpopulations

(b) F_{ST}—reduction in heterozygosity of a subpopulation due to drift

(c) F_{IT}—reduction in heterozygosity of an individual in relation to the total population.

Thus, F_{ST} is a measure of relative differentiation in allele frequences between subpopulations. G_{ST} is an extension of F_{ST} for the case of multiple alleles (Nei 1975; see Chapter 3). The difference between these two measures is usually negligible.

The effective population size, N_e, is a key parameter that determines the effect of drift rather than the census size, N. The effective size is the size of an ideal population which would give rise to the variance of change in gene frequency (variance effective size) or rate of inbreeding (inbreeding effective size) observed in the actual population. Inbreeding and variance effective sizes are usually very similar.

In the ideal population:

(a) gametes are drawn randomly from the parents and they unite randomly to form zygotes

(b) numbers of male and female parents should be equal

(c) variation in the number of progeny is Poisson distributed

(d) no inbreeding or fluctuations in the size of the population between generations occur.

The effective population size is usually lower than the census size because of deviations from the ideal population.

population, but rather on the genetically effective size. Populations with different reproductive and demographic systems, but where the consequences of genetic drift are similar, have the same effective population sizes (Box 7.1). Many demographic features of real populations tend to reduce the effective size compared to the census size, such as inbreeding or unequal fertility of trees. Longevity and overlapping generations usually reduce the effective size. However, seed banks in the soil may reduce the effect of bottlenecks. In clonally propagating species (clones) the effective size can be much smaller than the number of trees because there are fewer genets than ramets. The effective size can be estimated directly from the change in allele frequencies between generations. Demographic data, for example on fertility variation, can be used also (e.g. Crow & Denniston 1988; Caballero 1994).

There are few estimates of effective population sizes in trees. In the natural population of *Pinus radiata* in the Guadaloupe Islands, male fertility variation was measured based on paternity analysis using marker genes. The effective size was only 66% of the census size of the population (G.F. Moran, O. Savolainen & J.C. Bell, unpublished data). In many organisms the difference between the census and the effective size can be even more pronounced.

7.3 Low levels of variability

Population genetics theory predicts that genetic variability of neutral loci, measured as expected heterozygosity (H_e), depends on the product $N_e\mu$, where N_e is the effective population size and μ is the mutation rate. For quantitative traits, the amount of genetic variation is related to $\mathrm{Var}(m)N_e$, where $\mathrm{Var}(m)$ is the variance generated each generation by mutation. At equilibrium, the new variability generated by mutation equals that lost due to genetic drift. Mutation rates are considered to be rather similar in different taxa, and thus the level of variability should largely depend on the effective population size. We examine this relationship in some examples. A significant

a

Halocarpus bidwillii

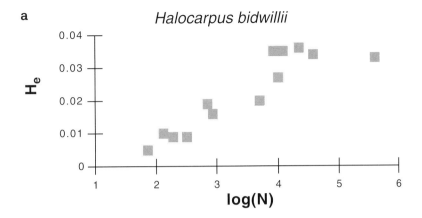

b

Washingtonia filifera

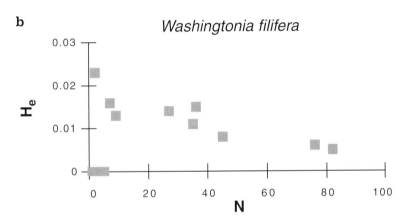

Figure 7.1   Expected heterozygosity (H_e) at allozyme loci in populations of different sizes. (a) Data for *Halocarpus bidwillii* are from Billington (1991). (b) Data for *Washingtonia filifera* are from McClenaghan and Beauchamp (1986).

correlation between population sizes and different measures of genetic diversity was found in a naturally rare dioecious tree, *Halocarpus bidwillii* (Hook. f. ex T. kirk) C.J. Quinn (Billington 1991) (Fig. 7.1a). Prober and Brown (1994) found that population size correlated with all measures of genetic diversity in fragmented populations of *Eucalyptus albens* Benth. In *Washingtonia filifera* H. Wendl. (McClenaghan & Beauchamp 1986) (Fig. 7.1b) and *Eucalyptus caesia* Benth. (Moran & Hopper 1983) there was a relationship between the number of alleles or the number of polymorphic loci and population size but not between heterozygosity and population size.

The lack of a relationship between population size and variability can be accounted for by several factors. First, the data compare census size, not effective population size. However, for outcrossing trees the differences are unlikely to be very important between populations. More importantly, we may not be examining an equilibrium situation—the present

population sizes may have prevailed for such short times that equilibrium has not been reached yet.

Hamrick and Godt (1989) examined the relationship of species distribution characteristics and genetic variability. Species with narrow distributions were found to have less variability than those with wide distributions. Localised species are thought to consist of smaller and more isolated populations than widespread species. The same trend was found in a group of *Eucalyptus* species (Moran 1992) and in tropical trees (Loveless 1992).

Marginal, isolated populations may have less variation than central populations of a species due to random processes in small populations. Marginal populations may have smaller effective population sizes than central populations because of increased selfing in a sparse population (Kärkkäinen *et al.* 1996) or increased selection pressures that may increase the variance in reproductive success between parents

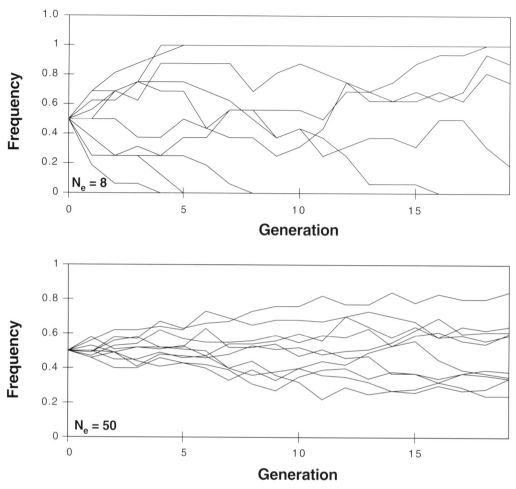

Figure 7.2 Allele frequency changes due to random sampling of gametes during 20 generations. In the simulation, the initial frequencies were 0.5 in each of 10 populations. The size of the populations were $N_e = 8$ in the upper figure and $N_e = 50$ in the lower figure.

(Savolainen 1996). Founder effects should be more frequent and gene flow restricted due to isolation of marginal populations. In *Pseudotsuga menziesii* (Mirb.) Franco (Douglas-fir) genetic diversity decreased towards the periphery of the distribution (Li & Adams 1989). However, in many cases, marginal populations receive sufficient gene flow to have equal amounts of variation compared to central ones, for example in *Picea abies* O. Deg. (Norway spruce) (Muona *et al.* 1990).

These general predictions account for only a small proportion of the differences in variability. Even after obvious characteristics have been taken into consideration, most differences between species remain unaccounted for.

Natural populations have reached their current levels of variability over long times, and may be in mutation–

drift equilibrium. For conservation genetics, we should consider what happens to genetic variation when an initially large population is divided into several small populations (Fig. 7.2, Table 7.1). This process is a highly simplified model of no selection, migration or mutation, which highlights the essential features of genetic drift. Each small population loses genetic variability at the rate of $\frac{1}{2N_e}$ per generation. If the small population size persists, this will result in a large reduction of expected heterozygosity over long times. The smaller populations in Figure 7.2 have nearly all become monomorphic in 20 generations, whereas the larger populations have retained some heterozygosity. The simulation of Figure 7.2 started with equal allelic frequencies of 0.5, but alleles with low frequencies would be lost more rapidly. The effect of genetic drift on the number of alleles (allelic richness) would be

TABLE 7.1 **Statistics for the simulation of changes in allele frequency due to random sampling of gametes over 20 generations (Figure 7.2)**

Single locus statistics are the average expected heterozygosity within populations, H_S (i.e. average H_e across populations); the expected heterozygosity in the total population, H_T; the mean allele frequency, p; and F_{ST}. The statistics for quantitative traits are calculated assuming phenotypic values −1 and 1 for the homozygotes, and 0 for the heterozygotes. V_A is the mean additive variance within populations; V_T, the total variance; X, the trait mean; and V_B, the variance among populations.

	N_e = infinite	N_e = 50	N_e = 8
Single locus			
H	0.5	0.44	0.09
H_T	0.5	0.50	0.50
p	0.5	0.49	0.48
F_{ST}	0	0.12	0.82
Quantitative trait			
V_A	0.5	0.46	0.07
V_T	0.5	0.59	0.89
X	0	−0.01	−0.04
V_B	0	0.13	0.82

larger than on expected heterozygosity itself. As the population size increases, recurrent mutation can restore the variability, but this takes place very slowly (Nei *et al.* 1975). Both the loss of genetic variability and restoration are slow processes. The number of alleles is reduced rapidly and recovered slowly, expected heterozygosity is retained for longer and is recovered more rapidly.

The long lasting influence of population history may account for cases where species with rather similar distributions and breeding systems, and thus presumably effective population sizes, seem to have very different levels of variability. A group of acacias with similar reproduction systems and fairly wide distribution show different levels of heterozygosity Moran *et al.* 1989c; Table 7.2). Similarly, in a group of conifers with wide continuous distribution there is considerable variation in the level of variability (Table 7.2). The nearly monomorphic species, *Acacia mangium* (Moran *et al.* 1989c), and *Pinus resinosa* (Mosseler *et al.* 1991) probably have experienced severe population bottlenecks, the effects of which can still be seen as reduced variability at marker loci. However, in rare species with relatively high variability, such as *Picea omorika* (Pancic) Bolle (Serbian spruce) (Kuittinen *et al.* 1991) or *Eucalyptus crucis* (Sampson *et al.* 1988), the population size either has remained rather large or the bottleneck has not lasted long. In the earlier examples (Fig. 7.1) there was a relationship between population size and number of alleles in *Washingtonia filifera* and *Eucalyptus caesia*, but not with expected heterozygosity. Perhaps this is related to how long the small population sizes have persisted.

The above discussion referred mainly to single locus marker genes. However, a similar reduction in genetic variability is expected for polygenic quantitative traits. In a simple model without selection the initial additive genetic variation (V_A) is reduced by the same proportion, $\frac{1}{2N_e}$, as expected heterozygosity. However, as natural selection may be more likely to influence variation in quantitative traits than molecular markers, the dynamics of change could be more complicated.

7.4 Subpopulation differentiation

Figure 7.2 displays clearly how, because of genetic drift in the absence of other evolutionary forces, populations become differentiated. Nearly all of the small populations have become fixed for one or other allele. The larger populations have also differentiated, but at a slower rate (Table 7.1). The reduction in heterozygosity of a subpopulation due to drift (F_{ST}) is increasing. For the single locus markers, the mean allele frequency, and the overall variability, measured as expected heterozygosity in the total population (H_T), remains nearly the same, but average expected heterozygosity within the populations (H_S) has decreased.

TABLE 7.2 Expected heterozygosities (H_e) in two groups of widespread species with continuous distribution, *Acacia* and *Pinus*

Species	Reference	No. of populations	No. of loci	H_e
Acacia auriculiformis A. Cunn. ex Benth.	Moran *et al.* (1989c)	1	30	0.084
Acacia crassicarpa A. Cunn. ex Benth.	Moran *et al.* (1989c)	1	30	0.081
Acacia dealbata Link	Moran *et al.* (1989c)	1	30	0.085
Acacia decurrens (H. Wendl.) Willd.	Moran *et al.* (1989c)	1	30	0.156
Acacia mangium	Moran *et al.* (1989c)	1	30	0.017
Acacia mearnsii De Wild.	Moran *et al.* (1989c)	1	30	0.206
Acacia melanoxylon R.Br.	Moran *et al.* (1989c)	1	30	0.300
Acacia parramattensis Tindale	Moran *et al.* (1989c)	1	30	0.179
Acacia trachyphloia Tindale	Moran *et al.* (1989c)	1	30	0.083
Pinus contorta ssp. *latifolia* (Douglas) Critchf.	Wheeler & Guries (1982)	24	42	0.118
Pinus ponderosa ssp. *ponderosa* Douglas ex Lawson & C. Lawson	Niebling & Conkle (1990)	3	23	0.155
Pinus pumila (Pall.) Regel	Goncharenko *et al.* (1993a)	5	22	0.255
Pinus resinosa	Mosseler *et al.* (1991)	Range-wide sample, $N = 96$	23	0
Pinus sibirica Du Tour	Goncharenko *et al.* (1993b)	8	20	0.173
Pinus sylvestris	Goncharenko *et al.* (1994)	18	21	0.276

These genetic diversity statistics depend on the allelic, not genotypic, frequencies. However, when dealing with quantitative traits, we assign each genotype a phenotypic value, and thus the population mean and its genetic variance will depend on the genotypic frequencies. Within individual populations the additive genetic variance decreased, but the among-population variance (V_B) increased and it will be $2FV_A$, where F is the fixation coefficient (F_{ST}). The overall additive genetic variance will thus increase, to $(1+F)V_A$. If there is no dominance, the overall mean of a quantitative trait will not change much.

The simple model displayed in Figure 7.2 assumes that there is no exchange of genes between populations after fragmentation. Often some gene flow between subpopulations may occur (see Chapters 9 & 10). Even a small amount of gene flow (1 migrant per generation regardless of the population size) prevents fixation of neutral alleles between subpopulations. In trees, migration can occur either through seed or pollen dispersal. The efficiency of seed and pollen dispersal may be very different, which can be seen in different geographical patterns of variation of nuclear versus organelle genes, that are mediated only through

seed or pollen (see Ennos 1994). For instance, seed-mediated chloroplast DNA variation in angiosperm trees is highly differentiated, while the same populations show little differentiation at nuclear markers.

Differentiation between isolated, small populations has been documented in many plant species. In conifers genetic differentiation is low due to wind pollination and effective seed dispersal. Average among-population differentiation measured as G_{ST} was 0.076 for 23 species (Muona 1990). In fragmented populations of conifers the G_{ST} values are sometimes higher (Table 7.3). This is associated with low levels of variability within the populations in *Pinus radiata* and *Pinus torreyana* Parry ex Carriere (Torrey pine). A short time since fragmentation in formerly continuous species and relatively large populations may explain why isolated populations of *Pinus sylvestris* (Kinloch *et al.* 1986; Prus-Głowacki & Stephan 1994), *Pinus leucodermis* Antoine (Boscherini *et al.* 1994) or *Pinus washoensis* H. Mason & Stockw. (Washoe pine) (Niebling & Conkle 1990) have not become differentiated. There are also rather high levels of variability in these species, consistent with much

TABLE 7.3 Among-population differentiation and level of polymorphism for sets of small disjunct populations in different tree species

Species	G_{ST}	H_e	No. of populations	Reference
Pinus leucodermis	0.054	0.123	7	Boscherini *et al.* (1994)
Pinus radiata	0.162	0.098	5	Moran *et al.* (1988)
Pinus sylvestris (Scotland)	0.028	0.303	41	Kinloch *et al.* (1986)
Pinus sylvestris (Spain)	0.040	0.325	7	Prus-Głowacki & Stephan (1994)
Pinus sylvestris (North and East Europe) wide distribution	0.025	0.363	16	Prus-Głowacki & Stephan (1994)
Pinus torreyana	1	0	2	Ledig & Conkle (1983)
Washingtonia filifera	0.023	0.009	16	McClenaghan Jr & Beauchamp (1986)

gene flow between populations. Isolated small populations of *Washingtonia filifera* (McClenaghan & Beauchamp 1986) lack genetic variation but are not different from each other. This suggests a recent colonisation event.

Sometimes the local habitat patches occupied by the species may experience repeated colonisation and extinction cycles. This kind of dynamics may be best described with metapopulation models (e.g. Hanski & Gilpin 1997). Long-lived forest trees often may not fit the metapopulation models, but colonising species that occupy frequently disturbed habitats, or those in ephemeral habitats at the margin of a species' distribution might be close to the idea of metapopulations. Metapopulation models may also apply to species under human disturbance with extremely fragmented populations. The mean variability in local populations and the distribution of variability among patches is governed by the carrying capacity of a patch, growth rate, rates of extinction and recolonisation, the number and source of founders, the number of patches and the level of gene flow between them. The effective size of a metapopulation can not be deduced from demographic data because it depends on extinction and recolonisation dynamics, and it may be considerably lower than the average census number (Hedrick & Gilpin 1997; Whitlock & Barton 1997).

The discussion above has related to those markers where selection is at most, weak. The distribution of variation at molecular markers is predominantly determined by a balance between drift and migration. Natural selection can be very strong, and adaptive traits may show contrasting patterns to neutral

markers (e.g. Karhu *et al.* 1996; Yang *et al.* 1996). Comparative quantitative genetics data are mainly available for trees with wide distributions, but species with smaller populations would be expected to also show contrasting patterns between neutral markers and adaptive traits. The effects of recent fragmentation of populations is discussed in more detail in Chapter 10.

7.5   Increase of homozygosity

We continue considering the situation where a large population is divided into many smaller units. As the subpopulations lose variability, they also become increasingly homozygous, even if mating is random. As shown in Table 7.1, the overall heterozygosity has decreased. This process resembles inbreeding due to non-random mating, as described in Chapter 5. For loci that are only weakly selected, or where the heterozygote is intermediate between the two homozygotes, the change in genotypic frequencies does not change the phenotypic means. However, many experiments have shown that inbreeding and increased homozygosity in outcrossing organisms results in low survival, growth and reproduction (inbreeding depression) (Charlesworth & Charlesworth 1987; for conifers, see Williams & Savolainen 1996). These effects are believed to be due to deleterious recessive genes that have become homozygous upon inbreeding. The harmful effects of the deleterious genes are normally not expressed in large outcrossing populations because they are hidden by the dominant 'wild type' allele in heterozygous genotypes. An alternative, probably less frequent, explanation for inbreeding depression is that the heterozygote at the locus is the best genotype, and the same fitness cannot be obtained by any homozygous genotype (overdominance).

The level of inbreeding depression varies between species. Predominantly outcrossing large populations, such as pine populations, can accumulate many deleterious alleles in the gene pool, which can then become homozygous. Often the consequences of homozygosity are rather severe, as for example in pine trees for which inbreeding effects prior to seed maturation are particularly marked, and subsequent effects on growth and reproduction can continue to be important (Williams & Savolainen 1996). In many other species deleterious effects are delayed, as for instance in some *Eucalyptus* (Eldridge & Griffin 1983). The long-term consequences of inbreeding depend on a several factors. If inbreeding depression results from a few very harmful genes (e.g. lethals), then drift will rapidly eliminate the homozygotes for these alleles, and the population will become purged of the deleterious alleles provided it has sufficient growth capacity. However, when inbreeding depression results from slightly deleterious alleles, such deleterious alleles cannot be purged by drift (Hedrick 1994). These alleles can be fixed in the population, and they may cause a permanent reduction in survival and reproduction. Futhermore, since selection is not strong, this can occur even in relatively large populations (Lande 1995). This phenomenon can be significant both in natural populations and in gene banks and during breeding. New deleterious mutations arise continuously, and they may also become fixed in small populations. The fate of new mutations depends on how deleterious they are (see also Caballero & Keightley 1996).

7.4 Conclusions

In small populations, genetic sampling results in a loss of variability within individual populations and an increase of genetic variation between subpopulations. There is also an increase in homozygosity in the overall population. The evaluation of the importance of small population phenomena depends on the ecological context. It is necessary to understand the role of ecological and demographic versus genetic threats to populations (Lande 1988, 1995). In practice, these will often interact; for example, demographic variation reduces effective population size and emphasises the role of drift and inbreeding. Further, it would be important to understand when variation is governed by metapopulation processes (Hedrick & Gilpin 1997).

The genetic effects of small population size take place slowly. Heterozygosity is reduced and inbreeding increased at a rate of $[1-1/(2N_e)]$. Thus, prolonged population bottlenecks may be needed for complete depletion of variability (as measured by heterozygosity, rare alleles are much more sensitive). However, once variability has been lost, recovery is slow (Nei *et al.* 1975). Thus, current effective sizes may be poor predictors of variability because equilibrium has not been reached.

Genetic drift can give rise to different patterns of variation at neutral loci than those that are influenced by selection. The nature of the differences will depend on the modes of selection. Thus, it is important to understand not only marker genes but also adaptively important quantitative variation (Lynch 1996).

In conclusion, large population sizes need to be maintained so that populations can keep responding to current selection pressures. Populations should be large enough so that mutation and recombination can generate enough variability for response to changing environmental conditions. Inbreeding and the fixation of deleterious alleles can be avoided only in large populations. The influence of population size has important consequences for conservation of natural populations, for planning of sampling strategies (see Chapter 14), for maintenance of collections and for tree breeding.

SELECTION

*Gene Namkoong, Mathew P. Koshy
and Sally Aitken*

SUMMARY

The power of selection in changing the genetic make-up of populations and species has
been demonstrated through both breeding of artificial populations and study of natural
populations. Genetic conservation of forest trees requires consideration of the process
of selection, not in isolation but in combination with the effects of other evolutionary
forces including mutation, migration and genetic drift and on how the interpopulational
genetic diversity is affected. Whether it is better for the conservation of genetic
diversity to promote, maintain or mute the effects of selection is situation-dependent,
as the former may increase the average fitness of populations under current
environmental conditions but will reduce the ability of populations to adapt to
future conditions.

8.1 Introduction

The fundamental goal of gene conservation is to maintain the genetic diversity necessary to allow populations to adapt to future environmental conditions and to ensure that populations are sufficiently well adapted to current conditions for adequate numbers of individuals to survive and reproduce. The capacity to adapt to selection pressures is a necessary element of conservation. Selection changes populations by favouring the survival and reproduction (called 'fitness') of some genotypes over others. The greater fitness of some genotypes over others may be imposed by relatively 'natural' environmental conditions, through intentional human interventions for ecological or economic reasons, or through unintentional side effects of human activity on the environment (see Chapters 9 & 10). Regardless of which agent is responsible, if the trait under selection is heritable, genotypic selection will change allele frequencies in the progeny, and hence move the population to a new state for the next generation. As such, selection should not be considered as a positive or negative factor in conservation, but simply as one of the major forces that affects evolution. Furthermore, since the response to selection can change the characteristics of a population, it can fundamentally alter the populations and species—the units of conservation.

The adaptive potential of a population depends upon the amount of genetic variation for traits under selection, which in turn is a product of both the dynamics which generate genetic variability and the selection process which acts upon the genetic variability. While selection is a normal feature of adaptive evolution and a natural response to environmental change, it can also alter a population in ways that decrease its capacity to respond favourably over the long term by decreasing genetic variability. An apparent paradox exists as genetic variation is the raw material allowing for natural selection to better adapt populations to environmental conditions, yet selection can simultaneously decrease the genetic variation for adaptive traits that are necessary for adaptation to future environments. Especially in rapidly changing or novel environments, the short-term response to selection may not be optimal for adaptability to long-term trends. Also, since selection can greatly reduce the number of potentially successful parents, the reduction in population size itself may impose unusually high risks of population extinction as well as loss of alleles

(see Chapter 7). The challenge for conservationists regarding selection is to simultaneously recognise it as a natural force of evolution while not hindering the dynamics, so as to maintain long-term adaptability. In this chapter we shall consider how selection can benefit or detract from conservation management. We will also consider that changes in population characteristics and in the structure of sets of populations may also change during evolution and, hence, also change the unit that is classified as the species and object of conservation.

8.2 Selection processes

Selection is the differential fitness of individual phenotypes whose performance is at least partially determined by their genotype. It may favour an intermediate phenotype and is then termed stabilising selection. When phenotypes for particular traits vary between individuals, some phenotypes may be more advantageous in a particular environment and such individuals will have a selective advantage (greater fitness) over others. The optimal phenotype may closely approximate the average phenotype of the population. Genotypes that depart from this optimum will be at a selective disadvantage. This will favour genotypes with smaller deviations from the optimal phenotypic value, which in turn will produce offspring with intermediate phenotypes. Consequently, this type of selection will not only stabilise phenotypic value around the mean, but also reduce variability around the mean leading to a reduction in genetic variance. For example, a trade-off exists between the duration of active growth and ability to withstand frost in some forest tree species in some environments (e.g. Eriksson 1982; Rehfeldt 1992). Trees that grow for a longer period will grow larger, providing some competitive advantage. However, if these trees initiate growth too early, or cease growth too late, injury or mortality may occur due to frost. Thus, stabilising selection can be expected to favour a phenotype with intermediate growing season length in a particular environment that balances competitive growth with risk of frost injury, and over time, reduces genetic variation for this trait.

If selection disfavours the intermediate phenotype, it is called disruptive selection. This may be a less common mode of selection than directional or stabilising selection. However, if an insect pest or pathogen was most successful in surviving and reproducing on the new growth of hosts with intermediate timing of bud burst, selection would favour phenotypes that burst bud relatively early or late relative to the population mean. When differences between environments are large enough, and if individuals tend to mate with individuals with similar phenology (assortative mating), the population will move towards two or more optima in different groups of individuals. A disruption in the continuity of variation expressed by a single smoothly varying optimum will evolve. Consequently, within each group, stabilising selection may prevail to move the population to an optimum within each group. Disruptive selection is thought to be the mode of selection under the multiple niche hypothesis, whereby genetic variation for adaptive traits is maintained in a population by differential selection among microhabitats associated with different optimal phenotypes (Emlen 1975; Campbell 1987).

Alternatively, selection could favour an extreme phenotype that may be outside the range of what is present in a population. This is called directional selection. In most breeding programs, the optimum phenotypes lie beyond the range of the character expression in the breeding population and, hence, selection directionally moves the mean of the population towards the higher value. However, eventually the population may stabilise at an optimum. Directional selection is a common form of natural selection. An example of this is selection for rapid height growth in a highly competitive environment, either due to interspecific or intraspecific competition. Faster growing genotypes will generally intercept more radiation for photosynthesis, grow faster, and thus reproduce at higher rates. Temperate trees from mild, moist environments exhibit rapid growth rates compared to populations from colder or drier environments (e.g. White 1987), providing evidence of selection for this trait in favourable environments.

While it is conceptually simplest to consider the fitness of an individual as an invariant property of its genotype regardless of any other condition in the population, in reality genotypes often have varying fitness depending upon the environment, population density and relative frequencies of other genotypes. When the fitness of a genotype depends on the frequency of other genotypes, it is termed frequency-dependent selection and if rarer genotypes are favoured, opportunities for maintaining polymorphism are increased. When the number of adults in a population is held constant at a carrying

capacity for that environment, soft selection occurs (Wallace 1968). With soft selection, it is not the absolute phenotypic values that are selected upon, but phenotypes relative to other phenotypes. Competition-induced mortality of trees in mild, productive environments would be an example of soft selection. However, when the number of seeds landing in each area is constant and natural selection is on absolute rather than relative phenotypes, resulting in different numbers of adults to remain, the process is called hard selection. For example, if a severe frost occurs and kills all seedlings except those with cold hardiness levels adequate to survive the minimum temperature (Timmis *et al.* 1994), the proportion of seedlings surviving the cold event is expected to be largely independent of seedling density. Hard selection in forest trees can be important in colonising species where adaptations to specific sites is necessary for survival.

Selection can also be expressed in the population in various other ways. When selection occurs as a result of differential mating success of adults, sexual selection is in effect. Differential production of gametes, called meiotic drive, can also cause selection for specific alleles. Differential success of gametes carrying specific alleles in accomplishing fertilisation leads to gametic selection. When different mating types generate different numbers of offspring, it is termed fecundity selection.

8.3 Selection effects

For selection on growth, viability and on general reproductive capacity, changes in phenotypic frequencies would be followed by some predictable changes in genotypic frequencies, and probably in allele frequencies as well. If other forces of evolution do not limit response in the heritable trait, the phenotype would either stabilise at the intermediate optimum, or would continue to change as allowed by gene action and as directed by selection. From generation to generation, a population's mean phenotype for a given trait is expected to change in the favoured direction resulting in a higher degree of adaptation to the environmental factor that is exerting the selection pressure.

To predict the effects of selection on allele frequencies for a single gene, the relative effects of different alleles on the phenotype, termed 'gene action', need to be considered. If two alleles are codominant, the phenotype of the heterozygote is intermediate to that of the two possible homozygotes. If one allele is dominant (and the other, by definition, recessive), the heterozygote will have the same phenotype as the dominant homozygote. If alleles are overdominant, the heterozygote's phenotype will be outside of the range of the two homozygotes. As selection acts on phenotypes, but alleles are inherited, gene action is an important factor in the maintenance or loss of alleles under selection.

Whether selection is stabilising, directional or disruptive, it could drive gene frequencies to an extreme in any finite population, resulting in the elimination of all but one allele from the population, an event termed gene fixation. If overdominance exists and the heterozygote has a more extreme expression than either homozygote, then directional selection for extreme expression will favour the heterozygote and selection will be a force to maintain both alleles in the population. However, if the heterozygote is not extreme in its behaviour, then stabilising selection for an intermediate optimum phenotype might be expected to maintain multiple alleles but would actually not be a strong force for allele conservation for multiple loci. Directional selection would also tend to promote more allele losses in a single population than if no selection was exercised.

While any one mode of selection may not explain satisfactorily the process of natural selection, an interplay of different evolutionary processes may be shaping the genetic diversity of populations or species. Hence, it is important to understand how conservation and management regimes interact with these processes. In large populations, disfavoured alleles may be present at low frequencies. Even strongly disfavoured alleles may be maintained in populations as a result of recurrent mutations, and mildly deleterious alleles or even alleles that are neutral with respect to fitness also may be present at low frequencies. While these alleles may have had negative or no significance for adaptation in the immediate past, these alleles may provide the necessary variation for future adaptational responses, and therefore should be conserved. While immediate selection response is due to alleles at relatively intermediate frequencies (e.g. 0.1–0.9), long-term response may arise from favourable alleles that exist at very low frequencies (e.g. below 0.1). Since these are also the alleles that are most likely to be lost by genetic drift, conservation strategies should focus on saving them.

As the loss of alleles due to genetic drift is much more likely for small than for large populations, large total population sizes are a key requirement for the conservation of future adaptability (Lynch 1996; see Chapter 7).

In large populations with greater than a few hundred mating adults, the response to selection in allele frequencies is strongly determined by the frequency of the alleles in the parent population. The one generation change in allele frequency (Δq) is expected to be:

$$\Delta q = \frac{1}{2}sq(1-q) \quad \text{(no dominance)} \qquad (8.1)$$

$$\Delta q = sq(1-q) \quad \text{(complete dominance)} \qquad (8.2)$$

where s is the selection differential, and q is the initial allele frequency.

From the above equations, it can be seen that only if the initial frequency of an allele is in the intermediate range, can the change in frequency be very large. In contrast, the change in phenotypic mean for quantitative traits can be very large as is evident from many tree breeding trials. The 'breeder's equation' describing the response to selection (R) in terms of change in phenotypic mean is:

$$R = h^2 s \qquad (8.3)$$

where h^2 is the heritability of the trait and s is the selection differential. The difference in the selection response for single-gene allele frequencies compared with the phenotypic response for quantitative traits lies in the cumulative effect that many small frequency changes at many loci can have on the phenotype in the latter case. Thus, even though we can observe large phenotypic changes without necessarily seeing large changes in allele frequencies, only the lowest frequency alleles are at risk as long as the population size is large. Even for loci that have very strong, qualitative effects on phenotypes, such as on stem form in conifers (Zobel & van Buijtenen 1989), alternate alleles can persist. Selection for the same, optimal phenotype for a quantitative trait in different populations or non-intermating subpopulations can result in the fixation of different alleles in these groups, as quantitative traits are determined by many loci, and different combinations of alleles can produce the same phenotypes (Holsinger 1993).

If the size of the parental population is small, additional factors affect the genetic dynamics (Barrett *et al*. 1991). Genetic drift is important in small populations (see Chapter 7). Genetic diversity is lost at a rate proportional to $\frac{1}{2N_e}$ per generation due to genetic drift, where N_e is the effective population size. Not only will disfavoured alleles be lost in small populations, but low frequency, favoured alleles can also be lost due to genetic drift, and the whole population can be at risk of losing genetic variation for non-genetic reasons. In large populations, selection is a much stronger evolutionary force than genetic drift, but in small populations, drift can cause alleles to go to extinction rapidly whether favoured by selection or not. Similarly, drift can result in the fixation of deleterious alleles in small populations (Lynch *et al* 1995). In addition, if the frequency of an allele that would be favoured for future fitness or gain is low, it is difficult to raise the frequency with any speed in large populations. Thus, even if the complete extinction of alleles due to selection may not be a problem, the rate of adaptation can be.

For slowly changing conditions, environmental selection is expected to be a positive force for increasing a population's adaptation to consistent trends in environmental pressures, at least in the short term (Lynch & Lande 1993). Thus, in large, intermating tree populations that are responding to gradual environmental changes, selection will continually shift population composition in the direction favoured by the shifting environment to increase adaptation. In long-lived tree species, most of the selection occurs during the juvenile phases when most of the mortality occurs. For example, Campbell (1979) estimated mortality rates in the life of coastal *Pseudotsuga menziesii* (Mirb.) Franco (Douglas-fir). From the 22 000 or so seed falling in the area typically occupied by one mature tree, about 5000 produce seedlings, of which half die due to random events. From the remaining seedlings, Campbell (1979) estimated that one mature tree will result. Thus, there is great opportunity for selection to act even in a single generation, particularly at the seedling stage. Selection is, therefore, expected to be followed often by an adaptational response in juvenile traits. However, as trees typically have very long life spans and populations regenerate many decades to centuries apart, an 'adaptational lag' may occur, whereby environments change considerably while trees are

mature and under little selection (Stettler & Bradshaw 1994). At the time of regeneration, seedlings may then be maladapted to the new regeneration environment.

If conditions are changing rapidly and involve intense selection in which relatively few parents survive to contribute gametes to the next generation, reductions in the mating pool may itself distort genotypic frequencies away from optimum and further expose the population to accidental extinction. In such cases, selection can limit the future capacity of populations to adapt. For example, selection can increase the rate at which alleles conferring low fitness are eliminated from populations above the rate that they may be lost due to genetic drift. In this case, selection reduces genetic variation as alleles are fixed. In addition, long-term evolution may require alleles that may be eliminated in short-term selection. Hence, whenever genetic means or variances change, there is a concern that even if selection may be operating to increase adaptation in any one generation, we may wish to moderate the loss of genetic variation necessary for long-term adaptability. Consequently, a dilemma exists in whether selection effects are to be muted or promoted, and how we can best conserve genes to meet both long-term and short-term conservation goals. Thus, selection is a force that not only increases the loss rate of some alleles, but also increases the survival rate of other alleles.

8.4 Levels of selection

So far we have been discussing selection which acts among individuals in a single population, and which can change allele and genotypic frequencies in that population from generation to generation. At this level, selection can result in an evolutionary series of population adaptations from generation to generation. For trees, we can also readily observe that large environmental changes can occur within the lifetime of individual trees and that individuals must be able to adapt their behaviour to that kind of variation. In such cases, viability and reproduction must be accommodated within an individual's lifetime and selection would be directed to an improvement of homeostasis of individuals. Trees must endure a larger variation in their physical environment than most plants because they persist for many years, and must endure seasonal as well as year-to-year climatic variation. They also endure more environmental variation than most animals because they are immobile. Although they may achieve a fitness homeostasis by absorbing environmental variations in changes in component growth traits, we are often concerned that tree size and functions would be affected so that their economic and ecological values may suffer even if they are capable of surviving. How individuals respond to environmental challenges is, therefore, important for conservation ecology as well as for conservation genetics. One of the consequences of an increased level of homeostasis for all individuals is a reduction of heritable differences among individuals in response to differences among those environments and a reduction in the intensity of selection among individuals in that population.

Similarly, if environmental changes occur among intermating individuals over a very short geographical distance, their progenies must be able to survive and reproduce in different environments. There would also then be a selective advantage for an individual level of homeostasis. One model of environmental variation in a single population is that it is random over a defined range within a lifetime, such that fitness is an average over that range (Lewontin 1972). If the environmental effects can be merely added up over a lifetime, then the arithmetic mean may be sufficient to represent lifetime fitness. If a series of critical survival steps have to be passed in order for an individual to survive and reproduce, then the geometric mean of fitness over that range of environments would suffice. Thus, even if allele effects are sometimes favoured in one state of the environment and disfavoured in others, as long as they exist within a single randomly mating population, a mean fitness can suffice for predicting evolutionary consequences of selection (Hedrick 1984).

However, if intermating is restricted to smaller distances than the scale of environmental differences, the different adaptabilities can be favoured among families. The forms of variability among environmental response functions, or their norms of reaction, would be subject to selection and would be particularly sensitive to the time and spatial scales of environmental variation. Hence, whenever populations are subdivided and have some limitation on the extent of intermating between subdivisions, we can expect some balance of between-individual and within-individual selection results. There are, therefore, several levels of selection, selection for individual homeostasis, selection among individuals within intermating populations and selection among populations.

Whenever genes have different effects on phenotypes or fitness in different environments, they must then be considered as functions of environmental variables instead of just constant effects. The analysis of selection effects on evolution then takes on another dimension. Gene effects now have to be changed from simple coefficients of fitness to functions of the environmental variables (Gregorius & Namkoong 1986). The outcome of selection then depends on the environmental variables and the magnitude and frequency of their occurrence.

8.5 Selection balances with other forces

Conservation management cannot be limited to only considering one evolutionary factor while ignoring the others. We need to consider the effect that other forces have on selection, before finally considering how the adaptability of populations can be managed for unknown and possibly novel future environmental conditions. Mutation, migration and genetic drift all interact with selection in the process of evolution (Lacy 1987). The relative strengths and effects of these forces are affected by both population size and mating system. Selection, as a strong evolutionary force, can cumulatively change population characteristics through determining the probability with which individuals in a population reproduce. Since selection operates on both individual and population characteristics, and these are affected by other forces, the effects of selection are moderated in the whole evolutionary system by population size, structure, migration and mutation.

Mutation is the force that generates new alleles, thus providing genetic variation on which other evolutionary forces act. Most mutations are deleterious or selectively neutral, with few mutations resulting in alleles conferring a fitness advantage. In large populations, selection will eliminate deleterious alleles or keep them at low frequency. Selection, however, may be ineffective at countering the effects of drift in small populations, and thus mildly deleterious alleles may accumulate in small populations. These accumulated, deleterious alleles, each with a relatively small effect on fitness, may cumulatively reduce population fitness to where selection can cause population extinction. This phenomenon is called mutational meltdown (Lynch & Gabriel 1990). Maintaining large population sizes is the best way to reduce the risk of such a meltdown (Lynch *et al*. 1995; Lande 1995).

Small populations can also experience random changes in allele frequencies, and thus the loss or fixation of alleles, through genetic drift. Small populations can also result in increased inbreeding within populations. Inbreeding may increase the efficiency of purging (see Chapter 7), but the population may go through an interim period of considerably reduced fitness due to inbreeding depression (Lynch 1996). Changing population structure can change the mating system in a population considerably. For example, an isolated population of *Pinus ponderosa* Douglas ex Lawson & C. Lawson (Ponderosa pine) that has been fragmented considerably by agriculture, forestry and urbanisation in the Willamette Valley of Oregon, USA, has an outcrossing rate (see Chapter 4) of around 60%. The outcrossing estimates elsewhere in the range of this species are around 90% (Gooding 1998). Seedlings resulting from self-pollination in this species have 10% lower survival and 21% slower height growth compared with outcrossed seedlings (Sorensen & Miles 1974). This increase in self-pollination of about 30% will result in considerably reduced population mean fitness. Selection will disfavour the inbred plants, reducing the effective inbreeding rate in mature trees, but this increased selection will come at a cost to the population. A more extreme example exists in *Pinus chihuahuana* Engelm. (Chihuahua spruce) in Mexico, where fragmentation in the Holocene has resulted in both strong divergence among populations and the evolution of a completely self-pollinating mating system in some populations (Ledig *et al*. 1997). A contrary case, however, has been found in *Acer saccharum* Marshall (sugar maple) in which fragmented populations have an apparently elevated outcrossing rate (Young *et al*. 1996).

If distinct populations exist in different environments for a lifetime, and there is restricted mating between sites, then the effects of selection in the separate environments are more complicated still. Evolution also depends on the sizes of the different populations and how they may intermate and contribute gametes to the next generation. At one extreme of population structure, where populations may be selected for survival in different environments for very different genotypes, but with little discrimination in mating, their evolution would be as described in the above paragraph. At the other extreme, selective differences may be slight but barriers to migration so strong that populations diverge (the first stage in speciation). In the latter case, any slight selective differences would result in adaptational divergence of these populations, ultimately resulting in fixation of different alleles. Migration via pollen or seed vectors can counter selective effects by moderating differences in allele frequencies between populations due to selection differences between the populations. Strong migration tends to homogenise allele frequencies among populations despite divergent selection among them. Conversely, convergent selection (similar selection pressures in different populations) can homogenise frequencies even if strong barriers exist to migration. Thus, similar patterns of variation among and within populations can result from different combinations of evolutionary forces.

When selection and migration have opposing effects, the strength and direction of selection in each population, as well as the strength and direction of migration between populations, will influence the outcome. If different alleles are favoured in adjacent populations, then the population sizes in different environments can have a strong effect on which allele persists to fixation. Therefore, differences in allele frequencies among populations can be reinforced or diminished by divergent versus convergent selection factors among populations.

The combined effects of these evolutionary forces on most wind-pollinated temperate and boreal tree species result in fairly strong genetic clines in population means for adaptive traits corresponding to environmental gradients in temperature and moisture. While natural selection is very difficult to detect and quantify in nature, the repeated concordance of genetic clines with environmental gradients is considered strong indirect evidence of differential selection pressures among environments (Endler 1986). Despite these strong geographical trends in average phenotypes, individual populations maintain high levels of genetic diversity for adaptive traits. Such high levels of within-population variation could reflect disruptive selection among microhabitats on a fine geographical scale, high levels of gene flow, large population sizes or probably some combination of all these factors.

Spatial patterns of genetic variation will vary among traits depending on the degree to which traits are under selection. As a result, patterns of variation in evidence for selectively neutral traits such as DNA markers will exhibit patterns of variation reflecting population size, history and gene flow, while traits under selection will exhibit divergence or convergence depending on differences in selection among environments in large populations, and reflect a combination of selection and drift effects in small populations (Lynch 1996).

The balance between the relative strengths of selection and migration, and the patterns of convergence between the mating neighborhood and the selection neighborhood are critical in determining the distribution of variation between and within populations (Eriksson *et al.* 1993). If species have been fragmented into sets of subpopulations such that evolution among populations requires conservation of a multiple population structure (Holsinger 1993), or if connectivity between populations is decreasing (e.g. due to changes in land use or broad-scale disturbances), then conservation will require understanding the historical divergence between populations, and managing both selection and migration.

8.6 Selection and population structure

Forest tree species exhibit a great variety in patterns of geographical distribution. Some species cover wide ecological zones with dense stands containing just one or a few tree species as in boreal coniferous species, while others have highly specialised ecological niches with sparsely distributed mature trees as in many tropical forests (Ashton 1969; Kageyama 1990). The causes of this variation in distribution may be historical (e.g. postglacial recolonisation by temperate and boreal conifers versus adaptive radiation and evolutionary specialisation over much longer periods of time for tropical species), or directed by physical or biotic selection factors. In any one species, the patterns of distribution may shift from generation to generation. In this shifting mosaic of species and populations that constitute forest ecosystems, the structure of interpopulational variation itself may require conservation to ensure the future evolution of the species.

Populations may exist in a single intermating neighbourhood; as a string of island populations with migration only between two adjacent islands of similar size in an 'archipelago'; as a network of similar sized 'stepping stones' with migration between nearest

neighbours in a two-dimensional array; in an oceanic or aerial Three- dimensional array, or between a large main or 'continental' body and a set of otherwise disconnected colonies; or any mixture of possible connections and sizes. For any given pattern of selection, the degree of connectedness and the degree of dominance in the genetic contributions of one or a few populations influence the existence of changes in allele frequencies. For example, clines in allele frequencies can be generated in the 'stepping stone' migration models even in the face of an abrupt and qualitative switch in allele selection coefficients. High levels of migration produce more nearly uniform frequencies among populations, but if fragmentation were to be imposed, then divergence in frequencies can be expected to develop if differences in selection coefficients between populations are strong or populations are small. As long as the separate populations are independently secure, the ensuing divergence may well imply that response to selection for local adaptability has increased local as well as species-wide probabilities of survival. The problem for conservation is that selection within each of the populations may reduce the population sizes so much that they are in danger of extinction due to a variety of demographic causes (Lande 1988). However, that selection can diversify populations does not in itself, constitute a threat of extinction, but may instead imply an adaptively advantageous shift.

Selection acts on whole individuals rather than on genes, and acts on multiple traits simultaneously, resulting in adaptive shifts in multitrait phenotypes. Selectively influenced divergence among populations will likely vary among traits and among the loci affecting them. Some traits will have a wide adaptive neighbourhood in which selection is for a common type of behaviour (e.g. photoperiod response, Eckberg *et al.* 1979), while other traits may be strongly influenced by local differences in soils, temperature and moisture regimes, or other factors. For example, local variation in *Pseudotsuga menziesii* cold and drought tolerance correlates strongly with temperature and moisture regimes (White 1987; Campbell 1979; Balduman 1995; Kavanagh *et al.* 1998). The spatial distribution of alleles can vary among loci depending on the relative strengths of selection and migration (Martinson 1996). In contrast, mating patterns are consistent for all of the genes in the parents being selected and, hence, the patterns of divergence between populations must be expected to differ according to the relative strengths of trait selection in each environment and the divergence of selection pressures between environments. Thus, for any given pattern of mating, the widest divergence in allele frequencies is likely to be found when environmental factors are at their most extreme within the species range.

Similarly, the connectedness between populations will vary among different stands of trees depending on their size, relative isolation, and pollen and seed movement patterns. Those populations that are isolated from other large populations are likely to diverge most in allele frequencies. Hence, remote populations may have alleles in relatively high frequencies that are rare on a species-wide basis. As a result of both selection and migration, rare alleles may be found in higher frequency in populations that are at the extremes of both environmental variables and in migrationally isolated locations. However, if populations in the core of a species range receive migrants from other populations, generally they will have higher levels of allelic diversity in terms of both numbers of alleles and number of heterozygous individuals than those at the margins of the species distribution. Thus there are good reasons for conserving both core and peripheral populations (Lesica & Allendorf 1995). Efficient sampling for gene conservation must consider the joint effects of selection with the other forces of evolution.

8.7   Forestry operations affecting selection and interactions

We often like to consider forests as existing in some steady state of birth and death, the maintenance of which can then be used as a goal of conservation. Since forests occur in different stages of succession and historically have been subjected to an array of selective events that favour one genotype or one population, or one species over another, attempts to establish a single type of forest (e.g. a hypothetical 'climax' type), would disrupt forest evolutionary processes. In most forests, selection events could be expected to occur at all hierarchical levels including the individual, stand and species levels, with a variety of selection intensities on different traits, and a variety of effects in different sets of populations. Therefore, conservation of the processes that allow forests to regenerate may require periodic disruptions of benign environments and shifts in the selective environment.

However, large and drastic changes that have a uniform effect, species-wide, such as large-scale conversions to agriculture, or massive fires and clearcuts, can pose a risk of local population extinction through destruction of regenerative capacities at both the individual and population levels. Indirect effects of disturbances on regeneration may occur through disruption of pollen or seed dispersal agents. Selective harvests of forests for just one or a few species may leave the forest environment physically undisturbed, but the availability of mating partners for the harvested species may be so diminished that regeneration is unlikely. Selection effects in these cases may be detrimental to the long-term evolution of populations or species if they reduce population size below the level required for self-regeneration. The larger the biotic scale of loss, the more difficult it is to restore a forest ecosystem, and hence, selection events that are involved in multi-level catastrophes are cause for major concern. If the new environments that may be available for regeneration are very different from the predisturbance environments, then adaptation to novel environments requires a large pool of potential parents available for reproduction. If adaptation is possible at all, the selection intensity may be so strong and the surviving proportion so small that the resulting population's size and amount of genetic diversity will both be very small. In such cases, genetic rescue operations using *ex situ* reserves or artificial migration may be necessary (Ledig & Kitzmiller 1992).

If mortality rates are high and local regeneration is poor for a particular population, but an essentially identical population nearby provides a source of recruitment, there may be no problem for conservation. However, if population differences are large between a declining population and a potential source of recruitment, and the recruitment pool for regeneration is ill-adapted, selectively caused differences may become a source of concern. The magnitude of the divergence between the populations and the availability of recruits will determine the magnitude of such an adaptive mismatch.

More subtle selection effects may also detrimentally shift population composition. Selective logging of vigorous trees can leave less vigorous, and economically less desirable, trees to reproduce and if the desirable traits of harvested trees are under some degree of genetic control (as measured by heritability), logging can reduce the ecological and economic value of the population regenerated from the remaining trees. Dysgenic selection effects by selective removal of vigorous, straight trees, can obviously diminish value. This effect may have occurred frequently and has been documented for *Swietenia macrophylla* King (mahogany) in Central America (Styles 1972) and for *Pinus brutia* Ten. (red pine) in the eastern Mediterranean (Palmberg 1975). Problems can also arise if tree breeding programs select for economic traits such as growth rate or wood quality that are unfavourably genetically correlated with adaptive traits such as insect, disease, cold or drought resistance (e.g. Aitken & Adams 1995; Andersson & Danell 1997). For example, for some species in environments with high degrees of cold or drought stress, selection for rapid growth rate may indirectly result in selection for genotypes more typical of warmer or wetter environments that are maladapted to planting environments (Rehfeldt 1992).

Stand structure manipulations can also result in selection effects. Opening stands for grazing can change the physical environment for reproduction of sensitive species without total forest removal. A gradual reduction in vigour of high value species may result from practices which affect tree health without tree removal, such as bark harvesting for medicinal or other purposes. Such a practice, which may target every tree of a sparsely distributed species, could lead to a sudden collapse of the population. Whole tree death or removal is not necessary to have effects on the parent pool since partial harvesting of individuals for flowers, seeds or fruits can result in selection at the gamete production or seed stages of the life cycle.

Gradual but large-scale changes in the selective environment, such as global climate change, also pose a risk of large-scale extinction because of the species-wide effects such changes could have. Even if individual trees are only moderately maladapted to new environments, as all individuals are simultaneously and similarly affected, species-wide threat is implied. In the event of large-scale environmental changes, natural selection may not be able to maintain or increase ecological security or economic productivity through increasing population adaptation to new conditions rapidly enough, and short-term responses may even hinder long-term evolution.

8.8 Conservation and selection

Since selection is a strong evolutionary force, it can be an effective tool for conservation management. At the individual level of selection, the possibilities of using artificial selection for homeostasis over environments of various types and magnitudes remains largely unexplored. The inheritance of genotypic responses to environments is poorly understood and rarely used (Gregorius & Namkoong 1987), either for individuals within populations, or for populations, but it may provide substantial economic gains if used in forest planning (Lindgren 1994).

One of the fundamental problems for conservation biology is the definition and identification of populations for conservation, when only a subset of extant populations can be protected *in situ*. Most efforts towards defining evolutionarily significant units (ESUs) have focused on using genetic distance measures based on molecular genetic markers to determine when populations are sufficiently distinct to warrant separate conservation (e.g. Moritz 1994). These methods, however, work poorly for many tree species as they are based on selectively neutral genetic markers and thus reflect mutation, genetic drift and migration, but not adaptive differences. The principles of selection can also be used to evaluate whether populations are sufficiently different to warrant separate protection. The potential of replacing the phenotypic distribution for adaptive traits (e.g. phenology, growth rate, cold or drought resistance) of one population through repeated selection on another population can be used to prioritise sets of populations for conservation. Under the concept of 'population replaceability', the difference between two populations (and potential for successfully recruiting individuals from one population into the other), could be quantified by the number of generations of selection and selection intensity required to reproduce the current phenotypic distribution of adaptive traits of a population not selected for conservation through either natural or artificial selection on another population (S. Aitken, in prep.). This approach requires information on variation and heritabilities of adaptive traits from common garden tests.

Considerable intervention may be required in natural processes to avoid species extinction if rates of environmental change are very rapid (Ledig & Kitzmiller 1992; Eriksson *et al.* 1993). The design of interpopulational divergence to conserve species has received very little attention. However, for threats of a magnitude and scope that threaten species existence and utility, where recruitment of similar individuals or populations may be difficult, novel environments may require adaptation to a wider range of environmental conditions than previously considered. Not only would this necessitate greater variation among populations, but the uncertainty of the nature of future changes might require greater within-population variation or adaptational response as well. Since populations can respond rapidly to selective breeding, divergent selection can be expected to generate large interpopulation differences in a relatively few generations. Hence, a multiple population selection system (Namkoong 1989) could be a genetic management tool for conservation purposes.

Selecting populations for adaptability to variable environments can be optimised with an array of divergent populations (Roberds & Namkoong 1989), and an optimum level of divergence between them can be derived. A finite number of such populations can be designed for a finite number of environmental variables. At least two strategies for selectively managing interpopulational diversity exist. One is to maximise the divergence and, hence, to maximise the mathematical hull (space enclosed by straight lines connecting the external points of a set) of the set of populations, and another is to maximise expected value by optimally distributing a set of populations (Namkoong 1984a). Selection at the interpopulation level can then be a tool to manage the total species adaptability.

The use of existing interpopulational diversity for an initial set of populations would constitute a means of using past selection–migration balance as a basis for future selection. This is an old tradition in forestry that is usually termed provenance selection. In terms of optimising a base set of populations, the extreme populations may be target for conservation, if a strategy of maximising divergence is followed. For future directions of selection, further diversification of those populations may be useful if the span of future conditions of adaptation is wide. However, total species-wide selection requires that selection within populations is not so drastic that the component populations are at increased risk due to low survival rates.

SECTION III
THREATS TO *IN SITU* GENETIC CONSERVATION

In some situations, for example when dealing with very rare or endangered species, *ex situ* conservation of genetic diversity, in seed collections or botanical gardens, is the only realistic management option. The sheer numbers of species of interest, however, means that this approach is not broadly applicable, and most management efforts are still concentrated on maintaining genetic diversity *in situ* within wild populations in natural or semi-natural forests. This approach also has the great advantage that it allows the continued operation of genetic processes such as selection and gene flow. This means that there is the possibility of conserving a dynamic gene pool capable of evolutionary response to changing environments, rather than a snap-shot of a species' genetic profile at a given time. While this represents a considerable advantage to the *ex situ* approach, the possible negative effects of continued environmental change on genetic processes and the variation they subtend need to be considered.

Continued deforestation, forest fragmentation and increased pollution as well as other ecosystem changes can all affect both demographic and genetic population processes and through this influence the genetic make-up of populations and ultimately species (see Section II).

In Chapter 9, Ratnam Wickneswari and Timothy Boyle examine the effects of logging and extraction of non-timber forest products on the gene pools of both tropical and temperate tree species. Andrew Young and Timothy Boyle then review in Chapter 10 how changes in population size, isolation and density following forest fragmentation affect patterns of mating and genetic variation in a range of forest species from herbs to trees. Thomas Geburek (Chapter 11) examines the variation in response of forest trees to gaseous and metal pollutants both among individuals within species and between species. This is followed by assessment of evidence for the heritability of plant responses to pollutants and how gene expression can be altered by environmental stress. This chapter also looks at whether it is possible to use genetic markers to assess population pollution status.

In Chapter 12, Margaret Byrne discusses the importance of disease for both *in situ* and *ex situ* conservation of forest genetic resources with specific regard to the preservation of coadaptation of host-pathogen systems and the importance of these interactions in shaping host resistance structure. The final topic on *in situ* conservation is

hybridisation and forest conservation—Chapter 13 by Shanna Carney, Diana Wolf and Loren Rieseberg. Here the possible genetic outcomes of hybridisation and the implications for gene conservation are considered in the light of several case studies. These include increases in genetic variation through the generation of hybrid genotypes, range expansion through adaptation to novel environments, genetic assimilation of one species by another and the formation of stable hybrid zones. Results from a computer simulation model are used to explore factors that effect the probability of species extinction due to hybridisation.

EFFECTS OF LOGGING AND OTHER FORMS OF HARVESTING ON GENETIC DIVERSITY IN HUMID TROPICAL FORESTS

Wickneswari Ratnam and
Timothy J. Boyle

SUMMARY

Commerical harvesting of timber through logging is a major form of disturbance in many areas, but the variety of products harvested from tropical forests is immense: wood, fruit and other foods, medicinal plants, construction materials and many more. Commercial logging has been separated from other types of human use to reflect the relative importance of logging in all types of forest.

In a comprehensive study of logging in tropical forests, genetic diversity (expected heterozygosity, H_e, for isozyme data and Shannon Diversity Index, H, for RAPD data) was reduced by 5.0–23.4% after logging for five species with different life history strategies. Generally, the order of reduction in number of individuals corresponded with the order of reduction in genetic diversity. *Daemonorops verticillaris* exhibited the highest loss of alleles (13.8%) most probably due to death of individuals following logging damage to supporting trees and being trampled during extraction of felled trees. No significant differences in outcrossing rates were observed between logged and unlogged stands for *Shorea leprosula* and *Dryobalanops aromatica*. No adverse changes in genetic diversity measures of *Parkia speciosa*, *Scaphium macropodum*, *Shorea leprosula*, *Garcinia malaccensis*, *Daemonorops verticillaris* and *Labisia pumila* were detected in two regenerated stands (13.5% and 40.7% disturbances) compared to an unlogged stand in the mixed dipterocarp lowland forest indicating maintenance of genetic diversity in regenerated stands after a single low intensity logging event.

In comprehensive studies on collection of non-timber forest products in the Biligiri Rangan Hills of south India, a significant effect on recruitment of *Phyllanthus emblica*, the fruits of which are harvested for food and as a base for numerous traditional medicines, was observed. Harvesting of fruits or flowers tend to have a more direct or dramatic effect on regeneration and genetic diversity than harvesting of leaves, bark or other non-reproductive parts of the plants.

9.1 Introduction

There are virtually no forests in the world today that are not directly affected by human activities. In the most remote parts of Siberia and northern Canada, where population densities are very low, the forests may be considered essentially free from the effects of human use, but in the tropics, even in the most inaccessible locations, forest-dwelling people clear patches for their dwellings and crops and collect non-timber forest products (NTFPs) in the forests themselves. Consequently conservation of forest genetic resources necessarily involves the management of these resources in the context of human use.

The philosophical debate about the role of humans in nature—whether as an integral part of natural systems, or separate from them, continues (e.g. McNeely & Wachtel 1988; Grumbine 1992; Visser 1992), but whatever the outcome of this debate, it is essential to understand the effects of human activities on forest genetic resources. Human activities in forests modify the size and age structure, both of species directly harvested and other associated species, thus potentially altering genetic structure and levels of genetic diversity.

Commerical harvesting of timber through logging is a major form of disturbance in many areas, but the variety of products harvested from tropical forests is immense: wood, fruit and other foods, medicinal plants, construction materials and many more. In this chapter we separate commercial logging from other types of human use, not because there is any essential difference among these activities in terms of genetic effects, but simply to reflect the relative importance of logging in all types of forest.

Figure 9.1 shows a generalised relationship between economic benefits from the forest and genetic diversity of its component species. As intensity of use increases, genetic diversity of many species will be affected, with some species that are unable to tolerate disturbance becoming locally extinct. The optimum level of management in economic terms may be substantially more intensive than the optimum for conservation of biodiversity. If, for political or policy reasons, a greater effort is considered necessary to conserve biodiversity (either at the genetic or species level), the need for compensation is inescapable, and research is required to develop an approach for establishing appropriate levels of compensation. Numerous examples show (e.g. Braatz 1992; Fairhead & Leach 1993; McNeely

1993; Pinedo-Vasquez & Padoch 1993) that unless local people and other affected forest users can gain economic benefits from protected areas, or are adequately compensated for their loss, there is little likelihood that effective conservation of long-term protected areas can be achieved (Vanclay 1992).

9.2 Effect of logging on genetic diversity

Annually, about four million hectares of the world's tropical forests are logged for timber (Jonsson & Lindgren 1990). Hence, sustainable management of forests has become a global issue. The International Tropical Timber Organization (ITTO) has stipulated that all timber from tropical forests is to be produced from sustainably managed forest by the year 2000—the so-called 'ITTO Objective Year 2000' (International Tropical Timber Organization 1992a). Similarly, the Helsinki (International Institute for Sustainable Development 1999) and Montreal Processes (Australian Department of Primary Industries & Energy 1997) have established guidelines for sustainable management of temperate forests.

Early studies on logging damage mainly concentrated on physical damage to trees (Nicholson 1958; Fox 1968) and regenerants (Wyatt-Smith & Foenander 1962) in an attempt to improve silvicultural practices before and after logging. More recent studies have reported the effect of logging on nutrient dynamics (Amir Husni et al. 1989), hydrology (Rahim & Harding 1992; Baharuddin & Rahim 1994; Zulkifli & Anhar 1994), animal distribution and abundance (Heydon & Bulloh 1996) and geometroid moth (Geometroidea family and subfamily) diversity (Intachat et al. 1997). Studies on logging damage to genetic diversity of forest plant or animal species are scanty.

Sustained management of natural forests depends on their ability to regenerate. Successful regeneration of forests accompanied by conservation of genetic diversity of its species is important for both short-term adaptation to environmental change and long-term effects on species and communities (Templeton 1995).

Maintenance of genetic diversity in forest tree populations that are undergoing population changes due to natural or human-induced events is seen to be instrumental for adaptability and continued evolution (Müller-Starck 1985; Gregorius et al. 1985; Ledig 1988; Namkoong 1991b; Bush & Smouse 1992;

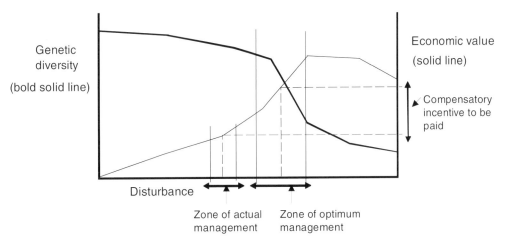

Figure 9.1 The relationship between genetic diversity (left axis, solid line) and economic value (right axis, solid line) as intensity of disturbance varies from low (left) to high (right). The actual shape of the lines will depend on the ecological characteristics of each species and their direct and indirect uses.

Müller-Starck *et al.* 1992). Hence, assessment of regeneration capacity and genetic diversity would be useful in developing sustainable forest management guidelines and effective conservation strategies. Estimation of genetic diversity of forest species can be rapidly done using isozyme markers (Yeh 1989).

Hamrick *et al.* (1979), who surveyed various schemes of species groupings, found that species with large ranges, high fecundities, an outcrossing mode of reproduction, wind-pollination, long generation time and from habitats representing later stages of succession have more genetic variation. Brown (1979) showed that patterns of interpopulation differentiation and heterozygosity within populations vary due to many complex factors affecting mating systems. Partial inbreeding due to restricted numbers of reproductive individuals or neighbourhood size is important to many outbreeding species. Hence, it would be interesting to assess how species with different life history strategies respond to disturbance or logging.

Most logging in temperate forests relies on clear cutting, followed by natural regeneration or, where a change of species is desired or where seed sources are considered inadequate for natural regeneration, by planting. For example, 90% of the timber harvested in Canada originates from clear cuts (Environment Canada 1991). Selective cutting in temperate forests is usually restricted to broadleaf forests managed for high value products such as veneer logs. The special case of clear cutting followed by planting obviously completely changes the genetic structure of the forest, and is not considered further here.

In many temperate forests, especially in the boreal region, natural disturbance regimes are characterised by large-scale disturbances such as fire or windthrow. Consequently, tree species are adapted for such conditions, and their reproductive systems enable long-range pollen and seed dispersal. For example, Young *et al.* (1993a) found that fragments of forest of *Acer saccharum* Marshall (sugar maple) in an agricultural landscape shared more alleles than similar-sized patches in an area of continuous forest. They attributed this result to long-distance pollen transfer through the agricultural matrix.

Buchert *et al.* (1997) reported a 12% increase in observed heterozygosity in *Pinus strobus* L. (white pine) in Ontario following harvesting. Despite this, appreciable losses of alleles (about 25%) occurred, and the authors concluded that these losses of alleles could imply that gene pools of naturally regenerated progeny stands may be different from the original parental stands (Box 9.1). They recommended silvicultural practices to ensure that gene pools of remaining pristine old-growth stands be reconstituted in the regenerating stands.

In the tropics the situation is reversed, with most logging being selective, except where land conversion is planned, and in special forest types such as mangroves. For example, logging in the production forests in Malaysia is managed under two harvesting

---

Box 9.1 Effects of logging in temperate forests

Two old-growth *Pinus strobus* stands were studied near Sault Ste-Marie in Ontario, Canada. White pine was the dominant upper canopy and emergent species, with a more species diverse understorey. Genotypes at 54 enzyme loci were inferred for 191 mature trees from the haplotypes of 8 to 10 megagametophytes per individual.

Even though a 'partial' cutting system was used, the lower species diversity and greater size conformity compared with most tropical forests (Box 9.2) meant that the reduction in population size and basal area was much greater, with 75% of the mature *Pinus strobus* trees being removed by logging. The resultant losses in allele numbers were dramatic, with 30% or more of all alleles being lost. However, these losses occurred only for low frequency and rare alleles, and in the latter category, 80% losses occurred. Mean numbers of alleles per locus and numbers or percentage of polymorphic loci showed similar losses of about 30%. The hypothetical multilocus gametic diversity was reduced by about 40% in each stand. However, observed heterozygosities did not follow the same trend, and in one of the two stands observed, heterozygosity increased after harvesting.

Nevertheless, it is clear both that genetic diversity is adversely affected by the high intensity of felling in most temperate forests and that gene frequencies in the resulting natural regeneration will be substantially different from those in the parent stand. For temperate conifers with reproductive systems adapted to high levels of gene flow, any adverse effects of harvesting could be at least partially offset by gene flow from neighbouring uncut stands. This implies the need for management of harvesting at the landscape level so as to provide for appropriate sources of gene flow (i.e. by by setting maximum interstand distances less than what is thought to be the average pollen dispersal range).

---

systems (i.e. the 30-year Selective Management System, SMS, cutting cycle and the 55-year Malayan Uniform System, MUS, cutting cycle; Wyatt-Smith 1963; Thang 1987). In the SMS practice, selective felling is carried out where the cutting limit for dipterocarp species should not be less than 50 cm diameter at breast height (dbh) and for non-dipterocarp species the cutting limit should not be less than 45 cm dbh. In this practice, sufficient adolescent trees (i.e. at least 32 commercial trees per hectare or equivalent; Thang 1988) should be left after logging. In the MUS practice, all commercial trees not less than 45 cm dbh are allowed to be logged if there are sufficient regenerants of the species.

In one of the most comprehensive studies of logging in tropical forests, Wickneswari *et al.* (1997a, 1997b) studied five species with different life histories (Box 9.2). Genetic diversity (expected heterozygosity, H_e, for isozyme data and Shannon Diversity Index, H, for RAPD data) was reduced by 5.0–23.4% after logging for all species examined. Generally, the order of reduction in number of individuals corresponded with

the order of reduction in genetic diversity. Among the non-timber species examined, *Daemonorops verticillaris* Mart. (climbing palm) exhibited the highest loss of alleles (13.8%) most probably due to death of individuals following logging damage to supporting trees and being trampled during extraction of felled trees.

Loss of alleles can lead to loss of adaptability (Gregorius *et al.* 1985) and environmental fitness (Müller-Starck 1985; Raddi *et al.* 1994). The existence of *Daemonorops verticillaris* may be further threatened as it is a dioecious species requiring sufficient individuals for natural regeneration. As it is quite common to allow extraction of palm species before logging is carried out, this prelogging practice needs to be re-examined to prevent further genetic erosion of these climbing plant species. However, the losses in genetic diversity were measured immediately after logging. These immediate losses in genetic diversity may be compensated by good seed or seedling bank in the forest management unit or in nearby undisturbed gene pools.

Box 9.2 Effects of logging in tropical forests

Genetic structure of plant species with different economic importance and life history strategies were examined in two humid tropical forest types: a ridge forest and a mixed dipterocarp lowland forest that had been logged less than one year ago and more than 40 years ago, respectively. About 30 mature individuals each of five to six species from each forest type were randomly sampled for isozyme and randomly amplified polymorphic DNA (RAPD) analysis.

The reduction in basal area of the forest management unit in the ridge forest was about 56% after logging. Genetic diversity (expected heterozygosity for isozyme data and Shannon Diversity Index for RAPD data) was reduced by 5.0–23.4% after logging for all species examined. Loss of alleles for the different species ranged from 7.7% to 25.0% after logging. Loss of genetic diversity immediately after logging was higher in commercial timber species of low abundance (9.4% loss in heterozygosity and 25.0% loss in alleles for *Shorea leprosula*) than that of high abundance (5.0% loss in heterozygosity and 7.7% loss in alleles for *Scaphium macropodum* Beumee ex K. Heyne) in the ridge forest management unit. This indicates the need to examine regeneration status for timber species of low abundance before permitting logging of these species. Among the non-timber species examined, the climbing palm, *Daemonorops verticillaris*, exhibited the highest loss of alleles (13.8%). Extraction procedures for felled trees may need to be improved to minimise damage to climbing plants such as rattans.

No adverse changes in genetic diversity measures of *Parkia speciosa* Hassk., *Scaphium macropodum*, *Shorea leprosula*, *Garcinia malaccensis* Hook. f., *Daemonorops verticillaris* and *Labisia pumila* Benth. & Hook. f. were detected in two regenerated stands (13.5% and 40.7% disturbances) compared to an unlogged stand in the mixed dipterocarp lowland forest. This indicates that genetic diversity is retained in regenerated stands after a single low intensity logging event.

Estimates of outcrossing rates were made for one dipterocarp species, *Shorea leprosula* Miq., and no significant differences were observed between logged and unlogged stands. Similarly, Kitamura *et al.* (1994) reported no significant difference in outcrossing rates between primary forest and secondary forest (logged selectively 20 years previously) for *Dryobalanops aromatica* Gaertn. (Dipterocarpaceae) in Brunei. They concluded that high levels of outcrossing rates were maintained by high flowering density regardless of size structure or the topography of the habitat. However, Murawski *et al.* (1994b) reported reduction of outcrossing rates in *Shorea megistophylla* P.S. Ashton (Dipterocarpaceae) in a selectively logged forest (about 20 years previously) in Sri Lanka, indicating inbreeding.

9.3 Effect of non-timber forest product collection on genetic diversity

The definition of what constitutes NTFPs can vary widely. In its broadest sense, it encompasses all products from the forest that are not timber, including animals and fish, charcoal and firewood, as well as fruit, nuts, medicinal plants and bark. More restricted uses of the term may exclude all wood products, and sometimes the products of hunting and fishing are also differentiated from other types of NTFPs.

The effects of NTFP collection often are considered to be more benign than those of logging (e.g. Panayatou & Ashton 1992), although Hall and Bawa (1993) concluded that there are few, if any, examples of sustainable NTFP collection systems. Since the nature of individual NTFPs varies so widely, as discussed above, and the form of harvesting can involve whole individuals, reproductive parts (e.g. flowers, fruits, nuts) or non-reproductive parts (leaves, barks, branches), the likely genetic effects will also be highly variable. Namkoong *et al.* (1996) considered the possible consequences of different forms of NTFP collection on genetic processes (Table 9.1). They concluded that the main effects of harvesting of whole individuals would be via genetic drift and indirect

TABLE 9.1 Genetic consequences of different types of utilisation of NTFP, non-timber forest product

Type of utilisation	Drift	Direct selection	Indirect selection	Mating system	Gene flow
NTFP—reproductive		X		X	X
NTFP—non-reproductive		X			
NTFP—whole individual	X		X		
Fire	X				X
Grazing	X				

selection. In contrast, harvesting of only reproductive structures would be most likely to affect gene flow, the mating system and direct selection. Thus, the major effects will differ depending on the form of harvesting.

There have been very few studies of the effects of NTFP collection, especially from the perspective of genetic diversity. Since the possible genetic consequences will differ depending on the type of the NTFP and harvesting, this paucity of empirical studies makes it very difficult to draw any conclusions concerning effects of NTFP collection.

In a series of comprehensive studies on NTFP collection in the Biligiri Rangan Hills of southern India, Hegde *et al.* (1996) demonstrated a negative correlation between the amount of time spent collecting NTFPs and the availability of salaried labour, while Murali *et al.* (1996) found a significant effect on recruitment of the harvested species, especially where harvest levels were highest (Box 9.3). The negative relationship between time spent collecting NTFPs and availability of salaried labour is a consequence of the seasonality and unreliability of NTFP collection, where crops in some years may be very low, and a consequence of the low value of the crop at the collection point. Uma Shanker *et al.* (1996) concluded that non-sustainable harvest levels were inevitable given these low value-added benefits accruing to the collectors. There are also numerous other examples of overexploitation of NTFPs (e.g. Browder 1992; Nepstad *et al.* 1992). While most of these studies focus on the direct effects on the harvested species, Boot and Gullison (1995) use the example of collection of brazil nut (*Bertholletia excelsa* Humb. & Bonpl.) to point out that there may be both direct and indirect effects on non-harvested species. High removal rates of *Bertholletia excelsa* may directly affect squirrels (Sciuridae) and agouti (Agoutidae)—the main consumers and seed distributors of the species, and this may have indirect

consequences on other plant species if these frugivores substitute other species into their diet.

These trends have led to an assumption that, increasingly, production systems and conservation goals will be integrated. Integration of production and conservation is a form of multiple-use management. Panayatou and Ashton (1992) examined multiple-use management from an economic perspective, and although they noted that greater management skills are required for multiple-use management, and that production costs may increase due to potentially higher harvesting costs (i.e. the loss of economies of scale) they concluded that 'the case for multiple-use management can often be based on financial analysis alone, without resorting to shadow pricing'.

Although multiple-use management may be justifiable on financial grounds, biodiversity is difficult to value. Van Noordwijk *et al.* (1996) constructed a model to evaluate the relative value, in terms of biodiversity conservation, of strategies to segregate or integrate production and conservation, and used data from agroforestry systems in Sumatra, Indonesia, to test the model. The model calculates the relative loss of biodiversity per unit increase in productivity for a range of possibilities from pure conservation to pure production. For systems in which the relative loss of biodiversity is higher at low or intermediate production levels, a segregation policy is optimal, in which production is concentrated on relatively small areas, allowing larger areas to be set aside for conservation. This situation may apply to most food crop-based production systems, due to the requirement of most crops for a high light, low competition environment, even at low production levels. In contrast, integration may be a valid strategy for 'jungle rubber' systems, involving the production of rubber and other crops in modified secondary forest.

Box 9.3 Effects of collection of non-timber forest products in tropical forests

In the Biligiri Rangan Hills of Karnataka State in southern India, the indigenous Soliga people harvest NTFPs from several dozens of species of plants and animals, both for subsistence reasons and for sale. Traditionally the Soligas were semi-nomadic, but have been increasingly encouraged by governments in the 20th century to adopt a settled lifestyle, and this became obligatory in 1972. Since that time, not only has exploitation of NTFPs been increasingly concentrated around the major settlements, but greater market availability has provided new incentives for harvesting, while access to salaried labour opportunities probably has had the opposite effect.

Intensive harvesting is carried out on relatively few of the NTFP species. One species that is heavily exploited is *Phyllanthus emblica* L., the fruit of which is collected and used for food and as the base of numerous traditional medicines. Although each collector will spend only five to six days per year harvesting *Phyllanthus emblica*, the synchrony of fruit maturation and ease of collection from this understorey tree mean that an average of more than 70 kg of fruits are collected per collector per year, for a total of nearly 500 t (Hegde *et al.* 1996). Furthermore, the fruit is an important seasonal food for various species, including monkeys (*Semnopithecus* spp., *Macaca* spp.) and *Cervus* spp. (deer). The genetic effects of such intense collection are easy to observe, since in the most heavily harvested areas, nearest the settlements, no *Phyllanthus emblica* regeneration was found.

Several other NTFP species were also studied. These included *Terminalia bellerica* (Gaertn.) Roxb. and *Terminalia chebula* Retz., for which the fruit is also used for medicinal purposes. The lower density of mature individuals of these species (10–25% of *Phyllanthus emblica* frequencies) means that there are lower returns per unit collection effort, and although comparable amount of time may be spent collecting fruits of these species, the total annual yield is only 2.6 t/y to 8.3 t/y. Seed set was observed for these two species even in locations subject to relatively heavy harvesting. In comparison with control (unharvested) sites, losses in allelic diversity at heavily harvested sites were only small (< 5%) for *Terminalia bellerica* and zero for *Terminalia chebula*.

Clearly, the genetic effects of NTFP collection are much affected by intensity of harvesting (as for logging), but also by the origin of the harvested product—harvesting of fruits or flowers will have a much more direct and dramatic impact on regeneration and genetic diversity in subsequent generations than harvesting of leaves, bark or other non-reproductive parts of the plant.

The model developed by Van Noordwijk *et al.* (1996) also has some limitations. For example, they point out that the effect of increasing productivity has a differential impact on different components of biodiversity. Data from Sumatra indicate that whereas for soil collembola, the impact may only be slight, for birds it is much more severe. If all groups of organisms are to be considered, then the choice to integrate or segregate will depend on the most sensitive group, and segregation will generally become much more attractive. Similarly, the choice depends on total productivity possible under pure production systems.

Whereas in the 1930s productivity of jungle rubber was comparable to monocultures, now it is only 25% to 30% of monoculture productivity, which again increases the tendency to segregate (Van Noordwijk *et al.* 1996).

Clearly, the decision of when to integrate and when to segregate production and conservation will depend on numerous factors, including not only relative biodiversity values and economic considerations, but also on land tenure systems, regulatory efficiency, administrative flexibility and so on.

Where segregation of conservation and production objectives is necessary, the identification of optimum areas for conservation relies increasingly on computer technologies, such as Geographical Information Systems (GIS) (e.g. Pressey *et al.* 1990; Margules *et al.* 1991; Vane-Wright *et al.* 1991; Ramesh *et al.* 1997). Such approaches may take the form of gap analyses (e.g. Scott *et al.* 1991) or the identification of minimum sets of conservation areas to meet a threshold value of some measurable variable (e.g. Margules *et al.* 1988; Kitching & Margules 1995). In most cases the measurable variable is species richness, and the threshold value is some arbitrarily fixed proportion of total species richness within a region. Raw measures of species richness may be supplemented with assessments of threat to prioritise area selection or conservation effort (Dickman *et al.* 1993; Smith *et al.* 1993).

Some studies that integrate socioeconomic and biophysical data by means of GIS have been undertaken in relation to biodiversity conservation. For example, Fox *et al.* (1994c) investigated the conflict between grazing and conservation of red panda habitats in Langtang National Park in Nepal. A model of grazing impact was developed, which showed that the strong preference for gentle slopes and open canopies in selection of grazing lands coincided with prime red panda habitat. Fox *et al.* (1994a, 1994b) also used GIS to study the farmer decision-making process in relation to biophysical and socioeconomic variables, and temporal change in land use in northern Thailand. There have been no examples of data on levels and pattern of biodiversity being combined with assessments of threat to determine a comprehensive conservation strategy.

9.4 Conclusion

Maintenance of genetic diversity in forest tree populations that are undergoing population changes due to natural or human-induced events is seen to be instrumental for adaptability and continued evolution. Immediate losses in genetic diversity due to logging may be compensated by good seed or seedling bank in the logged or nearby undisturbed stands. Silvicultural practices should ensure that gene pools of remaining pristine old-growth stands be reconstituted in the regenerating stands.

The main effects of harvesting of whole individuals would be effected via impacts on genetic drift and indirect selection. Harvesting of only reproductive structures would be most likely to affect gene flow, the mating system and direct selection. Thus, the major effects of harvesting of NTFPs will be different depending on the form of harvesting. These necessitate production systems and conservation goals to be integrated in an efficient manner to be economically viable.

FOREST FRAGMENTATION

*Andrew G. Young and
Timothy J. Boyle*

SUMMARY

Forest fragmentation is a pervasive threat to forest ecosystems throughout the world. Reduced overall numbers of individuals, decline in mean population sizes and the separation of forest remnants by non-forest land can affect the main genetic processes of genetic drift, gene flow, selection and mating. The three most obvious possible genetic effects of forest fragmentation are:

(a) loss of genetic diversity at the population and species level

(b) increased interpopulation structure

(c) increased inbreeding.

Responses are likely to vary depending on the ecological characteristics of the species involved, such as their breeding system, dispersal syndrome, longevity and ploidy level.

Although empirical data are sparse, results suggest that while genetic variation may decrease with reduced remnant population size, not all fragmentation events lead to genetic losses, with the interaction between local population size and isolation being important in determining levels of genetic diversity. Similarly, while inbreeding may increase in fragmented populations, sometimes even fairly dramatic shifts in pollinator guilds due to fragmentation do not appear to affect the rate of cross fertilisation. Overall, genetic effects of fragmentation are less easy to predict than expected, being determined, as they are, by the scale of fragmentation, the biology of the affected species and their associated pollinators and dispersers. More data are required in order to make sensible predictions regarding the likely response of a forest ecosystem to a given fragmentation event.

10.1 Introduction

Humans have been changing forest ecosystems for thousands of years as they seek to use forest species as resources for food, fuel and timber, or to clear forested land for agriculture, grazing and habitation. Throughout much of our history, human population densities have been low enough that these effects have been spread thinly across a wide geographical base although, as Ledig (1992) has pointed out, even early Greek and Roman civilisations precipitated some fairly dramatic local deforestation.

The rise in global human population over the last two centuries, however, has increased the rate of forest destruction and degradation. In North America, for example, forest cover was reduced significantly during the 19th century (Harris 1984). Similarly, in the southern temperate forests of New Zealand, the percentage of forest cover has been reduced from 66% to 22% since the beginning of European colonisation in the late 1830s (Hunt 1977). In addition, the range of forests exposed to intensive human disturbance has broadened, with particular pressure now on the species-rich tropical forest ecosystems. Between 1980 and 1990, 8.4% of the closed forest on the African continent underwent some form of land use conversion (Food and Agriculture Organization 1993), while in the Brazilian Amazon, between 1978 and 1988, the mean annual rate of deforestation and fragmentation has been estimated at 53 000 km^2 per year (Skole & Tucker 1993).

These rapid and large-scale changes in the extent and integrity of forest ecosystems have resulted in fragmented forests which often have little physical, and probably limited functional, resemblance to the original (Fig. 10.1). Changes in the species composition of forest communities have been observed to accompany forest fragmentation, especially at newly created edges (Ranney *et al.* 1981; Young & Mitchell 1994; Murcia 1995). While these primarily reflect ecological responses to fragmentation, due for instance to changes in disturbance regimes, or microclimate, they also represent genetic responses at the macro scale. More directly, however, at the intraspecific level, changes in population size, and isolation associated with reductions in forest cover will affect genetic processes such as mating, gene flow and selection that determine the levels of genetic diversity maintained within plant species (Billington 1991; Ledig 1992; Young *et al.*

1996). In the following sections we review the likely genetic effects of forest fragmentation and summarise the existing empirical data on such effects.

10.2 How can fragmentation affect genetic diversity?

Forest fragmentation has three main effects on species:

(a) reduction of overall numbers of individuals

(b) reduction of mean population sizes as individuals are restricted to small forest fragments

(c) spatial isolation of remaining populations within a non-forest-like land use matrix.

The genetic results of these changes are mediated by their effects on the main genetic processes of genetic drift, gene flow, selection and mating (see Section II). These processes determine the amount of genetic diversity found within a species and how it is partitioned among individuals and populations across its geographical range.

The three most obvious possible genetic effects of forest fragmentation are:

(a) loss of genetic diversity at the population and species level

(b) change in interpopulation structure

(c) increased inbreeding.

These effects suggest several causes for concern. In terms of *in situ* conservation, low genetic variation limits a species' ability to respond to changing environmental conditions through selection, while changes in interpopulation structure may alter the scale at which selective responses take place (e.g. individual-by-individual or population-by-population), as well as affecting the speed with which favourable mutations are spread. More immediately, loss of diversity at loci controlling disease resistance, or self-incompatibility, may have direct short-term effects on fitness and population demography (Barrett & Kohn 1991). For utilisation, a narrow genetic resource base limits the species' potential for domestication and reduces the efficiency of breeding programs.

Perhaps of more concern is inbreeding. Most forest trees, both temperate and tropical, are highly outcrossed, with self-fertilisation generally accounting for less than 20% of progeny (Boyle & Yeh 1988;

a

b

c

Figure 10.1 Fragmented forest ecosystems. (a) Beech maple forest in Ontario, Canada. Photograph: Kathy Freemark. (b) Dry tropical forest in Costa Rica. Photograph: David Boshier. (c) Mixed podocarp–broadleaf forest, New Zealand. Photograph: Neil Mitchell.

Figure 10.2 Fragments of mixed podocarp–broadleaf forest in northern New Zealand showing how the selective clearing of land for farming has restricted the forest ecosystem primarily to steep, south-facing slopes. Photograph: Neil Mitchell.

Loveless 1992; Mitton 1992; Murawski 1995). Because of this, trees maintain high levels of genetic load in the form of deleterious recessive alleles. If selfing or biparental inbreeding increases after fragmentation, due to reduced population sizes or effects of fragmentation on pollinator abundance or behaviour, the resulting increase in homozygosity may lead to severe inbreeding depression. For example, the effects of inbreeding on *Pinus* spp. (pines) include reduced seed yields, height growth and stress tolerance (reviewed by Ledig 1992). Such effects on fitness can reduce short-term population viability and limit the value of remnant populations either as foci for natural regeneration or seed sources for breeding.

Complicating these possibilities is that the response of different species to fragmentation is probably determined by their life history characteristics, in particular, their abundance, longevity, breeding system, dispersal syndrome and ploidy level. The same fragmentation event may have different genetic effects on different species, especially if responses are complicated by interactions with other species which are themselves affected by fragmentation. For example, for species involved in obligate relationships with pollinators or seed dispersers, such as the tight evolutionary relationships between figs and wasps, results of disturbance may be primarily mediated

through effects on the associated species rather than the plants themselves (Nason & Hamrick 1997).

10.3 Population-level genetic diversity

Initial effects on genetic diversity at the population level will occur at the time of fragmentation when alleles are lost due to generation of genetic bottlenecks. The magnitude of this loss depends on the pattern of forest clearing, and how much spatial genetic structure a species exhibits. If forest clearing is random, and there is little interpopulation genetic divergence, fragmentation can be viewed as a simple sampling effect. In this situation a logarithmic relationship between remaining numbers of plants and allelic richness is expected (Brown 1990; see Chapter 14).

However, many tree species, both temperate and tropical, show non-random distributions of genes among individuals and populations (e.g. Young *et al.* 1993b; Murawski & Bawa 1994; Dow & Ashley 1996), and patterns of forest clearing generally relate to environmental variables such as topography and drainage, such that remnant tree populations are an environmentally biased sample of the original. This is well illustrated by the fragmented podocarp–broadleaf forests of northern New Zealand, which now exist almost exclusively on steep, south-facing slopes at

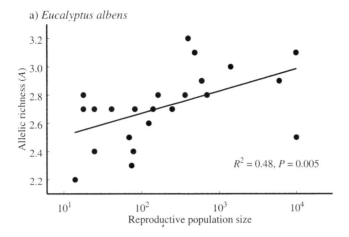

a) *Eucalyptus albens*

$R^2 = 0.48, P = 0.005$

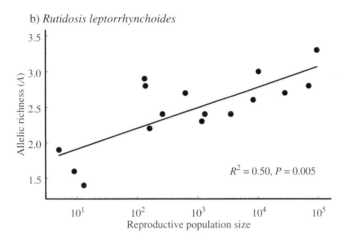

b) *Rutidosis leptorrhynchoides*

$R^2 = 0.50, P = 0.005$

Figure 10.3   Relationship between genetic diversity as measured by allelic richness at allozyme loci, and remnant population size. (a) *Eucalyptus albens* (Prober & Brown 1994). (b) *Rutidosis leptorrhynchoides* (from Young *et al.* 1999).

altitudes higher than 300 m (Fig. 10.2). In this situation, if there are significant environment–genotype relationships, such as those observed by Xie and Knowles between soil nutrients and allozyme genotypes for *Pinus banksiana* Lamb. (jack pine), remaining populations can no longer be viewed as random samples of the prefragmentation gene pool, and the potential for loss of genetic diversity is greater (Hamrick 1994b).

Subsequent to initial sampling effects, remnant populations that remain small for several generations will continue to lose alleles due to random genetic drift (Barrett & Kohn 1991; Ellstrand & Elam 1993), with the magnitude of the loss depending on generation time and population size. In the absence of immigration and selection the mean expected change in allele frequency (Δq) due to one generation of drift is:

$$\Delta q = \frac{q(1-q)}{2N_e} \qquad (10.1)$$

where N_e is the variance effective population size (Falconer 1989).

Two studies have examined erosion of genetic diversity due to forest fragmentation by comparing genetic diversity among different sized fragment populations using allozymes. Both were conducted in the grassland–woodland ecosystems of south-eastern Australia which have been reduced to only a small fraction of their original extent since the mid 1800s due to conversion of land to pasture for sheep grazing.

In the first of these, Prober and Brown (1994) found a significant positive relationship between population size and allelic richness in populations of *Eucalyptus albens* Benth., a large, long-lived, insect-pollinated tree species with a mixed-mating system (Fig. 10.3a). Similarly, looking at a species with a very different life history, Young *et al.* (1999) observed lower numbers of allozyme alleles with reduced population size in *Rutidosis leptorrhynchoides* F. Muell (daisy) (Fig. 10.3b), which is short-lived and self-incompatible. In both cases losses were due to the elimination of low frequency alleles in small populations, as might be expected for losses due to bottlenecks and genetic drift. For *Rutidosis leptorrhynchoides*, loss of diversity was associated with low seed production in small populations, probably relating to parallel reductions in the numbers of alleles present at loci controlling self-incompatibility (SI loci). Erosion of allelic richness at SI loci has also been observed in small populations of the rare lakeside daisy *Hymenoxys acaulis* var. *glabra* K.L. Parker (DeMauro 1993). Losses at such loci that control important ecological processes represent significant threats to population viability.

So far we have considered effects on genetic variation at single loci. However, quantitative genetic variation can also be eroded by reductions in population size. Although such effects are difficult to detect, requiring estimation of the genetic components of phenotypic variance, they are particularly important as much of the adaptively significant phenotypic variation in plant populations is under polygenic control. For a single isolated remnant population additive genetic variance (V_A) for neutral traits is lost due to genetic drift at a rate per generation of:

$$\Delta V_A = \frac{-V_A}{2N_e + V_M} \quad (10.2)$$

where V_M is the input of genetic variance by mutation and N_e is the variance effective population size (Lande & Barrowclough 1987). However, under some circumstances, additive variance may also increase as population size goes down. This is because fixation of alleles at some loci reduces the range of multilocus genotypes within which a polymorphism is expressed resulting in an apparent conversion of non-additive variance to additive variance (Goodnight 1988). As traits related to fitness often have high non-additive genetic variances, reduced population size associated with fragmentation may promote variation in individual fitness (Carson 1990), thus exposing

previously cryptic variation more directly to selection (Young *et al.* 1996). So far there are no studies of effects of fragmentation on quantitative variances within forest species, although both positive and negative effects of reduced population size on phenotypic variation and V_A have been observed for several grassland herbs (Ouberg *et al.* 1991; Widen & Andersson 1993).

10.4 Gene flow and interpopulation divergence

Indirect estimates of interpopulation gene flow for both temperate and tropical tree species are usually high, often with $N_e m > 5$, even over hundreds of kilometres (where $N_e m$ is the number of migrants per generation) (Govindaraju 1989; Hamrick & Loveless 1989). Similarly, direct estimates of pollen and seed dispersal distances in intact forests show that gene flow is extensive for species with a variety of pollination and seed dispersal syndromes (Loveless 1992; Boshier *et al.* 1995b; see Chapter 6). For example, Hamrick and Murawski (1990) observed average effective pollen dispersal distances of 374 m, 420 m and 369 m over three years in a population of the tropical angiosperm *Platypodium elegans* Vogel and commonly detected pollen movements of over 750 m. As a result, genetic neighbourhoods (Wright 1938; 1946; see Chapter 7) tend to be large, being about 20 ha to 50 ha even for low density tropical species (Hamrick & Murawski 1990). The result is that, subsequent to forest fragmentation, remaining populations may be smaller than original neighbourhoods. For example, Boshier *et al.* (1995b) showed that for *Cordia alliodora* (Ruiz & Pav.) Cham. neighbourhoods cover an area of about 7 ha, below the size of many forest fragments in which it now persists in Costa Rica.

Given this, the likelihood that original patterns of gene flow and genetic structure will be maintained depend on the species' ability to disperse across the interfragment matrix. For wind-pollinated and dispersed species this may be a simple function of interfragment distance. However, for species that rely on animals for pollination and dispersal, patterns of gene flow will depend on response of these associated species, in particular, their ability to cross non-forest habitat which is generally of low biomass and structural complexity relative to forest. There are now several studies showing that fragmentation affects the abundance, composition and behaviour of many

TABLE 10.1   Estimated rates of gene flow into fragmented *Spondias mombin* populations (from Nason & Hamrick 1997)

Population size	Isolation (m)	Apparent pollen flow[A]	Total pollen flow[B]
22	100	0.095	0.604
4	1000	0.400	0.891
2	150	0.579	1.013
1	300	1.000	1.020
8	80	0.857	1.071

[A]Apparent pollen flow is the actual proportion of progeny within a population that cannot have been sired by adults within the population and as such represents an underestimate of actual numbers due to the limitations on discrimination imposed by a finite set of marker loci (Ellstrand & Marshall 1985).
[B]Total pollen flow is an asymptotic estimator of the number of such individuals which takes account of the limitations imposed by marker diversity (Devlin & Ellstrand 1990).

pollinating species (see Didham *et al.* 1996 for a review). For example, Powell and Powell (1987) found that a clearing as small as 100 m formed an effective barrier to the movement of four euglossine bee species (*Euglossa* sp.) between remnants of Amazonian tropical forest.

Subsequent to fragmentation, if population structure is stable, interpopulation genetic divergence will increase due to random genetic drift, particularly when population sizes and gene flow are such that $N_e m < 1$ (Wright 1931; see Chapter 3). When population structure is unstable genetic divergence will depend on the frequency of colonisation and extinction events and the number of sources of colonists (Wade & McCauley 1988; Lande 1992; Whitlock 1992). Extinction and colonisation rates of more than twice the migration rate can substantially reduce genetic variance among subpopulations, whereas reducing the number of populations from which colonists are drawn increases differentiation.

Several recent studies have looked at patterns of interfragment gene flow and resulting genetic divergence. Nason and Hamrick (1997) examined gene flow into small, fragmented populations of the tropical tree *Spondias mombin* in Panama. *Spondias mombin* is an early successional, insect pollinated, hermaphrodite, canopy species. Using allozyme markers to identify progeny genotypes within fragments that could not have been sired by adults from that population, Nason and Hamrick (1997) estimated that pollen immigration generally accounted for at least half of the mating events occurring in fragment populations (Table 10.1) even with isolation distances of up to 1 km. This suggests that for

Spondias mombin gene flow events may not have been truncated by fragmentation.

A similar investigation by Foré *et al.* (1992) compared interpopulation genetic divergence between prefragmentation and postfragmentation cohorts of 15 *Acer saccharum* Marshall (sugar maple) populations in Ohio, USA. *Acer saccharum* is both insect and wind pollinated and has wind-dispersed samaras. Genetic divergence among post-fragmentation cohorts was less than half that among prefragmentation cohorts, indicating a reduction in genetic differentiation since fragmentation, and suggesting increased interpopulation gene flow which they attributed to increased wind velocity across the fields separating the fragments. Using similar methods to Nason and Hamrick (1997), more exhaustive intercohort comparisons on a subset of the same populations also revealed the presence of alleles in remnant population embryo cohorts that did not occur in their respective parental cohorts, providing clear evidence of gene flow into remnant populations (Ballal *et al.* 1994). Working on the same species in Canada, Young *et al.* (1993a) found a marginal increase in the spread of allozyme alleles among fragmented *Acer saccharum* populations relative to equivalently spaced sample sites in intact forest. Combined with higher mean polymorphism in fragmented populations, this also pointed to increased interpopulation gene flow as a consequence of fragmentation for this species.

Although these studies suggest that gene flow can be maintained among fragmented populations, this is unlikely always to be the case. Hall *et al.* (1996) identified significant reductions in genetic diversity and increased genetic divergence in fragment populations of the tropical tree *Pithecellobium elegans*

Box 10.1 Gene flow among *Pithecellobium elegans* trees isolated in pasture (Chase *et al.* 1996b)

Pithecellobium elegans is a large canopy tree in central American rainforests which has been subject to fragmentation of its forest habitat. It is pollinated by hawkmoths which are known to forage over long distances. Chase *et al.* (1996b) examined mating events among 28 adult trees left in pasture after forest clearing in the Atlantic lowlands of Costa Rica. Five microsatellite loci (simple sequence repeat, SSR—see Chapter 4) were used to conduct paternity analysis on 167 seed representing 72 mating events from six trees. The hypervariable nature of the loci, which had 5 to 14 alleles, allowed identification of a single pollen parent, or the exclusion of all known fathers, for 70 (97.2%) of these seed. Matings occurred between trees separated by as much as 350 m of pasture (Fig. 1.1a) and most mating events were not between nearest neighbours, but between individuals separated from each other by several other trees (Fig. 1.1b).

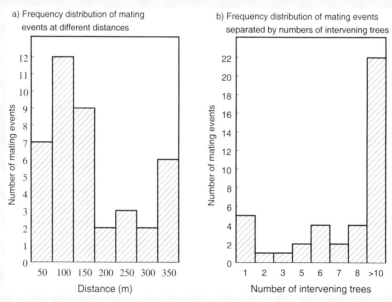

Figure 1 Frequency distributions of mating events among individuals of *Pithecellobium elegans* (after Chase *et al.* 1996b). (a) At different distances. (b) Separated by numbers of intervening trees.

with increasing distance from a large intact population in Costa Rica. *Pithecellobium elegans* is a large canopy emergent that is insect pollinated and bird dispersed. Similarly, Prober and Brown (1994) demonstrated apparent effects of population isolation on genetic variation. Small populations (< 500 reproductive individuals) of *Eucalyptus albens* that were less than 250 m from a larger population had high allelic richness, while more isolated populations of a similar size were genetically depauperate. These are both important results as they point to thresholds up to which gene flow from a large source population can maintain diversity, but beyond which genetic erosion occurs.

One factor that may be of considerable importance in determining the level of gene flow maintained among fragmented populations is the role of scattered trees that are often left in fields. The long-term ecological value of such individuals is debatable as the likelihood of regeneration is limited. However, recent work by Chase *et al.* (1996b) suggests that they may serve as stepping stones for gene migration between populations, at least for the duration of their life (Box 10.1). This may be either directly, if individuals receive pollen from an individual from one fragment and contribute seed to another, or indirectly, if they simply act as a corridor for pollinator movements.

Given the relatively low levels of gene dispersal required to prevent genetic divergence due to drift ($N_em > 1$), these individuals may be crucial in maintaining genetic cohesion among populations after fragmentation.

10.5 Mating and heterozygosity

The third major prediction about the genetic effects of fragmentation is that small populations occupying isolated habitat fragments will experience increased inbreeding. Such inbreeding may be due to either increased self-pollination, or through biparental inbreeding when mates are related through recent common ancestry in small populations. Inbreeding increases homozygosity and may result in inbreeding depression which can be severe in predominantly outbred species such as trees that carry high levels of genetic load.

As with gene flow, effects may be direct, due to changes in population size or indirect, owing to effects of fragmentation on pollinator abundance or behaviour. For example, it is well established that changes in plant density can effect the number and frequency of within-plant versus between-plant movements made by insects (e.g. Watkins & Levin 1990; Karron *et al.* 1995). Since such changes will immediately affect mating system parameters such as outcrossing rates, heterozygosity may respond more rapidly to fragmentation than allelic richness or gene diversity.

So far evidence for increased inbreeding following forest fragmentation is extremely limited. Prober and Brown (1994) found a positive relationship between remnant population size and heterozygosity in *Eucalyptus albens*, suggesting that decreases in heterozygosity accompany reductions in population size. The observed loss of variation was greater than expected due to bottlenecks and genetic drift alone, where the change in gene diversity is predicted to be $\Delta H_e = -H_e/S$ where S is the number of gametes (Frankel *et al.* 1995), and so presumably reflects increased inbreeding. However, no direct estimates of outcrossing rates were made.

Two other studies have made direct comparisons of mating patterns between fragmented and intact populations of forest species. In the first, Schmidt-Adam *et al.* (2000) compared outcrossing rates between two intact island populations and three fragmented mainland populations of the bird-pollinated tree *Metrosideros excelsa* Sol. ex Gaertn.

(pohutukawa) in New Zealand (Box 10.2). Significant increases in self-pollination were expected in the fragmented mainland populations as these only support one of the three major bird pollinators found in the intact populations. However, there was no consistent difference in levels of self-fertilisation between fragmented and intact populations, nor was there any decrease in progeny heterozygosity in forest fragment populations relative to that in the intact populations. Subsequent pollinator exclusion experiments suggest that exotic bird species may have taken over the roles of the two missing native species in the fragmented mainland populations—thus maintaining outcrossing rates.

In contrast to the apparent resilience of *Metrosideros excelsa*'s mating system to fragmentation, Young and Brown (1999) found significant changes in mating patterns with reduced size and increased isolation of remnant populations of the daisy *Rutidosis leptorrhynchoides*. Although outcrossing rate was not affected, due to the strong self-incompatibility in this species, there were clear changes in the patterns of paternity. Small populations (<10 plants), or medium-sized populations (100–200 plants) isolated by more than 5 km, exhibited increased likelihoods of producing full-sib families as measured by the correlation of paternity relative to either large populations (>10 000 plants), or medium-sized populations within 2 km of large ones (Table 10.2). Although not inbreeding in itself, this increased production of full-sibs, combined with limited seed dispersal in this species, foreshadows the possibility of significant biparental inbreeding in subsequent generations.

The demographic effects of inbreeding and associated inbreeding depression depend on the mode of gene expression, the stage of the life cycle affected and the intensity of selection against homozygotes. If inbreeding depression is severe, significant effects on population viability might be expected. However, under some circumstances, intense inbreeding coupled with selection against less fit homozygotes can eliminate deleterious alleles and restore fitness. This is known as purging genetic load (see Chapter 7) and is the subject of considerable research and debate (Barrett & Charlesworth 1991; Hedrick 1994; Lynch *et al.* 1995). However, comparative studies of genetic load in large and small plant populations have so far found little evidence of purging (van Treuren *et al.* 1993; Widen 1993).

Box 10.2 Outcrossing rates in fragmented *Metrosideros excelsa* populations (Schmidt-Adam *et al.* 2000)

Metrosideros excelsa is a mass-flowering tree of coastal forests in northern New Zealand. Present day populations are fragmented and represent only a small fraction of the original distribution, which in the 19th century formed an almost continuous coastal forest belt. In many mainland populations, habitat destruction and increased predation by feral animals has resulted in the local extinction of *Anthornis melanura* (bellbird) and *Notiomystis cincta* (stitchbird) which are two of the three honeyeating birds that are the primary pollinators, leaving only *Prosthemadera novaseelandiae* (tui).

To assess the effects of loss of pollinators on mating, Schmidt-Adam *et al.* (2000) compared outcrossing rates, heterozygosity and fixation coefficients between three mainland populations and two populations on offshore islands that maintain all three bird pollinators. On Little Barrier Island (Figure 1) the full suite of pollinators has always been present, while on Tiritiri Matangi Island, *Anthornis melanura* and *Notiomystis cincta* had previously become extinct, and have only recently been reintroduced.

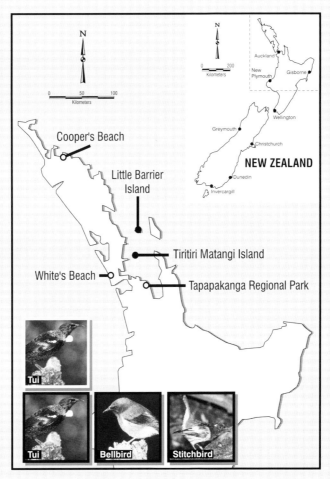

Figure 1   Location of *Metrosideros excelsa* island and mainland study populations and makeup of pollinator assemblages.

Table 1 shows that outcrossing rates of *Metrosideros excelsa* were uniformly low ($t_m = 0.22–0.53$) indicating that most seed set in all populations is the result of self-fertilisation. There was no obvious effect of the loss of *Anthornis melanura* and *Notiomystis cincta* in mainland populations on outcrossing rate. Similarly, progeny heterozygosity was uniformly low and fixation coefficients (F_{IS}) were high reflecting the predominance of self-fertilisation in all populations. Lower values for F_{IS} in adults suggest selection against selfed progeny occurs by reproductive maturity (7–8 years).

Table 1 **Multilocus outcrossing rates (t_m), heterozygosity (H_o) and fixation coefficients (F_{IS})**

Population	t_m (s.e.[A])	Progeny H_o	Progeny F_{IS}	Maternal F_{IS}
Islands				
Little Barrier	0.40 (0.08)	0.21	0.36	0.04
Tiritiri Matangi	0.41 (0.14)	0.15	0.46	0.16
Mainland				
Cooper's Beach	0.53 (0.09)	0.26	0.35	0.10
White's Beach	0.45 (0.08)	0.16	0.35	0.12
Tapapakanga	0.22 (0.09)	0.19	0.38	0.01

[A] Multilocus outcrossing rates (t_m) are based on five allozyme loci and 15 open-pollinated progeny from each of 15 maternal plants per population.

This result is quite surprising, showing what would be thought of as a major biological effect of fragmentation, loss of two pollinators, as having no apparent implications for patterns of mating. Current research is investigating how levels of cross-pollination are maintained in *Metrosideros excelsa* in the absence of *Anthornis melanura* and *Notiomystis cincta*. Initial results of pollinator exclusion experiments suggest that exotic birds may have increased pollination activity.

TABLE 10.2 **Mating system parameters for *Rutidosis leptorrhynchoides* (from Young & Brown 1999)**

Population	No. of flowering plants	Isolation distance (km)	Outcrossing rate (t_m) (s.e.)[A]	Correlation of outcrossed paternity (r_p)
Large				
Stirling Ridge	70 000	< 2	0.84 (0.06)	0.33 (0.08)
Queanbeyan	10 000	< 2	0.92 (0.06)	0.37 (0.08)
Medium				
Capital Circle	220	< 2	0.86 (0.06)	0.32 (0.09)
Barton	133	< 2	0.93 (0.06)	0.27 (0.06)
Captains Flat	161	> 5	0.92 (0.07)	0.53 (0.10)
St Albans	137	> 5	0.94 (0.06)	0.47 (0.07)
Small				
West Block	5	< 2	1.0 (0.01)	0.96 (0.054)

[A]Multilocus outcrossing rates (t_m) and correlation of outcrossed paternity (r_p) are based on eight allozyme loci.

10.6 Conclusions

Results so far show that forest fragmentation does affect population genetic processes and through these effects influences the amount and distribution of genetic diversity within forest species. However, these effects appear to be more varied than predicted from simple population genetics models. Certainly, fragmentation can erode allelic richness at the population level as fragments get smaller as shown in *Eucalyptus albens* and *Rutidosis leptorrhynchoides*. Given the relatively short time, and therefore small number of generations, since fragmentation for most species, these losses are more likely due to the initial formation of genetic bottlenecks rather than to subsequent random genetic drift. This means that present populations cannot be seen as being in mutation–drift balance, and current levels of genetic variation may provide little indication of a population's ability to retain variation in the long term.

Despite this, considerable among-fragment gene flow is possible, as shown indirectly by the work of Foré *et al.* (1992) and Young *et al.* (1993a) on *Acer saccharum* and Prober and Brown (1994) on *Eucalyptus albens*, and directly by Nason and Hamrick (1997) for *Spondias mombin* and Ballal *et al.* (1994) also for *Acer saccharum*. Given this, it seems possible that a group of fragment populations linked by gene flow, possibly through trees located in pasture between fragments, may sample and maintain a considerable proportion of the diversity within a species at the regional level in the long term. An issue regarding such a situation is the possible effect of a breakdown of local population genetic structure due to such gene flow. What are the possible effects of outbreeding depression? How will increased spatial scale and/or rates of gene flow affect the spatial and temporal dynamics of genetic interactions with parasites, pathogens and ecologically important symbionts such as mycorrhizal fungi or soil rhizobia? These questions are unanswered.

Given the potential importance of interfragment gene flow for determining effective population size, and through this affecting diversity within and divergence among populations, understanding related effects of fragmentation events on the abundance and behaviour of pollinators and seed dispersers is critical and this is a clear area for further research. The results from the study of *Metrosideros excelsa* already show that genetic effects can be difficult to predict even when changes to associated species are dramatic. The idea that different species may take greater or lesser roles in mating depending on fragmentation conditions means that effects may be subtle, involving changes in frequency and spatial scale rather than just differences in overall amount of cross-fertilisation. Such results highlight the value of empirical studies of real populations when examining the response of complex processes to disturbance.

Little can be said about the effects of fragmentation on plant fitness through increased inbreeding and reduced heterozygosity. The evidence from *Rutidosis leptorrhynchoides* that, for a given population size, increased isolation increases the correlation of paternity suggests a role for interpopulation gene flow in mitigating inbreeding. However, even in the smallest population of this species, where mating is essentially biparental, inbreeding has not as yet translated this into a significant increase in homozygosity. In terms of short-term population viability this is the crucial issue, and work examining interactions between inbreeding, heterozygosity, fitness and demography in remnant populations of forest species is a research priority.

Overall, the research conducted so far provides some initial insights into the genetic effects of habitat fragmentation. However, there is clearly a good deal more to learn. For example, changes in the biotic and abiotic environment that accompany forest fragmentation may lead to changes in local selection regimes. These changes may be direct effects of fragmentation, for example changes in air temperature, soil moisture and light due to the creation of new forest edges (Murcia 1995). They may also be indirect, if, for example, invasion of newly created forest edges by aggressive early successional species increases interspecific competition. Indeed some selection pressures may even moderate after fragmentation, as limiting resources such as light become more freely available in a habitat with a greater edge to area ratio. The effect of such changes on genetic diversity have so far not been considered.

11

EFFECTS OF ENVIRONMENTAL POLLUTION ON THE GENETICS OF FOREST TREES

Thomas Geburek

SUMMARY

Important aspects of environmental pollution on the genetics of forest trees are reviewed and include:

(a) a description of the variation in response to gaseous and metal pollutants among forest tree species and within-species variation on the family and clonal level

(b) heritability of plant response to gaseous pollutants

(c) a genetic assessment of polluted tree populations by means of gene markers

(d) altered gene expression triggered by environmental stress

(e) investigation of whether environmental pollution triggers evolutionary processes in forest tree populations.

In most forest tree populations (provenances), differences in resistance to gaseous and non-gaseous pollutants exist. Generally slow-growing and genetically highly variable tree populations seem to be less affected by different gaseous pollutants than fast-growing and genetically less variable tree populations. Intraspecific variation in resistance to pollutants increases from the provenance to the family level and is greatest among single trees (clones). Heritability estimates of resistance to gaseous air pollutants in forest trees are mainly high. Inheritance to pollution seems to be polygenic and in some cases is probably pleiotropic with respect to other stress factors. While genetic control of response to environmental stress (e.g. number of genes, gene dose, epistatic effects) is often known in herbaceous plants, knowledge in forest trees is still missing. Results of gene marker studies in polluted forest tree populations indicate that groups of trees (subpopulations, artificial subsets) with high genetic variation tend to be more tolerant to heterogeneous environmental stress than less variable ones. Environmental pollutants repress and/or induce plant genes.

Predictions of how forest tree populations will change genetically over time is complicated by the following facts: estimates of genetic variation derived from fumigation experiments cannot be generalised; response functions are unknown in detail; it is unclear how differences in tree vigour and foliar damage is related to fitness in natural tree populations; data on how pollution affects reproductive processes (pollen, ovule production) are scanty; it is still unknown how certain reproductive characteristics (e.g. serotiny) affect the evolutionary change to stress; intrinsically variable pollution over time and in space will cause episodic and space-dependent selection pressures; and data on 'net-fitness' regarding evolution of pollution resistance are missing. Nevertheless, there is circumstantial evidence that natural tree populations will change genetically because of pollution.

11.1 Introduction

Forest trees are subjected to many natural biotic and abiotic environmental factors, such as infections by pathogens, intraspecific and interspecific competition and physiochemical stresses such as cold, drought and salinity. Besides these natural factors, anthropogenic environmental pollution has become an important ecological issue. Environmental pollution is a multi-faceted phenomenon which places additional strains on forest ecosystems and has contributed to forest decline in Europe and other countries. While initially single causes such as acid rain, NO_x, ozone and pests have been blamed for the decline, today the destabilisation of forest ecosystems is seen as a result of several, partly interacting stress factors (Führer 1990).

From a genetic point of view environmental pollution has three important implications. First, evolutionary rates under environmental stress may differ from evolution under stress-free conditions. Phenotypic and genotypic variability tend to be high under severe physical and biological stress, especially for quantitative traits important in determining survival. Similarly, both recombination and mutation seem to be high. Experimental evidence suggests that a combination of high genetic variability of ecological, physiological and morphological phenotypes, high rates of mutation and recombination and some

protection from stress through shock proteins could collectively trigger accelerated and discontinuous evolutionary changes (Parsons 1987). Second, among various components of biodiversity the genetic patterns within species are important determinants of adaptive potential and thus have a basic significance for system resilience. Even accelerated evolutionary processes might not be sufficient to counteract severe induced genetic erosion or antibiotic interaction and consequently can destabilise forest ecosystems by reducing genetic adaptability (Gregorius 1995). Third, variations in silviculture, for instance a change in reforestation techniques or a shift towards more tolerant tree species, are being implemented to lessen detrimental effects. An altered management system, however, creates a 'new' environment. Probiotic interactions increase the adaptive potential of benefitting populations or, conversely, antibiotic ones decrease it.

Genetic variation in response to pollution is the pivotal point. In short-lived plants there is abundant evidence that pollution changes the genetic constitution of populations. This holds especially true when herbaceous or grassy plants are exposed to heavy metals (e.g. Bradshaw & McNeilly 1981; Roose et al. 1982; Macnair 1997). Since longevity is one of the main features of forest tree species, evolutionary effects are much more difficult to track down than in non-woody plants in which several generations can be studied easily. Therefore, this chapter portrays different levels of genetic variation of forest tree species in response to environmental pollution, mainly gaseous (i.e. NO_x, O_3, SO_2), summarises results of heritability studies and investigations based on gene markers, and discusses evidence of pollution-modified gene expression and evolutionary change in forest trees. As many different aspects are discussed, this chapter cannot claim to be comprehensive and many aspects remain untouched.

11.2 Genetic variation under pollution stress

The genetic constitution of tree populations will change if fitness of single trees, families or local populations (often called provenances) differ in a way that can be inherited. To assess whether and how environmental stress alters the genetics of forest trees, there must firstly be evidence of genetic variation and, in addition, that those genetic differences in sensitivity modify fitness. Genetic variation occurs at different levels: among different species, or within a single species among different populations, families and individuals (clones).

11.2.1 Interspecific variation to pollution

For among-species variation, differences are probably not the result of evolution in response to pollution since anthropogenic pollution has not been a factor over evolutionary time scales. Thus, differences in sensitivity are merely a result of differences in morphological and/or physiological traits that evolved during speciation.

Major differences among plant species in their response to pollution have been known for several decades. Kozlowski & Constantinidou (1986a) have throughly reviewed the response of woody plants to environmental pollution. In general, conifers are more susceptible to pollution than broadleaf species. However, sensitivity fluctuates greatly within each group and also different species within a genus vary considerably in their responses to certain pollutants (e.g. Winner & Mooney 1980a). Many lists ranking relative susceptibility of tree and shrub species are available, for instance concerning the response to sulfur dioxide (SO_2) (Hart 1973; Davis & Gerhold 1976; Garsed & Rutter 1982), ozone (O_3) (Davis & Gerhold 1976), fluoride (Weinstein 1977), aluminium (McCormick & Steiner 1978), heavy metals (Patterson & Olson 1983) or to combined pollution (Genys & Heggestad 1978).

11.2.2 Intraspecific variation to pollution

Many plant species including forest trees are intrinsically variable in their response to environmental pollution. Why does intraspecific genetic variation exist in forest tree populations? Mutation is the original source of variation and the rate of this evolutionary force can be altered erratically in forest tree species by environmental pollution (e.g. Kalchenko et al. 1993; Müller et al. 1994; Bakhtiyarova et al. 1995). However, this force is probably small because an exhaustive variation in tolerance to environmental stress usually already exists in forest trees. Certain genes, not those that are environmentally induced (mutated), which contribute to the genetic load of the population under non-stressed environments may become adaptive under stress. A hypothetical gene that is maladaptive through

decreasing vegetative growth may, in turn, reduce reproductive success. Under polluted conditions, however, this gene becomes highly adaptive because it avoids damage of leaves by early closure of the stomata in presence of gaseous pollutants.

Gene flow reduces genetic differentiation among populations and may allow both the spread of not only preadaptive genes but also genes that have not been selected by environmental stress and thus retards evolution. Overall, genetic effects from gene flow are likely to be small.

Effects by genetic drift are probably insignificant but can become important in severely damaged stands, when the effective population size is decreased to few survivors and they are subjected to stress.

Selection is the key factor that increases resistance to pollution in populations, depending mainly on the mode of inheritance and selection intensity. Anthropogenic pollution is a rather new environmental factor. Many genotypes (older trees) have had to cope with different selection regimes in both unpolluted and polluted environments causing high temporal environmental heterogeneity over a life span and thus slowing down evolutionary change. For younger trees, pollution in many areas is a 'normal' environmental factor. In addition, certain traits in forest trees exist that, for instance, cause drought tolerance in a non-polluted environment and are mutually beneficial under air pollution (Saxe & Murali 1989; Wellburn et al. 1996). Thus, even selection under non-polluted environments can maintain genetic variation important for gaining resistance under (later) polluted conditions. Studies in non-woody plants have shown that genes conferring tolerance to environmental stress have evolved independently within single species (Schat et al. 1996).

In forest trees abundant evidence for genetic variation in the response to pollutants comes from studies on intraspecific variation within provenances, families and clones.

11.2.2.1 Provenance level

Many studies that focus on the provenance level have been undertaken under differing field conditions (e.g. less versus heavily polluted areas) or under controlled environments in fumigation chambers, sand or solution cultures (Table 11.1). Certain field trials were established with the initial intention to assess provenance field performance (mainly growth traits)

TABLE 11.1 Intrapopulation (provenance) variation of forest tree species to environmental pollution

Al, aluminium; HF, hydrogen fluoride; O_3, ozone; SO_2, sulfur dioxide; NO_x, various nitrogen oxides

Material [No. and origin of provenances (N), No. of plants per provenance and treatment (P)]	Age (No. of years)	Pollutant(s)	Type of exposure and traits used for analysis	Remarks	References
				Abies alba Mill.	
12N; range-wide; 15P	3	SO_2	Fumigation; flushing, height, diameter, needle loss	Provenances from central Europe showed significant reduction both of height and diameter growth, provenances from eastern Europe only reduced height, and provenances from Calabria were not affected by SO_2-fumigation	Larsen & Friedrich (1988)
12N; range-wide provenances; 15P	?	SO_2	Fumigation; photosynthesis, transpiration	Provenances from north-eastern and central Europe were heavily affected, provenances from the south-eastern region and from Calabria were less damaged and had a pronounced regeneration ability after fumigation	Larsen *et al.* (1988)
6N; Italy (Apennines and Calabria); ?P	2	Biocide, tensioactive	Artificial fog: anatomical and biochemical characters	The most resistant provenance originated from Calabria	Pennacchini & Ducci (1991)
				Acer rubrum Lam.	
4N; range-wide; ?P	?	O_3	Fumigation; growth analysis	Seedlings from Alabama were the least susceptible and those originating from Pennsylvania the most susceptible to foliar damage	Townsend & Dochinger (1974)
				Betula alleghaniensis Britton	
8N; range-wide; 3–4P	—	Acidity	Pollen germination, LD_{50}	Very small variation among provenances in pollen germination response	Cox (1989)
				Betula papyrifera Marshall	
13N; range-wide; 18P	1	Al	Hydroponics; root growth, elements	Relative resistance (growth under stress/growth under control) varied from 10% to 37%, no clinal geographic pattern, more resistant provenances exhibited generally greater increases of foliar Al after exposure	Steiner *et al.* (1980)
8N; New Brunswick; 5–10P	—	Acidity	Pollen germination, LD_{50}	Very small variation among provenances in pollen germination response	Cox (1989)

Fraxinus americana L.

10N; range-wide, P10	1	O_3	Fumigation; foliar damage	Damage was not correlated to geographical seed origin	Steiner & Davis (1979)
10N; range-wide; 20P	2	O_3, SO_2	Fumigation; foliar damage	20% of variation in response to O_3 and 3% of variation in response to SO_2 were attributed to the provenance level; O_3 injury correlated with longitude of origin; resistant seed sources originated from near coastal locations (Maine, Mississippi, Conneticut), sensitive ones from central and western regions of the range, no geographical pattern was detected for SO_2 sensitivity	Karnosky & Steiner (1981)
16N; range-wide; ?P	2	O_3, SO_2	Fumigation; foliar damage	33% of variation in response to O_3 and 10% of variation in response to SO_2 were attributed to the provenance level; O_3 and SO_2 injury correlated with latitude of origin; tolerant seed sources originated from the Mississippi River Valley from central Illinois to Tennessee	Karnosky & Steiner (1981)

Picea abies (L.) H. Karst.

7N; Finland; P > 45 < 72	5–6	NO_x, SO_2	Field experiment; mortality, needle morphology, growth	Mortality varied from 3% to 94% among Finnish provenances	Huttunen (1978)
6N; Poland; ?P	4	SO_2	Fumigation; needle damage	Most resistant provenance originated from Puzcza Białowieska in east-central Poland, while most sensitive one was from the Polish Carpathians	Białobok et al. (1980b)
5N; Germany; ?P	1	Al	Sand culture; shoot and root growth	Reaction to Al exposure was similar among provenances	Makkonen-Spiecker (1985)
9N; Poland; ?P	13	Industrial emissions	5 field trials; mortality	Survival of provenance Istebna was poor	Rachwał & Oleksyn (1987)
15N; former CSSR, Romania; ?P	5–6	SO_2	Fumigation; needle damages, height	Variance of height growth before fumigation and visible damage after fumigation was mainly governed by genetic factors; increment in the year after fumigation poorly reflected genetic differences of pollution effects	Scholz & Venne (1989)

Material [No. and origin of provenances (N), No. of plants per provenance and treatment (P)]	Age (No. of years)	Pollutant(s)	Type of exposure and traits used for analysis	Remarks	References
18N; range-wide (Europe); 75P	1	Al	Hydroponics; needle damages, shoot and root growth	Relatively resistant provenances were derived from former CSSR, Poland, Harz Mt in Germany, while Romanian, Swedish and former USSR seed sources were more damaged. Al content of needle dry weight was not correlated to relative tolerance.	Geburek & Scholz (1989)
100N; range-wide; ?P	13	SO_2, HF	Field trials with differing pollution regimes; mortality growth	Relative resistance (growth under control site / growth under pollution) varied from 60% to nearly 90%; no details on geographical variation were reported	Knabe et al. (1990)
4N; Norway (?); 90P	1	O_3	Fumigation; dry matter (needles, stems, roots)	Concentration of 80–100 ppb over three months reduced total dry weight of two of the provenances, no visible damage was detected	Mortensen (1990)
6N; Belgium, former CSSR, Germany, Romania, Sweden; 75P	1	Al, SO_2, Al + SO_2	Fumigation + hydroponics; shoot and root growth	Plant response differed among provenances, different stress caused different ranking among provenances	Geburek & Scholz (1992)
64N; mainly from Hercyn-Sudetes region; 200P	15	Industrial emissions	Field experiment; mortality, health, height growth	Mortality of certain provenances from Beskides and the Bohemian Moravian Uplands was up to 50%, whereas only 9% mortality of the Harz Mt (German) provenance was found. The famous Polish provenance Istebna ranked among the worst seed sources.	Vančura (1993)
6N; Germany: Poland, Romania, provenance mixture; 300P	1	Heavy metals	Field experiment; mortality, isozymes	High mortality was found on toxic sites, variation of mortality among provenances varied from 58% to 92% and 49% to 59% at two polluted sites	Bergmann & Hosius (1996)
Pinus contorta Dougl.					
4N; Idaho, Oregon, Washington, Wyoming; 10P	1	SO_2	Fumigation; shoot and root weight	Coastal provenances (Washington and Oregon) were significantly more susceptible to SO_2 than continental ones (Idaho, Wyoming)	Lang et al. (1971)
Pinus nigra Arnold					
23N; Italy, Turkey, France, former Yugoslavia, Austria, Poland; ?P	11	Industrial pollution	Field experiments in pollution-free and heavily polluted area; mortality, height growth	Provenances from Austria and former Yugoslavia performed well under pollution compared to pollution-free sites	Rachwał & Oleksyn (1987)

Pinus strobus L.

Provenance	n	Pollutant	Experiment	Results	Reference
16N; range-wide; ?P	8–13	Industrial emissions	Field experiment	East-central provenances (Cent. W. Virginia, N. Ohio, E. Pennsylvania, N. New York) had high levels of tolerance, western provenances (E. Iowa, W. Lower Michigan, N. Minnesota, N. Wisconsin) were severely affected; significant correlation was detected between field sensitivity and estimated precipitation of seed source	Kriebel & Leben (1981)
9N; range-wide provenances; ?P	2	SO_2	Fumigation; needle damage	Most resistant provenance from Prince Edward Island (Canada), most sensitive ones were found from South Carolina	Genys & Heggestad (1983)
4N; New Brunswick, Ontario; 3–8P	—	Acidity	Fumigation; pollen germination, LD_{50}	Small, but significant differences among provenances were reported	Cox (1989)

Pinus sylvestris L.

Provenance	n	Pollutant	Experiment	Results	Reference
13N; Finland; provenances; 99P	5–6	NO_x, SO_2	Field experiment; mortality, needle morphology, growth	Provenances from more northern and eastern areas thrived best; mortality varied from 49% to 100%; damage was positively correlated to thickness of epidermal cells	Huttunen (1978)
3N; Finland;15P	4	NO_2, SO_2	Three field trials with low, medium, heavy pollution; needle damage, peroxidase activity	Most northern provenance was most resistant; most southern provenance was most sensitive; resistance was correlated to winter tolerance	Huttunen & Törmälehto (1982)
18N; range from 63°50' north (Arkhangelsk) to 40°30' north, (Turkey) and 63°00' east (Poland) to 21°08' east (Central Siberia); ?P	74	Primarily NO_x	Field experiment; mortality, growth	Least damage in provenances originating from Kiev, Jenisiejsk and Arkhangelsk was reported; severe damages in provenances from Tobolsk (Ural) and Kars (Turkey) were found; mortality over five years varied from 23% to 57%	Oleksyn (1987)
19N; Europe; 80P	5	SO_2, fluorides	Field experiment; mortality, height growth, needle damage	Least productive provenances from northern part of the range (Sweden, former USSR) survived best; mortality varied from 25% to 95%; damage was positively correlated with height growth	Oleksyn (1988)

Material [No. and origin of provenances (N), No. of plants per provenance and treatment (P)]	Age (No. of years)	Pollutant (s)	Type of exposure and traits used for analysis	Remarks	References
20N; Europe; 3P	1–3	1. SO_2, 2. NO_2, 3. $SO_2 + NO_2$, 4. HF, 5. SO_2, fluorides, 6. SO_2, heavy metals	1.-4. fumigation, 5.-6. field tests in polluted area; comparison between field and controlled experiments; needle damage	Northern (Sweden, former USSR) and southern (Turkey, former Yugoslavia) provenances had low damage compared with trees from central regions in Europe; plant reactions under field test conditions (SO_2 and fluorides) correlated to reaction under SO_2 and HF fumigation, and reactions under field test conditions (SO_2 and heavy metals) correlated to SO_2 under controlled conditions	Oleksyn et al. (1988)
18N; Europe; 14–23 P (SO_2), 3P (NO_2, $SO_2 + NO_2$), 4P (HF)	1–2	1. SO_2, 2. NO_2, 3. $SO_2 + NO_2$, 4. HF	Fumigation; analysis of needle prolin content	Gaseous stress induced higher prolin content; variation of prolin content among provenances before stress was not correlated with resistance	Karolewski (1989)
20N; Europe; 80P	9	SO_2, fluorides	Field experiment; height growth, flowering	Slow-growing provenances from southern (40° north to 45° north) and northern (58° north to 61° north) range were lesser damaged (pollution induced growth reductions) than those from central part of the range. Flowering of resistant provenances was more affected by pollution than in sensitive ones.	Oleksyn et al. (1992)
15N; range from 63°50' north (Arkhangelsk) to 40°30' north (Turkey) and 63°00' east (Poland) to 21°08' east (Central Siberia); 20P	78	Primarily NO_x	Field experiment; growth during unpolluted and polluted periods by tree ring analyis.	Pollution induced growth reductions in provenance Kharkov of 6% and in provenance Kurland of 16%, while most sensitive provenances originating from Kars and Akhangelsk had both a 60% growth reduction	Oleksyn et al. (1993)
8N; Europe, north-to-south latitudinal range; 10P	—	SO_2, fluorides	Field experiment; photosynthesis, needle element content	No latitudinal trend in relation to photoynthesis or needle element content was reported.	Reich et al. (1994)

Populus tremuloides Michx.

Material [No. and origin of provenances (N), No. of plants per provenance and treatment (P)]	Age (No. of years)	Pollutant (s)	Type of exposure and traits used for analysis	Remarks	References
15N; range-wide; 7–15P	1	O_3	Fumigation	Most resistant provenances were found from the West Coast and the north-east and industrialised proportion of the Great Lakes; significant negative correlation was detected between the amount of injury and the maximum daily ozone average at the localities where the plant samples were taken	Berrang et al. (1991)

and were later re-evaluated in view of pollution effects (e.g. Kriebel & Leben 1981). Controlled experiments, however, generally have been undertaken on young plant material normally exposed to high toxic doses of pollutants over short times. Hence, different methodological reasons complicate a synoptic review of the results. Most studies were done in Europe.

In *Abies alba* (silver fir) controlled field experiments indicate that Calabrian provenances are more tolerant to different pollution regimes than seed sources from other regions of the native range. Decline in indigenous *Abies alba* populations is detectable over huge areas in central Europe, while firs in south-eastern Europe and Calabria (southern Italy) still thrive under polluted conditions. Larsen (1986) and Bergmann *et al.* (1990) related genetic variation to decline. Calabrian and south-eastern *Abies alba* provenances are genetically highly variable and probably can better cope with environmental stress. Estimates of how much of the variation as a result of environmental stress is governed by provenance effects are still missing.

For *Picea abies* (Norway spruce) it is much more difficult to draw a conclusive picture for two reasons. First, in this conifer there has been significant transfer of seed material in European forestry since the 19th century. Many populations and thus, in due course, material used in field trials and controlled experiments are not necessarily autochthonous. Second, the range is much more extensive than, for instance, *Abies alba*, and studies that sufficiently represent autochthonous populations over a wide range are lacking. Nevertheless, about 10% of the variance in 'needle damage' is due to provenance effects, as expected much less than the variation in families or even siblings (Scholz & Venne 1989). However, these estimates are only indicative at the species level since provenances of the Hercyn-Carpathian range only were studied. In general, it appears that slow-growing provenances that avoid much uptake of pollutants are more tolerant of pollution than rapidly growing ones (Saxe & Murali 1989). An example is the famous Polish seed source Istebna. This provenance has become one of the most universal provenances in different trials in Europe and the USA (Giertych 1984). In polluted areas, however, it normally ranks poorly (Vančura 1993). An exception to the simplified rule that slow growers are more tolerant are Alpine spruce provenances. These

seed sources are normally slow growing when planted outside their range, but are heavily damaged when exposed to environmental pollution (Geburek & Scholz 1989; Vančura 1993). Alpine provenances are genetically less variable compared with spruces from other regions of the native range, while Baltic and south-eastern populations are characterised by high genetic variation (Lagercrantz & Ryman 1990). This may explain the comparably low tolerance of Alpine spruce populations.

In *Pinus sylvestris* (Scots pine) the general trend exists that slow growing (Oleksyn 1988) but genetically highly variable provenances (Oleksyn *et al.* 1994) are less sensitive to air pollution. Mainly northern and eastern provenances thrive best under environmental pollution. Presumably traits such as xeromorphy of needles or thickness of epidermis, slow growth and a diverse gene pool of a population confer tolerance under gaseous pollution. Since mortality of *Pinus sylvestris* in heavily polluted areas varied from 20% to 95% among provenances (Oleksyn 1987, 1988), ongoing pollution will doubtless change the genetic composition of this species.

Compared with the exhaustive collection of data in *Pinus sylvestris*, provenance data of other pine species in response to environmental pollution are small. In *Pinus strobus* (eastern white pine), provenances originating from pollution-free sources suffered more under pollution than those that were long exposed to pollution in the eastern and central States of the USA (Kriebel & Leben 1981). *Pinus nigra* seed sources from Austria and former Yugoslavia performed well compared with French, Turkish or Polish provenances (Rachwał & Oleksyn 1987).

Among broadleaf species only few data are available. In *Fraxinus americana*, 20% of the variation in the response to O_3 was attributed to the provenance level. Ozone injury was significantly correlated with the longitude of seed origin and near east coastal provenances were tolerant whereas provenances from central or western regions of the native range in North America were more sensitive. Unexpectedly only 3% of the response variation was explained by provenance differences to SO_2 and no geographical pattern was discernible (Karnosky & Steiner 1981). In *Fraxinus pennsylvanica* Marshall, 33% and 10% of the variation to O_3 and SO_2, respectively, were attributed to geographical origin and for both pollutants damage was correlated with latitude (Karnosky & Steiner 1981).

TABLE 11.2 Clonal and family variation in resistance based on visible foliar damage to gaseous pollutants in forest trees

Many other fumigation studies in forest trees not listed in this table have employed family or clonal material (e.g. Kainulainen et al. 1995). Only results derived from studies based on at least 10 clones or families are presented.

Tree species	Concentration[A] ($\mu g/m^3$)	Time of exposure (hours)	Material[B]	Results (%)[C]	References
Hydrogen fluoride (HF)					
Larix decidua Mill.	817	5	20 C	5–70 F*	Karolewski & Białobok (1979)
Pinus sylvestris	408	12	17 C	1–37 F*	Karolewski & Białobok (1978)
	408	12	17 F	20–55 F*	Białobok et al. (1980a)
Nitrogen dioxide (NO_2)					
Pinus sylvestris	5375	60	17 C	1–18 F*	Białobok et al. (1980a)
	5375	60	17 F	0–5 F*	Białobok et al. (1980a)
Ozone (O_3)					
Fraxinus americana	980	7.5	50 F	0–200 I**	Karnosky & Steiner (1981)
Fraxinus pennsylvanica	490	6	10 F	2–33 F	Steiner & Davis (1978)
	980	7.5	59 F	0–230 I**	Karnosky & Steiner (1981)
Larix decidua	1960	25	20 C	0–5 F*	Karolewski & Białobok (1979)
Pinus sylvestris	1960	66	17 F	2–38 F*	Białobok et al. (1980a)
Pinus taeda L.	490	8	18 F	2–49 F*	Kress et al. (1982)
Populus tremuloides	392	3	11 C	7–56 F*	Karnosky (1977)
Ozone and sulfur dioxide ($O_3 + SO_2$)					
Larix decidua	980/5220	15	20 C	7–65 F*	Karolewski & Białobok (1979)
Pinus sylvestris	980/1300	13	17 C	0–5 F*	Białobok et al. (1980a)
	980/1300	36	17 F	1–14 F*	Białobok et al. (1980a)
Populus tremuloides	98/914	3	10 C	0–20 F	Karnosky (1977)

			Sulfur dioxide (SO$_2$)		
Fraxinus americana	7.5	1305	50 F	13–213 I*	Karnosky & Steiner (1981)
Fraxinus pennsylvanica	7.5	1305	55 F	1–116 I**	Karnosky & Steiner (1981)
Larix decidua	5	10440	21 F	3–67 F*	Karolewski & Białobok (1979)
Pinus sylvestris	24	2610	17 C	0–18 F*	Białobok et al. (1980a)
	24	2610	17 F	0–41 F*	Białobok et al. (1980a)
Populus tremuloides	3	1305	10 C	1–46 F*	Karnosky (1977)

[A] Conversion factors used: ppm O$_3$ x 1960 = mg/m^3; ppm SO$_2$ x 2610 = mg/m^3 (25°C); ppm HF x 817 = mg/m^3; ppm NO$_2$ x 10750 = mg/m^3 (20°C). Where 2 figures are given, the first refers to O$_3$, and the second to SO$_2$.
[B] Material: C, clone (genotype); F, open-pollinated families.
[C] Results: F, percentage of foliar damage; R, injury rating on scale from 0 to 90% foliar damage; I, injury index (% leaves injured in plot) x (average degree of severity per leaf), degree of severity followed a scale from 0 to 5 (i.e. 0, no injury; 1, 1–20%, 2, 21–40% etc.); significance level, *$P < 0.05$, **$P > 0.01$.

In *Populus tremuloides* (aspen) a significant correlation was found between foliar damage and O_3 impact at the location of the source population (Berrang *et al.* 1991). When provenances of *Betula papyrifera* (Steiner *et al.* 1980) and *Fraxinus pennsylvanica* (Steiner & Davis 1979) were exposed to aluminium in hydroponics, no geographical pattern was found.

11.2.2.3   Family and clonal level

Differences among families derived from controlled and open pollination, and among different clones provide evidence for a genetic basis to resistance to air pollutants and metal exposure (Table 11.2). Reports of a range of foliar damage from nearly 0% to about 50% have been published, for instance in family studies on the variation to O_3 in the genus *Fraxinus* (Karnosky & Steiner 1981), and *Pinus taeda* (Kress *et al.* 1982), to SO_2 in *Larix decidua* (Karolewski & Białobok 1979), to hydrogen fluoride in *Larix decidua* (Karolewski & Białobok 1979) and *Pinus sylvestris* (Białobok *et al.* 1980a) and to a combination of O_3 and SO_2 in *Larix decidua* (Karolewski & Białobok 1979). Conversely, several studies have reported no statistically significant pattern in plant response. For example, no difference in visible needle injuries was found in *Pinus taeda* in O_3-fumigated open-pollinated families (Wiselogel *et al.* 1991) or full siblings (Weir 1977). In many field and laboratory studies SO_2 needle damage is negatively correlated with growth (e.g. *Pinus sylvestris*, Oleksyn 1988), whereas visible foliar damage caused by O_3 is often not (e.g. *Pinus taeda*, Winner *et al.* 1987). In *Picea abies*, the variance components for SO_2-damage varied from 10% to 23% for families (Scholz & Venne 1989).

Variation among different trees in resistance to environmental stress is even higher among clones than in families as evidenced by high variation in clonal studies (Table 11.2). While certain genotypes are resistant to certain pollutants, others suffer severely. In *Pinus sylvestris*, for instance, pollen tube growth under O_3 pollution [340 $\mu g/m^3$] for 48 hours resulted in a 35% reduction in resistant individuals, while pollen tube growth of sensitive ones was ceased completely (Venne *et al.* 1989).

11.3 Heritability studies

Whereas studies discussed in Chapter 11 on the provenance, family or clonal level already provide evidence that resistance is genetically controlled, the question is how much phenotypic variation is genetic and how much is environmental. This leads us to heritability studies, which estimate the genetic component of tolerance to pollutants. If the genetic component of the variance in resistance is high, evolutionary response to stress adaptation will be rapid. If the environmental component is high, mainly environmental effects of resistance would slow down a genetic change of tree populations. Resistance can be hereditary in the sense of being determined by the genotype (clone) or being transmitted from the parental trees to their progenies. These two different ways are reflected by heritability in a broad and a narrow sense, respectively.

Broad sense heritability (H^2) is defined by:

$$H^2 = \frac{V_G}{V_P} \tag{11.1}$$

where V_G is genotypic variance and V_P is phenotypic variance. Narrow sense heritability (h^2) expresses the extent to which phenotypes are determined by transmitted genes and is defined by:

$$h^2 = \frac{V_A}{V_P} \tag{11.2}$$

where V_A is additive genetic variance. When in clonal studies the environmental variance, instead of genetic variance, is partitioned, the parameter 'clonal repeatability' is often used. Although not a genetic parameter in a strict sense, clonal repeatability neither equals H^2 nor h^2, it sets the upper limits of both heritability estimates (see Falconer 1981, pp. 124–33). All three parameters are greatly influenced by the success of the investigators to reduce environmental variance and, hence, estimates obtained apply only to the specific experimental conditions (e.g duration and toxicity of fumigation or exposure under field conditions, plant material, traits measured). In this context highly significant results are from one of the earliest heritability studies regarding the response to pollution. Weir (1977) was the first to investigate genetic variation in O_3 response within open-pollinated and control-pollinated *Pinus taeda* families. While narrow sense heritabilities (h^2) were estimated in 40 open-pollinated families to be 0.27 (North Coast region) and 0.13 (South Coast region), less than 0.09% of the variation was attributed to additive genetic effects in 32 full sibs and the author provided evidence for non-nuclear inheritance to O_3 resistance.

TABLE 11.3 Broad sense heritability (H^2), narrow sense heritability (h^2) and clonal repeatability (R) for response to different gaseous air pollutants in forest tree species

Species	Pollutant(s)	Parameter	Trait	Estimate	Reference
Picea abies	Unspecific	h^2	Buffering capacity	0.78	Scholz & Reck (1977)
	HF	H^2	Damage (%) in current-year needles	0.60	Scholz et al. (1979)
	HF	h^2	Damage (%) in current-year needles	0.34	Scholz et al. (1979)
Pinus strobus	$SO_2 + O_3$	R	Damage (%) in current-year needles	0.82	Houston & Stairs (1973)
	$SO_2 + O_3$	R	Scoring of needle necrosis on a 6-value scale	0.82	Houston & Stairs (1973)
Pinus sylvestris	SO_2	H^2	Scoring of needle necrosis on a 9-value scale	0.60	Tzschacksch (1982)
		h^2	Scoring of needle necrosis on a 9-value scale	0.60	Tzschacksch (1982)
Pinus taeda	O_3	h^2	Visible needle damage	0.13–0.27 (half sibs) $\leq$ 0.09 (full sibs)	Weir (1977)
	O_3	h^2	Visible needle damage	0.43	Ward (1980, cited after Taylor (1994))
Populus tremuloides	SO_2	R	Counts of leaves damaged per plant	0.64	Karnosky (1977)
	O_3	R	Counts of leaves damaged per plant	0.62	Karnosky (1977)
	$SO_2 + O_3$	R	Counts of leaves damaged per plant	0.46	Karnosky (1977)
Pseudotsuga menziesii	SO_2	H^2	Scoring of needle necrosis on a 9-value scale	0.79	Tzschacksch (1982)
(Mirb.) Franco	SO_2	h^2	Scoring of needle necrosis on a 9-value scale	0.66	Tzschacksch (1982)

Most estimates of heritabilities and repeatability are high (Table 11.3). However, it is difficult to question a profound genetic component in tree response to pollution. Thus, in view of an evolutionary change, high estimates of these parameters for resistance make genetic changes due to selection in natural populations very likely. Estimates of additive gene effects are the most useful in this context.

Pleiotropic effects such as the effects on carbon and nitrogen metabolism, resistance to drought or pests, and stomatal control are important in understanding inheritance of pollution resistance. There is ample evidence that resistance to pollution is pleiotropic. For example, O_3 resistance in *Pinus taeda* is related to cold hardiness (Chappelka *et al.* 1990), carbon (Wiselogel *et al.* 1991) and nitrogen metabolism (Manderscheid *et al.* 1992).

The genetic control of response to environmental stress (e.g. number of genes, gene dose, epistatic effects) is often known in detail in herbaceous plants (Macnair 1993, 1997) whereas knowledge in forest trees is still missing. For instance, data on how the degree of foliar damage in heritability studies is related to fitness are not available. Predictions regarding how fast and to what extent populations will change genetically based on the results of heritability studies must remain speculative.

11.4 Gene marker studies

Especially in Europe gene markers have been used to assess mainly populational genetic effects of pollutants in forest tree species, such as *Abies alba*, *Fagus sylvatica* L. (European beech), *Picea abies* and *Pinus sylvestris*. Studies include comparisons between:

(a) different ontogenetic stages, for instance embryonic and seedling pools at polluted sites to assess viability selection (Prus-Głowacki & Godzik 1991)

(b) relatively tolerant and sensitive groups under controlled and specified stress (Scholz & Bergmann 1984)

(c) relatively tolerant and sensitive groups under long-term influence of unknown environmental stress (Müller-Starck 1989) (Table 11.4).

Although early studies suffer from small sample sizes and were rightly criticised (Barrett & Bush 1991), many results are now available that indicate that

pollutants change the genetic structure in forest tree species. These changes are indicated by shifts in genotype and/or gene frequencies or by different heterozygosities, genetic diversities or genetic multiplicities between resistant and sensitive groups. In certain cases, viability selection can be detected at specific gene loci (e.g. Müller-Starck & Ziehe 1991) (Fig. 11.1). Tolerant groups of trees usually are genetically more variable at marker genes than more sensitive ones (e.g. Raddi *et al.* 1994).

Viability selection by environmental stress in forest trees can be studied best by exposing plant material to stress and by comparing genetic structures before and after the size of the population has been reduced. This approach is restricted, however, to young plant material and selection is probably different during later ontogenetic stages. Such experiments have been done in *Fagus sylvatica*, *Picea abies* and *Pinus sylvestris*. In *Fagus sylvatica*, germinating acorns and two-year-old plants were compared genetically at five environmentally stressed sites and observed heterozygosity was 22.9% (acorns) and 24.8% (survivors) (Müller-Starck & Ziehe 1991). That the increase of heterozygotes in *Fagus sylvatica* does not simply reflect a pruning effect of inbreds is demonstrated by a reduced number of alleles per locus (3.04 versus 2.56). Contrary results were reported in *Pinus sylvestris* at a site polluted by a zinc smelter and a control site in Poland where observed heterozygosity of surviving seedlings on the control site was greater than under zinc [24% (embryos) versus 22% (seedlings, control) versus 18% (seedling, zinc stressed)] (Prus-Głowacki & Godzik 1991). However, significance of these estimates of both studies have not been tested by bootstrapping or other appropriate methods. Bergmann and Hosius (1996) analysed *Picea abies* seedlings that originated from six provenances before and after the exposure to heavy metal stress. Significant differences were detected at certain isozyme loci [Lap-B, Pepca-A, 6-phosphate-gluconate-dehydrogenase (6Pgd-B) and 6Pgd-C]. In surviving plants observed heterozygosity was greater than in the embryonic pool (Pepca-A: 18% versus 28%; 6Pgd-B: 35% versus 40%; 6Pgd-C: 42% versus 52%).

Although selection effects could be determined in the above studies, there are long-term experiments that compare groups of trees that have visible damage (e.g. foliar discolouration, foliar loss, crown structure) with unaffected trees. Selective response and thus evolutionary effects between these groups must remain

TABLE 11.4 Gene marker studies in polluted forest tree populations

HF, hydrogen fluoride; NH$_3$, ammonia; NO$_x$, nitrogen oxides; O$_3$, ozone; SO$_2$, sulfur dioxide

Tree species, stress, material & methods	Significant differences in genetic structure (resistant v. sensitive trees)[A]	Heterozygosity (expected = H_e, observed = H_o; conditional = H_c)	Genetic multiplicity or diversity (resistant v. sensitive trees)	Remarks	References
Abies alba; unknown, different factors likely, 27 permanent observation plots; comparison resistant v. sensitive trees	Aap-A, Got-C, Idh-A, Idh-B, Mdh-B, 6Pgd-A	No difference between resistant and sensitive trees	No difference between resistant and sensitive trees	Pollution induced selection against Got-C$_3$	Konnert (1992)
Acer saccharum; unknown different factors likely; 6 stands; comparison resistant v. sensitive trees	Not reported	No difference	Gametic multilocus diversity: 10% shortage of diversity in resistant subsets	No clear pattern of whether more genetic variation is haboured in resistant trees was found	Geburek & Knowles (1992)
Fagus sylvatica; unknown, different factors likely; 1 adult stand; comparison resistant v. sensitive trees	Acp-B, Got-B, Mdh-B, Skd-A	Increased H_o in resistant plants (31% v. 27%)	Genic multilocus diversity:553 v. 233	Greater variation in resistant group	Müller-Starck (1985)
Fagus sylvatica; unknown, different factors likely; 6 adult stands; comparison resistant v. sensitive trees	Aco-A, Aco-B, Got-B, Mdh-B, Mdh-C, 6Pgd-A, Pgm-A, Skd-A	Increased H_o in resistant plants 28% v. 22%	Gametic multilocus diversity: 90% excess of diversity in resistant subsets; number of alleles 2.64 v. 2.48	Greater variation in resistant groups, averaged over stands, 10.5 private alleles were found in the tolerant, while 5.1 private alleles were found in the sensitive subset	Müller-Starck (1989)
Fagus sylvatica; unknown, different factors likely; 1 adult stand; comparison resistant v. sensitive trees	Amyl, Lap, Mdh, 6Pgd	Increased H_o in resistant plants 16% v. 11%	Gametic multilocus diversity: 161% excess of diversity in resistant subsets	Greater variation was reported in resistant group	Hertel & Zander (1991)
Fagus sylvatica; unknown, different factors likely; comparison seeds v. 2-year-old survivors at 4 sites	Not reported	Increased heterozygosities in survivors v. seeds— H_o: 25% v. 23%; H_c: 62% v. 57%	Survivors v. seeds, no. of alleles per locus 3.04 v. 2.56	Drastic increase of Lap-A$_4$ in survivors at all sites	Müller-Starck & Ziehe (1991)
Picea abies; unknown, different factors likely; 2 adult stands; comparison resistant v. sensitive trees	Grd-A	Increased in resistant plants 31% v. 25%	Gene pool diversity: 1.31 v. 1.34; no. of alleles: 2.33 v. 2.22	Greater variation in resistant groups	Bergmann & Scholz (1987)

Tree species, stress, material & methods	Significant differences in genetic structure (resistant v. sensitive trees)[A]	Heterozygosity (expected = H_e; observed = H_o; conditional = H_C)	Genetic multiplicity or diversity (resistant v. sensitive trees)	Remarks	References
Picea abies; unknown, different factors likely; 4 adult stands; fumigation of clones with O_3, SO_2, HF; comparison resistant v. sensitive trees	Got-B, Grd-A, Pepca-A	Increased H_o in resistant plants 31% v. 27%	Not reported	While the resistant groups of 4 stands exhibited higher variation than sensitive trees, there was not such a trend in the fumigated material	Bergmann & Scholz (1989)
Picea abies; unknown, different factors likely; 24 permanent observation plots; comparison resistant v. sensitive trees	Got-C, NADdh-B, 6Pgd-C, Pgi-A	No difference between resistant and sensitive trees	No difference between resistant and sensitive trees	$6Pgd$-C_0 and $6Pgd$-C_2, and Idh-B_3 were private alleles in tolerant trees, while Skd-A_4 was exclusively found in sensitive trees	Löchelt (1994)
Picea abies; unknown; 3 regions; comparison resistant v. sensitive trees; logistic regression analysis	Not reported	Increased hereozygosities in resistant trees— H_o:14% v. 12% H_C: 86% v. 80%	No. of alleles per locus: 2.4 v. 2.3	Four marker loci (Lap-B, Pgm-A, Pgi-B, Mnr-B) contributed significantly to a logistic regression model, which predicted probability of trees belonging to the resistant or sensitive group	Raddi *et al.* (1994)
Picea abies; unknown, mainly SO_2 and NO_x; 1 stand, comparison resistant v. sensitive trees	Fest, G6Pd, Got-B	Increased H_o in resistant population 26% v. 19%	No differences in no. of alleles per locus	Genotypic polymorphism index in polluted population was slightly higher than in reference populations	Prus-Głowacki & Godzik (1995)
Picea abies; zinc, lead; comparions of embryonic pool with surviving seedling at 2 sites	Lap-B, Pepca-A, 6Pgd-B, 6Pgd-C	Pooled data not reported	Pooled data not reported	Tendency of increased number of heterozygotes in survivors and a pronounced shift towards Pepca-A_1A_2 in survivors were reported	Bergmann & Hosius (1996)
Pinus sylvestris; pollution (SO_2, fluorides) from phoshorus fertiliser plant; 1 adult stand; comparison resistant v. sensitive trees	Acp-B	Increased H_e in resistant plants at Acp-B (83% v. 79%)	Not reported	Decreasing frequency of Acp-B_6 due to emission	Mejnartowicz (1983)
Pinus sylvestris; unknown, different factors likely; field test with family material comparison resistant v. sensitive trees	Gdh, Got-B	Increased H_o in resistant plants 41% v. 31%	Genic multi-locus diversity: 124 v. 52	Greater variation in resistant group	Geburek *et al.* (1987)

Pinus sylvestris; 1 population influenced by heavy metals, 1 unpolluted population; comparisons of different ontogenetic stages	Got-A, Est, Skd-A	In both populations H_O increased from embryonic, seedling, young to adult trees; no pronounced difference between polluted and unpolluted population	Not reported	With age increasing genotypic polymorphism index was detected in polluted population, increase in unpolluted population was less pronounced	Prus-Glowacki & Nowak-Bzowy (1989)
Pinus sylvestris; 1 seedling population predominantly influenced by lead and zinc, 1 unpolluted population; comparisons of different ontogenetic stages (embryos v. 1-year-old survivors)	Mdh-A, Skd-A	Decreased H_O from embryonic to seedling pool at both sites; decrease more pronounced under pollution	Not reported	Fixation index increased from embryonic to seedling population at both sites (more pronounced under pollution)	Prus-Glowacki & Godzik (1991)
Pinus sylvestris; Pb, Zn, SO_2, NO_{xx}; comparison of natural regenerated trees with natural regeneration in 9 (unpolluted) reference populations; comparison resistant v. sensitive trees	Gdh, Skd-B	Decreased H_O in resistant population 24% v. 29%, no differences in H_e	No. of alleles: 2.6 v. 2.4	Genotypic polymorphism index in polluted population was slightly higher than in reference populations; fixation index in polluted populations was higher than in unpolluted ones	Prus-Glowacki & Nowak-Bzowy (1992)
Pinus sylvestris; toxic dust, SO_2, NH_3, NO_x; provenance trials; correlation analysis between growth decline and genetic polymorphism	Not applicable	Not applicable	Not applicable	Significant relations between several genetic indices and pollution-induced growth decline existed	Oleksyn et al. (1994)

[A] Aap, alanine-aminopeptidase; Aco, aconitase; Acp, acid-phosphatase; Amyl, amylase; Est, esterase; Fest, (fluorescent) esterase; Gdh, glutamate-dehydrogenase; Got, glutamate-oxalacetate-transaminase; G6Pd, glucose-6-phosphate-dehydrogenase; Grd, glutathione-reductase; Idh, isocitrate-dehydrogenase; Lap, leucine-aminopeptidase; Mdh, malate-dehydrogenase; NADdh, NADH-dehydrogenase; Pepca, phosphoenolpyruvate-carboxylase; Pgm, phosphoglucomutase; 6Pgd, 6-phoshate-gluconate-dehydrogenase; Pgi, phosphoglucose isomerase; Skd, shikimate-dehydrogenase.

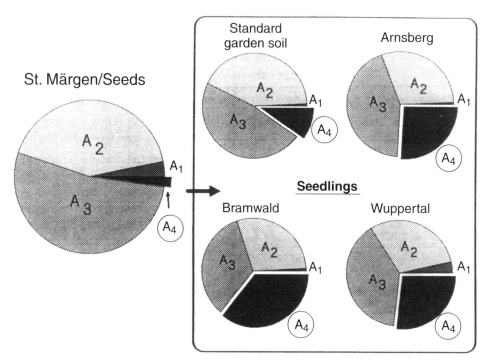

Figure 11.1 Environment-induced viability selection in *Fagus sylvatica* populations. Frequencies of the allozymes Lap-A1, A2, A3 and A4 in the embryonic pool of the provenance 'St Märgen' (Germany) and in four two-year-old seedling populations in standard garden soil (stress-free environment) and three different natural stress environments (pollutants unknown in detail) originated from this source (from Müller-Starck & Ziehe 1991, © Sauerländer's Verlag, reproduced with kind permission).

speculative. However, it is intriguing that groups resistant to varied environmental stress tend to be more genetically diverse than sensitive ones. This is reported for different tree species, such as *Picea abies*, *Pinus sylvestris* and *Fagus sylvatica* (Müller-Starck 1985, 1989; Bergmann & Scholz 1987; Geburek *et al.* 1987; Oleksyn *et al.* 1994; Raddi *et al.* 1994; Prus-Głowacki & Godzik 1995). This is also observed by DeHayes & Hawley (1992) when studying the fitness of stressed *Picea rubens* Sarg. populations. However, genetic diversity did not differ between sensitive and tolerant trees exposed to varied air pollutants in certain field studies (Geburek & Knowles 1992; Konnert 1992; Löchelt 1994) and, expectedly, in studies that focus on a single stress, no pronounced difference in genetic diversity between sensitive and tolerant trees was reported (SO$_2$: Scholz & Bergmann 1984; zinc smelter: Prus-Głowacki & Nowak-Bzowy 1989, 1992).

The question that arises is 'What mechanistic explanation can be given for associations between allozyme heterozygosity and fitness-related traits?'

Mitton and Grant (1984) and Bush and Smouse (1992) have discussed the heterozygosity–fitness phenomenon, which besides forest trees is also well documented in marine organisms (see e.g. Nevo *et al.* 1983). This phenomenon could be due to homozygosity for deleterious alleles, to differing degrees of heterozygote advantage or to a combination of both. The first answer to the question, the dominance hypothesis, holds that allozymes at polymorphic loci are selectively neutral and that allozyme heterozygosity is reflecting genomic heterozygosity. Whether allozyme variation can reflect genomic variation sufficiently is questionable as pointed out by Mitton and Pierce (1980) and Chakraborty (1981). When sampled specimens are not unrelated, however, allozyme variation can indicate genomic heterozygosity (Smouse 1986). The second answer, the overdominance hypothesis, presupposes that allozymes themselves influence fitness or closely linked genes are adaptive. This explanation is supported by results from viability experiments (e.g. Müller-Starck & Ziehe 1991) and field comparisons of tolerant and sensitive trees (e.g. Mopper *et al.* 1991) indicating that selection is triggered by environmental stress on single (hitchhiking) genes. Nothing is known about the physiology, however.

11.5 Alteration of gene expression in response to environmental pollution

Response mechanisms to stress also have metabolic components, such as changes in proteins, enzymes, lipids or different acids (e.g. Kozlowski & Constantinidou 1986b; Newton *et al.* 1991; Wellburn *et al.* 1996). However, so far the influence of pollutants on the structure and function of plant genes have been studied only scarcely. In *Nicotiana tabacum* L. (tobacco) for instance, gaseous pollutants repress photosynthesis genes, induce pathogen defence genes and nuclear anti-oxidative defence genes are transcribed at higher rates (Bahl *et al.* 1993). In *Picea abies* seedlings, the enzyme short-chain alcohol dehydrogenase (scADH) may protect against abiotic stress. In O_3-fumigated seedlings the expression of a novel gene is induced that belongs to the broad short-chain alcohol dehydrogenase gene family (Bauer *et al.* 1993). Cinnamyl alcohol dehydrogenase (CAD) is an enzyme involved in lignin biosynthesis and also in abiotic plant defence. Hence, CAD mRNA accumulation can be expected in polluted plants and this was shown in O_3-treated *Picea abies* seedlings (Galliano *et al.* 1993a). To get better insights into the relationship between environmental pollution and gene expression in forest trees, *in vitro* translated products can be studied. Distinct changes of the poly(A)+ RNA were caused by O_3 fumigation and induced polypeptides ranging from 12 kD to 50 kD after gaseous exposure were found in *Pinus sylvestris*. However, identities of these mRNAs and functions of translation products are still unclear (Großkopf *et al.* 1994). Ozone also has increased the mRNA of extensin, a specific hydroxyproline-rich cell wall glycoprotein. In *Fagus sylvatica*, *Picea abies* and *Pinus sylvestris* this pollutant caused a strong increase of extensin mRNA within a few hours of fumigation (Schneiderbauer *et al.* 1995).

11.6 Does environmental pollution trigger an evolutionary change in forest trees?

Evolution of resistance to pollution will occur if, over generations, the gene pool of forest tree changes. Thus,

fitness rates of genotypes are to be switched by the action of pollution. Sensitive genotypes may be eliminated by selection before reaching maturity or selection-driven reproductive processes may occur. Certain gene marker studies (see Chapter 11.4) have already furnished evidence for viability selection and changes in the genetic composition over generations in pollution-stressed tree populations are very likely.

Figure 11.2 summarises the potential influences of pollution during different developmental stages. The question now in view of a pollution-driven evolutionary change in forest trees is how much genetic variation can be attributed to the respective potential detrimental effects. An extensive literature has been accumulated on effects of various pollutants on tree growth and foliar damage. Although there is substantial evidence derived from heritability studies, field tests and fumigation experiments that genetic variation in foliar damage in resistance to pollutants exists, the following points (a–h) complicate the prediction of evolutionary effects.

(a) Estimates of genetic variation (e.g. heritability parameters) derived from fumigation experiments predominantly done under short-term conditions, with high dosages and young plant material cannot be generalised. It is simply not known how foliar damage of seedings or cloned material under fumigation is related to damage under long-term exposure to the pollutant in the field. Results of fumigation experiments are biologically useful if identical plant material is exposed to air pollutants both under field and laboratory conditions (Oleksyn *et al.* 1988).

(b) Since the response function (i.e. variation in tolerance over a wide range of toxic dosages) is unknown in detail, selection, and thus evolutionary change, is difficult to predict. At the lower and upper bounds of environmental stress, the range of genetic variation in tolerance probably diminishes. For instance, in highly polluted areas where topodemes are on the brink of extinction, selection is unlikely to be a major factor in evolutionary change and genetic drift effects will dominate instead. Since high phenotypic variation is needed for the selection-driven process, this evolutionary force is probably more effective under moderate environmental stress.

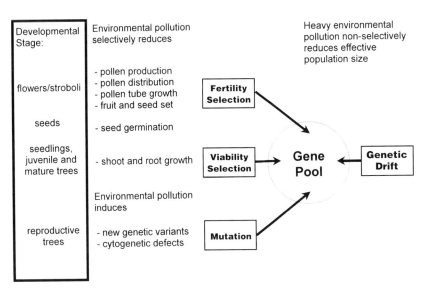

Figure 11.2 Environmental pollution affects the gene pool of forest trees.

(c) How are differences in tree vigour and foliar damage related to fitness in natural tree populations? Open-grown trees or dominant trees are the most prolific seeders. For example, Fowells and Schubert (1956) showed that in *Pinus lambertiana* Douglas and *Pinus ponderosa* Douglas ex Lawson & C. Lawson cones were found on more than 98% of the dominant trees and it is feasible that by means of growth-induced differences in polluted areas, fitness rates of pollution-affected genotypes are changed. Even small growth differences may be advantageous for certain trees especially during early life stages in view of competition for light and other vital resources. Less obvious is the relationship between fitness and visible damage (e.g. leaves, needles, crown architecture).

If certain genotypes are eliminated before reaching maturity because of pollution, evolutionary effects are obvious. However, less conspicuous are potential effects, if damaged trees are already spreading seeds and pollen. Negligible seed production in conifers was reported to occur in polluted areas as early as 1912 (Hedgcock 1912). Pollution has been shown since to affect various reproductive traits. For example, gaseous pollution (fluorides, SO_2) affects the number of seeds per cone, cone size and cone abortion in *Pinus sylvestris* (Roques *et al.* 1980) whereas acid rain reduces seed

production per cone as shown in *Picea rubens* (Feret *et al.* 1990). However, it is still an open issue how differently damaged genotypes contribute their genes to the next generation. A study on tree fruiting in Great Britain over three years has shown that cone production of *Picea abies*, *Picea sitchensis* (Bong.) Carrière and *Pinus sylvestris* was greater in trees with the least transparent (damaged) crowns whereas in *Fagus sylvatica* cupule production was greater in highly stress-affected trees (Innes 1994).

(d) Pollution can directly affect reproduction and it is well known that pollen germination and subsequent pollen tube growth are among the most sensitive pollution indicators (e.g. Feder 1968; Krug 1990; Hughes & Cox 1994). However, sufficient genetic data on pollution-affected reproductive processes are still missing— only very few studies have been done to evaluate family or clonal variation of pollution-affected reproduction in forest trees (Cox 1989; Venne *et al.* 1989).

(e) Longevity and thus long generation cycles as well as certain reproductive characteristics (e.g. serotiny) of trees slow down evolutionary response to pollution.

(f) Pollution is intrinsically highly variable in its components and interactions, resulting in episodic selection pressures.

(g) Strictly speaking, evolution of pollution resistance is governed by 'net fitness' as shown by Roose *et al.* (1982). Evolutionary changes in populations normally entail some trade-offs. In herbaceous plants, for instance, there is substantial evidence that metal tolerance is disadvantageous in the absence of the selective metal (see Macnair 1997). Slower growth rates, and thus lesser competitive ability of pollution-tolerant plants, may be disadvantageous in unpolluted environments. However, certain mechanisms of tolerance to pollution also involve physiological or morphological properties of trees which may be simultaneously of adaptive value in unpolluted environments (pleiotropy). A xeromorphy of needles, for instance, can be beneficial under air pollution (Huttunen 1978) and under water stress (Newton *et al.* 1991), or the avoidance of plant injuries by pollution (Winner & Mooney 1980a) or drought by stress-induced higher concentration of abscisic acid (Zeevart & Creelman 1988), which triggers early closure of stomata, are examples of mutual physiological benefits under differing environmental conditions.

Encoding 'pollution-resistant' genes may, therefore, be found at moderate to high frequencies in polluted and unpolluted environments. However, known mechanisms of tolerance to pollution also involve properties that are only beneficial under pollution and, consequently, opposite selection forces may act in natural tree populations. For example, genotype-dependent energy consumption for an increased buffering capacity for detoxification under pollution (Scholz & Reck 1977) may be detrimental under non-polluted environments. The frequency of genes coding such energy-requiring resistance mechanisms probably is low in unpolluted populations due to natural selection. In herbaceous plants, genes conferring resistance are generally in low or very low frequencies as indicated by a very small number of survivors at polluted sites derived from unpolluted plant populations (Gartside & McNeilly 1974; Ingram 1988). So far no studies have examined the energy needs to establish resistance mechanisms to environmental pollution in forest tree species.

(h) Evolutionary response to environmental pollution is determined by the magnitude of stress-induced (net) fitness differences (mentioned above) and the mode of inheritance (e.g. dominance versus recessivity, number of adaptive genes, pleiotropic hitch-hiking effects). For instance, if resistance to pollution is governed by dominance and pronounced fitness differences between sensitive and resistant genotypes exist, rate of evolution is rapid, whereas recessivity and a minor fitness difference slow down evolutionary change. Whereas in certain annual plant species mode of inheritance of pollution resistance has been long known (see examples in Roose *et al.* 1982; Macnair 1993, 1997), genetic studies in forest trees do not furnish enough information to compute the rate of evolution for resistance.

11.7 Conclusions

Field resistance to environmental pollution and also resistance to specific pollutants in controlled experiments are governed by high amounts of genetic variation and for certain environmental stress and species, geographical patterns in resistance to certain pollutants exist. However, data to assess reaction norms to single 'classical' pollutants, such as NO_x, SO_2 or O_3, lack completeness and genetic data on peroxyalkyl nitrates, hydrocarbons and also many non-gaseous pollutants are entirely missing. Studies of genetic control (i.e. number of genes involved in resistance mechanisms, their interactions and environmentally triggered modification) are still in their infancy, but fortunately scientific development is brisk in this field. However, even based on a rather incomplete knowledge, limited data already indicate that genetics of forest trees are doubtless altered in polluted environments. It seems very likely that over generations genetic shifts will occur (i.e. that rate and magnitude of different evolutionary factors are modified and alteration in genetic architectures over generations will result). On the single gene level viability selection in polluted environments may be already discernible within very few years, but on the polygenic level significant differences in field resistance to pollution in forest trees are more likely to emerge over longer time scales. There is no scientific doubt that resistance to specific pollutants or specific mixtures thereof can be artificially selected and thus breeding for more tolerant populations, families or clones is theoretically possible. However, selection for certain phenotypic traits may increase the possibility of reducing the potential genetic adaptability in forest trees (Ziehe & Hattemer 1989, but see also the multiple populations breeding system introduced by Namkoong *et al.* 1980). Furthermore, modified pollution regimes (factors, concentrations) and additional anthropogenic effects (e.g. CO_2) in the field are likely, especially in view of the longevity of forest trees and their typical long rotation times in managed forests. Hence, future growing conditions are to a great extent not predictable, which makes traditional breeding goals evasive and questionable. Tree breeding for increased resistance to specific pollutants may be better substituted by tree improvement for increased homeostatic capacity by promoting high physiological adaptive potential (Larsen 1991). Unfortunately, appropriate methodical tools for assessing physiological adaptive potential in forest trees are not yet available.

Box 11.1 *Pinus strobus* and *Populus tremuloides*—two case studies of pollution-driven evolution

It is elusive how environmental pollution affects evolutionary processes in forest tree species. However, in *Pinus strobus*, permanent changes in the proportions of tolerant and sensitive trees (i.e. shifts in the genetic structure) can be concluded from pollution-affected field experiments (Kriebel & Leben 1981). Seeds were harvested in 1955 and 1966 from high-pollution localities in central West Virginia, northern Georgia, northern Ohio and from pollution-free localities (at the time of seed harvest), for instance from eastern Iowa, USA. The seedlings were planted in a heavily polluted region in northern Ohio. In each of the three pollution-affected States, nearly all pines derived from the 1964 or 1966 seed harvest were free of symptoms (Figure 1). Although seed was collected from different trees in 1955 and 1966, consistency of shifts among different seed sources strongly indicate a pronounced loss of sensitive trees in the native pine stands. Results from Karnosky's (1989) study are in the same line. He has shown a 10-fold increase in the mortality of O_3-sensitive *Pinus strobus* genotypes compared to tolerant ones in a 15-year study in southern Wisconsin which documents the elimination of air pollution-sensitive genotypes as an early stage of natural selection.

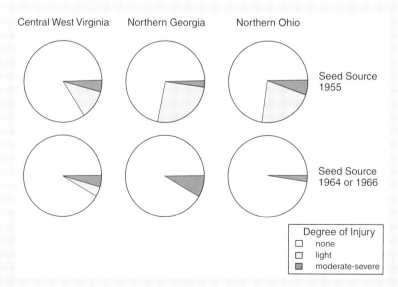

Figure 1   Changes in population structure of *Pinus strobus* at three heavily polluted sites over one decade as indicated by differing sensitivity of offspring to gaseous pollution in field trials in northern Ohio, USA (based on data from Kriebel & Leben 1981).

Berrang *et al.* (1991) concluded that pollution triggers an evolutionary change from a study in *Populus tremuloides*. In an O_3-fumigation experiment of cloned material originated from the whole species' range, populations sampled from areas that failed to achieve the US National Ambient Air Quality Standard for ozone had significantly less injuries than source populations grown up in a less ozone-polluted environment. A significant negative correlation exists between *Populus tremuloides* leaf injury and the maximum daily O_3 averages at the population localities (Figure 2).

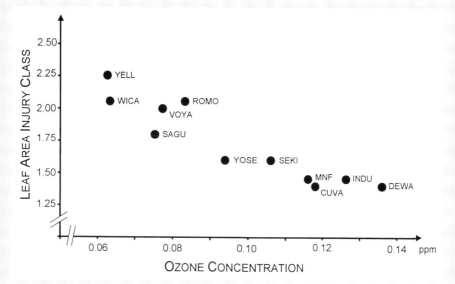

Figure 2   Association between natural ambient ozone (O_3) concentration (average maximum from 0900 hours to 1700 hours) at population sites and induced foliar damage after controlled O_3 fumigation (from Berrang *et al.* 1991, slightly modified). ***Data are significant at $P = 0.001$. CUVA, Cuyahoga Valley National Recreation Area (Ohio); DEWA, Delaware Water Gap National Recreation Area (Pennsylvania); INDU, Indiana Dunes National Lakeshore (Indiana); MNF, Monongahela National Park (West Virginia); ROMO, Rocky Mountains National Park (Colorado); SAGU, Saguaro National Monument (Arizona); SEKI, Sequia National Park (California); VOYA, Voyageurs National Park (Minnesota); WICA, Wind Cave National Park (South Dakota); YELL, Yellowstone National Park (Wyoming); YOSE, Yosemite National Park (California) (after Berrang *et al.* 1991).

DISEASE THREATS AND THE CONSERVATION GENETICS OF FOREST TREES

Margaret Byrne

SUMMARY

Disease has had major impacts on forest trees and the ecosystems they support. Disease expression is a result of complex plant-pathogen interactions and disease severity is often related to whether the pathogen is endemic or has been introduced. The genetic effects of disease include changes in the distribution and frequency of resistance genes, fragmentation and reduction in population size. Disease control involves containment and quarantine in the field, and increasing the resistance of populations. Disease effects have implications for conservation, both *in situ* and *ex situ*. *In situ* conservation preserves the coadaption of host-pathogen systems and allows the continued evolution of resistance. *Ex situ* conservation collections provide the gene pool from which disease resistance selection and breeding will be achieved, and therefore should encompass as wide a range of genotypes as possible.

12.1 Introduction

Disease is an important, although sometimes overlooked, factor in the dynamics of forest populations. The dramatic effects that disease can have on trees has been evident in several major disease epidemics. For example, Dutch elm disease caused by *Ophiostoma ulmi* (Buism.) Nannf. has affected a range of *Ulmus* sp. (elm) in Europe and North America, chestnut blight caused by *Cryphonectria parasitica* (Murr.) Barr has resulted in the near extinction of *Castanea dentata* (Marsh.) Borkh. (American chestnut) once a major hardwood species in North America, and the root pathogen *Phytophthora cinnamomi* Rands has caused devastation in forest ecosystems in Australia, as many of the native flora are susceptible. Although the severe effects of these major epidemics are obvious, many disease effects in forest communities are less conspicuous and go unnoticed. These less obvious effects have important consequences on the ecology and demographics of the forest, at the community level and for individual species.

The effects of disease can be wide ranging and depend on the particular interaction between pathogens and their hosts. These interactions are dynamic processes that vary spatially and temporally, and lead to changes in population structure, both in physical and genetic terms. Effective conservation strategies should take into consideration the changes that are wrought through host-pathogen interactions. The main interactions of pathogens and forest trees and their effects on forest populations will be outlined and implications for the conservation genetics of forest communities discussed.

12.2 Plant–pathogen interactions

The interactions between plants and pathogens are complex and in this chapter have been placed into two groups depending on whether the organisms have evolved together through coexistence in communities, or whether the pathogen has been introduced into communities with no previous exposure to it.

12.2.1 Exotic pathogens

Most plants are resistant to infection by most pathogens. However, when a pathogen is introduced to a community with susceptible species, these species are generally widely affected and show easily discernible symptoms, such as widespread death.

Although epidemics are rare in non-agricultural plants, they generally occur when exotic diseases are introduced into natural communities. Exotic diseases can have specific effects where the pathogen has a restricted host range. For example, *Cryphonectria parasitica* only affects *Castanea dentata* and the introduction of the pathogen into mixed hardwood forests has resulted in the replacement of chestnut with other codominant species (Woods & Shanks 1959; Stephenson 1986). In this case the effects on community structure are subtle, the loss of one species in a mixed forest does not have significant effects on the forest as a whole. However, where many species are susceptible to an introduced pathogen, the effects on the community structure can be widespread and dramatic. For example, the introduction of *Phytophthora cinnamomi* into Western Australia and Victoria has resulted in forest and heathland communities being replaced by resistant communities dominated by sedges (Weste 1981; Shearer & Tippett 1989). The complex forest community, as it was prior to infection, has been eliminated and replaced by a floristically depauparate, non-forest community.

Susceptible species are often severely affected by exotic diseases because epidemics can spread quickly due to the ample supply of susceptible material. If resistance exists in the species, it is usually at low frequency, for example, resistance to white pine blister rust (caused by *Cronartium ribicola* J.C. Fisch. ex Rabenh) in *Pinus lambertiana* Douglas (sugar pine) (Kinloch 1991) and resistance to *Ophiostoma ulmi* in *Ulmus americana* L. (American elm) (Smalley & Guries 1993). The rate of spread of infection depends on several factors relating to both plant and pathogen populations. Plant factors that affect disease spread include: density, which is related to the dominance of the species in the community; age and stage of the tree, such as seedlings or ageing trees; and whether the species has some level of resistance or is uniformly susceptible. Pathogen factors that affect disease spread include: means of dispersal of inoculum (e.g. wind, water or insect vectors); environmental conditions that favour pathogen survival and reproduction; and generation time of the pathogen.

For species with widespread distributions, disease may be a threat to the survival of populations but not a threat to the survival of species as a whole, particularly if the range of the species encompasses environmental conditions that are unsuitable for the pathogen. For species with a limited distribution, however, disease can pose a significant threat of extinction. Many of the Proteaceae are highly susceptible to *Phytophthora cinnamomi*, and the survival of several species endemic to Western Australia, for example, *Banksia brownii* Baxter ex R. Br. (feather-leaved banksia) and *Banksia verticillata* R. Br. (granite banksia) are severely threatened by *Phytophthora* infection (Kelly & Coates 1995). Although *Pinus radiata* D. Don (Monterey pine) is the most widely planted pine species in the world, it has a limited natural distribution in three mainland Californian populations and two island populations off the coast of Mexico (Moran *et al.* 1988). Pine pitch canker, caused by the fungus *Fusarium subglutinans* (Wollenweb. & Reinking) P.E. Nelson, T.A. Toussoun & Marasas f. sp. *pini*, is threatening the native Californian populations of *Pinus radiata* (Smith & Ferlito 1997).

12.2.2 Indigenous pathogens

The effects of pathogens endemic to plant communities are often less obvious than those that are introduced. Where indigenous pathogens have coevolved with their plant hosts they establish a degree of balanced interactions which enable populations of each to survive. The basic theoretical models for coevolving pathogens and hosts propose that genes for resistance in the host are matched by genes for virulence in the pathogen. Both resistance and virulence genes exert a selective force in the population, and both have fitness costs associated with them (see Burdon 1987 for details of models). The balanced interactions that result from these gene-for-gene models do not mean the populations are static, rather they are dynamic processes where local extinction and recolonisation events lead to continual change within communities. The dynamics of this change is affected by the distribution of hosts, distribution and dispersal of pathogen, distribution and frequency of resistance and virulence genes. An example of the effects an endemic pathogen has on forest communities is *Phellinus weirii* (Murrill) R.L. Gilbertson, a root-rot fungus. This fungus is endemic to the conifer forests of the Pacific Northwest of the United States of America where the dominant species, *Tsuga mertensiana* (Bong.) Carrière (mountain hemlock), *Abies amabilis* Douglas ex Forbes (Pacific silver fir), *Pinus contorta* Douglas ex Loudon (lodgepole pine) and *Pinus monticola* Douglas ex D. Don (western white pine), vary in their susceptibility. Combinations of the differential susceptibility of species and their environmental requirements lead to changes in the structure and succession of forests attacked by *Phellinus weirii* compared to uninfected forests (Cook *et al.* 1989; Burdon 1994).

Disease expression may be strongly influenced by environmental conditions, such as water availability, disturbance, temperature, nutrient deficiency or toxicity. Environmental effects work in two directions. Poor environmental conditions place stress on plants which make them more susceptible to biotic infections; and diseased plants are generally less able to withstand unfavourable environmental conditions (Burdon 1987). Drought stress is often associated with higher disease expression. For example, *Quercus cerris* L. (oak) and *Populus tremuloides* Michx. (aspen) are more susceptible to Hypoxylon canker (*Hypoxylon* ssp.) when water stressed (Manion & Griffin 1986; Vannini & Scarascia Mugnozza 1991). Drought and *Phytophthora cinnamomi* infection are major factors in the decline of oak in the Mediterranean (Brasier 1996), but the direction of cause and effect is less clear. Drought stress on trees already subject to low levels of infection may reduce their tolerance of the fungus; alternatively, infection with the fungus may make trees less tolerant to the effects of water stress. The stress induced in trees by non-optimal environmental conditions, such as air pollution or competition from introduced species, can lead to increases in pathogen infection. Reduced light intensity on north-facing slopes increases *Brunchorstia pinea* (Karst.) Hoehnel infection on *Pinus nigra* var. *maritima* (Aiton) Melville (Corsican pine) in England (Read 1968).

Disease occurrence can also be influenced by biotic factors other than infection by the primary pathogen. In heteroecious rusts (those that require the presence of two hosts for the completion of the life cycle of the pathogen), the numbers and distribution of the alternate host will affect the level of disease expression. For example, the build up of oaks in southern pine forests of North America due to fire suppression has been a factor in the increase of *Cronartium fusiforme* Hedgc. & Hunt ex Cumm. (fusiform rust) infection on *Pinus taeda* L. (loblolly pine) and *Pinus elliottii* Engelm. (slash pine) (Zobel 1982). Many fungal pathogens are spread by insect vectors [e.g. *Ophiostoma ulmi* and oak wilt caused by *Ceratocystis fagacearum* (Bretz) Hunt], therefore, spread of the disease depends on the level of insect activity in the forest, which in turn can be affected by environmental conditions.

There are a range of pathogenetic effects associated with disease expression. The most obvious disease effects are those that result in the death of the tree. However, many diseases do not threaten the survival of the tree but cause a reduction in fitness. For example, severe growth reduction occurs with *Lophodermella sulcigena* (Rostr.) v. Hohn (grey needle cast) on *Pinus silvestris* L. (Scots pine) in Finland (Jalkanen 1982). This reduction in fitness will have long-term genetic effects on populations if it also affects fecundity of trees.

12.3 Genetic effects of disease

12.3.1 Resistance genes

The distribution and frequency of resistance genes across species will vary, and often geographical differences in resistance frequency are correlated with differences in environmental conditions which influence the pathogen (Burdon 1982). Such differences have often been observed in provenance trials. For example, the northern Scandinavian and Russian provenances of *Pinus silvestris* L. are more resistant to snow blight (caused by *Phacidium infestans* P. Karstens) than southern populations (Roll-Hansen 1989). Provenances from the periphery of the natural range of *Pinus taeda* L. are more resistant to *Cronartium fusiforme* than other provenances (Wells & Wakeley 1966). Geographical distribution of resistance will also be correlated with distribution of alternate hosts for heteroecious fungi. For example, the frequency of resistance to *Cronartium comptoniae* Arth. (sweet fern rust) in *Pinus banksiana* Lamb. (jack pine) and *Pinus contorta* Dougl. was correlated with the distribution of the alternate hosts [*Comptonia peregrina* (L.) Coult., sweet fern, and *Myrica gale* L., sweet gale] (Hunt & Van Sickle 1984). The distribution of resistance genes suggests they will evolve, or be maintained, where there is selective pressure from the pathogen, that is, where the environmental conditions are most favourable for the pathogen, or where the presence of the pathogen is maintained by the occurrence of both hosts.

The frequency and distribution of resistance genes will be a major determinant on the change in population structure of a species following cycles of disease activity. It may also be responsible for the distribution of species themselves, as species may be restricted to areas where environmental conditions are suitable for the host, but not so favourable to the pathogen (Burdon 1982). It has been suggested that the natural restriction of *Larix decidua* Mill. (European larch) to the Alps is due to the devastation caused by the pathogen *Trichoscyphella willkommii* (Hart.) Nannf. when *Larix decidua* is grown in lowlands (Burdon 1982).

12.3.2 Fragmentation and reduction in population size

Two of the main effects of disease on populations are reduction in size and fragmentation of populations. Small population size and fragmentation reduce genetic diversity through changes in the effects of drift, gene flow, selection and mating (see Chapters 7 and 10). Epidemics, such as those that occur from introduced pathogens, often result in substantial reductions in population size or bottlenecks. Migration is beneficial to populations that have undergone a bottleneck. However, migration into a population that is small because of a disease epidemic will only be beneficial if it brings disease resistance genes into the population. Disease outbreaks in endemic pathogen systems may not have substantial effects on population size but are likely to result in localised reduction in plant density and patchy distribution. This may result in reduction in effective population size leading to increase in neighbourhood structure and increased inbreeding (Murawski & Hamrick 1991), particularly for insect-pollinated species.

Fragmentation of populations, or an increase in distance between them, results in isolation and divergence between populations or fragments. In endemic systems local adaptation between pathogen and host will be established, and fragmentation of populations from disease outbreaks may lead to divergence of host-pathogen adaptations between populations. Migration from divergent populations will not be beneficial since these migrants will not be adapted to the selective pressures exerted by the local pathogen population.

Diseases that reduce fecundity rather than survival may not result in a physical reduction in population size or density, but will cause a reduction in the number of trees contributing to the breeding pool. This results in a reduction in the effective population size. Spatial variation in fecundity has effects on the behaviour of insect and bird pollinators, leading to increased inbreeding and biparental breeding

(Sampson *et al.* 1996). Dioecy is not uncommon in tree species, and reduction in population size or density may result in a bias in the sex ratio of trees within populations. Such an imbalance could lead to overrepresentation of particular individuals in the gamete pools of the population and consequent reduction in diversity in later generations.

12.4 Disease control

12.4.1 Containment

Controlling disease or reducing its effects usually involves minimisation of its spread through the isolation and containment of affected areas. Containment of infected areas involves strict quarantine measures, such as disinfection of vehicles, equipment and people leaving an infected site, restriction of movement within infected sites and prevention of transport of infected material out of infected sites. The spread of *Phytophthora cinnamomi* in *Eucalyptus marginata* G. Don ex Sm. (jarrah) forest in Western Australia was significantly reduced following the implementation of quarantine and hygiene practices (Shea 1979). Quarantine and hygiene practices have been implemented in California to control the spread of pine pitch canker, including the cessation of landscape plantings of *Pinus radiata* in California, and prohibition of planting of any pine species in close proximity to the Torrey Pine State Park (Storer *et al.* 1994). For diseases that affect species that are widely planted outside their natural habitat, quarantine measures may also need to be applied at an international level. For example, regulations are in place to prevent the spread of *Ceratocystis fagacearum* into Europe from North America (Delatour 1996) and to prevent the introduction of the rust pathogen *Puccinia psidii* Wint. into Australia where it has the potential to attack many eucalypt species (Frankel *et al.* 1995). Uninfected forests should be isolated from any disturbance where possible. Disturbance will often spread the pathogen or predispose the forest to infection should one be introduced. Quarantine, hygiene and isolation will not completely prevent the spread of the disease or lead to its eradication, but it does slow its progress, and provide time for other control measures to be developed (Shea 1979).

Management of disease may also include control of vectors that transmit pathogens from tree to tree, either by insecticides, or by removing insect breeding and feeding areas, such as removal of dead material

from elms and oaks (Manion 1981). Root grafting in oaks results in rapid local spread of *Ceratocystis fagacearum* and containment can be achieved by destroying the contact between roots of healthy and infected trees by poisoning healthy trees around the perimeter of an infected site (Nair 1996).

Application of chemicals, such as fungicides and insecticides, can help to slow or prevent disease progression. Benomyl and thiabendazole have been used as foliar sprays, trunk injection and soil treatment in the control of *Ophiostoma ulmi* (Sinclair & Campana 1978). Phosphite has been shown to be effective in the control of *Phytophthora cinnamomi* in *Eucalyptus marginata* and a range of other native plant species (Komorek *et al.* 1994) where the chemical has fungicidal properties but also appears to induce resistance effects in host plants. The large-scale application of chemicals to forests is generally impractical. However, chemical applications can be useful in the protection of trees of high value, such as amenity plantings of elms and chestnut, and in the protection of rare or threatened species, such as in Western Australia where a major phosphite application program is now underway to protect native flora against *Phytophthora cinnamomi* infection (K. Vear, pers. comm. 1998).

12.5 Disease resistance

Genetic resistance is part of the plant's natural method of disease control, and can be manipulated to increase the frequency of resistance genes in a population. Disease resistance can be controlled by major genes, where one, or a few genes have a large effect on resistance, for example against *Cronartium ribicola* in *Pinus lambertiana* (Kinloch & Littlefield 1977) and *Cryphonectria parasitica* in *Castanea dentata* (Bernatzky & Mulcahy 1992). Alternatively, resistance can be quantitative, being controlled by a number of genes of small individual effect, which is the case for disease resistance in most trees. Increasing the resistance of species can be achieved through breeding. This involves selection of resistant individuals and families, and incorporation of these individuals into breeding populations, through crossing programs or seed orchards. Reliable assessment of disease susceptibility and resistance is critical to resistance breeding, and is often more difficult than scoring for other breeding characters. It is important that resistance is expressed in the field, but scoring this is complicated by environmental effects. Controlled trials in the glasshouse give better repeatability in phenotype scoring. However, the correlation with field-based resistance, and resistance at different ages, needs to be established.

Breeding for resistance is an on-going activity as the plant-pathogen interaction is constantly evolving. Selection for resistance in the host favours selection for virulence in the pathogen (Namkoong 1991a). Hence, tree breeding requires the incorporation of multiple forms and sources of resistance that can be manipulated to endure into future generations (Namkoong 1991a). Diversity of resistance genes is the most effective means of stabilising pathogen populations (Kinloch 1982). Breeding for resistance has been undertaken in several diseases of forest trees [e.g. Dutch elm disease, white pine blister rust, jarrah dieback (caused by *Phytophthora cinnamomi* Rands) and chestnut blight]. Numerous elm cultivars resistant to *Ophiostoma ulmi* have been developed in the Netherlands and the USA, and many are commercially available (Smalley & Guries 1993). *Phytophthora cinnamomi* rootrot resistant lines have been developed in *Eucalyptus marginata* (Stukely & Crane 1994). Many blight-resistant *Castanea sativa* Mill. (European chestnut) trees have been incorporated into a gene bank for reforestation in Switzerland (Bazzigher 1982). *Pinus taeda* and *Pinus elliottii* seedlings from *Cronartium fusiforme* -resistant seed orchards are being planted in the southern USA (Powers 1984).

Molecular biological approaches to breeding are now being undertaken in some tree species. For example, molecular markers are being used to select for individuals carrying disease resistance genes (marker-aided selection). This involves identifying markers linked to the resistance genes through analysis of segregation patterns in families. Identification of tightly linked markers flanking each side of the gene will assist in minimising problems associated with linkage equilibrium in family selections. Resistance to *Cronartium ribicola* is conferred by a single gene and markers linked to the gene have been identified in *Pinus lambertiana* (Devey et al. 1995). Identification of markers that tag resistance genes is being carried out for leaf rust in poplars caused by *Melampsora laricini-populina* Kleb. (Cervera et al. 1996b; Villar et al. 1996), leaf spot in poplars caused by *Septoria populicola* Peck. (Newcombe & Bradshaw 1996), black leaf spot [caused by *Stegophora ulmea* (Schw.: Fries) Sydow & Sydow] in *Ulmus parvifolia* Jacq. (Chinese elm) (Benet et al. 1995) and dieback cause by

Phytophthora cinnamomi in *Eucalyptus marginata* (Byrne et al. 1997). Where clonal propagation is feasible, selected individuals can be incorporated directly into existing breeding and deployment programs.

An alternative approach to marker-aided selection is to clone genes for resistance and then transform them into susceptible individuals. Cloning of genes is usually achieved through use of heterologous genes from other species or through map-based cloning. No forest disease resistance genes have been cloned to date, although research is being carried out to clone the resistance gene against *Cronartium ribicola* from *Pinus lambertiana* (D. Neale, pers. comm. 1997). The recent identification of similarities in disease resistance genes in agricultural crop species holds promise for the identification of resistance genes in forest trees. Sequencing of cloned resistance genes has identified high levels of homology in some common domains within the genes (Michelmore 1996). These genes have been identified in a range of plants and confer resistance to fungal, bacterial, viral and nematode pathogens (Kanazin et al. 1996; Lagudah et al. 1997). Use of conserved sequences from the common domains is being evaluated for the isolation of resistance gene analogues in other species (Kanazin et al. 1996; Yu et al. 1996).

12.6 Implications for conservation

Conservation of forest genetic resources is important for three reasons:

(a) genetic uniformity increases vulnerability to pests and climatic extremes

(b) genetic variants are important for their potential breeding value some time in the future

(c) the loss of diversity by local or global extinction of a species may reduce the stability of entire ecosystems (Ledig 1986).

The maintenance of genetic variation, both as gene diversity and coadapted gene complexes, is a key component of conservation strategies. Maintaining genetic variation guards against overall vulnerability to disease, both in natural populations of tree species and in plantings, whether they are ornamental or commercial. Vulnerability increases with genetic uniformity; plantings of a single, or a few, clones could be highly vulnerable to a new pathogen or to a more virulent race of an existing pathogen (Libby 1982).

Natural populations that have reduced variability through fragmentation, bottlenecks or selection would also have increased vulnerability to disease. This is of particular importance in trees compared to crop species, because of the long generation time of most tree species. Pathogens have multiple generations for each generation of a tree; therefore, virulence in the pathogen population can evolve more rapidly than resistance in the tree population (Ledig 1986). Populations with reduced variability may also be more vulnerable to stresses resulting from climatic or other environmental factors (see Chapter 11). This may lead to increased susceptibility to disease in itself due to the interaction of environmental conditions and disease expression.

Maintenance of genetic variation for its breeding value is relevant to both commercial plantings and natural populations. The base breeding populations for commercial species should be maintained with sufficient diversity for selection of new forms of disease resistance to be undertaken in the future. Maintenance of genetic variation in native populations will enable resistant individuals to be naturally selected when disease pressure is applied in a population. These naturally selected resistant trees will be the source of seed for future regeneration of disease sites, either by natural regeneration or through planting and reafforestation programs. Reafforestation using resistant trees in natural populations may be the only major defence against severe exotic diseases where broad-scale control measures often have limited effectiveness or are impractical. Natural regeneration from resistant trees will produce trees adapted to the site. However, if there are only a few resistant trees left in a population and they are scattered, regeneration will suffer from inbreeding and bottleneck effects. The diversity in resistance may be limited in small, disease-affected populations. Effective regeneration of natural populations may require active revegetation with disease-resistant stock from a breeding program. Breeding for resistance aims to identify multiple resistance genes to maximise variability in the resistance response. This will produce enduring resistance by avoiding selection pressure for virulence in the pathogen population. Breeding for resistance should enable a larger gene pool to be created from which resistant planting stock is generated across a range of genetic backgrounds. However, some partitioning of the breeding populations may be

required in order to generate provenance-correct planting stock for natural populations (see Chapter 8). An active combination of conservation and breeding is necessary for the use of resistance in ameliorating disease threats (Namkoong 1991a). Such breeding and conservation are being combined in breeding programs for *Quercus suber* L. (cork oak) in Portugal (Varela & Eriksson 1995), and species in the United States Pacific South West Tree Improvement Program (Kitzmiller 1990).

In situ conservation is often implemented so no manipulation or disturbance of the populations occurs. This is because disturbance can cause changes in ecological interactions resulting in increased disease activity in endemic systems, and can lead to the introduction of exotic pathogens (Burdon 1987). However, in conservation against disease threats some management may be necessary. Thus, control of pathogen vectors may be a disturbance that could be permitted in conserved populations in order to reduce the occurrence and spread of a disease. In other cases, the maintenance of multiple age-classes within the populations could be important since old or young trees may be more vulnerable to pathogen infection (Koski 1996). Fire can also be an important disturbance that is beneficial in disease cycles. Fire has been suggested as a conservation measure against pitch canker in *Pinus radiata* forests in California in order to promote regeneration and kill inoculum sources in the soil (Cylinder 1997; Gordon *et al.* 1997).

Ex situ conservation of germplasm is particularly important as reservoirs of disease resistance diversity. *Ex situ* collections will not lead to natural selection or identification of disease resistant individuals, but do provide a gene pool from which selection for disease resistance can be carried out. Programs of selection for resistance will need to be initiated as new diseases evolve or are introduced, or as different races of the pathogen evolve. Therefore, continued access to *ex situ* collections is of high priority. It is often not desirable to test for disease resistance by inoculating or releasing pathogens into planted collections, therefore forms of tissue storage are an important aspect of protection against disease threats. Testing for disease resistance needs to be carried out on plants generated from seed collections or tissue cultures that are dedicated for that purpose, as there is the potential that plants will be lost in the process of disease testing. Factors that improve the genetic diversity captured in the collections, such as appropriate sampling and storage strategies, will increase the chance of successfully using the collections to identify disease resistance when required. Representative habitats across the species range are usually sampled for *ex situ* collections, but marginal habitats should also be sampled because selection may have favoured novel variants (Ledig 1986). For example, two unusual sites of *Pinus lambertiana* were sampled and one was found to have a high frequency of a major gene for blister rust resistance (Ledig 1988).

Ex situ collections may themselves be subject to disease threats. Domestication and *ex situ* conservation plantings can disturb the balance between plant and pathogen populations. This may result in increased disease severity in managed plantings from pathogens that are endemic, but of minor concern, in natural populations. For example, *Cronartium fusiforme* is endemic to southern pine populations of North America and is only of minor concern in native stands. However, it has become a major problem in plantations of *Pinus taeda* and *Pinus elliottii* (Zobel 1982). *Mycosphaerella* leaf disease is endemic in native eucalypt forests but can become severe in plantations of *Eucalyptus globulus* Labill. (blue gum) in Australia (Keane 1997). Planting trees in mixed stands rather than pure stands would be beneficial since disease effects are likely to be lower in mixed stands (Libby 1982). Mixtures of host and non-host plants reduce epidemic development through dilution of susceptible tissue, reduction in source of inoculum, increased distance between susceptible plants, reduction in inoculum dispersal and interception of inoculum by non-host plants (Burdon

1987). The rate of increase of disease tends to be greater at high than at low plant density; therefore, the effect of pathogens in mixed stands is likely to be frequency dependent (Burdon 1982). Whereas most evidence of the effects of mixed plantings in reducing disease come from agricultural systems, there are some other examples. Planting clonal mixtures of *Salix* sp. (willow) has been shown to reduce the effects of rust infection from *Melampsora epitea* var. *epitea* Thum. (McCracken & Dawson 1997).

The cloning and characterisation of resistance genes is forecast to be a major activity in crop plants (Michelmore 1995) which is likely to flow through to forestry. The increasing ability to characterise and manipulate resistance genes will lead to greater prospects for protection of trees against disease threats. For crop species, it is proposed to use markers that tag resistance genes to screen wild germplasm so that collections can be made to encompass resistance gene diversity as well as general genetic diversity (Michelmore 1995). Developing disease resistance in a species is not confined to identification and selection of disease resistance in the species itself. Resistance is often present in one species and can be transferred into a susceptible species through hybridisation or genetic engineering. Therefore, we should not limit our efforts to the conservation of disease threatened species only. Genetic engineering will proceed by selecting from the diversity which is available, not by creating genes *de novo*. The genetic diversity must be preserved now, however, so that it can be utilised when needed (Ledig 1986).

HYBRIDISATION AND FOREST CONSERVATION

*Shanna E. Carney, Diana E. Wolf and
Loren H. Rieseberg*

SUMMARY

Despite the presence of many barriers to interspecific reproduction, hybridisation is common among plants and in some groups of animals. It is especially common in plants living in disturbed habitats, on islands, and when introduced species are involved. Numerous outcomes are possible, including an increase in genetic variation, adaptation to novel environments, the merging of species and the formation and maintenance of stable hybrid zones. Several models regarding the formation and maintenance of hybrid zones have been proposed, and they differ in whether selection is environment-dependent or environment-independent, the fitness of hybrids and the importance of dispersal.

There are many rare plants and animals that are threatened by hybridisation with relatives. In these cases there are two possible outcomes that are relevant to conservation biology: outcrossing depression and genetic assimilation. Factors that affect the probability of extinction due to hybridisation are discussed, as well as management implications. In rare cases, hybridisation may also be used to save species on the brink of extinction.

13.1 Introduction

Hybridisation can be defined as cross-fertilisation between individuals of different species or more broadly as the 'crossing between individuals belonging to separate populations which have different adaptive norms' (Stebbins 1959). This broader definition is more widely accepted by evolutionary biologists because it is independent of species definitions, which are often controversial. For example, many biologists employ the biological species concept (BSC) (Dobzhansky 1937b) which defines species as groups of interbreeding plants or animals that are reproductively isolated from all other such groups. The emphasis of the BSC on reproductive isolation and the maintenance of species integrity makes it incompatible with a definition of hybridisation as the interbreeding of species unless hybrids are sterile, hybrid breakdown occurs, or hybridisation is extremely localised without significant introgression. Strong proponents of the BSC often view hybridisation as unimportant and hybrids as evolutionary dead ends. Botanists are less likely to employ a strict biological species definition because the importance of hybridisation in the evolution of plants and its high frequency have long been recognised. Groups of organisms with high rates of hybridisation can be taxonomic quagmires because hybridisation tends to blur species boundaries and confound species identification and classification.

Despite the many fields of study that are affected by hybridisation, it was not often considered as a factor relevant to species conservation until very recently. The main factors that previously were cited as a threat to rare species include habitat destruction, over harvesting, extinction chains and competition—especially from introduced species (Diamond 1989). During the 1990s, however, several discussions of the role of hybridisation in extinction and conservation have been published (Rieseberg 1991; Whitham *et al.* 1991; Ellstrand 1992a; Ellstrand & Elam 1993; Levin *et al.* 1996; Rhymer & Simberloff 1996; Whitham & Maschinski 1996), illustrating an increased awareness of the importance of hybridisation as a factor that can influence extinction rates.

While in some circumstances hybridisation between a common species and its rare congener (species from the same genus) can be beneficial as a final effort to maintain some of the genetic information present in a species on the brink of extinction, mostly it is detrimental to the persistence of the less frequent species. The threat may be due to decreased fitness of the rare parent because of reduced viability or fertility of hybrid progeny (i.e. outbreeding depression; Price & Waser 1979; Templeton 1986). If hybrid fitness is not greatly reduced relative to the parental taxa, frequent interspecific matings may lead to genetic swamping or assimilation of the rare species by its common congener. Hybridisation involving rare plant taxa also has legal ramifications because of the importance of reproductive isolation in widely accepted species definitions such as the BSC and the emphasis of many legal measures on protecting species.

13.2 The process of natural hybridisation

For hybridisation to occur, many criteria need to be met. In plants, pollen needs to be transferred to the stigma of another taxon, and seeds must be produced. The first requirement for this to occur is that the species must be parapatric or sympatric to each other. They must flower at the same time and utilise the same pollinators, at least occasionally. Pollen must be transferred interspecifically, and it must be able to germinate on the foreign stigma. Once germination has occurred, the pollen tubes must be able to grow down the foreign style to the ovary and successfully fertilise ovules. The hybrid embryos must then develop into mature seeds. Finally, the seeds must be able to germinate and seedlings must successfully compete with 'pure' progeny in order to form adult hybrid plants that are in turn able to reproduce. Any step in this pathway may be blocked by a reproductive barrier, thus preventing hybridisation. In many cases, more than one reproductive barrier isolates a species pair (Dobzhansky 1937a, 1937b). These barriers probably evolve as a byproduct of diversification of the two species, although it is possible that in some cases where hybrids have low fitness, reproductive isolation is reinforced through natural selection against individuals that hybridise (Dobzhansky 1940; Mayr 1963).

13.3 Frequency and outcomes of hybridisation

Despite the numerous mechanisms that can prevent hybridisation, it appears to occur commonly in plants. It has been estimated that as many as 70% of plant species experienced hybridisation at some point in their evolutionary history (Whitham *et al.* 1991; Masterson 1994). However, this does not mean that

70% of extant species are currently hybridising. Analysis of the taxonomic distribution of hybrids in a tropical flora, two European floras and floras of two regions of North America (Ellstrand *et al.* 1996) revealed that the frequency of natural hybrids ranged from about 22% for the British flora to 5.8% for the intermountain flora of North America. They also found that hybridisation is not randomly distributed among taxa. Rather, it is concentrated in certain families and genera. About 16% to 34% of families in these floras have at least one hybrid member, whereas only 6% to 16% of genera have at least one hybrid. Furthermore, most hybrids reported were perennials with some mode of clonal growth, although this may be because it is easier to detect hybridisation in perennials.

Hybridisation occurs less frequently in animals than in plants and often has been considered unimportant by zoologists (Dobzhansky 1937b, 1940; Mayr 1942). However, some animal groups experience relatively high levels of hybridisation. For example, Grant and Grant (1992) reported that about 9% of bird species hybridise, and the hybridisation frequency in families of North American freshwater fish ranges from 3% to 17% (Hubbs 1955).

The habitat in which a species occurs appears to be important in determining whether the species will hybridise. Hybridisation is frequently associated with disturbed habitats. In many cases, disturbance disrupts geographic barriers between species, and competition with other species may not be as strong in disturbed habitats (Anderson 1948). Anderson (1948) suggested that hybrids are more likely to find a suitable niche in disturbed areas because niches intermediate to those of the parental species are likely to exist. Similarly, hybridisation is observed in areas where two niches meet and create a mosaic of habitats. Hybridisation also appears to be especially common among island flora. This is due to: small niche size; small population sizes; extensive sympatry of congeners; colonisation of islands by closely related exotics; increasing disturbance of island habitats; and the close genetic relationship and lack of strong reproductive barriers often seen due to relatively short divergence times (e.g. following adaptive radiations) (Rieseberg *et al.* 1989; Rieseberg 1991; Rieseberg & Swenson 1996). Hybridisation also may be encouraged because of human-mediated introductions or natural species movements, such as those caused by global climate changes (e.g. glaciations) or changes in sea levels that isolate and rejoin land masses (Box 13.1).

The possible results of hybridisation are diverse and depend on the degree of compatibility between the parental species, the fertility and viability of the hybrids, habitat preferences of the parents and hybrids, and numerous other factors. Anderson (1948) states that 'the commonest result of hybridisation is introgression' or the movement of genes from one species into another by repeated backcrossing. Although that observation is probably biased by introgression being the most easily observed result of hybridisation, introgression is nonetheless an important evolutionary phenomenon (Rieseberg & Wendel 1993). It can lead to an increase in genetic diversity of each of the parental species because of the presence of genes from the alternate species or the formation of new alleles through intragenic recombination (Golding & Strobeck 1983). Similarly, adaptations may be transferred between species via hybridisation, or new adaptations not present in either of the parental species may arise. These new phenotypes can allow the hybrid individuals to utilise different habitats than the parental species, thus increasing the probability that they will persist. Hybrids that are able to persist in this way, by avoiding competition from the parental species, may eventually become reproductively isolated from their progenitors, forming a new species. This process can occur rapidly through polyploidisation or more slowly in the case of homoploid hybrid speciation. In contrast, hybridisation can be seen to decrease biodiversity when it leads to the genetic assimilation of one of the species in a hybridising pair, causing its extinction, or when it leads to the merging of species due to the breakdown of reproductive barriers. Additional outcomes of hybridisation include the formation of a stabilised hybrid zone or the reinforcement of reproductive barriers leading to a decrease in the frequency of hybridisation. Finally, when cultivated plants hybridise with wild relatives, natural or engineered genes such as those that confer pest or herbicide resistance to the crop can be transferred to wild or weedy relatives possibly exacerbating existing weed problems or creating new ones. For this reason, crop × wild hybridisation is a topic of intense interest and study (e.g. Ellstrand 1988; Langevin *et al.* 1990; Baranger *et al.* 1995; Arriola & Ellstrand 1996, 1997; Bing *et al.* 1996; Lefol *et al.* 1996; Landbo & Jorgensen 1997; Van Raamsdonk & Schouten 1997; Whitton *et al.* 1997).

Box 13.1 The effects of global climatic changes on hybridisation

During the earth's history, there have been many dramatic changes in climate. These changes in temperature and rainfall patterns often caused species' ranges to expand and contract, at times isolating them and at others bringing them together. During periods of extreme heat or cold, species retracted into refugia. Occasionally, hybridisation resulted from normally allopatric taxa coming into sympatry in refugia, such as among fir species of the genus *Abies* Mill. (Fady *et al.* 1992). In other cases, populations of the same species were isolated in different refugia where they diverged from one another. When the divergence was not great enough for reproductive isolation to be complete, secondary contact of the divergent groups following population expansion often resulted in hybridisation.

Hybridisation in *Pinus* L. (the pine genus) due to climate-induced range changes (Millar 1993) is an example of this process. During the Eocene, pines shifted away from the middle latitudes, where temperatures and humidity were highest, into cooler refugia. At the end of the Eocene the earth experienced a major cooling trend and pines migrated out of the refugia. During this period, hybridisation occurred between formerly isolated lineages. Population expansions and contractions during the Pleistocene glaciations also allowed some *Pinus* species to hybridise.

13.4 Hybrid zone theory

It is vital for conservationists to understand the contrasting theories regarding the nature and maintenance of hybrid zones, allowing them to effectively manage hybridising endangered species. These theories or models can be differentiated based on the type of selection that is invoked, the importance of dispersal, the fitness of hybrids and the stability of the hybrid zone. This subject is reviewed in detail by Arnold (1997) and is summarised here. Selection pressure on hybrids in a hybrid zone can be either intrinsic (environment-independent) or extrinsic (environment-dependent). Barton and Hewitt's dynamic equilibrium or tension zone model (Barton & Hewitt 1985) applies when selection is independent of the environment. In their model, hybrids exhibit reduced viability or fertility due to disruptions of coadapted gene combinations. These tension zones (Key 1968) are maintained by a balance between dispersal of parental individuals into the zone and intrinsic selection against the hybrids that are produced. This on-going process allows tension zones to be stable for long periods of time, thus the name dynamic equilibrium. Similarly, Harrison's mosaic model (Harrison 1986) predicts that selection and dispersal are both important in maintaining the hybrid zone, and that hybrids are less fit than parental individuals. However, in Harrison's model, selection against hybrids is due to environmental heterogeneity such that the one parental species is fit in one habitat,

the other species is fit in another and hybrids are less fit in both. Thus, in contrast to the clinal patterns predicted for tension zones, mosaic hybrid zones have a patchy distribution of parental and hybrid genotypes that tracks habitat variation. A third hybrid zone model is Moore's hybrid superiority model (Moore 1977). As suggested by the name, this theory states that hybrids are more fit than parental taxa in intermediate habitats. Thus, selection is thought to be environment-dependent. In contrast to the theories described previously, dispersal is thought to be of little importance for hybrid zone maintenance. Instead, hybrid zones of this type may be stable for long times due to the fitness advantage of hybrids in intermediate habitats, often along ecotones. Finally, Arnold's evolutionary novelty model (Arnold 1997) combines characters of several previous models. He suggests that hybrid zones are structured by both intrinsic and extrinsic selection. After an initial hybrid is established (which in his model is a rare event), hybrid zones may be long lived with mosaic distributions of genotypes. He stresses that some hybrid genotypes are unfit while others may exceed that of parental genotypes, and hybrids may be associated with the same or different habitats as the parental species.

The tension zone model is the most frequently cited hybrid zone model (Arnold 1997). However, the information necessary to correctly identify the model that best fits a given hybrid zone is often equivocal. For example, whether a hybrid zone appears clinal or

TABLE 13.1   The effects of increases in 13 different ecological parameters on the time to extinction (T_E) of a hybridising rare species (from Wolf & Rieseberg, in prep; see Chapter 13.5.2)

Model parameter	Effect on T_E of rare species
Frequency of rare species (Population size constant)	Increase[A]
Total population size, frequencies of rare and common species constant	Increase
Total population size, variation in	Decrease
Pollen grains, number of per stigma	Increase
Seeds, number of per patch of habitat	Species with faster growth rate benefits
Selfing rate of rare species	Increase
Pollen production of rare species	Increase
Pollen tube, growth rate of rare sp. on rare sp. flower	Increase
Pollen tube, variation in growth rate of rare species	Increase
Number of ovules of rare species	Increase
Ovules, variation in number of rare species	None
Seedling growth rate of rare species (competitive ability)	Increase[B]
Competitive ability variation in rare species	Increase

[A]See text Figure 13.2. [B]See text Figure 13.3.

mosaic depends on the scale at which it is sampled. The *Gryllus firmus* Scudder × *Gryllus pennsylvanicus* Burmeister hybrid zone in the north-eastern United States of America exhibits clinal variation on a broad scale, but on a local scale, individuals are distributed in a mosaic fashion (Rand & Harrison 1989). At even finer scales along ecotones, variation is once again clinal (C. Ross, pers. comm., 30 July 1997). Another obstacle to identifying the correct hybrid zone model is the difficulty of obtaining lifetime fitness estimates for long-lived hybrid plants. To date, most fitness estimates for plant hybrids have been based on vegetative growth and/or seed set from a single season and seem unlikely to be valid for perennial herbs or trees. Ideally, to determine which hybrid zone model best fits the patterns of a particular set of hybridising taxa, one would: genetically sample individuals across the zone using both large-scale and small-scale sampling regimes; determine habitat dependence versus independence; and compare the lifetime fitness of hybrids to that of the parental species in a variety of habitats. An ideal way to measure fitness in plants is to determine the parentage of all individuals in a population with molecular markers rather than estimating fitness indirectly from vegetative growth rates or seed sets. Parentage analysis takes into account both male and female components of fertility and that selection may eliminate some of the progeny before they reach maturity. Dispersal of individuals within

the hybrid zone would also be measured. Finally, historical information relating to the stability of the zone would be obtained. Needless to say, it will often be impossible to obtain all of this information, and in many situations, robust estimates of fitness may be sufficient for the study in question.

13.5 Natural hybridisation and conservation

Plant evolutionary biologists historically have stressed the creative nature of hybridisation; individuals may be formed whose novel genotypes allow them to utilise different niches than their parents (e.g. Anderson 1948; Stebbins 1950). In such situations, hybridisation can lead to race formation, hybrid speciation or to the formation of stabilised introgressants (Grant 1981; Rieseberg & Wendel 1993). However, hybridisation can also lead to a loss of biodiversity. There are two ways in which hybridisation of a rare species can further endanger that species. When the hybrids that are produced have low fitness, the result can be outbreeding depression (Price & Waser 1979; Templeton 1986). Rare populations suffer reduced fitness caused by wasting their gametes on hybrid individuals that have low viability and/or fertility. This problem may be manifested as the low levels of seed set sometimes seen when rare plants and their congeners are sympatric or manifested as sterile or weak hybrids

---

Box 13.2 Genetic assimilation through hybridisation: the Catalina Island mahogany

One of the rarest trees in North America is *Cercocarpus traskiae* Eastw. (Catalina Island mahogany), whose population size has dwindled to six pure adult trees (Rieseberg 1996). This distinctive species is restricted to wild boar gully on the south-west side of Santa Catalina Island off the coast of California, USA. When the population was first discovered in 1897, it consisted of more than 40 trees, but it has declined rapidly over the past century. Two factors appear to have caused this decline: grazing and rooting by introduced herbivores; and interspecific hybridisation with its more abundant congener, *Cercocarpus betuloides* ssp. *blancheae* (C. Schneider) Little (mountain mahogany).

Although *Cercocarpus betuloides* ssp. *blancheae* is not found in wild boar gully, hybridisation between the two mahogany species appears to be frequent. In addition to the six pure *Cercocarpus traskiae* trees in wild boar gully, five other adult trees and at least 7% of newly established seedlings are of hybrid origin (Rieseberg & Gerber 1995). Presumably, wind pollination allows *Cercocarpus betuloides* ssp. *blancheae* from nearby canyons to sire hybrid plants in wild boar gully.

Because nearly half of the global genetic diversity in *Cercocarpus traskiae* would be lost by the removal of hybrid or introgressed adult and juvenile individuals, it would be risky to eliminate these individuals from wild boar gully. The other alternative, which is being implemented by the Catalina Island Conservancy, is the transplantation of cuttings of pure *Cercocarpus traskiae* individuals to other locations on the Island where hybridisation is unlikely.

(Ellstrand & Elam 1993). When populations are small or when hybridisation is common, inadequate numbers of non-hybrid seeds may be produced to maintain the rare species. If hybrids exhibit little or no reduction in fitness, extinction through genetic assimilation rather than outbreeding depression can occur. In each successive generation of hybridisation, increasing numbers of individuals are hybrid and there is a corresponding decrease in the number of pure individuals of the rare species. Factors affecting the rate of genetic assimilation (time to extinction) include the strength of reproductive barriers, the fitness of hybrids and population sizes (Table 13.1).

13.5.1 Examples of hybridising rare species

There are many plant and animal species that are threatened with extinction due to hybridisation with more common relatives. Some of the more well-known examples are *Canis rufus* Audubon & Bachman (red wolf; Wayne & Jenks 1991), *Cercocarpus traskiae* Eastw. (Catalina Island mahogany) (Box 13.2; Rieseberg *et al.* 1989; Rieseberg & Gerber 1995), several duck endemics that hybridise with mallards

(*Anas platyrhynchos* L.) (Lever 1987; Griffin *et al.* 1989; Mazourek & Gray 1994; Rhymer *et al.* 1994), *Felis silvestris silvestris* Schreber (wildcat) (Hubbard *et al.* 1992), *Cervus elaphus* L. (Scottish red deer) (Abernethy 1994), and *Oncorhynchus clarki* Richardson (cutthroat trout) (Martin *et al.* 1985; Allendorf & Leary 1988). Other rare taxa that hybridise are shown in Table 13.2 that lists rare woody plant taxa from California, Hawaii and the British Isles that hybridise. These floras were surveyed because they are among the few that provide data on both rarity and hybridisation, or where other resources are available from which the necessary information can be obtained easily (e.g. the *RareFind* database for California; California Department of Fish and Game, Natural Diversity Data Base 1997a). In the British Isles, a total of 130 rare plant taxa are involved in hybridisation, while in California there are 39 and in Hawaii, 38. The greatly elevated figure for the British Isles relative to the other two floras could be partly because of the degree to which the plants of the three regions are known. However, it is unlikely that this explanation accounts completely for the difference. Historical factors such as species

TABLE 13.2 Examples of hybridisation involving rare and endangered woody plants from three floras

Underlined taxa are rare. A '?' after the first species' name means that hybridisation is suggested but not certain between the pair.

Family	Species	Species	Species
	California, USA[A]		
Ericaceae	_Arctostaphylos pallida_ Eastw. (2)	× ? Species not specified	
Fagaceae	_Quercus tomentella_ Engelm. (3)	× _Quercus chrysolepis_ Liebm.	
Fagaceae	_Quercus englemannii_ E. Greene (3)	× _Quercus dumosa_ Nutt.	
Juglandaceae	_Juglans californica_ var. _hindsii_ Jeps. (1)	× _Juglans regia_ L.	
Rhamnaceae	_Ceanothus ophiochilus_ S. Boyd, T. Ross, & Arnseth (1)	× _Ceanothus crassifolius_ Torrey	
Rosaceae	_Cercocarpus traskiae_ Eastw. (1)	× _Cercocarpus betuloides_ var. _blanchae_ (C. Schneider) Little	
Salicaceae	_Salix delnortensis_ C. Schneider (4) ?	× _Salix lasiolepis_ Benth.	
	Hawaii, USA[B]		
Amaranthaceae	_Charpentiera densiflora_ Sohmer (V)	× _Charpientiera elliptica_ (Hillebr.) A. Heller	
Asteraceae	_Argyroxiphium kauense_ (Rock & M. Neal) O. Deg. & I. Deg. (E)	× _Dubautia dolosa_ (O. Deg. & Sherff) G.D. Carr	
Asteraceae	_Argyroxiphium kauense_ (E)	× _Dubautia ciliolata_ ssp. _ciliolata_ (DC.) Keck	
Asteraceae	_Argyroxiphium sandwicense_ ssp. _macrocephalum_ DC. (E)	× _Dubautia menziesii_ (A. Gray) Keck	
Asteraceae	_Argyroxiphium sandwicense_ ssp. _macrocephalum_ (E)	× _Argyroxiphium virescens_ Hillebr. (Ex?)	
Asteraceae	_Argyroxiphium virescens_ (Ex?)	× _Argyroxiphium grayanum_ (Hillebr.) O. Deg.	
Asteraceae	_Bidens micrantha_ ssp. _ctenophylla_ Gaudich. (R)	× _Bidens menziesii_ (A. Gray) Sherff	
Asteraceae	_Dubautia arborea_ (A. Gray) Keck (V)	× _Dubautia ciliolata_ ssp. _glutinosa_ (DC.) Keck	
Asteraceae	_Dubautia arborea_ (V)	× _Dubautia linearis_ ssp. _hillebrandii_ (Gaudich.) Keck	
Asteraceae	_Dubautia imbricata_ St. John & G.D. Carr (R)	× _Dubautia laevigata_ A. Gray	
Asteraceae	_Dubautia imbricata_ ssp. _acronaea_ St. John & G.D. Carr (R)	× _Dubautia paleata_ A. Gray	
Asteraceae	_Dubautia imbricata_ ssp. _acronaea_ (R)	× _Dubautia waialealae_ Rock	
Asteraceae	_Dubautia microcephala_ Skottsb. (R)	× _Dubautia plantaginea_ ssp. _magnifolia_ Gaudich.	
Asteraceae	_Dubautia sherffiana_ Fosberg (V)	× _Dubautia plantaginea_ ssp. _plantaginea_ Gaudich.	
Asteraceae	_Tetramolopium rockii_ var. _calcisabulorum_ Sherff (V)	× _Tetramolopium rockii_ var. _rockii_ Sherff (V)	
Asteraceae	_Wilkesia hobdyi_ St. John? (E)	× _Dubautia_ spp.	
Asteraceae	_Wilkesia hobdyi_ ? (E)	_Argyroxiphium_ spp.	

173

Family	Species	Species	Species
Campanulaceae	*Cyanea obtusa* (A. Gray) Hillebr. (Ex?)	× *Cyanea elliptica* (Rock) Lammers	
Euphorbiaceae	*Chamaesyce halemanui* (Sherff) Croz. & O. Deg.? (E)	× *Chamaesyce remyi* (A. Gray ex Boiss.) Croz. & O. Deg. (R)	
Euphorbiaceae	*Chamaesyce remyi* (R)	× *Chamaesyce sparsiflora* (A. Heller) Koutnik (R)	
Gesneriaceae	*Cyrtandra dentata* St. John & Storey (E)	× *Cyrtandra laxiflora* H. Mann	
Gesneriaceae	*Cyrtandra giffardii* Rock (E)	× *Cyrtandra platyphylla* A. Gray	
Gesneriaceae	*Cyrtandra kohalae* Rock (Ex?)	× *Cyrtandra platyphylla*	
Gesneriaceae	*Cyrtandra munroi* C. N. Forbes (E)	× *Cyrtandra grayana* Hillebr.	
Gesneriaceae	*Cyrtandra oxybapha* W. L. Wagner & Herbst (R)	× *Cyrtandra grayana*	
Gesneriaceae	*Cyrtandra polyantha* C. B. Clarke (E)	× *Cyrtandra paludosa* var. *paludosa* Gaudich.	
Gesneriaceae	*Cyrtandra sandwicensis* (H. Lév.) St. John & Storey (V)	× *Cyrtandra cordifolia* Gaudich.	
Gesneriaceae	*Cyrtandra sandwicensis* (V)	× *Cyrtandra grandiflora* Gaudich.	
Gesneriaceae	*Cyrtandra subumbellata* (Hillebr.) St. John & Storey (E)	× *Cyrtandra garnotiana* Gaudich.	
Gesneriaceae	*Cyrtandra subumbellata* (E)	× *Cyrtandra kalihii* Wawra	
Malvaceae	*Hibiscadelphus giffardianus* Rock (E)	× *Hibiscadelphus hualalaiensis* Rock (E)	
Malvaceae	*Hibiscus arnottianus* ssp. *immaculatus* A. Gray (E)	× *Hibiscus rosa-sinensis* L. (ornamental)	
Pittosporaceae	*Pittosporum napaliense* Sherff (R)	× *Pittosporum kauaiense* Hillebr.	
Pittosporaceae	*Pittosporum napaliense* (R)?	× *Pittosporum glabrum* Hook. & Arnott	
Plantaginaceae	*Plantago princeps* Cham. & Schlechtend. var. *longibracteata* Mann.? (R)	× *Plantago pachyphylla* A. Gray	
Ranunculaceae	*Ranunculus hawaiensis* A. Gray? (E)	× *Ranunculus mauiensis* A. Gray (R)	
Rubiaceae	*Hedyotis fluviatilis* (C. Forbes) Fosberg (R)	× *Hedyotis acuminata* (Cham. & Schltdl.) Steud.	

British Isles[C]

Family	Species	Species	Species
Ulmaceae	*Ulmus plotii* Druce (R)	× *Ulmus glabra* Huds.	
Ulmaceae	*Ulmus plotii* (R) ?	× *Ulmus glabra*	× *Ulmus minor* Mill.
Ulmaceae	*Ulmus plotii* (R)	× *Ulmus procera* Salisb.	
Ulmaceae	*Ulmus plotii* (R)	× *Ulmus minor*	
Betulaceae	*Betula nana* L. (R)	× *Betula pubescens* Ehrh.	
Tiliaceae	*Tilia platyphyllos* Scop. (R)	× *Tilia cordata* Mill.	
Cistaceae	*Helianthemum apenninim* (L.) Mill. (RR)	× *Helianthemum nummularium* (L.) Mill.	
Salicaceae	*Salix lapponum* L. (R)	× *Salix caprea* L.	

Family	Species	Species	Species
Salicaceae	_Salix lapponum_ L. (R)	× _Salix aurita_ L.	
Salicaceae	_Salix lapponum_ L. (R)	× _Salix myrsinifolia_ Salisb.	
Salicaceae	_Salix lapponum_ L. (R)	× _Salix myrsinifolia_	× _Salix phylicifolia_ L.
Salicaceae	_Salix lapponum_ L. (R)	× _Salix phylicifolia_	
Salicaceae	_Salix lapponum_ L. (R)	× _Salix repens_ L.	
Salicaceae	_Salix lapponum_ L. (R)	× _Salix lanata_ L. (RR)	
Salicaceae	_Salix lapponum_ L. (R)	× _Salix arbuscula_ L. (R)	
Salicaceae	_Salix lapponum_ L. (R)	× _Salix herbacea_ L.	
Salicaceae	_Salix lapponum_ L. (R)	× _Salix reticulata_ L. (R)	
Salicaceae	_Salix lanata_ (RR)	× _Salix herbacea_	
Salicaceae	_Salix arbuscula_ (R)	× _Salix repens_	
Salicaceae	_Salix arbuscula_ (R)	× _Salix herbacea_	
Salicaceae	_Salix arbuscula_ (R)	× _Salix reticulata_ (R)	
Salicaceae	_Salix myrsinites_ L. (R)	× _Salix caprea_	
Salicaceae	_Salix myrsinites_ (R)	× _Salix myrsinifolia_	
Salicaceae	_Salix myrsinites_ (R)	× _Salix myrsinifolia_	× _Salix phylicifolia_
Salicaceae	_Salix myrsinites_ (R) ?	× _Salix phylicifolia_	
Salicaceae	_Salix myrsinites_ (R)	× _Salix herbacea_	
Ericaceae	_Erica ciliaris_ L. (RR)	× _Erica tetralix_ L.	
Ericaceae	_Erica mackaiana_ Bab. (RR)	× _Erica tetralix_	
Rosaceae	_Rosa agrestis_ Savi (RR)	× _Rosa pimpinellifolia_ L.	
Rosaceae	_Rosa agrestis_ (RR)	× _Rosa stylosa_ Desv.	
Rosaceae	_Rosa agrestis_ (RR)	× _Rosa canina_ L.	
Rosaceae	_Rosa agrestis_ (RR)	× _Rosa tomentosa_ Sm.	
Rosaceae	_Rosa agrestis_ (RR)	× _Rosa sherardii_ Davies	
Rosaceae	_Rosa agrestis_ (RR)	× _Rosa micrantha_ Borrer ex Sm.	
Rosaceae	_Sorbus rupicola_ (Syme) Hedl. (R)	× _Sorbus aria_ Crantz	
Rosaceae	_Sorbus rupicola_ (R)	× _Sorbus torminalis_ (L.) Crantz	
Rosaceae	_Sorbus porrigentiformis_ E. F. Warb. (R)	× _Sorbus torminalis_	
Thymelaeaceae	_Daphne mezereum_ L. (R)	× _Daphne laureola_ L.	

A The flora of California, USA. Hybridisation data are from California's _RareFind_ program (California Department of Fish and Game, Natural Diversity Data Base 1997a), and rarity information is from the California Special Plant List (California Department of Fish and Game, Natural Diversity Data Base 1997b). Species status: 1 , fewer than 1000 individuals, or fewer than 2000 acres (1 ha = 2.4 acres); 2, 1000–3000 individuals or 2000–10 000 acres; 3, 3000–10 000 individuals or 10 000–50 000 acres; 4, apparently secure, but factors exist to cause some concern; 5, population demonstrably secure.
B The flora of Hawaii, USA (Wagner _et al._ 1990). Species status: R, rare; V, vulnerable; E, endangered; Ex, extinct.
C The flora of the British Isles (Stace 1997). 'R's following a species name indicate the degree of rarity: R, scarce, found in no more than 100, 10 by 10 km grid-squares in the British Isles; RR, rare, found in no more than 15 grid-squares (these plants are listed in the British Red Data Book; Perring & Farrell 1983); RRR, rare and endangered or vulnerable (these plants are listed in Schedule 8, 'Plants Which are Protected', of the Wildlife and Countryside Act; United Kingdom 1981).

introductions and habitat disturbance probably also contribute. It is also interesting to note the large fraction of Hawaii's rare hybridisers that are woody (32 of 38). This is most likely because several of the most speciose genera in Hawaii are woody.

For rare taxa to hybridise, they must occur within mating distance of a closely related species. In California, where there are many plant species threatened by hybridisation (Table 13.2), 90% of the threatened or endangered plants occur sympatrically or parapatrically with one or more congeners (Ellstrand 1992a). Hybridisation is also more likely if one species is present in larger numbers than another. The common species acts as a source of pollen and the rare species as a sink (Ellstrand 1992a; Ellstrand & Elam 1993). If only a small number of flowering individuals of a rare species is present at the same time that many congeners are present, most pollen available to the rarer species will be heterospecific. A higher proportion of its progeny will be hybrid than will those of the common species since most of the pollen available to the common species will be conspecific. The greater the difference in numbers, the greater the detrimental effect to the rare species (Table 13.1).

13.5.2 Theoretical work

Very little theoretical work has been directed to determining rates of extinction due to hybridisation. Ellstrand and Elam (1993) suggested that absolute and relative population sizes, the proximity of related species and rates of hybridisation affect the time to genetic assimilation. Other factors that have been predicted to affect rates of genetic assimilation include the fitness of hybrids and the availability of suitable habitats for hybrids (Levin et al. 1996).

The first detailed theoretical model to directly estimate rates of extinction due to hybridisation is presented (Fig. 13.1; Wolf and Rieseberg, in prep.). These include the parameters suggested by Ellstrand and Elam (1993) and Levin et al. (1996), as well as numerous additional parameters such as selfing rate, pollen production and competitive ability. The life cycles of two hybridising annual plants were explicitly simulated in C, using Metrowerks CodeWarrior Academic Pro 11 for MacOS. Fifty rare plants and 100 common plants produced pollen, selfed at the prescribed rate and contributed outcross pollen to a common pollen pool. Pollen was distributed randomly among stigmas leading to non-selfed ovules, and the fastest pollen grain on each stigma fertilised the one

ovule in the associated ovary. Seeds were distributed randomly among the 150 patches, seedlings competed within patches and the fastest-growing seedling in each patch flowered. Values for most parameters were chosen randomly from a normal distribution, although the number of pollen grains per stigma and the number of seeds per patch were chosen from a Poisson distribution, and selfing-rate was a fixed value. Values for the remaining parameters (Table 13.1) were the same in all simulations except when a particular parameter was under investigation. Total population size was held constant at 150 individuals. For all other parameters, the rare species, common species and hybrids had equal values, except that conspecific pollen tubes were assumed to grow twice as fast as foreign pollen. All individuals were classified as 'rare', 'hybrid' or 'common'. An offspring resulting from a hybrid parent could never be classified as rare or common, even if it was a late-generation backcross. This may not be completely realistic, as late-generation backcross individuals may be nearly indistinguishable from pure parental individuals. However, this assumption seems justified for two reasons: hybrids are more likely to backcross with the common species than the rare species, so the assumption should not affect the probability of extinction of the rare species very much; and it is important to understand the probability that the rare taxon can survive as a distinct species without contamination from a foreign genome. However, this does not mean that 'impure' populations of rare species should not be protected. Nearly all of the variables examined were important in determining the mean time to extinction (T_E; Table 13.1) in a non-linear fashion (Figs 13.2 & 13.3). In general, favourable values for only one parameter could not prevent extinction. However, the rare species could be protected by the synergistic action of favourable values for several parameters.

Although the simulations showed that total and relative population sizes (Fig. 13.2), relative reproductive values and competitive ability (Fig. 13.3) affect the extinction rate of the rare species, as predicted previously (Rieseberg 1991; Ellstrand & Elam 1993; Levin et al. 1996), there were also some unexpected results. The mean number of pollen grains per stigma and the mean number of seeds per habitat patch (in which only 1 seedling will mature to flowering) strongly influenced the number of generations to extinction of the rare species. If, on

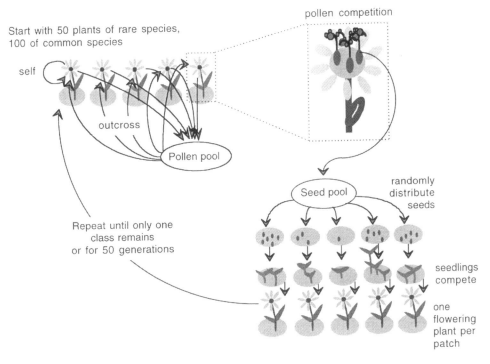

Figure 13.1 Simulation of life cycles of two hybridising annual plants in order to estimate rates of extinction due to hybridisation.

average, rare pollen grew faster on the stigmas of rare plants than foreign pollen, as would be expected (i.e. prepotency of conspecific pollen; Darwin 1876), then increases in the number of pollen grains per stigma increased the probability that one of the pollen grains was a fast, rare grain. The rare pollen grain would subsequently fertilise the single ovule to which the stigma led, and a seed of the rare species would be produced. Thus, an increase in the number of pollen grains per stigma increased the T_E. Similarly, if the mean growth rate of rare seedlings was faster than that of the common species (i.e. they have a greater competitive ability), an increase in the number of seeds per patch would benefit the rare species. However, if the rare species has slower growth rates, it benefited from fewer seeds per patch. This was because seeds were likely to land in a patch with low levels of competition, thus increasing the probability that rare seeds would mature and flower.

Furthermore, the variances associated with parameter means were just as important as the means themselves for several parameters: population size, pollen tube growth rate and seedling growth rate. Increased variation in the total number of available patches (i.e. population size) decreased T_E because of increased demographic stochasticity. When variation in rate of rare pollen tube growth increased, the rare species generally benefited. When variation was high, a few very fast pollen grains were produced. If many pollen grains were produced by each plant, many grains would land on each stigma. Then the chance that a stigma had a fast, rare pollen grain was increased, resulting in greater production of pure rare seeds. Some very slow pollen grains would also be produced, but the loss of these was less important than the benefit from the fast pollen. Similarly, increased variation in seedling growth rate was beneficial to the slower-growing seedlings when there were many seeds in each patch; and variation was detrimental when there were few seeds per patch. Preliminary analyses suggested that rare seed production and competitive ability (both mean and variance) were the most important variables in determining the probability that the rare species would become extinct. However, these results may have been because of the assumption of only one flowering plant per patch.

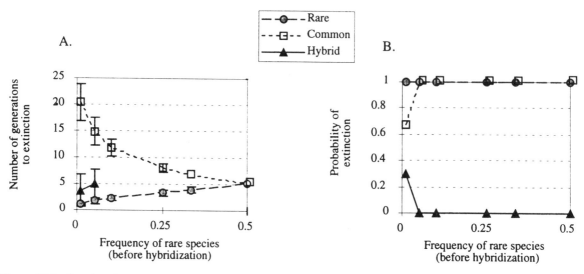

Figure 13.2 Results of simulations (see text and Fig. 13.1) were tabulated by recording (a) the time to extinction and (b) the probability of extinction of a rare species that hybridises with a common species. In this run, the relative population sizes of the two species were varied, with the total population size held constant at 150 individuals. (a) Time to extinction. The rare species becomes extinct within one to five generations, depending on the rare species' population size relative to the common species. Hybrids only become extinct when the frequency of the rare species is extremely low; thus, the 'hybrid' line ends at the point where their extinction no longer occurs. (b) Probability of extinction. Under these conditions, both the rare and common species became extinct and hybrids persisted (i.e. assimilation of the rare species occurred) except when the frequency of the rare species was extremely low.

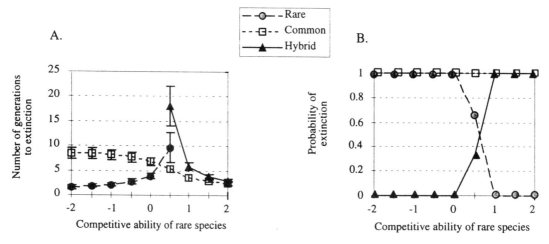

Figure 13.3 Results of simulations (see text and Fig. 13.1), varying the relative competitive advantage of the rare species over the common species. The average competitive ability, or seedling growth rate, of the rare species increases, while the average competitive abilities of hybrids and the common species are constant with a relative value of zero and a standard deviation of one. (a) Time to extinction. When the competitive value of the rare species is one standard deviation greater than that of the hybrids and common species, the 'rare' species never becomes extinct, thus the 'rare' line stops at this point. (b) Probability of extinction. The rare species can persist only if it has a greater competitive ability than the common species or hybrids.

13.5.3 Management implications

Our model verifies the potential risk of hybridisation when a rare species comes in contact with a congener. Extinction through hybridisation can happen very quickly—in one generation in extreme cases and easily within five generations, especially if the threatened species is less vigorous or produces fewer seeds or pollen grains than the more common species. If the common species has expanded recently into the range of the threatened species and hybrids are observed, the common species and hybrids should be eradicated quickly before pure rare individuals are lost. Alternatively, the rare species can be transplanted to a remote location where the other hybridising taxon does not occur. However, if the two species have been co-occurring for many generations without a noticeable decline in the rare species, hybridisation is probably not an imminent threat.

To determine the imminency of the hybrid threat, we suggest the simple approach of monitoring population structure because the various parameters interact synergistically, and all of the parameters are important. Thus, it is probably not sufficient to measure a single parameter, such as the probability of first generation (F_1) formation in a glasshouse cross, or the fertility of F_1s. These studies are useful in estimating the hybrid threat, and can help managers learn to identify the variety of hybrid morphologies. However, low rates of F_1 formation in the glasshouse do not guarantee that there is no risk, as interspecific pollen grains or tubes on the rare plants could block intraspecific pollination (Levin et al. 1996). Low F_1 fertility also fails to predict the hybrid threat, as seen in sunflowers. Only 5.6% of pollen is viable in hybrids between *Helianthus annuus* L. and *Helianthus petiolaris* Nutt. (Ungerer et al. 2000). Yet hybrid zones between the two species are persistent and common, and late-generation hybrids regain nearly full fertility (Rieseberg et al. 1999). Furthermore, a lack of F_1s in the field is not sufficient to guarantee that hybridisation is not a threat. The formation of F_1s is extremely low in natural populations of *Iris fulva*, *Iris hexagona* and *Iris brevicaulis* (Louisiana irises) (< 1%; Arnold et al. 1993; Cruzan & Arnold 1993, 1994; Hodges et al. 1996), yet later-generation hybrids are abundant. Thus, tests of F_1 formation, vigour and fertility should be combined with careful observations of changes in the number of rare, common and hybrid individuals from one generation to the next.

13.5.4 Controlled hybridisation of rare species

In some cases, controlled hybridisation of an extremely rare species with a common congener is the only way to preserve the species. Hybridisation between subspecies or congeners can be used to prevent inbreeding depression from further threatening the rare species and to increase its chance of survival in a changing environment. For example, O'Brien et al. (1990) suggested that introgression of *Felis concolor* L. (South American panther) genes into *Felis concolor coryi* L. (Florida panther) might help eliminate the genetic defects that were caused by extensive inbreeding. Similarly, *Panthera leo persica* L. (Asiatic lions) involved in a captive breeding program in the United States' zoos demonstrated increased vigour when two *Panthera leo* L. (African lions) were unintentionally included in the program (Wildt et al. 1987). Hybridisation may also be used to transfer resistance genes into a species threatened by pests or disease, such as *Castanea dentata* (Marshall) Borkh. (American chestnut) (Box 13.3). Finally, when very few or only a single individual of a species remains, hybridisation with a congener may be employed so that its unique genome will not be lost entirely. The *Trochetiopsis erythroxylon* (Forst. f.) Marais (St Helena redwood) became extinct in the wild in the 1950s. Seed was gathered prior to its extinction, but the resulting trees were self-pollinated and the species soon exhibited symptoms of catastrophic inbreeding depression. In 1980, two individuals of *Trochetiopsis ebenus* Cronk (St Helena ebony), also presumed extinct, were rediscovered on a cliff (Cronk 1986). The two species were crossed and the hybrids, known as *Trochetiopsis* × *benjamini* Cronk, exhibited extreme heterosis. Therefore, genes of each of these two species were preserved (Cronk 1995; Rowe 1995).

13.5.5 Taxonomic and legal issues

Hybridisation creates taxonomic ambiguity because it blurs boundaries between otherwise distinctive sexual species (such as in *Eucalyptus*; Potts & Wiltshire 1997) and, in conjunction with polyploidy and asexual reproduction, can generate flocks of clonal microspecies that span the range of morphological variability between parental taxa. This not only causes problems in species delimitation but it also provides a challenge to species concepts based on sexual isolation such as the BSC. Until recently, scientific controversy relating to species classification was of academic

Box 13.3   Using hybridisation to save the American chestnut

Castanea dentata (Marshall) Borkh. (Fagaceae) (American chestnut) was previously a widespread species in the USA ranging from southern Maine to Alabama and west to Illinois and the Appalachians, often making up 25% of forests (Exum 1992). Individual trees reached up to 100 ft (30.5 m) in height and 10 ft (3 m) in diameter (Figure 1). It was an economically important tree, with its nuts a holiday favourite, its bark providing tannins important to the leather trade and its timber prized for its straight grain, light weight and resistance to decay and warping. It also participated in complex ecological associations with many other species, and the nuts were a major food source for many animals.

Figure 1   This early 20th century photograph of lumberjacks standing among *Castanea dentata* (American chestnut) trees illustrates the large size commonly reached by these trees before the introduction of chestnut blight. Photo courtesy of the American Chestnut Foundation.

In 1904, *Cryphonectria parasitica* (Murr.) Barr (formerly *Endothia parasitica* (Murrill) P.J. Anderson & H.W. Anderson) pathogen (chestnut blight) was introduced into the USA on a *Castanea crenata* Sieb. & Zucc. (Japanese Asian chestnut) or a *Castanea mollissima* Blume (Chinese Asian chestnut). The disease was first discovered in the Bronx Zoo (Merkel 1905), and quickly spread throughout the range of *Castanea dentata*, at 20 to 50 miles per year (32–80 km/y) (Brooks 1937; Exum 1992). By the 1950s the tree had essentially disappeared from the forests, a loss of roughly 3.5 billion trees (Wink 1984; Exum 1992).

Chestnut blight is a disease caused by a fungus. The fungus causes galls or orange cankers to form on the limbs and trunks of chestnuts, killing them above the point of infection. However, the roots are not killed by the fungus and some continue to send up shoots that later are infected by the blight (West 1988). These remaining shoots serve as a genetic reservoir for the species, as there is no significant reproduction in the wild.

Castanea mollissima is resistant to the chestnut blight and shows few symptoms when infected (Shear & Stevens 1913). Therefore, scientists have been working for several decades to transfer these resistance genes into *Castanea dentata* through a backcrossing regime (i.e. through controlled introgression). Hybrids (F_1) were produced and then crossed to *Castanea dentata* for several generations, interspersed with sibling matings. In each generation, those trees that exhibit resistance are selected for use in crosses. The eventual result should be a tree that is phenotypically a *Castanea dentata* but possesses resistance from *Castanea mollissima* to chestnut blight (Wallace 1984). This is the same process that plant breeders use to transfer desired traits among cultivars. However, a study of genetic diversity in *Castanea dentata* (Huang *et al*. 1998) suggests that more plants with diverse genotypes should be used as backcross parents to ensure that the resulting hybrids will be able to survive in the numerous habitats that comprise the natural range of *Castanea dentata*.

interest only. However, due to the current emphasis of many conservation strategies and legal measures on protecting species, these arcane scientific issues have become important legal and management concerns. Questions include:

(a) What is the legal status of species that hybridise? and

(b) What is the legal status of hybrid products such as F_1 hybrids, hybrid swarms, introgressive races and hybrid species?

In the USA, endangered species laws do not provide protection for most products of hybridisation such as hybrid zones or introgressive races. However, the law does appear to protect hybridising rare species such as *Cercocarpus traskiae* (Box 13.2), and legal attempts to exclude hybridising rare species from protection (e.g. Fergus 1991) have been unsuccessful. Whitham *et al*. (Whitham *et al*. 1991; Martinsen & Whitham 1994; Whitham *et al*. 1994; Whitham & Maschinski 1996) have

argued that legal protection should be extended to hybrid zones for the sake of preserving important centres of diversity and speciation. However, this may not be practical given the limited resources of management agencies and might result in the preservation of hybrid swarms that were actually contributing to the extinction of the rare taxon through outbreeding depression or genetic assimilation. However, strong arguments can be made for the preservation of *bona fide* hybrid species, since they often are morphologically and ecologically distinctive and sometimes found dynamic and speciose evolutionary lineages (Rieseberg 1991).

13.6 Conclusions

Despite the many isolating mechanisms that can prevent interspecific matings at any stage of the reproductive process, hybridisation is a common occurrence. However, it is not evenly distributed among organisms. Plants hybridise at higher frequencies than animals, and certain families and genera have higher incidences of hybridisation than others. Additionally, the outcome of hybridisation events varies greatly. In some cases, hybrids are sterile or inviable, while in others, hybridisation has led to the formation of new species or hybrid zones that are stable for long periods of time.

Hybridisation can also lead to a decrease in biodiversity. It is, indeed, a real threat to the survival of many plant and animal species. Ironically, their very rarity makes them more likely to hybridise than common species whose reproduction is not limited by a lack of available conspecific mates. Hybridisation involving rare species can be a threat in two ways. The fitness of the rare species may be decreased because of outbreeding depression, or there may be a cascade effect with more hybrids being formed each generation

until only individuals of the common species and hybrids exist (genetic assimilation). The number of generations necessary for the assimilation of the rare species may be very small and depends on population size and various reproductive and competitive factors. When hybridisation threatens a rare species, it is important that the more common species and the hybrids be eradicated quickly before all pure individuals of the rare species are lost.

In rare cases, hybridisation with a common congener may be the only way to save a species from extinction. Controlled hybridisation can be beneficial when genetic variation is extremely low, when genetic resistance to a threatening disease or pest is lacking, or when mating is no longer possible within the species.

SECTION IV
GENETIC DOMESTICATION AND
EX SITU CONSERVATION

The types of human disturbance described in Section III may have a variety of effects on forest gene pools and in some cases may necessitate interim or long-term *ex situ* measures to ensure conservation of genetic diversity. Planting trees clearly has the potential to conserve tree genetic resources *ex situ*, but may also modify gene pools in a variety of ways—these effects form the focus of Section IV. Collecting seed for *ex situ* gene conservation is a common practice which can complement *in situ* efforts when resources are limited, act as a primary conservation method when the long-term security of remaining populations within a provenance is tenuous or provide a genetic base for breeding programs. In making such collections, however, the opportunity exists for selection, either conscious or unconscious, the possibility that the sample is unrepresentative of the targeted gene pool, and as a consequence the conservation objectives are immediately compromised. In Chapter 14, Anthony Brown and Craig Hardner look at sampling strategies to ensure maximum representation of a species' gene pool and discuss issues that relate to:

- maximising sampling efficiency

- maintaining diversity within collections

- the importance of propagation for long-term *ex situ* gene conservation

- the role of core collections.

The process of domestication, by definition, involves selection and an associated narrowing of the natural gene pool. However, the extent to which commercial planting and associated breeding programs narrow gene pools varies greatly depending on the awareness of the issues and risks involved. While well-planned breeding programs inevitably narrow natural gene pools, they are still much wider than many poorly planned plantation programs. In some cases plantations have been established with seed from very few trees, with subsequent reductions in yields over generations through inbreeding (e.g. *Acacia mangium* in Sabah). In Chapter 15, Yousry El-Kassaby reviews the effects of domestication on gene pools and shows how the stages of domestication differ in their potential to conserve or erode genetic variation. He then focuses on approaches to conservation through domestication and the factors to be taken into account to meet conservation goals.

While tree breeders have, for many years, been concerned with the issue of 'contamination' of breeding or production populations by pollen from 'wild' populations, the complementary issue of gene flow from artificial populations into natural populations has recently also come under investigation. In particular, with the first commercial plantations of genetically modified trees imminent, the possibilities for escape into closely related natural stands become a critical issue. Similarly with an increasing interest in commercial plantations of species within their natural ranges and their use in ecological restoration (e.g. corridors), the degree to which locally adapted allelic complexes become diluted may compromise the adaptive ability of particular populations. In Chapter 16, W. Thomas Adams and Jaroslaw Burczyk look at interactions between artificial and natural populations and discuss the mode and consequences of gene exchange and their implications for conservation of genetic resources.

SAMPLING THE GENE POOLS OF FOREST TREES FOR *EX SITU* CONSERVATION

Anthony H.D. Brown and
Craig M. Hardner

SUMMARY

A population approach to sampling the genetic resources of forest species is inherent in the prominence of the 'provenance' concept in forest management. It is known from sampling theory for selectively neutral alleles that the main quantity of interest, the allelic richness or number of alleles per locus actually present in a sample, is expected to increase linearly with increasing population size. In contrast, the allelic richness of the sample increases in proportion to the logarithm of the sample size. This implies that the reward for extra sampling effort at a site follows the law of diminishing returns. An adequate sample is defined as one that includes with 95% certainty, at least one copy of an allele with arbitrary frequency 0.05 and this requires 59 unrelated gametes. Larger samples are needed when there are several alleles at the boundary frequency— the worst-case scenario of 20 equally frequent alleles requires a doubling of the size. Fruits are often the unit of sampling in trees species and we examine the effect of single versus multiple paternity of fruit on the number of fruit required. When paternity is unknown, a worst-case scenario is that all the fruit from one individual form a single full-sib array. It follows that fruit should be sampled from at least 15 trees to be sure of an adequate sample.

Strategies for the sampling of many populations depend on the amount of genetic diversity in each population and the genetic divergence between them. The best strategy accords priority to so-called 'local' common alleles, or those that exceed 0.10 in frequency in only one region or population. Such alleles make up an appreciable fraction (10–20%) of the genetic resources of trees. This strategy is in line with the provenance concept and contrasts with samples of 172-plants-per-species, which is an approach recently advocated by Lawrence and colleagues.

14.1 Introduction

Sampling is crucial to the conservation and use of the genetic resources of wild or cultivated plant populations. The collector needs to avoid three major pitfalls in sampling; that is, the sample is:

(a) too limited in coverage of the variation existing in the population

(b) biased in content

(c) too large to deal with.

Excessively small samples risk omitting significant genetic variation, and the source population may not survive for later resampling. Bias in samples can arise at two levels, the choice of populations to sample and the choice of individuals within populations. Whereas some degree of departure from naive random sampling may be sensible, excessive bias risks the omission of variants that may have a future use. Excessive numbers of samples are wasteful of effort and imply that other species or fewer populations will not be sampled.

Several authors have addressed the theory and the rationale of optimal sampling strategies (e.g. Allard 1970; Marshall & Brown 1975; Crossa 1989; Brown & Marshall 1995; Lawrence & Marshall 1995 and references therein). Much of this literature has had herbaceous plants in mind (but see Namkoong 1988; Krusche & Geburek 1991; Forest Resources Division, FAO 1995). Tree species, however, possess several distinctive features that affect sampling in the field (Box 14.1).

Perhaps the most important feature of Box 14.1 is the central concept of provenance in forest trees. The provenance concept has had a long history in forestry (Matyas 1996). It signals that geographical features, like climate, topography, soils and spatial isolation, shape the patterns of genetic variation in tree species. Differences in growth and phenology tend to relate to origin of samples in adaptive ways. Many studies have demonstrated genetic divergence among provenances of the same species (e.g. Stern & Roche 1974; Mikola 1982).

Thus, geographically different environments are likely to be home to suitably adapted genotypes. The provenance concept forms an essential link between *in situ* and *ex situ* conservation of forest trees. In order to restore forests after logging, the reseeding process should use seed originating from the same environment as the one to be restored. When species are introduced and tested in other countries, the provenance of individual samples becomes a pointer to their performance. For example, the proven provenance for planting *Eucalyptus camaldulensis* Dehnh. (river red gum) in tropical areas is one from Petford, northern Queensland, whereas the provenance sought for introduction to temperate regions is one from Lake Albacutya from

---

Box 14.1 Features that most forest tree species possess and that affect sampling (after Forest Resources Division, FAO 1995)

Features of most forest trees that affect sampling are:

1 adaptive divergence recognised as provenances

2 predominantly or preferentially outcrossing with high levels of intraspecific variation

3 tendency to asynchronous and irregular annual flowering and fruiting

4 fruiting often in inaccessible parts of the canopy

5 high ratios of fruit to seed mass, requiring the processing of samples in the field

6 collection of vegetative samples which sometimes is more efficient than seed sampling

7 coevolution with and dependence upon root microsymbionts, which may be essential for growth and long-term survival

8 large, long-lived organisms with complex age structure and great variation in fecundity.

north-western Victoria (Eldridge *et al.* 1993). The provenance tag is then an essential descriptor for the repeated sampling of germplasm of known performance.

Collecting strategies should therefore recognise the centrality of provenances as units of genetic resources. In this chapter, we will outline a basic sampling strategy that adheres to this principle, based on previous work on this topic in other plant species. Lawrence *et al.* (1995a, 1995b, 1997) have advocated an alternative approach to sampling—one that reduces the importance of population divergence. It is, therefore, important to examine their proposal with respect to forest trees.

14.2 Theoretical considerations

14.2.1 Assumptions

The following are the framework and assumptions that underlie our basic sampling strategy and its development.

(a) The collector aims to sample propagating material (e.g. seed, cuttings) for one or more purposes. These purposes are to: (i) provide germplasm (e.g. for breeding programs, seed orchards); (ii) provide samples for long-term storage as a method of conservation *ex situ* together with programs *in situ*; or (iii) provide planting material for direct use in the restoration of logged areas, or for introduction elsewhere. This chapter concentrates on (i) and (ii).

(b) The collector has limited time and resources and can collect and use only up to a fixed total amount of material.

(c) The species is sufficiently numerous that the collector has choices for the number and composition of the samples collected. Rare species, with few populations or individuals, may require modified strategies (Brown & Briggs 1991).

(d) The target species contains substantial stores of genetic variability—each individual is genetically unique. In some tree species, individuals can form extensive clones of a single genotype. Such species require care in sampling to avoid biased sampling of one or a few abundant clones.

(e) Within the target region, little or nothing is known of the amount of variation in each

population, nor the pattern of genetic divergence between them. The collector then begins with the assumption that each population is of equal value, and that all populations have the same level of genetic variance, with no 'hot-spots' of variation. This assumption often does not apply to inbreeding species (Schoen & Brown 1991).

(f) The collector's task is to capture the maximum amount of useful variation within a limited number of samples. Unfortunately the specific variants that would meet future needs are usually unknown, so a subsidiary aim is to preserve as many variants as possible at each locus. This implies that allelic richness, or the average number of alleles per locus, is the primary relevant parameter of genetic diversity.

14.2.2 Allelic richness of a sample from a population

Levels and patterns of genetic variation in populations can be summed up as a set of parameters. Strictly speaking, estimates of genetic diversity parameters cannot be known ahead of sampling. Estimates of allelic richness in particular are difficult to derive because their sample values steadily increase with sample size. Yet knowledge of the relationship between sample size and richness would assist decisions about sampling strategy.

The neutral allele theory of Kimura and Crow (1964) generates a family of allele frequency distributions for which the sampling theory is well developed. The equilibrium distributions are a function of the single combined parameter $\Theta = 4N\mu$ where N is the population size and μ is the mutation rate. The number of selectively neutral alleles, k, at a locus in a sample of size S random gametes is approximately:

$$k = \theta\log_e[(S+\theta)/\theta]+0.6 \qquad (14.1)$$

where $4N\mu > 0.1$ and $S > 10$ (Brown & Moran 1981). Table 14.1 lists values of allelic richness (k), for various values of polymorphism (Θ) and geometrically increasing values of S. If the rows of Table 14.1 represent a sequence of comparable model populations with the same μ that differ only in N and hence Θ, then by comparing the values within any one column, it is evident that the number of alleles present in a series of samples of the same fixed size increases almost linearly with increasing population size. In contrast, allelic richness in a series of samples from the

TABLE 14.1	The number of neutral alleles (*k*) expected in a random sample of size *S* gametes from a population in neutral equilibrium for various levels of polymorphism			
Θ	*S* = 25	*S* = 50	*S* = 100	*S* = 200
0.1	1.15	1.22	1.29	1.36
0.2	1.57	1.71	1.84	1.98
0.3	1.93	2.14	2.34	2.55
0.4	2.26	2.53	2.81	3.09
1.0	3.86	4.53	5.22	5.90
2.0	5.81	7.12	8.46	9.83
3.0	7.30	9.22	11.2	13.2
4.0	8.52	11.0	13.6	16.3
10.0	13.1	18.5	24.6	31.1

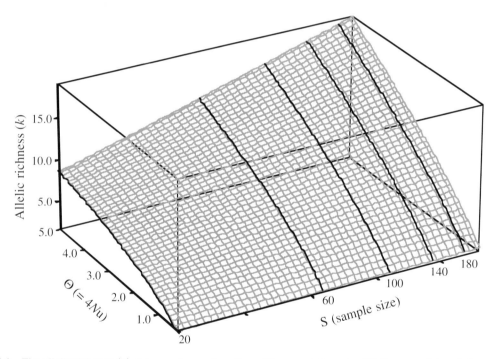

Figure 14.1 The allelic richness (*k*) of a sample as a function of the population size (*N*, linear scale) and the sample size (log$_e$*S*, logarithm scale).

same population increases in proportion to the logarithm of *S*. Figure 14.1 displays the interaction between population size and sample size in determining the allelic richness of a sample.

Results for allozyme alleles in natural populations of a species, however, tend to show a logarithmic relationship between population size and richness (Young *et al.* 1996), rather than the linear one that the neutral model predicts. Three possible reasons are as follows.

(a) Populations frequently depart from the theoretical assumption of equilibrium under the neutral allele model. In particular, population size is not stable for long. Many populations that are small may have been recently much larger. This implies that the allelic richness of these small populations will continue to decline. Their allelic richness will take many generations to fall to that expected at neutral equilibrium for the new size.

(b) Populations are not completely isolated over long periods as assumed in the model. Natural populations commonly form a dynamic network, or metapopulation, with cycles of migration, local extinction and recolonisation. Selection may inflate the effective migration rate (Hardner *et al.* 1998). Thus, it is difficult to estimate effective population size of a local population from its actual size.

(c) Allozyme electrophoresis fails to resolve all mutation as unique events and this affects the allele frequency distributions (Brown *et al.* 1981). Thus, in the infinite allele model, where every new mutation is to a unique allele, the number of alleles increase without limit as population size increases. In contrast, in the stepwise mutation model to which allozyme variation tends to conform, a unique mutation at the DNA level may be a recurrent mutation for net charge on the protein. In such a model, the actual number of alleles tends to approach a limit with increasing population size (Kimura & Ohta 1975). The same general tendency might be expected for allele frequencies at microsatellite loci (Shriver *et al.* 1993).

14.2.3 Sufficient size for a sample

Our present concern, however, is with the logarithmic relationship between S and k, which implies that the richness per individual declines as S increases. In contrast, the resources required to collect the sample at a site increase linearly with S, once the site is reached. Thus, each unit of cost incurred in increasing S is met with a diminishing return of collecting additional alleles. Further, excessive sampling at any one site limits an explorer's opportunity to collect new variants at other sites. It is, therefore, logical to define a criterion for an adequately sized sample. Marshall and Brown (1975) suggested that the sample should be sufficient to include at least one copy of 95% of the alleles that occur in the target population with $q > 0.05$. A random sample of 59 unrelated gametes is sufficient to achieve this objective. In a species mating completely at random this equates to about 30 individuals, whereas in a completely selfing species, this target requires 60 individuals. For simplicity, we convert this to 50 random unrelated individuals to

allow for differences in breeding system among plants. For other values of q, for example $q < 0.2$, S is given by:

$$S = \frac{-3}{\log_e(1-q)} \approx \frac{3}{q} \qquad (14.2)$$

If at least r copies of an allele are required, rather than just a single copy, sample size is approximately (Brown & Marshall 1995):

$$S \approx \frac{r + 1.645\sqrt{r+0.5}}{q} \qquad (14.3)$$

Rather than considering the numbers required for a single rare allele, Gregorius (1980) investigated the sample size required to ensure that all the alleles at a locus with frequency $(q) \geq 0.05$ are included with probability σ. For example, three alleles with $q = 0.05$ require $S = 80$ gametes (Marshall & Brown 1975). With a lower bound of target allele of $q = 0.05$, the most extreme case would be a locus with 20 alleles all of equal frequency. Gregorius (1980) computed the size required to ensure at least one copy of all 20 alleles ($\sigma = 0.95$ surety) to be 117. In this extreme case in which all the target alleles are equally frequent, an approximate general formula ($q < 0.2$) for the S required is:

$$S \approx \frac{3 - \log_e q}{q} \qquad (14.4)$$

The arbitrary criterion of 95% alleles with $q > 0.05$ has led to considerable debate (see Marshall 1989; Krusche & Geburek 1991; Lawrence & Marshall 1995). Lawrence and Marshall (1995) suggest either increasing the surety level to greater than 95%, or lowering the critical allele frequency. Two points need stressing. First, such changes dramatically and steeply increase the required sample size. This increase imposes a substantial cost particularly in forgoing the sampling of other sites. Second, the nomination of a specific general benchmark is an initial theoretical proposal and does not preclude the collector from varying the sample size because of species attributes or ecological conditions (Brown & Marshall 1995).

When open-pollinated fruits are to be harvested, the pollination system is an important variable. Wind pollination is more likely than animal pollination to generate progeny arrays that are half-sibs. Hence, 30 individual seeds might in theory yield a minimum similar number of independent male gametes. Animal pollination is generally more likely than wind pollination to produce fruit with a limited paternity. In extreme cases, a single sire might be the source of all the pollen involved in producing one or more fruit. In addition, a significant proportion of open-pollinated species may derive from selfing, with proportions varying between families and sites (Hardner *et al.* 1996).

14.2.4 Full-sib versus half-sib arrays

The effect on sampling of paternal relationship within progeny arrays of an outbreeding species is illustrated by comparing full-sib arrays with half-sib arrays in a simple model, where n progeny are sampled from each of m maternal plants for a total of $S = m \times n$ seeds (Brown *et al.* 1975). Consider a rare allele (A_1) with frequency q such that in a large, fully outbred population this allele occurs mostly in the heterozygous state (A_1A_x where A_x denotes any and all other alleles at that locus) with frequency $2q$ and all other genotypes collectively sum to $1 - 2q$.

In the full-sib case (FS), the m mothers correspond to m randomly chosen mating events. Let M_1 ($M_1 = [1 - 2q]^2$) denote the frequency of the first kind of mating ($A_xA_x \times A_xA_x$), M_2 ($M_2 = 4q[1 - 2q]$) that of the second kind ($A_1A_x \times A_xA_x$) and M_3 ($M_3 = 4q^2$) that of the third kind ($A_1A_x \times A_1A_x$). The joint probability that the sample consists of h matings of the first kind, i of the second kind and j of the third kind ($h = 0,...,$ m; $i = 0, ..., m - h$; $j = m - h - i$); and that allele A_1 is lost from all subsamples of k progeny from each mating is:

$$\frac{m! M_1^h M_2^i M_3^j (1/2^n)^i (1/4^n)^j}{h! i! j!} \quad (14.5)$$

Therefore, the overall probability that the allele will be lost ($P_L[FS]$) follows from summing over all permitted values of h, i and j as:

$$P_L[FS] = \left\{ (1 - 2q)^2 + \frac{4q(1 - 2q)}{2^n} + \frac{4q^2}{4^n} \right\}^m$$

$$= \left\{ 1 - 2q \left(1 - \frac{1}{2^n} \right) \right\}^{2m} \quad (14.6)$$

An interesting limiting case is when the number of seeds per mating (maternal plant) is extremely large, such as in numerous tree species with abundant tiny seed (*Eucalyptus* spp., *Melaleuca* spp.). With large n, limit of $P_L[FS]$:

$$P_L[FS] \to (1 - 2q)^{2m} \quad (14.7)$$

In the half-sib case (HS), the joint probability is conditioned on the two kinds of maternal genotypes (A_1A_x and A_xA_x). Analogous reasoning leads to the overall probability that the allele will be lost from m half-sib arrays each of size n as $P_L[FS]$:

$$P_L[HS] = (1 - q)^{mn} [1 - 2q(1 - 1/2^n)]^m \quad (14.8)$$

Note that:

$$\frac{P_L[FS]}{P_L[HS]} = \frac{P_L[HS]}{P_L} = \left[\frac{1 - 2q(1 - 1/2^n)}{(1 - q)^n} \right]^m \quad (14.9)$$

where P_L is the probability that the allele is loss in an unstructured random sample of nm offspring from the population. The fraction (14.9) exceeds one when $n > 1$. Thus $P_L[FS] \geq P_L[HS] \geq P_L$.

Table 14.2 lists values for $P_L[FS]$ and $P_L[HS]$ for various values of rare allele frequency, array number (m) and array size (n). The first set of examples considers our minimum target allele frequency of $q = 0.05$ with a constant total effort of $m \times n = 30$ ($= S/2$). This total number (30) is the number of unrelated outbred zygotes required for assuring sampling of such alleles with probability greater than 0.95. As expected, deploying the sampling effort within arrays of partially related progeny (n increasing) at the expense of fewer arrays (m) increases the chance of loss of rare alleles. Half-sib arrays are more powerful than full-sib arrays in retaining (i.e. lower P_L) rare alleles.

The second set of examples are for determining the number of arrays (of a given limited size n) required to reduce P_L below the chosen value of 0.05. Thus, for a fixed array size of three, 17 full-sib arrays are needed to meet this target, whereas only 13 arrays are needed in the case of half sibs. *The limit formula above (Equation 14.7) establishes that a minimum of 15 trees should be sampled when large amounts of seed are available from each maternal tree and when one wishes to guard against the extreme case of a single male paternity (full sibs) for all the seed on any one tree.* The third set of results in Table 14.2 answers the same questions as the second set, but with minimum allele frequency of 0.025, half the former case. As expected, the values of n required are roughly double those for $q = 0.05$.

TABLE 14.2 The probability that a given rare allele with population frequency q is not included in a sample of m arrays (maternal plants) each of array size n

The progeny arrays are either full sibs (*FS*) or half sibs (*HS*)

m	n	$P_L[FS]$	$P_L[HS]$
Values for $m \times n = 30$, $q = 0.05$			
5	6	0.355	0.128
6	5	0.294	0.116
7.5	4	0.228	0.103
10	3	0.160	0.086
15	2	0.096	0.067
30	1	0.046	0.046
Values for $q = 0.05$, $P_L < 0.05$			
20	2	0.044	0.027
17	2	0.071	0.046
17	3	0.044	0.015
13	3	0.092	0.041
16	4	0.043	0.008
10	4	0.140	0.046
15	500	0.042	0.0
Values for $q = 0.025$, $P_L < 0.05$			
60	1	0.048	0.048
40	2	0.047	0.029
34	2	0.074	0.049
34	3	0.048	0.017
25	3	0.107	0.048
32	4	0.046	0.008
21	4	0.133	0.044
30	500	0.046	0.0

Another effect of limited or biased paternity arises when only a few plants are the effective sources of all the pollen. The limiting worst case occurs when the A_1 allele happens to be absent from the sampled pollen pool, and ovules are the only source of the rare allele in question. In this case, the probability of loss of A_1 is:

$$P_L[HS \text{ no } A_1 \text{ in pollen}]$$
$$= [1 - 2q(1 - 1/2^n)]^m \qquad (14.10)$$

As expected, with large n, the limit is:

$$P_L[HS \text{ no } A_1 \text{ in pollen}] \to [1 - 2q]^m \qquad (14.11)$$

In such a case, a sample (m) of between 29 maternal plants (n large) and 59 plants (when $n = 1$) is needed for the rare allele ($q > 0.05$); and 59 to 119 for $q > 0.025$.

14.2.5 Sufficient number of sites

Perhaps more controversially, Brown & Marshall (1995) suggested that these statistical and cost-benefit arguments extend analogously to answering the question of how many sites to sample. A justifiable sample on such a basis consists of a set of 50 sites. This would ensure the sampling of at least one copy of all common alleles ($P > 0.05$) that occur in more than 5% of populations with about 90% surety.

Admittedly the analogy gives only a weak guide. Unlike the members of a population, potential sites in a region can differ completely from one another and yield no redundant samples. Yet it is helpful to start with a specific target of 50 sites per species per region in mind, and alter this target to meet specific situations. For example, it may prove very difficult to locate 50 populations of a rare species. However, it may be evident from data, or from observation during the mission, that the species is found in many distinct and diverse habitats in the region.

14.2.6 Kinds of alleles

The optimal strategy for sampling more than one population depends crucially on the amount of genetic diversity in each population and the extent of genetic divergence between them. In an attempt to define priorities and 'useful variation', Marshall and Brown (1975) defined four conceptual classes of alleles based on their allele frequency and their distribution within and between populations. Each allele was classed into whether it only ever occurred as a rare allele (frequency always < 5% or 10%) in contrast to when it was common and exceeded that frequency at least once. These two classes were each then divided into two subclasses on their geographic distribution, recognising whether the allele occurred in many populations (widespread), or in only one or a few adjacent populations (local).

This yields four classes of alleles:

1. common, widespread

2. common, local

3. rare, widespread

4. rare, local.

For common alleles, the division into widespread versus local is based only on the pattern of their common occurrences (i.e. an allele that reaches a frequency of 20% in only one population, and occurs as a very rare allele in several other places, is classed as a 'common, local'—its common occurrences are highly localised).

Of the four classes of alleles, Marshall and Brown (1975) argued that the 'common, local' alleles merit priority in sampling and hence in the devising of sampling strategies, for the following reasons. This second class presumably includes the alleles that confer specific adaptation to the local conditions. Collecting the first class—'common, widespread', which includes broadly adapted alleles—presents no problem as they are inevitably collected regardless of

strategy. Members of the third class are rare but in many populations. Their capture will depend on the total collecting effort and not on how numbers are deployed among populations. Specific members of the fourth class are extremely difficult to collect. This class includes variants that are very rare in the whole species and include recent or deleterious mutants. A fraction of this class of alleles will be included, but the pursuit of every existing specific rare localised variant is beyond the resources available.

Table 14.3 lists some recent examples of allozyme frequency distributions in tree species where more than five populations have been studied. These more recent data support the similar earlier analysis of Brown and Moran (1981) and demonstrate the reality and extent of the priority class (common, local) of alleles (about 10–20% of alleles in these studies).

14.3 Basic sampling strategy

From the above outline of the theory and rationale of sampling, the basic strategy is as follows.

14.3.1 Number and distribution of sites

The optimal procedure for coverage of the genetic diversity within a species is a structured random sample of sites. The first step, therefore, is to assemble the available physiographic, climatic, edaphic, botanical, genetic and land use information for the target region. The region is then zoned into a limited number of areas to emphasise environmental differences between zones. The total effort is then divided among these zones according to occurrence of the target species and any possible differences in genetic diversity among zones. Zones where the target species is more frequent, or apparently more variable, merit more sampling effort.

The target is about 50 sites per species. This is to be varied according to circumstance (e.g. species abundance, species ecological diversity, collecting resources, number of species to consider, value of the material).

14.3.2 Number of plants per site

The number of plants required to reach the benchmark minimum of 59 unrelated gametes varies with many factors (e.g. breeding system, pollination system, flowering determinacy, seed quality, losses during handling). In the case of fully autogamous species, the sample—either vegetative or seed—should come from

TABLE 14.3 Distribution of allozyme variation within and among populations of several forest tree species

P_n, the number of populations; R_g, the number of geographical regions; N, the average number of individuals sampled per population per locus; L, the number of loci studied; k, the average number of alleles per locus. The percentage is shown of variant alleles occurring at frequencies greater than 0.1 (common) or less than 0.1 (rare) in more than one region (widespread), or only in one region (local), and the % of gene diversity between populations (G_{ST} or F_{ST}). The common-widespread alleles are split into those with overall species frequency > 0.05, and the remainder. 'Extensive' species are defined as those that cover a range of 600 km or more; 'regional' as having a range of 150 km to 600 km, and 'restricted' a range of 150 km or less.

Species	P_n	R_g	N	L	k	Common Widespread $q>0.05$	Common Widespread $q<0.05$	Common Local	Rare Widespread	Rare Local	G_{ST} or F_{ST}	Source
Angiosperms – Extensive												
Acer saccharum Marshall		9	50	13	3.5	21	3	18	34	24	5	Young et al. (1993b)
Carapa guianensis Aubl.	9		52	6	3.0	58	—	8	17	17	5	Hall et al. (1994)
Castanea sativa Mill.	15	3	22	7	2.4	80	—	20	—	—	9	Villani et al. (1991)
Casuarina cunninghamiana Miq.				19	4.3	25	9	27	12	27	29	Moran & Hopper (1987) and unpublished
Eucalyptus albens Benth.	25	5	35	18	5.1	24	16	16	28	16	6	Prober & Brown (1994) and unpublished
Eucalyptus urophylla S.T. Blake	25	9	50	13	3.9	24	10	16	42	8	12	House & Bell (1994)
Grevillea robusta A. Cunn.	23	2	52	9	2.7	47	—	13	7	33	18	Harwood et al. (1997)
Regional												
Eucalyptus caesia Benth.	13		38	11	2.5	41	—	12	29	18	61	Moran & Hopper (1983)
Eucalyptus diversicolor F. Muell.	13	3	27	16	2.4	45	9	23	—	23	9	Coates & Sokolowski (1989)
Restricted												
Eucalyptus pendens Brooker	7		26	16	2.9	28	—	24	14	34	8	Moran & Hopper (1987) and unpublished
Ulmus minor Mill.		5	33	8	3.0	50	—	12.5	25	12.5	12	Machon et al. (1997)
Conifers												
Pinus radiata D. Don	5		200	27	2.9	26	—	7	33	35	16	Moran et al. (1988)
Taxus brevifolia Nutt.	9	3	25	13	2.8	44	—	13	17	26	0.08	El Kassaby & Yanchuk (1994)

about 60 randomly sampled plants in the population. Species of outbreeding panmictic forest trees first divide into those where vegetative material is targeted, and those where seed is available and preferred. A sample of cuttings from 30 random plants would reach the benchmark we have set.

For open-pollinated seed collections of outbreeding species, an important parameter in determining the number of maternal plants to sample is the pollination system, as it determines the paternity of fruit. The results in Table 14.2 suggest that when large numbers of seed per plant (> 30) are available, at least 15 maternal plants should be sampled. If the number of seed per plant is restricted, more maternal plants should be sampled. In general, a limited number of seed from many plants is a more efficient sample of the diversity within a population than many seeds from a limited number of parents. Individuals should be sampled at random at each site. However, when the habitat at the site is heterogeneous, a structured sample consisting of a random sample from each distinct local microenvironment is worthwhile. If possible, it is beneficial to keep different maternal samples separate, as bulking of the seed into populations will reduce the effective population size.

Generally, it is desirable to choose strategies that lessen the genetic relationship among the individuals in a sample and diversify the seed source as much as possible, both in space and time (seeds produced at different times of the year or ages in the life cycle of a perennial species). The sampling of neighbours that may be related due to limited seed or pollen dispersal (e.g. Hardner *et al.* 1998) should be avoided. In addition, the collector should be aware that isolated trees of self-compatible species are prone to yield seed from self fertilisation (Hardner *et al.* 1996). Sufficient seeds or vegetative material per plant should be collected to assure representation of each original plant in all duplicates.

Plant species differ in many distributional, life history, ecological and genetic attributes and sampling objectives and practicalities that will require amendment to this basic strategy. Difference in such points as geographical range, local abundance, life cycle duration, population age structure, determinacy of the flowering season, breeding system and fecundity affect the amount of material available and the level and pattern of genetic diversity. Brown and

Marshall (1995) and Marshall and Brown (1998) discuss in detail such modifications to the basic sampling strategy.

14.4 The Lawrence and Marshall alternative

The above theory and practice of sampling for genetic diversity rests on the concept that the population is the unit of sampling. This is particularly relevant to forest trees where the notion of provenance is central in forest genetic resources. Lawrence *et al.* (1995a, 1995b, 1997) have put forward an alternative approach to sampling based on 'virtual certainty' of sampling 'all' the variation within a species. Most notable is their contention that all the genetic variants of the whole target species can be conserved with a random sample of only 172 plants. If this contention is correct, it is a serious challenge to germplasm collections that consist of thousands of accessions, and labels as extremely wasteful those collection that comprise tens of thousands of accessions.

A striking difference between these papers and the earlier literature is the idea of aiming for 'virtual certainty' of sampling outcome by using very high levels of assurance probability (i.e. 0.9999 or low values of P_L = 0.0001). 'A sample of about 172 plants, drawn at random from a population of a target species, is of sufficient size to conserve at very high probability, all or very nearly all of the polymorphic genes ...' with allele frequencies > 0.05. However, their concept of 'virtual certainty' depends on an arbitrary value of P_L. As the value of P_L approaches zero, it has a larger and larger effect on the required sample size. The problem with the terminology of certainty and an exact number such as 172, is that it implies a fully deterministic outcome from what appears to be a modest sample.

Another issue that is important to contest is their claim that '172 seed taken from a single plant should give the same probability of conservation as a sample ... (composed of) a single seed from 172 randomly chosen plants' (Lawrence *et al.* 1995b). Many tree species such as *Eucalyptus* spp. and *Pinus* spp. present precisely this very tempting possibility—to collect the whole sample from a single individual. Seed collectors of such species might interpret Lawrence and colleagues claim as vindication for such a sample. The claim is, however, extremely misleading. This is because:

(a) such a sample is not as efficient in collecting allelic diversity due to the multiple redundant sampling on the maternal contribution

(b) it results in biased allele frequency—the variance effective population size of such a sample cannot exceed four (Frankel *et al.* 1995).

Such a strategy would be disastrous for reforestation programs.

More problematic, however, is their contention that, when the species is the unit of genetic conservation, the same computations apply directly at the species level as for a single population. This assertion leads to their conclusion that 'when samples are taken from a number of populations, the size from each need be no larger than 172 divided by the number of populations visited' (Lawrence *et al.* 1995a). In a later paper, they state that 'a random sample of 172 seeds, collected from the constituent populations of a species, is sufficient to conserve all or very nearly all of its genetic variation' (Lawrence & Marshall 1997, p. 112). This surprising conclusion is potentially seriously misleading. It assumes that all populations are equivalent in both the kinds and the levels of genetic variation they contain.

This assumption, namely that population differentiation is absolutely zero ($G_{ST} = 0.0$), is at odds with the overwhelming bulk of empirical evidence from plant population genetics (e.g. Briggs & Walters 1997). Cultivars, or populations of cultivated or wild plants, have diverged from one another genetically (Frankel *et al.* 1995). As Bengtsson *et al.* (1995) state, 'In no natural situation would one expect an allele to be perfectly evenly spread over the entire species range'. Population divergence is, thus, the basic starting point and key variable of genetic conservation (Moran 1992). Particularly in forest trees, local ecotypes with special adaptations or superior performance are the objective of collecting. The basic problem of the 172-plants-per-species approach to sampling is that it focuses on the 'common, widespread' alleles, as defined above. It has long been known that common, widespread alleles are:

(a) easiest to collect

(b) included in a sample at high frequency regardless of strategy

(c) most likely to be present in any collections already to hand (Marshall & Brown 1998).

Indeed the approach fails to guarantee the recovery of all the alleles in this class. Some alleles classed as common, widespread (with frequencies > 0.05 in more than one region) have an overall species frequency of < 0.05. The data in Table 14.3 give examples of studies where the proportion of such alleles is appreciable (e.g. *Eucalyptus urophylla* S.T. Blake, about 10%). These few 'widespread common' but species-rare alleles will not have the surety levels claimed for the 172-plants-per-species approach. Indeed of the 375 variant alleles ($\Sigma g[k-1]$) included in the 13 examples (Table 14.3), only 121 variant alleles (or 32%) have a species-average allele frequency of > 0.05 and thus form the target of the Lawrence and Marshall approach.

Further, this approach to sampling a whole species or region pays no attention to alleles that may be common, local and rare in the whole species ($q < 0.05$), but may occur and predominate in only one or a few populations on the margins of a species range. Table 14.3 gives 11 examples to show that common, local alleles make up more than 10% of variant alleles detected in tree species. Despite their present low frequency in the species as a whole, such alleles may be crucial in adapting a species to future changing environments.

14.5 Conclusions

The objective of sampling plant gene pools for conservation *ex situ* is to obtain the maximum amount of useful variation while keeping the number and size of samples within practical limits (Allard 1970). This objective is relevant for forest trees and is met by dividing the target region into distinct ecogeographical areas and collecting a limited number (about 50) of random individuals from many populations within each area totalling about 50 populations for the mission. When open-pollinated arrays are the collected material, seed should be sampled from at least 15 maternal trees. These approximate guidelines should be adapted to take account of biological features among the species such as breeding system, population numbers and sizes (Brown & Marshall 1995). When the mission has specific genetic variants as goals, it is appropriate to increase sample size at any site containing such variants. For example, when the mission is to repeat the sampling of a region, measures of genetic divergence among populations are useful in dividing the region into more homogeneous areas. Measures of genetic richness are useful in pointing to richer sites for larger samples (Brown & Schoen 1994).

The strategy we recommend stresses the importance of population divergence in the sampling of forest tree gene pools and the targeting of alleles that are rare in the whole species but common in a locality. A growing body of evidence from isozyme markers shows that such alleles make up an appreciable fraction (10–20%) of alleles that exist in tree species. The collection of such alleles is the most appropriate conceptual focus for sampling the genetic resources of forest trees.

14.6 Acknowledgments

Our thanks are due to H.-R. Gregorius, L. Guarino, D.J. Schoen, P.H. Thrall who read and commented on previous drafts. P. Thrall prepared the figure and H.-R. Gregorius pointed out formal relationship between P_L [FS], P_L [HS] and P_L. This chapter was prepared while AHDB was Honorary Research Fellow with International Plant Genetic Resources Institute.

Effect of Forest Tree Domestication on Gene Pools

Yousry A. El-Kassaby

Summary

The little available information on the domestication of forest tree species is mainly devoted to the genetic consequences of domestication rather than the process itself. Domestication of forest tree species is fragmented and follows the classical horizontal organisation of management systems, thus making evaluation difficult. However, systematic monitoring of the various steps of the domestication process (selection-breeding methods and strategies, and seed and seedling production) has identified cases of unintentional directional selection where genetic diversity could be affected. Phenotypic selection has been proven effective in capturing most of the genetic variation existing in natural populations. The progress from breeding population to production populations did not substantially reduce genetic variability; however, the chance for rare, endemic allelic loss increased with a reduction in number of parental trees in seed orchards. Use of a breeding strategy that provides for sampling over several geographical locations and selection for various goals, as well as adaptation to different environments (i.e. Multiple Population Breeding System) was demonstrated to be effective. Generally, understanding the species biology and the role of biology on genetic diversity is of great importance. The need for change and/or the implementation of management practices is effective in maintaining the genetic variability.

15.1 Introduction

The domestication of forest tree species can be described as the process whereby plants are taken from their wild (natural), undomesticated state through a series of sampling and selection stages, with each stage curtailing the genetic variation, and ultimately resulting in the production of somewhat genetically uniform plantations. There is a systematic hierarchical structure to the populations that descend from large collections from natural populations, to undergo selection, testing, breeding and finally to advanced generation breeding populations that are characterised by their small numbers of parents (Fig. 15.1).

As economic forces act to increase pressure for the rapid culling of genotypes to attain higher genetic gain, the plant breeding response to this genetic erosion has focused on gene conservation through seed collections (Plucknett *et al.* 1987). However, this has not been an entirely successful strategy (Brown & Briggs 1991). For forest tree species, breeding to narrowly based

advanced generations has been seen as a particularly dangerous precedent to follow for two reasons:

(a) each tree plantation must endure more variable environments than do traditional agricultural crops

(b) breeding cycles are usually long, forcing an extended response time for genetic infusion.

The amount of genetic data relating to the study of forest tree domestication is very limited. Furthermore, these studies were mainly devoted to the genetic consequences of domestication rather than evaluation of the process itself. In this chapter, I examine the steps involved in tree breeding programs and their delivery systems, indicate where problems may occur and make recommendations for managing the genetic resource for forest tree plantations (Fig. 15.2). First, I consider that the process is multi-generational and distinguish, in the structure of genetic variation, between-population versus within-population variation, and the initial state of that variation versus the condition

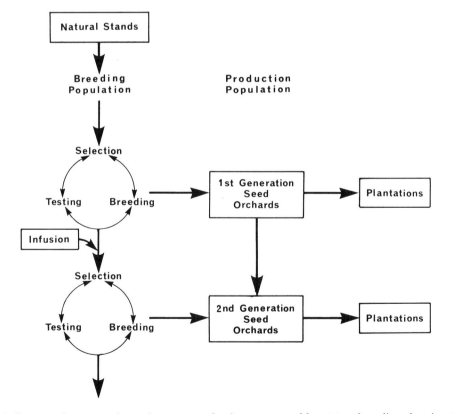

Figure 15.1 A diagram of a commonly used recurrent selection program of forest tree breeding showing the relationship between natural populations and first-generation and second-generation seed orchards as well as the infusion population (from El-Kassaby & Ritland 1995b).

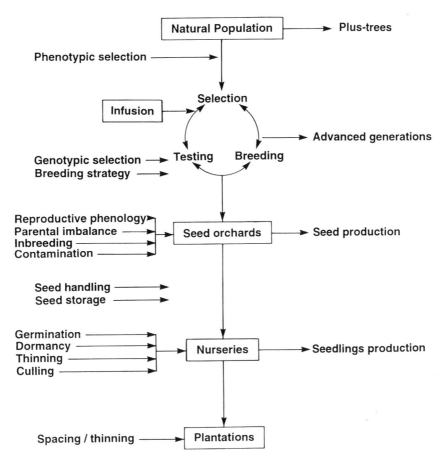

Figure 15.2   A diagram depicting the tree improvement delivery system with its associated activities (after El-Kassaby 1992).

that could be predicted for a managed resource. I then consider the effects that occur within a single generation and the problems that may exist during seed and seedling production.

15.2 Breeding

15.2.1 Phenotypic selection

The genetic variation of a natural population is a result of its evolution and involves factors such as mutation, migration and inbreeding, recombination events, and various forms of natural and human influenced selection. The genetic variation of natural and artificial (plantations) populations can be assessed and characterised by morphological, physiological and quantitative attributes as well as biochemical and molecular markers. To be able to compare the genetic diversity and its organisation between natural populations and the product of phenotypic selection (i.e. breeding and production populations) ideally

requires the use of a combination of these attributes due to the apparent limitation of using only one type. However, due to the neutral, random nature of allozyme markers and the relative simplicity of the technique, allozyme markers have been used extensively in assessing the levels of genetic variation in several populations of forest tree species (Hamrick & Godt 1989). Therefore, these studies provide common ground for comparison; allozyme markers alone, however, do not provide a complete picture of the genetic variation.

Comparison of heterozygosity parameters (percentage of polymorphic loci, average number of alleles per locus, mean expected heterozygosity; Hartl 1980) between the product of phenotypic selection, represented by first-generation populations of seed orchards, with their corresponding natural populations have been conducted for five species. These species are: *Picea abies* (L.) H. Karst. (Norway

TABLE 15.1 Heterozygosity parameters comparison between phenotypic selection output (seed orchards) and natural populations

N = the number of seed orchards or natural populations studied.

Species		% Polymorphic loci	No. of alleles/ locus	Mean expected heterozygosity
Picea abies	Seed orchard (N = 1)	88.0	2.4	0.320
Bergmann & Ruetz (1991)	Natural populations (N = 3)	100.0	2.5	0.230
Thuja plicata	Seed orchard (N = 1)	11.1	1.2	0.058
El-Kassaby et al. (1993c)	Natural populations (N = 3)	11.1	1.1	0.055
Picea sitchensis	Seed orchard (N = 1)	100.0	2.8	0.229
Chaisurisri & El-Kassaby (1994)	Natural populations (N = 10)	66.9	1.8	0.183
Pseudotsuga menziesii	Seed orchard (N = 12	62.5	2.3	0.172
El-Kassaby & Ritland (1995a,1995b)	Natural populations (N = 49)	52.6	2.1	0.171
Picea glauca × *engelmannii*	Seed orchard (N = 1)	64.7	2.4	0.194
Stoehr & El-Kassaby (1997)	Natural populations (N = 9)	64.7	2.7	0.189

spruce) (Bergmann & Ruetz 1991), *Thuja plicata* Donn ex D. Don (western red cedar) (El-Kassaby *et al.* 1993c), *Picea sitchensis* (Bong.) Carrière (Sitka spruce) (Chaisurisri & El-Kassaby 1994), *Pseudotsuga menziesii* (Mirb.) Franco (Douglas-fir) (El-Kassaby & Ritland 1995a, 1995b) and *Picea glauca* × *engelmannii* ('interior spruce') (Stoehr & El-Kassaby 1997) (Table 15.1). In general, the comparison of heterozygosity parameters between the seed orchards and pooled data from their corresponding natural populations indicated that the amount of genetic variation present in natural populations was retained or even increased during the phenotypic selection step (Table 15.1). The difference between the sampling methods (i.e. sampling breadth) of natural populations and individual was attributed to be responsible for the retention of most of the genetic variation observed in natural populations. Sampling of natural populations is usually restricted to a relatively small area (i.e. the local gene pool); however, the selection of many individual trees (i.e. plus-trees) from several populations represents a situation whereby the sampled trees for breeding may include other alleles not originally sampled from the natural populations. In other words, the sampling of natural populations is usually restricted to relatively few locations due to experimental limitations but phenotypic selection with its restriction on the number of plus-trees that can be included in the breeding program per location allows more populations to be sampled. The frequently observed reduction of genetic diversity in agricultural

crop plants was not observed in the species studied above. This may be because conifers with their known high genetic variability are still in their natural state and that early domestication does not cause appreciable reduction to genetic variability. Additionally, the seed orchard population represents an assemblage of the intrapopulation and interpopulation variation.

15.2.2 Breeding versus production populations

As distinct from the genetic variation that lies in the breeding populations, the production populations (seed orchards) usually contain a subset of parents that represent the entire breeding population. Thus, comparing heterozygosity parameters between breeding and production populations is of great importance in gauging the magnitude and direction of genetic change. This type of comparison sheds light on how tree improvement deployment systems could affect genetic variability and more importantly it provides guidelines for the implementation of various breeding and conservation strategies.

El-Kassaby and Ritland (1995a, 1995b) monitored the heterozygosity parameters through the progressive process of British Columbia's low-elevation, coastal *Pseudotsuga menziesii* tree improvement program (Fig. 15.1). Genetic variation was monitored and evaluated in natural populations, then in breeding populations and finally in production populations

TABLE 15.2 A comparison of heterozygosity parameters between natural, breeding, production and infusion populations and first-generation (1st gen. S.O.) and second-generation (2nd gen. S.O.) seed orchards in *Pseudotsuga menziesii* (El-Kassaby & Ritland 1995a,1995b)

Population	Heterozygosity parameters		
	% polymorphic loci	No. of alleles/ locus	Mean expected heterozygosity
Natural	53	2.14	0.171
Breeding	65	2.50	0.176
Production	65	2.65	0.167
Infusion	65	2.30	0.151
1st gen. S.O.	63	2.28	0.172
2nd gen. S.O.	56	2.25	0.163

(Table 15.2). In addition, the first-generation and second-generation seed orchards of the production populations were compared, and the amount of genetic heterozygosity introduced to the breeding population through the infusion process of breeding was determined (Table 15.2). This study has confirmed the ability of phenotypic selection to capture most of the genetic diversity in the species' natural range, as well as maintenance of the breadth of this variability in the production population (the aggregate of the 12 first-generation seed orchards) (Table 15.2). The comparison between the first-generation and second-generation seed orchards was of interest: although the reduction in percentage of polymorphic loci (from 63 to 56) and mean heterozygosity (from 0.172 to 0.163) was not substantial, the average number of alleles per locus (from 2.28 to 2.25) indicated permanent loss of some rare localised or private alleles (Table 15.2). This study has included the only four established second-generation seed orchards in 1995 and the eventual inclusion of the additional new seed orchards in the

deployment program is expected to decrease the chance of allelic loss.

15.2.3 Breeding strategy

Breeding strategy, which is the management and structure of breeding population(s), is able to affect the structure of genetic variation of the breeding population and, thus, genetically alter its production populations and subsequently the entire tree improvement delivery system. For example, the main objective of the Williams *et al.* (1995) study on *Pinus taeda* L. (loblolly pine) was to compare the Multiple Population Breeding System (MPBS) (Namkoong 1984a) with that of the Hierarchical Open Ended system (HOPE) (Kannenberg 1983) (Table 15.3). MBPS combines short-term gain, germplasm enhancement and genetic conservation objectives into the same breeding population, while HOPE is characterised by a trade-off between short-term gain and the maintenance of long-term genetic diversity. In the Williams *et al.* (1995) study, genetic diversity differences between the two breeding strategies were not significant. Most of the genetic variation (94%)

TABLE 15.3 Heterozygosity parameters comparison between natural, Multiple Population Breeding System (MPBS) with that of the Hierarchical Open Ended system (HOPE) in *Pinus taeda* (from Williams *et al.* 1995)

Population	Heterozygosity parameters		
	% Polymorphic loci	No. of alleles/ locus	Mean expected heterozygosity
Natural	100	2.63	0.213
MPBS	90	2.50	0.180
HOPE	85	2.40	0.181

resided within populations of the two breeding systems; however, loss of rare alleles was observed. Williams *et al.* (1995) suggested that MPBS has the potential of increasing the genetic variability through the genetic fixation of rare alleles in the smaller breeding populations. In addition, these populations represent different geographical locations that might be associated with different genetic and adaptive backgrounds and are bred for different selection objectives. The aggregate of these populations, however, will produce greater variation than that of HOPE. In the HOPE strategy, the selection objectives are the same; thus, any genetic loss that is associated with one selection method will not be recovered in other populations as in MPBS.

15.3 Seed production

Seed orchards represent the link between tree improvement programs and reforestation through the delivery of consistent, abundant yields of genetically improved seed. It is important to translate the genetic gain and diversity present in breeding programs to reforestation efforts. Seed orchards usually consist of a restricted number of genotypes; therefore, the potential for loss in genetic variability is high. The attainment of genetic gain through selection and breeding depends on maintaining the frequency of desirable genes in the seed orchard crops at the level present in the selected parent population (i.e. the seed orchard). This goal requires the attainment of the random mating assumption of the Hardy-Weinberg theorem. Panmictic equilibrium is achieved only when seed orchards function as closed, perfect populations. Reaching perfection in seed orchard populations requires that all the seed orchard's clones are in reproductive synchrony and contribute similarly to the gene pool (i.e. equal reproductive output). In addition, the satisfaction of this condition assumes lack of or minimum levels of:

(a) outside pollen migration (i.e. contamination)

(b) inbreeding

(c) selection among the gametes of various clones.

In most cases, these conditions are not fully met and seed orchard management practices are always relied upon to correct for these deviations. The following section summarises the various biological factors affecting seed orchards.

15.3.1 Reproductive phenology

Failure to achieve reproductive phenology synchrony among a seed orchard's clones (genotypes) affects all seed orchards' panmictic-equilibrium requirements (El-Kassaby *et al.* 1986). Both time and temperature, summarised as heat sum, have a significant effect on reproductive phenology of forest tree species (Worrall 1983). Because seed orchards consist of several genotypes that are usually adapted to different environmental conditions with various heat sum requirements for bud burst, time differences in reproductive bud phenology among orchard clones will occur. When reproductive phenology differences occur, the orchard clones form several temporally isolated breeding subpopulations, thus increasing the rate of inbreeding, reducing panmixis and exposing the early and late 'flowering' clones to increased levels of non-orchard pollen contamination (for review see El-Kassaby 1989). The reproductive phenology dynamics of seed orchards of several species were investigated, and, with the exception of *Thuja plicata* (Fig. 15.3), all displayed an extended pollination season and subdivision of the seed orchard population into several temporally isolated breeding subpopulations (El-Kassaby 1992). This temporal isolation also affected seed yield (El-Kassaby & Reynolds 1990), rate of pollen contamination (El-Kassaby *et al.* 1988) and mating pattern (selfing rate, level of correlated matings) (El-Kassaby & Davidson 1991).

Reproductive phenology differences also affect the gametic contribution by every clone to the resultant seed crop. A high seed-producing or pollen- producing clone in the early or late phenology classes could contribute the same as or less than the amount contributed by a low output or medium reproductive-output clone during the peak of pollination (El-Kassaby & Askew 1991; Roberds *et al.* 1991).

The frequency and timing of receptive females and pollen-shedding males also affect the rate of pollen contamination. If background pollen is high when early or late phenology classes are receptive, most of the receptive trees will be pollinated by background pollen (El-Kassaby *et al.* 1988). This will reduce the observed rate of inbreeding (El-Kassaby & Ritland 1986b), consequently increasing the effective population size (i.e. increasing the genetic variability). However, pollen contamination is often associated with reduction of the expected genetic gain (Askew

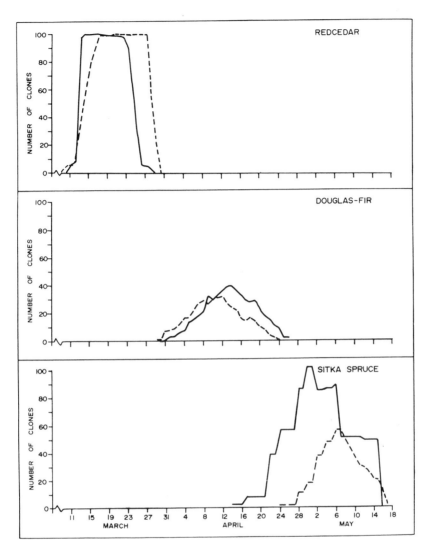

Figure 15.3   Reproductive phenology diagram showing the extent of the pollination season for *Thuja plicata* (western red cedar), *Pseudotsuga menziesii* (Douglas-fir) and *Picea sitchensis* (Sitka spruce) (from El-Kassaby 1992). The solid line represents females, dotted, males.

1986), as well as increased maladaptation, especially if the seed orchard is located outside its natural range (Johnsen 1989a, 1989b, 1989c).

15.3.2 Reproductive output

The maintenance of similar allelic frequencies between a seed orchard and its seed crop depends on the presence of balanced gametic production among the orchard's clones and absence of effective selection involving specific alleles. Parental-balance estimates of several studied seed orchards have revealed that the parental contribution to seed-cone/seed crops and pollen varied substantially among a seed orchard's

clones (for review see El-Kassaby 1995). Mostly, a small proportion of clones contributed the majority of male and/or female gametes, thus reducing the expected genetic variation and producing crops with unpredictable allelic frequencies (Muona & Harju 1989; Roberds *et al.* 1991; El-Kassaby & Ritland 1992; Nakamura & Wheeler 1992).

15.3.2.1   Maternal contribution

Because it is easy to estimate clonal seed-cone production, the parental balance of seed orchards often is based solely on seed-parent contributions. The '20/80' rule (i.e. 20% of the clones produce 80% of the

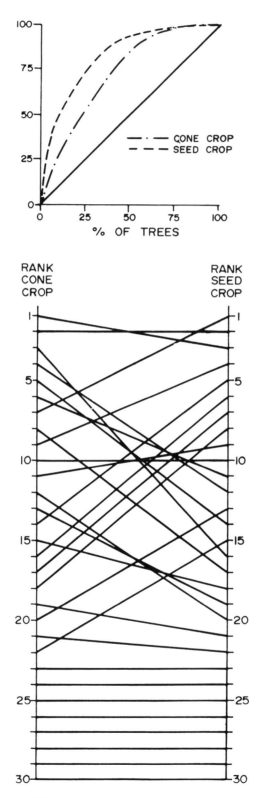

Figure 15.4 Cumulative seed-cone and filled-seed production curves for 30 *Pseudotsuga menziesii* trees and their rank order change between seed-cone and filled-seed production (from Reynolds & El-Kassaby 1990).

cone crop) was coined by the North Carolina State Tree Improvement Co-operative in 1976 to describe the parental balance in seed orchards (Anonymous 1976). Since then, parental-balance values that are solely based on seed-cone parent contributions have become an accepted method to evaluate seed-orchard crops.

Parental-balance estimates of several *Pseudotsuga menziesii* seed orchards have revealed that the parental contribution to cone-seed crops varied substantially among the seed orchard's clones (El-Kassaby *et al.* 1986; Reynolds & El-Kassaby 1990; Copes & Sniezko 1991; El-Kassaby & Askew 1991; El-Kassaby & Cook 1994). Furthermore, placing greater value on determining maternal contribution on seed-cone production produces inaccurate information due to the observed differences between reproductive energy and reproductive success. Reynolds and El-Kassaby (1990) and El-Kassaby and Cook (1994) have demonstrated that the relationship between reproductive energy and reproductive success is different among clones. In these two studies the maternal contribution based on seed-cone production differed substantially from that obtained from filled-seed count (Fig. 15.4).

The concept of effective female or male population number (Crow & Kimura 1970) can be used to demonstrate deviation from the ideal case of equal contribution. The proportion of female effective and actual numbers ($\frac{N_e}{N}$) would serve as a good estimate to determine the expected reduction in genetic diversity ($\frac{N_e}{N} = 0.45$) produced for a *Picea sitchensis* seed crop by Chaisurisri & El-Kassaby 1993). Furthermore, El-Kassaby and Cook (1994) have demonstrated that the reliance upon either female effective population number estimates as shown by Chaisurisri and El-Kassaby (1993) or parental-balance curves that are based on seed-cone and filled-seed contribution which was introduced by Reynolds and El-Kassaby (1990) provided only a static description of the genetic representation and failed to depict the dynamics of maternal reproductive output.

15.3.2.2 Paternal contribution

Paternal contribution to seed crops is difficult to assess due to the interaction between male reproductive output and reproductive phenology (i.e. quantity of pollen production, time of female receptivity) and the male–female complementarities (Schoen *et al.* 1986;

Schoen & Stewart 1986, 1987; Denti & Schoen 1988; Apsit *et al.* 1989; El-Kassaby & Ritland 1992). Even if the quantity of pollen production is being estimated, the accuracy of the estimate depends upon the viability of the pollen produced and its success in uniting with ovules. Roberds *et al.* (1991) developed a method for determining the paternal contribution of seed orchard clones to the seed crop utilising the concept of Wright's (1931) effective number of pollen parents. These effective numbers are variance effective numbers for populations of male gametes that are successful in uniting with ovules to produce viable seed. In this method, Roberds *et al.* (1991) considered the allele frequencies of genes in the orchard's pollen pool to be exactly equal to the allelic frequencies of the orchard's ramet population. Furthermore, they used this pollen pool as a reference population from which binomial samples of genes in the male gametes were drawn. Such sampling generates the allele frequency drift that forms the basis for defining the effective numbers of pollen parents for three levels (orchard, clone, ramet).

Roberds *et al.* (1991) applied the above-mentioned methods to estimate the effective number of pollen parents in a *Pseudotsuga menziesii* seed orchard utilising allozyme data of the embryo/gametophyte of individual ramets. The results demonstrated that the effective number of pollen parents varied considerably among clones and among ramets within clones. The results were in agreement with the mating pattern observed in the seed orchard (Ritland & El-Kassaby 1985). Early flowering clones produced lower effective number of pollen parents when compared with the number of intermediate ones (El-Kassaby & Ritland 1986b). In addition, the low effective number of pollen parents estimated for the seed orchard was also in agreement with the low effective contamination observed by El-Kassaby and Ritland (1986a) who studied the same seed crop.

Roberds *et al.* (1991) also indicated that the ramet, clone and orchard estimates of the effective number of pollen parents provide insight into the mating patterns in seed orchards. It is expected that the effective number of pollen parents for an individual ramet should be smaller than that of an individual clone, and that of an individual clone should be smaller than the number for the entire seed orchard. Roberds *et al.* (1991) presented five cases that deviated from the expected relationship among the three levels (i.e. ramet, clone, orchard) and provided an interpretation of possible mating pattern.

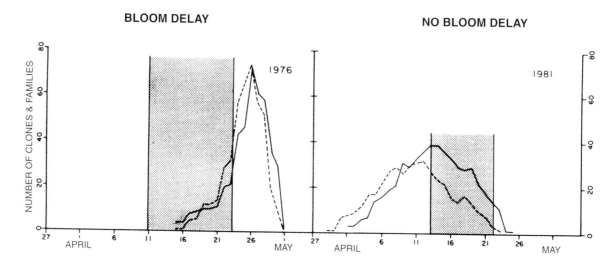

Figure 15.5 Reproductive phenology comparison for bloom delay treatment for a *Pseudotsuga menziesii* seed orchard showing the extent of the pollination season and its relation to background pollen flight (from Fashler & El-Kassaby 1987). The shaded area represents the local pollen flight peak; the solid line represents females, and the dotted line males.

15.3.3 Inbreeding

The mating system, the union of male and female gametes and their genetic relationship, is crucial in determining subsequent population structure and how genetic variation is being transmitted and distributed from one generation to another. With few exceptions, the natural estimates of outcrossing in conifers have been found to be quite high ($t > 0.9$; Adams & Birkes 1991). Studies of mating systems in seed orchards indicated that the levels of inbreeding are low; however, its importance to forestry practices requires added attention, since most forestry programs rely on non-competitive plantings in both nursery-production and plantation-establishment phases.

In general, estimates of outcrossing rate obtained from seed orchards are higher than those reported for natural stands of the same species (Shaw & Allard 1982; Rudin *et al.* 1986). This suggests that population structure (i.e. the physical arrangement of related individuals within a population) affects the rate of outcrossing. If orchard population arrangement can increase the rate of outcrossing, then additional manipulation of pollen dynamics (i.e. pollination effectiveness) could also further reduce inbreeding and, consequently, increase heterozygosity in the seed crop. The use of the overhead cooling treatment (i.e. bloom delay; Silen & Keane 1969) in coastal *Pseudotsuga menziesii* seed orchards was a successful

contamination prevention method (El-Kassaby & Ritland 1986a) and produced several additional benefits, including drastic reduction in reproductive phenology differences (Fashler & El-Kassaby 1987). The pollination season in cooled seed orchards was shorter, yielding higher overall outcrossing than that obtained from uncooled ones (Fig. 15.5) (El-Kassaby & Davidson 1990, 1991). In addition, the use of supplemental mass-pollination as a crop management practice has reduced the level of inbreeding in *Pseudotsuga menziesii* seed orchards (El-Kassaby & Davidson 1990), thereby increasing the level of genetic variability.

In summary, the expected levels of genetic variability in the crops of seed orchards are greatly affected by the interaction among reproductive phenology, reproductive output and the management practices applied to every seed crop.

15.3.4 Cost of reproduction

The practice of managing crop trees (clones with consistently high seed-cone production) in seed orchards is common. This practice makes sense from the seed orchard manager's point of view since it concentrates the management efforts on fewer clones. However, there is an increasing amount of theoretical literature, most commonly called the theory of life history evolution, that casts doubt on the validity of this approach. The theory maintains that since there are finite resources available to an organism, selection that involves increasing the resource allocation to one trait will of necessity decrease resource availability elsewhere. More specifically, selection for higher reproductive output should concomitantly select for slower growth. Stearns (1977, 1980) has presented mounting empirical and experimental evidence that demonstrates such trade-offs when selection is imposed on a population. This trade-off has been observed for *Pseudotsuga menziesii* where negative genetic correlations between seed-cone production and annual growth increments were detected for six out of eight years investigated (El-Kassaby & Barclay 1992). These negative correlations imply the existence of genetic variability in the proportional allocation of photosynthate to reproduction and growth, and that selection for seed-cone production (i.e. managing crop trees only) will result in negative selection for growth. Restricting seed-cone harvesting to fewer clones reduces the number of maternal genotypes contributing to the seed crop, hence reducing genetic variability. Pollen management practices that favour the use of pollen from low seed-cone producing clones in the pollen mixes used for supplemental mass-pollination are expected to increase the genetic variability and counteract this unintentional selection. Naturally, the reliance on supplemental mass-pollination for increasing the parental contribution of orchard's clones depends on how this method is being practised to ensure high success rate of the applied pollen (El-Kassaby *et al.* 1993a).

15.4 Seed biology

Commonly, seed orchard managers collect their seed crops, extract the seeds and then store them as a single bulk lot. Bulk seed lots represent the sum of all seed-producing parents within the orchard, but it has been established (see Chapter 15.3.3.1) that not all individual parents contribute equally in terms of seed production. Bulking the seeds also explicitly assumes that all seeds will respond uniformly to storage, dormancy breaking treatment, germination and emergence in the seed bed.

15.4.1 Seed size

Seed size varies within and among genotypes. These variations are caused by both environmental factors (i.e. position within the seed-cone, height and aspect of the cone in the crown) and genetic factors. For example, Silen and Osterhaus (1979) and Chaisurisri

et al. (1992) demonstrated significant differences in seed weight among *Pseudotsuga menziesii* and *Picea sitchensis* clones, respectively. Both studies revealed that the smallest seeds within a clone could be as large or larger than the largest seeds in another clone. Seed sizing during seed extraction and processing is common. If this practice is being used for these two species, seed sizing is liable to cause the loss of a major proportion, or the entire removal, of a specific clone, resulting in reduced genetic representation as well as the expected genetic gain.

15.4.2 Germination parameters

Germination parameters (germination capacity, peak value, germination value) of coniferous seeds are under strong genetic control (Fig. 15.6) (El-Kassaby *et al.* 1993b). El-Kassaby *et al.* (1992) conducted a study to determine the genetic control of germination parameters in *Pseudotsuga menziesii* using seed from 19 different clones and found that genotypic differences represented a major and important component of the total variation observed for all germination parameters. Broad-sense heritability estimates of germination parameters ranged between 0.91 and 0.93 for peak value (a mathematical expression of the break of a sigmoid curve representing a typical course of germination) and germination value (a combination of speed and completeness of germination, Czabator 1962), respectively. The anatomical structure of conifer seeds with its dominant maternal contribution ($2n$ seed coat, $1n$ megagametophte, $1n$ embryo, total $4n$) compared with the paternal contribution ($1n$ embryo) explains this phenomenon (El-Kassaby *et al.* 1992). Thus, it can be expected that bulk seed lots will vary in their germination behaviour reflecting both the genetic control of germination, as well as their parental contribution, when given a uniform treatment such as a standard germination test and storage. Furthermore, Edwards and El-Kassaby (1995) showed that marked differences in seed dormancy (i.e. the need for seed pretreatment such as prechilling) were present among different genotypes of *Pseudotsuga menziesii*, but that these genetic differences can be minimised by modifying the commonly prescribed prechilling routine (increased from 3 to 5 weeks). Alternatively, if seeds from individual parents with their known differences in germination parameters are kept separate, then they can be given specific treatments to

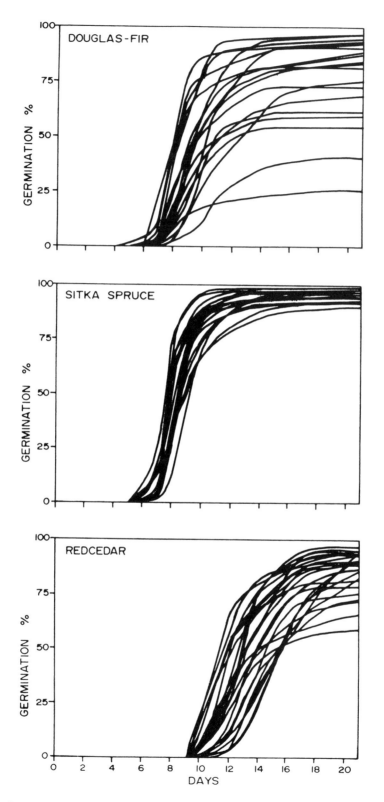

Figure 15.6 Germination curves for different *Pseudotsuga menziesii*, *Picea sitchensis* and *Thuja plicata* genotypes (from El-Kassaby *et al.* 1993b).

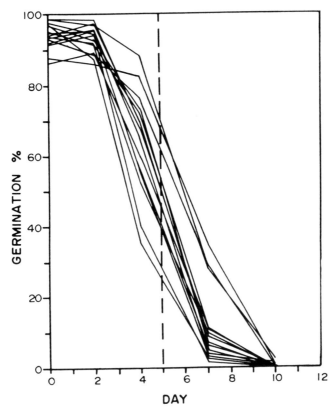

Figure 15.7   Germination curves of different *Pseudotsuga menziesii* genotypes after simulated long-term storage
(i.e. accelerated ageing) (vertical dashed line shows the difference in the rate of deterioration in seed
germination among genotypes).

maximise their germination performance, thus
minimising differences in their response.

15.4.3 Seed storage

Seed viability is highest at cone harvest and declines
with age. The processes of seed deterioration in a
population of seeds are independent among the
individual seeds and the time for deterioration ranges
from days to years. In agricultural crops, genotypes
differ in the degree of seed viability loss during storage
(Delouche & Baskin 1973). If conifer seed crops are to
be collected, extracted and stored in bulk, then seed
viability losses by a specific genotype cannot be traced,
and a reduction in the total genetic diversity passes
undetected (El-Kassaby 1992). The effects of viability
loss in several species [*Picea sitchensis*, *Pseudotsuga
menziesii* (Fig. 15.7), *Tsuga heterophylla* (Raf.) Sarge.
(western hemlock), *Tsuga mertensiana* Carrière
(mountain hemlock), *Picea glauca* (Moench) Voss
(white spruce) and *Pinus contorta* Douglas ex Loudon
(lodgepole pine)] seed were assessed with a simulated

ageing technique known as accelerated ageing
(Delouche & Baskin 1973). In this test, seeds were
exposed to a range of temperatures and humidities for
varying times. In all the species investigated,
germination rate declined steadily with time of
exposure; however, the rate of decline varied among
genotypes, indicating differential viability losses (Fig.
15.7) (El-Kassaby 1992; Chaisurisri *et al.* 1993). These
differences alter the genetic makeup of stored, bulked
seed lots from collection to sowing. However, if seed
lots were harvested, extracted and stored on a clonal
basis then reductions in genetic variability caused by
viability losses could be identified by periodic tests.

15.5 Seedling production

Container nurseries provide an alternative, secure
source of seedling production by avoiding the
uncertainties typical of traditional bare-root nurseries.
A common assumption is that container-nursery
seedling crops grown from seed-orchard seeds

maintain the same genetic composition as that present in the orchard. As has already been discussed:

(a) bulked seed orchard seed lots are usually composed of undetermined proportions of seeds contributed by many seed parents

(b) germination parameters are under genetic control

(c) the application of uniform seed handling (extraction, storage, germination) protocols will not produce a uniform response from all genotypes.

Thus, the germination response of a seed orchard seed lot in the nursery represents the weighted average of the germination parameters of the individual parents contributing to that seed lot.

The economic production of seedlings in container nurseries depends upon maximising the number of plantable seedlings per unit area of the nursery. Therefore, the presence of empty cavities represents a loss in productivity. To overcome this and ensure that there is at least one germinant present in all cavities, multiple sowing has become a common practice (Vyse & Rudd 1974). This frequently results in multiple germinants in each cavity and these must be thinned to leave a single germinant. When bulk seed lots are sown, the biological factors discussed above affect the genetic representation of the seedling crop. For example, it should be expected that differences in dormancy level as well as germination speed will cause differences in germinant size at the time of thinning. Thus, thinning to leave the largest germinant probably will result in unintentional directional selection favouring the parents that produced less-dormant, fast-germinating seeds.

At the end of the nursery rotation, seedlings that do not meet 'standards' are usually culled and discarded. This not only represents a further loss in nursery productivity, but it could also affect parental genetic representation if the culled seedlings are from parents that are characterised by slow seedling development. This situation was corroborated by Piesch (1987), who compared the rank in height-performance among 45 *Pseudotsuga menziesii* families from the time of lifting from the nursery (1 year) to eight years in the field, and found that some families that ranked the lowest during the nursery phase ended up in the highest rank in the field. Thus, culling would have removed superior parents that were hidden in the bulk-sown seed lot. For the nursery manager, culling of substandard seedlings

may maintain a uniform, high quality stock, but this practice might harbour the means by which slow-developing genotypes are excluded from the reforestation efforts.

Three recent studies have evaluated the cumulative effects of seed biology (by looking at differences in parental reproductive output and parental variation in germination parameters) and container nursery practices of thinning and culling, on the genetic representation of parents in the resultant seedling crops. These studies included:

(a) a stochastic simulation designed to estimate the consequences of seed biology on various scenarios of sowing (El-Kassaby & Thomson 1990)

(b) an experimental container nursery trial to determine the interaction between seed biology and nursery management (El-Kassaby & Thomson 1996)

(c) an operational nursery trial to evaluate the findings of El-Kassaby & Thomson (1996) (Edwards & El-Kassaby 1996).

In the simulation study, El-Kassaby and Thomson (1990) demonstrated that there was increased selection pressure in favour of fast-germinating parents when multiple sowing was performed. They concluded that the selection pressure was greater when three seeds were sown per cavity compared to two seeds per cavity (i.e. the probability of one or more parents being selected against during thinning was greater the higher the number of seeds per cavity). Single seed sowing represented the genetic ideal, but was associated with a nursery productivity loss.

In the experimental container nursery trial, El-Kassaby and Thomson (1996) presented empirical evidence that seed biology and nursery practices affected the genetic representation of the seedling crop. This study, which was based on 19 *Pseudotsuga menziesii* parents represented by 42 000 seeds that were sown three per cavity, demonstrated that 66% of the observed variation in the genetic representation was attributed to seed germination, 20% to germinant thinning and 14% to seedling culling. To ensure the maintenance of genetic diversity, they recommended the use of either single seed sowing or individual parent sowing.

To test the recommendation for individual parent sowing, a recent study (Y.A. El-Kassaby, unpublished

data) was conducted on 10 *Tsuga heterophylla* parents. The genetic representation of seedling crops of these 10 parents was compared when seeds were sown either by individual parents, or in a mix (i.e. simulating sowing a bulked seed lot). This design permitted the evaluation of intraparent (single parent sowing) and interparent (mixed) competition and its effect on postgermination seedling growth. The overlapping seedling growth habit of *Tsuga heterophylla*
provided an opportunity to study such competition. The comparative study included 11 000 seeds representing the 10 parents with a sowing factor of three seeds per cavity in both parts of the trials.

The effects of germination parameters and thinning on the genetic representation of the 10 parents should be observable only in the mixed portion of the trial, where thinning was among the three seeds of the different parents within each cavity. By contrast, in the single-parent sowing portion, the effects of germination parameters should not affect the genetic representation of parents since thinning was conducted among seeds belonging to the same parent. If variation in seedling growth habit exists, then different competitive abilities among seedlings of the 10 parents (i.e. interparent competition) should be manifested in the mixed portion of the trial, as opposed to competition among seedlings of the same parent (i.e. intraparent competition).

The parental representation in the mixed and the single-parent portions of the trial produced highly contrasting results. After thinning, as a percentage of their seed contributions, two parents contributed less than 15% and 17% germinants to their parental representation when grown in the mix, compared to 33% germinants in the single-parent portions of the trial. That is, the contribution as germinants remaining after thinning was almost double when seeds of these two parents were grown in the single-parent portion, implying that variation in germination parameters affected their genetic representation. However, the remaining eight parents showed no striking differences in the percentage of germinants left after thinning. They averaged 42% (range: 29–54), as compared to 33% (range: 30–33) germinants of their parental representation in the mix as compared to the single-parent portions, respectively. In other words, their genetic representation as germinants was 1.3 to 2.5 times higher than the first two parents.

For the two parents discussed above, the culling rate was higher (18% and 35%) when seedlings of mixed parents were grown together rather than when they were grown in the single-parent portion of the trial (12% and 24%). At the same time, the culling percentage for the other eight parents averaged 9% (range: 4–14) when grown in a parental mix and 10% (range 8–13) when grown as single-parents (i.e. there was no difference in culling percentage for those eight parents using either growing regime). However, culling percentages for the first two parents ranged from 2 to 4 times higher when grown in a mix, and 1.2 to 2.4 times higher when grown as single-parents, compared to the remaining eight parents. The results confirmed the prediction that the genetic representation for the entire seedling crop could differ from their original seed lot representation when seeds were sown in a parental mix (i.e. simulating sowing of a bulk seed lots).

These nursery trials demonstrated the interplay between seed biology and nursery systems and their potential effects on the genetic constitution of seedling crops. Based on these observations the use of individual seed or single-parent sowings represents the optimal means for maintaining the genetic representation of the original seed lots. The choice between these two options should be considered in light of:

(a) the economic loss incurred by the presence of empty cavities

(b) the cost of thinning when seeds are multiple sown

(c) the availability and cost of extra seeds used in multiple sowings

(d) the added logistical work associated with single-parent sowing

(e) the desire to capture the potential genetic diversity of the seed lots.

15.6 Plantation management

Since forest tree plantations are subject to various harvesting and silvicultural activities, the effect of these activities on tree species and their genetic diversity needs to be rigorously evaluated. Commercial thinning is one of these silvicultural activities and is being commonly practiced by plantation foresters to:

(a) reduce stand density

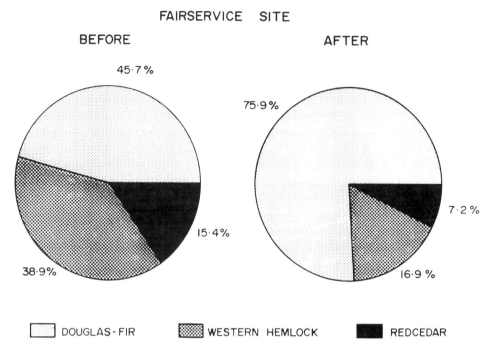

FAIRSERVICE SITE

BEFORE   AFTER

DOUGLAS-FIR   WESTERN HEMLOCK   REDCEDAR

Figure 15.8 Percentage distribution of *Pseudotsuga menziesii* (light stipple), *Tsuga heterophylla* (medium stipple) and *Thuja plicata* (solid) before and after commercial thinning.

(b) remove disease infected or damaged trees

(c) modify stand species composition

(d) improve the overall stand productivity.

In most cases, the opportunity to significantly alter the species or genetic diversity of the thinned stand exists.

Y.A. El-Kassaby (unpublished data) conducted a study to evaluate the effect of commercial thinning on the biological diversity (species or genetic) of forest trees. The study plot was located in a 60-year-old *Pseudotsuga menziesii* plantation (Fig. 15.8). In the study plot, all tree species were stem mapped (a map in which the spatial arrangement of each living tree is plotted) and vegetative buds or foliar tissue from every individual tree was collected for isozyme analysis. The genetic data provided a baseline for the degree of genetic variation within and among the various species in the study plot. A commercial thinning treatment was applied to the study plot area and the thinned trees were identified. The genetic data for the thinned trees were removed from the prethinning data, and a new assessment of the level of genetic variation within and among the various species was determined.

The results indicated that the species composition differed from the original plantation. Originally the site was planted with *Pseudotsuga menziesii* (i.e. 100%); however, at the time of thinning the species composition was 46%, 39% and 15%, *Pseudotsuga menziesii*, *Tsuga heterophylla* and *Thuja plicata*, respectively (Fig. 15.8). The thinning altered the species composition to 76%, 17% and 7%, of *Pseudotsuga menziesii*, *Tsuga heterophylla* and *Thuja plicata*, respectively. The Shannon index (Pielou 1975) estimate of species diversity showed a reduction from 1.01 to 0.70. All three species remained after the thinning; however, the reduction in the Shannon index clearly reflected the change in the three species proportional representation. Estimates of heterozygosity parameters for the study plot before and after thinning are presented in Table 15.4. In general, average number of alleles per locus, percentage polymorphic loci and expected heterozygosities did not differ significantly. Thinning, however, resulted in a loss of eight alleles (Table 15.4). Condensing the species or genetic diversity of the three different species in one index was done by estimating weighted heterozygosity across species according to their proportional representation. Following this concept, weighted heterozygosities of 0.110 and 0.133

TABLE 15.4 Number of loci, total number of alleles, number of alleles per locus (N_a), percentage of polymorphic loci (PLP95%), expected heterozygosity (H_e), and allelic loss for *Thuja plicata* (Cwr), *Tsuga heterophylla* (Hw) and *Pseudotsuga menziesii* (Df) before and after commercial thinning (before and after) in a 60-year-old *Pseudotsuga menziesii* plantation

Species	Thinning	No. of loci	No. of alleles	N_a	PLP95 %	H_e	Allelic loss
Cwr	Before	21	27	1.29 (0.12)	19.05	0.068 (0.039)	—
	After	21	27	1.29 (0.12)	23.80	0.087 (0.041)	—
Hw	Before	20	31	1.55 (0.15)	15.00	0.051 (0.026)	—
	After	20	24	1.20 (0.09)	15.00	0.043 (0.027)	7
Df	Before	22	56	2.55 (0.18)	54.55	0.153 (0.031)	—
	After	22	55	2.50 (0.19)	59.10	0.158 (0.033)	1

for before and after commercial thinning were obtained, indicating that the overall level of heterozygosity was increased following thinning. The observed increase was due to the increased proportion of *Pseudotsuga menziesii*, the species with the highest expected heterozygosity (Table 15.4).

The present review on the effect of forest tree domestication on gene pools illustrates the necessity of understanding the biology of the species and the biological ramifications of the various processes and practices used during domestication to avoid the effect of some specific practices that may have unintentional long-term consequences on genetic diversity.

MAGNITUDE AND IMPLICATIONS OF GENE FLOW IN GENE CONSERVATION RESERVES

W. Thomas Adams and
Jaroslaw Burczyk

SUMMARY

A practical means of long-term genetic conservation in forest trees is to establish natural (*in situ*) populations (i.e. gene resource management units, GRMUs) or *ex situ* plantings as gene conservation reserves. Results from pollen contamination studies in conifer seed orchards, however, indicate that gene flow in such reserves could be extensive. Although gene flow can be beneficial in terms of introducing new genetic variants, immigration of genes from domesticated populations is likely to reduce total genetic diversity within reserves and potentially lower their fitness. The prudent position on gene flow in reserves, therefore, is to limit it as much as possible. Pollen gene flow can be eliminated in *ex situ* plantings by controlled mating. If controlled mating is not feasible, applying pollen management techniques to increase pollen production within the plantings relative to external sources can minimise gene flow. Strategies for reducing gene flow in GRMUs are to make reserves as large as possible, include buffer zones around reserves (perhaps planted with an exotic species), and to assure that natural regeneration, or seed collection for artificial regeneration, occurs in heavy seed-crop years. Gene flow in forest trees is still poorly understood and the effectiveness of various approaches to limit gene flow in reserves, especially in GRMUs, cannot be quantified with any accuracy. Thus, research on gene flow in forest trees should receive high priority.

16.1 Introduction

Establishment of gene resource management units (GRMUs) is often proposed as a primary means of preserving genetic diversity in forest trees (Ledig 1988; Millar & Libby 1991). GRMUs are defined as parcels of land chosen to include a representative sample of the genetic diversity of the target species in a region, and designated for long-term genetic management. *In situ* reserves are superior to *ex situ* methods of genetic conservation because target species can continue to evolve in their native habitats, and because entire ecosystems are conserved, including other targeted and non-targeted species. In addition, as long as regeneration by local, native, seed sources is assured, timber harvest within GRMUs is compatible with the primary goal of preserving genetic resources.

In intensely managed species it is expected that GRMUs will eventually be surrounded by plantations of improved varieties. A question then arises as to the extent to which gene flow (i.e. immigration of genes via pollen or seeds from surrounding plantations) will influence the integrity of GRMUs. Because the size of GRMUs is central to their acceptance by managers, much has been written about minimum population numbers required to prevent loss of genetic diversity or adaptive potential due to inbreeding and genetic drift in small populations (National Research Council 1991; Frankel *et al.* 1995). Little attention, however, has been directed to potential negative effects of gene flow (Ellstrand 1992a; Ellstrand & Elam 1993). Given large differences in the genetic composition of GRMUs and surrounding plantations, even small amounts of gene flow in GRMUs could influence profoundly their genetic composition and adaptation potential. The magnitude of gene flow and factors influencing this magnitude, therefore, are of much interest.

A problem in evaluating potential effects of gene flow in GRMUs is the paucity of information on the magnitude of gene flow in native populations. Studies of patterns of genetic diversity among and within populations of forest trees using isozyme genetic markers (allozymes), indicate that gene flow has been strong enough in most cases to prevent genetic divergence among populations within regions (see Chapter 6; Ellstrand 1992b; Hamrick *et al.* 1992). Nevertheless, since only a small amount of gene flow is needed to arrest divergence among populations, it is impossible to discern from these data what the magnitude of gene flow might be in any one generation (Frankel *et al.* 1995).

One source of information on the potential magnitude of pollen gene flow in GRMUs are studies of pollen contamination in forest tree seed orchards. Seed orchards are important in forestry because they are the primary source of genetically improved seeds used in reforestation (Zobel & Talbert 1984). Orchards consist of either clones (i.e. grafted cuttings) or offspring (families) of parent trees selected for desirable characteristics (e.g. fast growth, disease resistance, favourable wood quality) in breeding programs. The number of clones (or families) typically ranges from 50 to several hundred, each replicated many times in the orchard. For efficient management, orchard sites often contain more than one orchard block, with the parents of each block derived from a separate geographical region. Because orchard blocks rely primarily on open (i.e. wind-mediated or animal-mediated) pollination, the potential for fertilisation by pollen sources outside blocks (i.e. pollen contamination) is always possible. Pollen contamination is detrimental because it reduces the genetic gains achieved by breeding, and if contaminant pollen comes from trees not adapted to the intended planting region, the adaptability of orchard seed is also negatively affected.

Estimates of pollen contamination (i.e. proportion of offspring sired by contaminant pollen) have been obtained recently for a variety of species and orchard management conditions with the aid of allozymes. In this chapter, we summarise the results of pollen contamination studies and evaluate their relevance to predicting levels of gene flow in GRMUs. We then discuss implications of gene flow in terms of effects on the integrity and fitness of GRMUs, and ways in which gene flow might be limited. *Ex situ* plantings of genetic resources (e.g. clone banks, arboreta, provenance trials) are also important in forest genetic conservation programs (National Research Council 1991; Rogers & Ledig 1996), and may even be more susceptible to gene flow. Thus, we also discuss implications of gene flow in *ex situ* plantings. We refer to GRMUs and *ex situ* gene conservation plantings collectively in this chapter as 'gene conservation reserves'.

16.2 Pollen contamination in seed orchards

Because pollen contamination (m) is the proportion of orchard seeds resulting from fertilisation by background stands (versus fertilisation by pollen produced within the orchard), it ranges in magnitude

---

Box 16.1 Statistical estimation of pollen contamination

Several procedures are used to estimate pollen contamination from genotypes observed in the offspring of mother trees (Smith & Adams 1983; Friedman & Adams 1985; El-Kassaby & Ritland 1986a; Devlin & Ellstrand 1990; Xie *et al*. 1991; Stewart 1994; Adams *et al*. 1997). The most commonly used approach is based on simple paternity exclusion (Smith & Adams 1983; Devlin & Ellstrand 1990). The first step is to determine the multilocus genotypes of all parents in the recipient population. Multilocus genotypes of seed offspring are then compared to parental genotypes and the proportion of offspring that could not have been sired by males in the recipient population are determined (detected immigrants). The proportion of detected immigrants (*b*) provides only a minimum estimate of pollen contamination because some immigrants are likely to have multilocus genotypes that are indistinguishable from those that can be produced by parents within the recipient population. To estimate the true proportion of immigrants (*m*), *b* must be adjusted by the probability that an immigrant offspring has a detectable genotype. This process is relatively straightforward in conifers where the genotype of a pollen gamete can be directly inferred by comparing the genotype of the megagametophyte (equivalent to the female gamete) to that of the embryo in the same seed. The detection probability, *d*, can then be estimated from allele frequencies in surrounding stands, such that:

$$m = \frac{b}{d} \tag{1}$$

In angiosperms, estimating *m* by paternity exclusion is more complicated because genotypes in pollen gametes rarely can be determined directly and detection probabilities vary depending upon the genotype of the mother (Devlin & Ellstrand 1990). As an example of estimating *m*, 16 out of 200 seeds sampled from an orchard crop had pollen gametes with multilocus genotypes that could not have been produced by parents in the orchard. Thus, a minimum estimation of pollen contamination is *b* = 16/200 = 0.08. The probability that stands surrounding this orchard produce pollen gametes with genotypes different from those produced within the orchard is *d* = 0.19. The estimate of *m* is, therefore, 0.08/0.19 = 0.42.

---

from 0 (no contamination) to 1 (total contamination). We list pollen contamination estimates (*m*, Box 16.1) for seed orchards of six conifer species, but include only cases where orchards have not been subjected to special management conditions to limit contamination (Table 16.1). The estimates, which cover a wide range of orchard sizes, ages and isolation (distance to nearest stands of the same species), range widely (0.01–0.91), but on average are quite high (mean *m* = 0.45). In some cases, only a minimum estimate of *m*, the proportion of detected contaminants (*b*), was available (Box 16.1). Typically only 0.25 to 0.5 of contaminated seeds are detected genetically using allozymes (see references in Table 16.1); thus, these minimum estimates probably underestimate true *m* two-fold to four-fold, supporting the high levels of pollen contamination observed in the other orchards.

Among factors which influence the magnitude of pollen contamination in seed orchards are:

(a) degree of isolation from background stands

(b) orchard size

(c) pollen production within the orchard

(d) synchrony of flowering in the orchard with flowering in background stands (although not botanically correct, in this chapter we refer to mating structures in conifers as flowers).

In cases where isolation distances are not reported, it is likely that there is no separation between the orchard and stands of the same species. Thus, cases of more than nominal isolation of orchards in Table 16.1 are few. Nevertheless, it is clear from the estimates of *m*, that isolation of less than a few hundred metres affords little protection from pollen contamination. The lowest *m* among all reported was for a *Picea glauca* (Moench) Voss (white spruce) orchard with isolation of 1000 m (Stewart 1994). In addition, the lowest

TABLE 16.1 Estimates of proportion of detectable immigrants (b) and pollen contamination (m) in conifer seed orchards

Species	Orchard				b	m (s.e.)	Reference
	Location	Size (ha)	Isolation (m)[A]	Age			
Larix decidua	Slovakia	—[B]	—	—	0.05	—	Paule & Gomory (1992)
Picea abies	Sweden	—	—	—	0.10	—	Paule *et al.* (1991)
	Sweden	—	—	—	0.17		Paule *et al.* (1991)
Picea glauca	Canada	—	1000	11–12	—	0.01 (0.01)	Stewart (1994)
Pinus sylvestris	Germany	—	—	—	0.02	—	Müller-Starck (1982)
	Finland	3.0	—	27	—	0.33 (0.03)	Harju & Muona (1989)
	Finland	3.2	—	29–33	—	0.26[C] (0.02)	Harju & Muona (1989)
	Finland	22.9	2000	20–23	—	0.48 (0.06)	Harju & Nikkanen (1996)
	Finland	22.7	—	31–33	0.18	0.67[D]	Pakkanen & Pulkkinen (1991)
	Finland	13.7	—	20–22	0.06	0.49[D]	Pakkanen & Pulkkinen (1991)
	Finland	6.0	—	26	—	0.53 (0.10)	Pakkanen *et al.* (1991)
	Poland	3.0	1000	16–18	0.15[D]	—	Burczyk (1992)
	Slovakia	—	—	11–12	0.11	—	Paule & Gomory (1992)
	Sweden	6.0	500	18–25	0.38	—	Nagasaka & Szmidt (1985)
	Sweden	16	—	14–18	0.21	—	El-Kassaby *et al.* (1989b)
	Sweden	12.5	—	17–18	0.36	—	El-Kassaby *et al.* (1989b)
	Sweden	13.8	—	18–21	0.35[E]	—	Paule (1991)
	Sweden	12.5	100+	17–18	—	0.72[F]	Yazdani & Lindgren (1991)
	Sweden	16	100	29–31	0.29	0.56	Lindgren (1991)
	Sweden	16	100	25–27	0.30	0.55	Wang *et al.* (1991)
Pinus taeda	S. Carolina	2.0	100	15–17	—	0.36[G] (0.03)	Friedman & Adams (1985)
	Texas	—	—	—	—	0.51 (0.05)	Wiselogel (1986)
Pseudotsuga menziesii	Oregon	1.8	None	14	—	0.52[H] (0.06)	Smith & Adams (1983)
	Oregon	—	None	20	—	0.29	Smith & Adams (1983)
	Oregon	3.3	None	8–9	—	0.91 (0.08)	Adams & Birkes (1989)
	Oregon	2.0	None	14–24	—	0.49[I] (0.05)	Adams *et al.* (1997)
	Washington	20	500	15	—	0.11	Wheeler & Jech (1986)
	B.C. Canada	—	None	11	—	0.34	Xie *et al.* (1991)

[A]Distance to nearest stand of the same species. [B]A dash means no information or estimate available. [C]Mean over four crop years. [D]Mean over three crop years. [E]Mean for two orchard blocks in one crop year. [F]Mean for three orchard blocks in two crop years. [G]Mean over two orchard blocks in three crop years. [H]Mean for 10 orchard blocks in one crop year. [I]Mean for one orchard block in five crop years.

among several *Pseudotsuga menziesii* (Mirb.) Franco (Douglas-fir) estimates ($m = 0.11$), was for a Washington orchard with isolation of 500 m (Wheeler & Jech 1986). A Finnish *Pinus sylvestris* L. (Scots pine) orchard was separated by 2000 m from the nearest stands of the same species, yet m was very high (0.48, Harju & Nikkanen 1996). Some individual *Pinus sylvestris* trees, however, were scattered in the *Picea abies* (L.) H. Karst. (Norway spruce) stands immediately surrounding the orchard. The value of isolation in limiting pollen contamination in seed orchards is unclear. Isolation distances of at least 500 m to 1000 m appear necessary for at least some protection. Nevertheless, large amounts of pollen can be dispersed into seed orchards from stands 50 km to 60 km away (Di-Giovanni *et al.* 1996). If this far-distant source of pollen is an important component of contamination, isolation zones within the natural range of species may be ineffectual.

Although gene flow is expected to decrease as recipient populations become larger (Ellstrand 1992a; Ellstrand & Elam 1993), no relationship between pollen contamination and orchard size is evident in Table 16.1. A complicating factor is that the magnitude of orchard flowering can vary widely from year to year, and estimates of orchard pollen production are often not available. Fertilisation and stem girdling were applied in a 15-year-old *Pseudotsuga menziesii* orchard to increase flowering (Wheeler & Jech 1986). As pollen production within orchard blocks increased, m decreased dramatically. Levels of pollen contamination, however, did not vary significantly over six crop years in another *Pseudotsuga menziesii* orchard where pollen production ranged six-fold (Adams *et al.* 1997). Likewise, there was no relation between the magnitude of pollen crops within orchards and contamination over three years in two *Pinus sylvestris* seed orchards (Pakkanen & Pulkkinen 1991). The influence of within-orchard flowering on contamination cannot be completely assessed without information on levels of pollen produced by background sources, because it is the relative concentration of orchard versus background pollen that largely determines the success of contaminant pollen (Adams 1992). By placing pollen traps inside the orchard to measure the sum of pollen produced by orchard and background sources, and in open fields nearby to measure background pollen levels, pollen contamination can be estimated as the ratio of background to orchard pollen cloud densities

(Greenwood & Rucker 1985; Webber & Painter 1996). Estimates of pollen contamination based on relative pollen cloud densities have been found to roughly approximate m derived from genetic markers (Greenwood & Rucker 1985; Wheeler & Jech 1986).

The degree of synchrony in floral phenology between the orchard and background stands is another factor influencing relative fertilisation success of background versus orchard pollen. Normally, if an orchard is located in the same region as its parents there is large overlap in floral phenology with background stands. Flowering within *Pseudotsuga menziesii* orchards, however, can be delayed, relative to background stands, by slowing flower development in late winter and spring with cooling water mists. This technique, called 'bloom delay' can be very successful in reducing pollen contamination in this species. For example, in a 15-year-old orchard in the State of Washington, m in one orchard block was reduced by more than 50% from 1983 ($m = 0.56$), when there was no bloom delay, to 1985 ($m = 0.26$), when bloom delay was applied (Wheeler & Jech 1986). Pollen density within this block increased three-fold during this period, suggesting that the reduction in m may also be due to an increased concentration of within-block pollen. Nevertheless, in an adjacent orchard block with a similar increase in pollen production from 1983 to 1985, but where bloom delay was not applied, m decreased only from 0.43 to 0.33. Bloom delay also appears to have had a dramatic effect on pollen contamination in a British Columbia seed orchard. In a year when bloom delay was applied, m was nearly zero (0.002) (El-Kassaby & Ritland 1986a). In the following year, with no bloom delay (but, with poorer within-orchard flowering), $m = 0.12$ (El-Kassaby & Ritland 1986b). Although a promising tool, application of bloom delay requires very special circumstances (e.g. overhead irrigation systems, good soil drainage) and may not work in all cases (Webber & Painter 1996). Indeed, few *Pseudotsuga menziesii* orchards use this methodology, either because they have no overhead irrigation system, or because trees have become too large, making misting impractical.

16.3 Expected magnitudes of gene flow in gene conservation reserves

How relevant are estimates of pollen gene flow in conifer seed orchards to predicting gene flow in gene conservation reserves? Levels of pollen contamination in *ex situ* gene conservation plantings of conifers probably will be similar to those observed for seed orchards. *Ex situ* plantings resemble orchards in terms of containing a single species, and having limited size (2–20 ha), uniform spacing of individuals and intensive management (e.g. competition control, fertilisation). Evidence that pollen gene flow can also be large in small populations of insect-pollinated angiosperm trees comes from several studies of effective pollen dispersal in natural populations (see following paragraphs).

GRMUs, however, could differ from seed orchards in several ways. First, one might expect GRMUs to be much larger than seed orchards. GRMUs must be large enough to ensure with reasonable probability that the reserve will survive and evolve in perpetuity (Ledig 1986; Millar & Libby 1991; National Research Council 1991; Rogers & Ledig 1996). The minimum number of individuals required for this purpose is called the minimum viable population (MVP) size. MVP size depends on several demographic, genetic and environmental factors that are likely to vary unpredictably over time (National Research Council 1991; Frankel *et al.* 1995). The MVP size most often suggested as a minimum necessary to maintain evolutionary potential (i.e. genetic diversity) is 500 (Frankel *et al.* 1995), but Lande (1995) argues that this number is an order of magnitude too small. Furthermore, MVP sizes refer to effective population size (N_e). Frankel *et al.* (1995) suggest that N_e is 20% to 10% of the actual census number (N) in forest trees. Thus, actual population sizes that are needed lie somewhere between 2500 and 50 000. Orchards cited in Table 16.1 range from a few hundred individuals to over 8000, so numbers in seed orchards are at the lower end of recommended minimum population sizes for GRMUs. Because trees often show interpopulation variation within regions, several small GRMUs may conserve total genetic diversity more effectively than a single large reserve (National Research Council 1991). This is basically the approach taken by the Washington Department of Natural Resources in establishing, perhaps, the only extensive GRMU system for a forest tree in North America (Wilson 1990). They have designated over 100 reserves for *Pseudotsuga menziesii* in western Washington—each about 10 ha and containing more than 400 dominant or codominant trees. Certainly the size of these

GRMUs are well within population sizes typical for seed orchards.

Stand structure is likely to be much more complex in GRMUs, with unevenness in spacing and tree size, and presence of multiple tree and shrub species. The influence of complex stand structure on pollen gene flow is unknown, but the few estimates of effective pollen dispersal in natural populations of both conifers and angiosperm tree species suggest that the levels of gene flow observed in seed orchards are representative of what occurs in similarly sized natural stands (Hamrick & Murawski 1990; Adams 1992; Boshier *et al.* 1995a; Schnabel & Hamrick 1995; Burczyk *et al.* 1996; Chase *et al.* 1996b; Dawson *et al.* 1997). For example, minimum estimates of pollen gene flow (i.e. proportion of detected pollen immigrants) in two shelterwood stands of *Pseudotsuga menziesii*, each 2.4 ha and containing 36 to 43 old growth (> 200 years) trees, were 20% and 27%, respectively (Adams 1992). Minimum estimates of pollen gene flow in two stands of the dioecious, insect-pollinated *Gleditsia triacanthos* L. (honeylocust) (each stand about 3 ha and containing 60 males), were 17% to 19% in good seed years and 28% to 30% in poor seed years (Schnabel & Hamrick 1995). In both these examples, the studied populations were not isolated spatially from background sources of pollen.

Only pollen gene flow is of concern in seed orchards, but gene flow by seed (or fruits or vegetative propagules) is an additional possibility in GRMUs, where seedlings can become established naturally and ultimately interact genetically with the population. Seed dispersal is usually more restricted than that of pollen (Levin & Kerster 1974; Ellstrand 1992a), but gene flow by seed, especially from nearby populations, can be substantial (Adams 1992; Dow & Ashley 1996). Because seed immigrants carry twice the number of genes than pollen gametes, they have twice the effect on gene flow.

Based on evidence from pollen contamination studies in seed orchards, and pollen and seed gene flow in natural stands, gene flow in *ex situ* plantations and GRMUs could be extensive, even when reserves are relatively large. Until further data become available, it is prudent to expect significant gene flow in all situations, except when gene conservation reserves are exceptionally well isolated (i.e. by several thousand metres) from populations of the same species.

Box 16.2   An example of the strong force of gene flow

As an illustration of the potential for gene flow to maintain maladapted genes in populations, we plot change in the frequency (p) of an undesirable gene (A_2) in a recipient population, using a migration-selection model (Hartl & Clark 1989) (Figure 1). We assume A_2 is fixed in the donor population, all immigration (m) is by pollen, A_2 has an additive effect on fitness (with selection coefficient s), and all selection occurs in offspring after random mating. Notice that when selection against allele A_2 is of the same magnitude as the rate of immigration (0.05), change in p (Δp) is always positive, such that A_2 will eventually become fixed in the recipient population. Even when selection against allele A_2 is relatively strong ($s = 0.10$ or 0.20), the limited gene flow we have assumed causes p to be positive whenever the frequency of A_2 is low, such that A_2 is never completely purged from the recipient population.

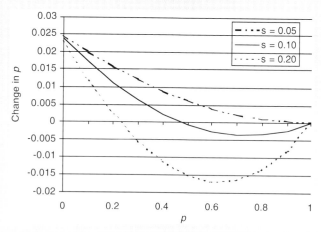

Figure 1   Change in the frequency (p) of an undesirable allele (A_2) under a migration–selection model, where the loss of A_2 due to selection (selection coefficient, s, is 0.05, 0.10 or 0.20) is countered by gene flow ($m = 0.05$) from a donor population where A_2 is fixed.

16.4   Significance of gene flow in gene conservation reserves

The effect of gene flow depends on the magnitude of immigration and the degree of genetic differentiation between donor and recipient populations (Hartl & Clark 1989). Studies based on allozymes have revealed little differentiation between domesticated populations (e.g. seed orchards, breeding populations) and natural stands in early generations of tree improvement programs (Adams 1981; Szmidt & Muona 1985; Chaisurisri & El-Kassaby 1994; Williams *et al*. 1995; El-Kassaby & Ritland 1995b; Stoehr & El-Kassaby 1997). Thus, gene flow may initially have little influence on the genetic integrity of reserves. These effectively neutral genetic markers, however, may underestimate changes that occur at loci under selection in breeding programs (Williams *et al*. 1995). In addition, genetic differentiation between

surrounding stands and reserves is bound to increase in future generations as surrounding stands are replaced by increasingly domesticated varieties.

Gene flow into reserves may be beneficial if the immigrants are well adapted to local environmental conditions, because the immigrants may be a source of new genetic variants (Ellstrand 1992a; Frankel *et al*. 1995). Indeed, in early generations of breeding, genetic variation in domesticated populations may be greater than within local natural stands because seed orchards contain selections from broad areas (i.e. many stands) (Chaisurisri & El-Kassaby 1994; El-Kassaby & Ritland 1995b; Stoehr & El-Kassaby 1997). Nevertheless, as domestication proceeds over generations, genetic variation in domesticated varieties is expected to be reduced to only a fraction of the amount in reserves, such that substantial gene flow from surrounding stands will reduce the amount of genetic diversity within reserves (Ellstrand & Elam 1993).

Potentially even more damaging to the genetic integrity of reserves is if domesticated varieties are less well adapted to current or changing local environments. If so, gene flow may disrupt or prevent local adaptation and reduce the fitness of reserves (Millar & Libby 1991; Ellstrand 1992a; Ellstrand & Elam 1993; Frankel *et al.* 1995). Some forest geneticists believe that immigration of maladapted genes into GRMUs is not a problem because with dense regeneration, natural selection will remove unfit individuals (Ledig 1986; Millar & Libby 1991). Even modest gene flow, however, can be a potent force in counteracting relatively strong selection and in maintaining maladapted genotypes in populations (Box 16.2). Immigration of maladapted genes may even be more of a problem in *ex situ* plantings that are regenerated artificially. Some of the strongest selection occurs at early seedling stages in field environments (Campbell 1979), but maladapted genotypes may undergo little selection, or perhaps may even be favoured, when raised under mild nursery conditions (Campbell & Sorensen 1984). Thus, genotypes that otherwise may not have made it beyond the seedling stage could be planted, and perhaps survive to reproduction.

Gene flow can be a severe detriment to the integrity and survival of gene conservation reserves. We agree with Millar & Libby (1991), that the most appropriate strategy in managing these reserves is to do all that is possible to limit immigration of foreign genes.

16.5 Strategies for limiting gene flow in gene conservation reserves

Three general approaches, used independently or combined, can be applied to limit gene flow in reserves:

(a) control the location, size and isolation of reserves

(b) increase the ratio of pollen production within reserves relative to that in surrounding populations

(c) control mating.

16.5.1 Controlling location, size and isolation of reserves

When feasible, reserves should be located where gene flow, when it occurs, will have the least negative effect

(Frankel *et al.* 1995). These are areas where surrounding populations have not been domesticated (e.g. locations in, or adjacent to, parks or other natural reserves), or where harvesting is followed by natural regeneration or accomplished artificially using seed from local, native stands. Unfortunately, the areas in greatest need of gene conservation reserves are those where management is the most intense and where populations surrounding reserves are under the most rapid domestication. Thus, optimal conditions for reserve placement often are not available.

Despite ambiguous results from pollen contamination studies in seed orchards, increasing the size of reserve populations should decrease gene flow, whether genes are transported by wind or by pollinators (Ellstrand 1992a; Ellstrand & Elam 1993). Most *ex situ* reserves are not likely to be large, simply because population sizes do not need to be great to capture most of the useful genetic diversity within a region (i.e. N in hundreds, or less, is adequate; Frankel *et al.* 1995). MVP size sets the minimum numbers necessary to ensure long-term evolutionary potential of GRMUs, but because harvesting is compatible with gene conservation, GRMUs much larger than the minimums required may be acceptable to managers (Ledig 1988).

As indicated earlier, spatial separation of reserves from plantations of the same species may need to be substantial for isolation to have a reasonable effect on limiting contamination. Spatial isolation is easiest to achieve in *ex situ* plantings that could be located in areas far removed from pollen sources of the same species. A drawback with this strategy, however, is that if the planting environment differs from the one in the source location, the genetic composition of the reserve population could be altered significantly by selection. Isolation of both *in situ* and *ex situ* reserves is probably best accomplished by surrounding them with buffer zones or wind breaks, to provide a physical barrier to foreign pollen and seeds (Di-Giovanni & Kevan 1991; Millar & Libby 1991). Buffer zones could consist of natural stands, or planted trees of local origin, of the same species in the reserve. An alternative is to plant buffers with an exotic species known to be adapted to the local environment, but not invasive. The advantage of using an exotic is that seed immigrants can be easily identified, and the buffer would absorb, not generate, contaminating pollen. The width of buffer needed for adequate protection is unclear, but will depend on the species, size of the reserve and meteorological factors (Di-Giovanni & Kevan 1991), and is crucial to the total land area required. For example, with a buffer of 200 m around a square reserve of 100 ha, a total land area (reserve plus buffer) of nearly 200 ha would be required.

16.5.2 Maximising pollen production within reserves relative to surrounding stands

Artificial manipulation of floral phenology and magnitude is probably feasible only in *ex situ* gene conservation plantings where tree location, spacing and size can be controlled. The magnitude of flowering and pollen production in *ex situ* collections can be increased greatly using floral stimulation treatments developed for seed orchards (Wheeler & Jech 1986; Wheeler & Bramlett 1991). Increasing flower production within reserves will also reduce gene flow in tree species with animal-mediated pollination, because pollinators will be encouraged to forage exclusively within reserve populations (Levin & Kerster 1974). Only one stimulated crop should be needed for regenerating an *ex situ* population. But, if by chance, pollen production in surrounding stands is large relative to that within the reserve, seed collection can be deferred to another crop year when flower stimulation produces a more favourable ratio of reserve to background pollen density. Bloom delay could be used to offset timing of peak flowering within *ex situ* plantings relative to surrounding natural stands, but it is unlikely that the necessary irrigation systems would be available in most cases.

Manipulation of the ratio of pollen production within GRMUs relative to surrounding stands can be done silviculturally by controlling harvesting. Timing of harvests to promote natural regeneration (e.g. seed tree or shelterwood cuts) should coincide with heavy seed crops within the reserves. Surrounding stands should be harvested several years before regeneration within reserves is anticipated and adult trees removed (e.g. by clearcutting or by overstory removal after seed tree or shelterwood regeneration) to reduce background pollen (Millar & Libby 1991). If the GRMU is to be regenerated artificially, seed should be collected from scattered trees within the GRMU only in good seed years. In addition, seedlings should be raised under conditions that are least likely to promote artificial selection (e.g. sow at wide spacing in nursery beds or single sow in containers; El-Kassaby & Thomson 1996). Planting in the field, however,

should be at relatively high density to promote early competition and natural culling of maladapted genotypes.

16.5.3 Controlling mating

The optimum solution to pollen contamination is to control mating completely by applying pollen artificially to bagged flowers. It is possible to make controlled crosses among trees in GRMUs by climbing the trees on site or by conducting crosses in clone banks on grafted cuttings of the trees (Wilson 1990). It is hard to imagine, however, that these expensive alternatives could be justified except under very special circumstances, such as might occur with highly valued or endangered species. Controlled pollination is most feasible in *ex situ* plantings, but even here, costs may be prohibitive. An alternative to completely controlled crossing is the broadcast application of pollen onto unbagged female flowers (called supplemental mass pollination, SMP, Bridgwater *et al.* 1993). SMP has a lot of promise, but reported success rates have been variable, with often less than 50% of offspring resulting from the applied pollen (Bridgwater *et al.* 1993; Eriksson *et al.* 1994). In concert with other tools, such as isolation and flower stimulation, SMP may be helpful in reducing pollen contamination, but if SMP is to have a primary role for this purpose, more work will be needed to perfect the technique.

16.6 Need for research and monitoring

We have a lot to learn about gene flow in forest trees, its effect on the integrity of gene conservation reserves and the methods by which it can be curtailed or controlled. Much continues to be learnt about pollen management in seed orchards (Bramlett *et al.* 1993; Webber & Painter 1996), and as this technology becomes available, it will be applicable to more intensely managed *ex situ* gene conservation plantings. Control of gene flow in GRMUs may be difficult to achieve, and the relative roles of reserve size, isolation, buffer zones and background versus reserve pollen production, on levels of pollen contamination are nearly impossible to quantify with any accuracy. Research on gene flow in GRMUs, or other natural populations with properties similar to GRMUs, must be given high priority. Methods similar to those employed in measuring pollen contamination in seed orchards can be used, but the reliability of these methods is highly dependent on the ability to genetically discriminate pollen from local and foreign sources. Application of these methods using allozymes has been possible in seed orchards only because of the relatively small number of genotypes involved (Adams *et al.* 1992a). Estimation of gene flow in larger populations, like GRMUs, requires more polymorphic markers than available with allozymes. Hopefully, hypervariable microsatellites (or simple sequence repeats) will prove more suitable (Frankel *et al.* 1995). Some early applications of these molecular markers to gene dispersal patterns in forest trees are promising (Chase *et al.* 1996b; Dow & Ashley 1996; Dawson *et al.* 1997).

Regardless of the approaches taken to limit gene flow in gene conservation reserves, it will be important to monitor their success in at least a representative sample. Only in this way can the validity of the approaches be confirmed and improvements designed.

SECTION V
MONITORING, SOCIOECONOMICS AND POLICY

Concern for the world's forests has prompted many initiatives to promote conservation of forest genetic resources. We have seen, in earlier sections, that the levels and patterns of genetic diversity are due to a balance of evolutionary forces that operate at different spatial and temporal scales, but that human impacts, both current and future, are also influential in shaping the patterns and levels of genetic diversity. Indeed others have shown that humans have always influenced forest genetic resources to a greater or lesser degree (e.g. Ledig 1992; Fairhead & Leach 1998). As Westoby (1989) said 'forestry is about people, not about trees' and any practical consideration of conservation must be made within the context of human involvement at all levels, while also recognising a range of, often conflicting, rights and priorities from diverse stakeholder groups. Indeed the extreme diversity of human (economic) activities and the cultural and socioeconomic contexts in which they occur lead to highly complex interactions. This last section of the book therefore examines the socioeconomic and policy context within which practical initiatives for the conservation, sustainable management and monitoring of forest genetic diversity are set.

Since most threats to forest genetic resources come from human activities, managers need to be able to recognise when and under what circumstances species or populations may be under genetic threat. In Chapter 17, Gerhard Müller-Stark and Roland Schubert outline the role of genetic markers as bioindicators. Recent work has seen the development of genetic criteria and indicators (C&I) to provide guidance on what genetic processes may be affected by which forest practices and forming part of a more general set of biological, economic, and social C&I for use in the assessment of forest sustainability. Timothy Boyle (Chapter 18) describes potential indicators of genetic processes, their relationships to different forest practices and their use in monitoring the current genetic status of populations.

Stakeholders in conservation genetics may be manyfold, depending on the type of forest, the socioeconomic and cultural circumstances of the resource to be conserved. Stakeholders in the conservation of temperate rainforest systems on the west coast of North America will be very different, in many senses, from stakeholders in dry deciduous tropical forest, and this should affect the definition of conservation goals, criteria and their ultimate implementation. Conservation models developed for one situation may be entirely inappropriate in other circumstances and yet the blanket application of particular models has been a common scenario. In some instances, although policy makers may consider conservation desirable, owing to potential benefits at some future date or to non-local stakeholders, such benefits are often not tangible to those who most directly utilise the forest. Where conservation results in some stakeholders forgoing benefits they would otherwise reasonably expect to enjoy, some form of compensation may have to be established. The nature of this compensation may vary, with achievement of the correct balance between conservation and development critical to the former's success; but conservation genetics inevitably involves an economic dimension, both direct and indirect. In Chapter 19, McNeely discusses the economics of genetic conservation, including political issues such as intellectual property rights. Whatever the situation, however, experience suggests that conservation can only be effective with the active involvement, commitment and ideally participation of a range of stakeholders throughout the process. Thus, Thomas Enters, in Chapter 20, looks at means of involving stakeholders in genetic conservation.

Peter Kanowski completes Section V, in Chapter 21, by examining current policy issues and initiatives related to forest genetic diversity within a historical context of concern, and

action, for its conservation. He discusses the relative roles and emphasis on *in situ* and *ex situ* conservation, incorporation of genetic conservation criteria into forest and farm management practices and the tensions between strategies that emphasise conservation of diversity at different scales. Finally he explores how the conservation and use of forest genetic diversity might best be advanced within the current policy environment.

GENETIC MARKERS AS A TOOL FOR BIOINDICATION IN FOREST ECOSYSTEMS

Gerhard Müller-Starck and
Roland Schubert

SUMMARY

Bioindication measures the response of individuals and populations to stress with main emphasis on anthropogenically induced change of environment. Forest ecosystems reveal a high indicative potential for the verification of long-term environmental stress. This study aims at a greater efficiency of bioindication via integration of genetic markers as indicators, which are a response variable not currently being employed for this purpose.

Principles of adaptation to stress are outlined and influences of environmental stress on genetic structures are classified. A set of selected genetic parameters is suggested for utilisation in bioindication. It includes indicators for climate change with respect to thermic impacts and indicators for environmental stress governed by air pollution. Chances of molecular tools for bioindication are described with respect to such genes encoding heat shock proteins, stilbene synthase as well as cinnamyl alcohol dehydrogenase enzymes, and the species-specific detection and quantification of pathogens. Five major problems in bioindication are discussed and the importance of genetic tools in bioindication is pointed out.

17.1 Introduction

17.1.1 Bioindication

As a consequence of increasing influences of industrial pollution on the biosphere, various attempts were made to measure the response of organisms or communities of organisms to anthropogenically induced change of environment. The verification of such responses is commonly termed 'bioindication' and covers many phenomena. A major problem arises because anthropogenic effects are complex and cannot be observed independently from the dynamics of 'natural' environmental factors (i.e. the variety of primarily non-anthropogenic processes and corresponding stress conditions). Consequently specified stress scenarios were established under controlled conditions in the phytotron or *in vitro*.

Generally, bioindication does not refer exclusively to anthropogenic effects. Bioindication *per se* refers to the response of biological systems to any environmental impact and can be defined in many different ways. In the present study, bioindication is used to verify the response of biological systems to environmental change directly or indirectly induced anthropogenically.

17.1.2 Bioindicators

Bioindication requires variable parameters, bioindicators, which allow the monitoring of the effects of specified environmental impacts on a time scale (cause–effect relationship). Such parameters are designated as 'bioindicators'. Generally, bioindicators are organisms whose occurrence (or behavioural trait) can be so closely related with certain environmental conditions that it can be utilised as a pointer or quantitative test (Ellenberg *et al.* 1991). With respect to pollution, bioindicators are defined as organisms or communities of organisms which respond to stress by modifications of essential life functions and accumulations of pollutants—the corresponding indicators are designated as reacting indicators ('Reaktionsindikatoren') and accumulating indicators ('Akkumulationsindikatoren'), respectively (Arndt *et al.* 1987). The term 'biomonitoring' is used when an area is continuously observed by means of bioindicators, that is, 'biomonitors' (Wittig 1993).

Bioindicators refer to various levels of biological systems—individuals (cell, tissue, organ, organism), populations, multi-species communities and ecosystems. Morphological and physiological characters are used (e.g. Arndt *et al.* 1995). Genetic markers are not yet integrated in the classical concepts of bioindication.

17.1.3 Environmental stress

'Environmental stress' stands for a variety of abiotic and biotic stress factors that affect the biosphere in terms of death of individuals and corresponding reductions of density and size of populations. Particularly for long-lived tree species, environmental stress is a complex phenomenon which is highly variable in space and time. In the case of 'forest decline', environmental stress is considered to be governed by anthropogenic stress. Predisposing factors are not only climate, chemical composition of the atmosphere and dynamics within soil but also genetic characteristics such as non-adapted populations, genetic erosion and inbreeding depression. Disposing factors are, for instance, stress by extreme temperatures, air pollution in high concentrations, pathogen stress and damage of root systems and crowns following acidification.

17.1.4 Forest ecosystems

Forest tree species are significant members of terrestrial ecosystems. About one-third of the ice-free land surface of the globe is covered by closed forests or other wooded areas. This area is decreasing—between 1980 and 1990 average global losses were about 4.4%, mainly in tropical zones (8.9%). There was no reduction in the boreal zone and a slight increase of 1.4% in the temperate zones (Food and Agriculture Organization of the United Nations 1992). If the average rate of extinction (4.4% per 10 years) is projected in time, by the year 2207 all forests would be extinguised. This time period is smaller than the natural life span of most of the forest tree species in one generation.

Tree populations provide substantial benefits both ecologically and economically by harbouring many plant and animal species, by protecting landscapes against soil erosion and flood, by providing various recreational and social venues and by supplying many materials for industrial use, for fuel wood and food. The release of oxygen and the biochemical fixation of carbon dioxide are essential for life on our planet and are needed to compensate for the increase of global temperatures following greenhouse effects.

In contrast to agricultural systems, most forest ecosystems can still be considered to contain biodiversity in a non-domesticated status. Tree populations are long lived and are exposed to many biotic and abiotic components of environmental stress. Control of stress is not possible for exposure to extreme temperatures or air pollutants and is not very effective in other cases such as long-term improvement of site conditions via fertilisation or disease management (without cost-effective pest control). Future environmental stress cannot be forecasted clearly because of complex interactions among stress variables. Global warming will affect various abiotic and biotic stress conditions and possibly furthermore increase challenges to adaptability and survival abilities of forest tree populations after environmental change.

In addition to exploitation and devastation, the vulnerability of forest ecosystems has two major causes: first, the large potential of parasites to adapt successfully to hosts because of the discrepancy between extremely short generation cycles of parasites in contrast to long generation cycles of immobile trees; and second, increasing anthropogenic effects on environments and climate which increase heterogeneity of stress conditions and may substantially reduce chances for adaptation and survival.

Because forest ecosystems are long lived and predominantly non-domesticated, tree populations can be considered to indicate efficiently long-term anthropogenic environmental stress. Genetic variability within tree populations is an essential part of the high indicative potential of forest ecosystems because it is a major factor in the determination of the adaptive potential of tree populations and its survival abilities under various stress scenarios (e.g. Scholz *et al.* 1989; Kim & Hattemer 1994; Baradat *et al.* 1995a; see Chapter 5).

17.2 Genetic response of tree populations to environmental stress

17.2.1 Genetic variation in forest tree species

Genetic variation is defined here as the within-species part of 'biodiversity'. Other levels of biodiversity are the diversity with respect to different species or to ecosystems. Within species, genetic variation quantifies the variation of genetic types (genotypes, gametes—single or multilocus) within and between populations (e.g. for methods see Gregorius &

Roberds 1986; Müller-Starck & Gregorius 1986; El-Kassaby 1991).

Most of the tree species tested reveal remarkably great intrapopulational genetic variation (for survey see Mitton 1983; Hamrick & Godt 1989; Fineschi *et al.* 1991; Müller-Starck & Ziehe 1991; Adams *et al.* 1992b). In a comparison of such investigations with those that used isoenzyme gene markers (Müller-Starck 1995), the average observed heterozygosities of tree species amounted to 24.25% and of reference species to 13.90% (dicotyledons 11.3%, monocotyledons 16.5%). The corresponding ratio is 1.74:1. The average number of alleles per locus was 2.45 for tree species and 1.55 for reference species (dicotyledons 1.4, monocotyledons 1.7). The corresponding ratio is 1.58:1. Increasing numbers of alleles coincide with an exponential increase of the potential of populations to form a maximum number of genetically different multilocus genotypes in the next generation and thus to respond to environmental challenges via genetically variable offspring. If 18 polymorphic gene loci are included in the calculation of such maximum potential, the values for tree species are 1.68×10^6 times greater than those of the reference species (Müller-Starck 1995).

The great genetic variability (i.e. potential to form genetically different gametes or genotypes) belongs to the predominant characteristics of tree populations as compared to populations of other plant species which have a shorter life span and are exposed in most cases to less heterogeneous environmental conditions. This is interpreted as a response of the genetic system of species and populations to the complex selection pressure following highly variable stress conditions.

17.2.2 Principles of adaptation

The following scheme (Fig. 17.1) demonstrates in which way genetic variability is crucial for genetic adaptation and survival of populations, species and entire ecosystems under changing environmental conditions.

The state of being adapted (adaptedness) describes the condition in which populations can survive and reproduce without modifications of its integrity. Adaptation is the process which results in the state of adaptedness. In most cases, adaptation coincides with structural changes at the population level (i.e. selective processes which eliminate certain genotypes and thus modify frequency distributions). In addition,

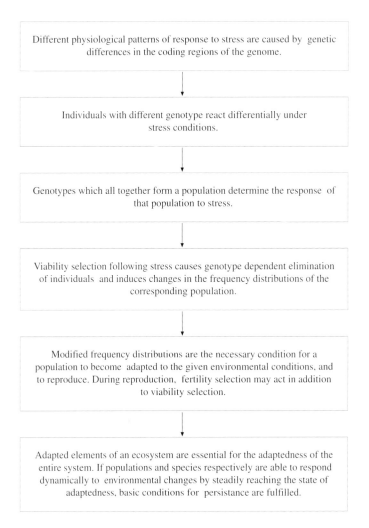

Different physiological patterns of response to stress are caused by genetic differences in the coding regions of the genome.

↓

Individuals with different genotype react differentially under stress conditions.

↓

Genotypes which all together form a population determine the response of that population to stress.

↓

Viability selection following stress causes genotype dependent elimination of individuals and induces changes in the frequency distributions of the corresponding population.

↓

Modified frequency distributions are the necessary condition for a population to become adapted to the given environmental conditions, and to reproduce. During reproduction, fertility selection may act in addition to viability selection.

↓

Adapted elements of an ecosystem are essential for the adaptedness of the entire system. If populations and species respectively are able to respond dynamically to environmental changes by steadily reaching the state of adaptedness, basic conditions for persistance are fulfilled.

Figure 17.1 Principles of genetic adaptation following viability selection.

adaptation may occur exclusively by regulatory processes at the individual level.

A system is called adaptive if it responds to environmental changes by reaching again the state of being adapted. From a genetic point of view, the adaptive potential of a population describes its genetic variation with respect to a specified census stage and the corresponding reaction potential, that is, the ability to create genetic variation (genetic variability).

Dynamics of environmental stress conditions create selection pressure on populations which need to react via adaptation in order to avoid successive extinction. The genetic reaction is successful, that is, populations adapt and survive if they contain a sufficiently large numbers of adapted genotypes after each modification

of stress conditions. The remaining population must be able to reproduce under the given environmental conditions without being endangered by genetic drift processes.

Physiological adaptation, that is, adaptation by means of regulatory mechanisms at the individual level, is limited in its efficiency because the chance is very low that the metabolic function of one and the same individual responds successfully to any abiotic or biotic stress. Genetic adaptation primarily affects populations and not individuals. Such adaptation coincides with losses of non-adapted individuals (i.e. structural changes) and a modification of the adaptive potential. These costs of adaptation are the precondition for any genetic change in response to the dynamics of environmental stress. Modified genetic structures of populations will result in corresponding changes in the offspring populations.

Genetic changes occur at different levels. The genetic structure of populations is modified particularly by viability selection, reproduction with inclusion of fertility selection, by gene flow via gametes and/or genotypes and by genetic drift. At the individual level, genetic change is established by means of mutations of the nuclear and the extra-nuclear genetic information. Genetic changes are part of any evolutionary process. Populations (i.e. reproductive communities) are elements of evolution and thus of adaptation to the dynamics of environmental conditions.

17.3 Problems related to bioindication

17.3.1 Cause–effect relationships

Under complex field conditions, cause–effect relationships cannot be considered isolated with respect to specific stress factors. The existing networks of abiotic and biotic stress components allow general statements on predominant stress characteristics but no on-going interpretations with respect to single components. Under conditions *in vitro* or in the phytotron, cause–effect relationships can be studied with respect to single stress factors or selected combinations of factors. Evidently, the more complex the application of stress condition, the less precise are the resulting statements of such experiments.

17.3.2 Representativeness of response to stress

Response of a populations of a certain tree species at one location cannot be considered to be representative for conditions at other locations. The local characteristics of the stress scenario and the genetic structure of the responding population interact and result in location-specific and population-specific responses. Even in cases of genetic identity between populations at two locations (e.g. clonal material), the specific local arrangement of stress components will create different selective forces and thus result in differential responses at the individual and the population level.

17.3.3 Time scale

Bioindication can be related to seconds, months, vegetation periods or much longer time scales. Observations performed within short intervals cannot represent the whole life span of a tree. Therefore, if possible, bioindication should cover a large part of the life span of respective individuals. Bioindication with respect to short-lived individuals will not result in adequate prognoses of long-term responses of complex ecosystems. Particularly with respect to long-lived forest ecosystems, permanent experimental plots are essential tools in bioindication.

17.3.4 Levels of bioindication

Bioindication can be related to entire ecosystems, to certain sets of species or populations, to groups or to specified individuals. Even within one individual, various levels can be distinguished such as organs, tissues, cells, effects of gene regulation or mutations of the nuclear or the extra-nuclear genetic information. The dilemma is that most experiments and resulting information is focused on individuals (i.e. a low level of organisation) (e.g. Arndt *et al.* 1992) which is not representative for higher levels of organisations such as populations and ecosystems. The higher the organisation level, the more representative are findings by bioindication but the smaller is the current information base.

17.3.5 Dose–effect relationships

Abiotic and biotic stress mostly affect metabolic functions of individuals in a complex way. Linear dose–effect relationships are the exception—even *in vitro* or in phytotrons. At the population level, complex interactions can be expected with respect to fluxes of matter and energy which do not fit linear dose–effect relationships. Indicators which reveal an accumulation of effects more than simple reactions are important particularly in long-lived forest ecosystems.

Genetic markers reveal a new category of dose–effect relationships. They allow the monitoring of different intensities of viability selection and various other forms of genetic responses to environmental stress (see Chapter 4). Genetic markers are part of the gene pool of populations and species and reflect differentially the effects of stress on genetic resources.

17.4 Genetic markers as bioindicators

Genetic markers are part of the genetic resources of species and can characterise efficiently genetic changes following selection pressure by various environmental stress factors. In long-lived forest ecosystems, biomonitoring via genetic markers can reveal long-term trends in the response to environmental stress independent from non-representative short-term fluctuations with respect to specific environmental stress factors.

The frequency distributions of nuclear genetic markers in forest tree populations measured by means of isoenzymes fit into the category 'minor polymorphism' (for survey see e.g. Fineschi *et al.* 1991; Müller-Starck & Ziehe 1991; Adams *et al.* 1992b; Kremer *et al.* 1993; Baradat *et al.* 1995a). In most cases one or two alleles are predominant while others are present in low frequencies (i.e. 10% or much lower) (e.g. Finkeldey & Gregorius 1994). The occurrence of these genes is interpreted as a latent genetic potential (Bergmann *et al.* 1990) which has a buffer function in so far that it facilitates adaptation to the future under certain stress combinations. Consequently, such rare alleles can be classified as 'pre-adaptive'.

Environmental stress can be expected to induce changes in the frequency distribution of genetic types in populations, particularly by affecting the following processes:

(a) *loss of genetic variation*—the loss of rare alleles following severe reductions of population sizes (genetic drift) mainly concerns that part of the adaptive potential which is classified as pre-adaptive and is important as a tool for future adaptation

(b) *viability selection*—certain combinations of genes (single-locus or multilocus) are subject to selection and induce an individually different response to environmental stress at individual level—individuals with sensitive genotypes will be eliminated while the tolerant individuals remain in the population

(c) *fertility selection*—the formation of female and male gametes and of zygotes can be substantially influenced by means of selective elimination—air pollution in particular is expected to interfere in fertility and following reproductive processes

(d) *inbreeding depression*—the reduction of population sizes and preferential elimination of reproducing individuals will reduce neighbourhoods and increase the presence of full sibs and half sibs as well as offspring from self fertilisation—for dioecious species, the proportion of male and female individuals can be

substantially modified which again reduces the effective size of neighbourhoods

(e) *gene regulation and mutation*—stress components that affect gene regulation will substantially determine the genetic response of individuals and populations to the actual and future environmental situation, respectively—the acceleration of mutations by means of mutagenic stress factors can both increase or decrease the chances of elimination of respective individuals.

For survey of the results of experimental investigations and of case studies see Müller-Starck (1994), Degen (1997), and Chapter 11. Based on these studies which primarily utilised isoenzymes markers and on findings from recent molecular studies (Schubert *et al.* 1997a, 1997c; Schubert *et al.* 1998, 1999), the following genetic markers are suggested as bioindicators.

17.4.1 Isoenzyme markers for climate change with respect to thermic influences

The study of the thermostability of the system of the isocitrate dehydrogenase (IDH, EC 1.1.1.42) proved that certain isozymes at the gene locus IDH-B show different sensitivities to summer temperature regimes (Bergmann & Gregorius 1993; see Fig. 17.2). The allele B_2, significantly more frequent in the southern parts of Europe, reveals a higher thermostability than the allele B_1, significantly more frequent in populations in northern Europe. The frequencies of these two alleles in *Abies alba* populations apparently respond to climatic conditions and may function as indicators for climate change with respect to local or regional increase of temperature gradients

17.4.2 Isoenzyme markers for environmental stress governed by air pollution

17.4.2.1 Viability advantages of specific alleles—example 1

In the case of *Fagus sylvatica* L. (European beech), the genetic comparison between germinating seeds and corresponding samples of survivors under field stress dominated by severe acidification in the upper layers of the soils resulted in a highly significantly increase of carriers of the allele leucine aminopeptidase-A_4 (LAP, EC 3.4.11.1) (Müller-Starck & Ziehe 1991). The mean frequency of this allele was 8.9% in all five sets

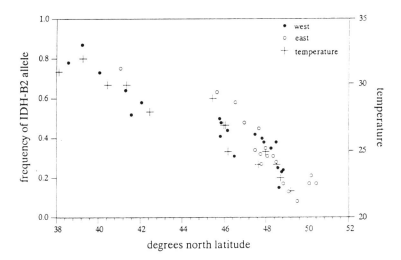

Figure 17.2   Latitudinal variation in IDH-B2 allele frequencies along an eastern and a western north–south transect of *Abies alba* (silver fir) populations (Bergmann & Gregorius 1993).

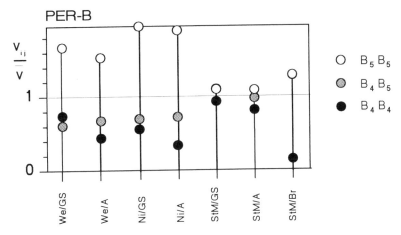

Figure 17.3   Survey of viability coefficients (V_{ij}) for single genotypes at the gene locus PER-B in relation to the population mean. Data refer to three samples of germinating seeds of *Fagus sylvatica* (We, Ni, StM) and corresponding survivors at two or three locations (A, GS, Br) (for details see Müller-Starck 1993).

of germinating seeds and 25.8% in the corresponding nine sets of survivors (i.e. a ratio of 1:2.9). This relative increase of frequencies of this allele under different field conditions of acidification cannot be interpreted as other than selective advantage compared to other alleles at this gene locus. For demonstration of the increase of the allele LAP-A$_4$ see Chapter 11.

17.4.2.2 Viability advantages for single genotypes or classes of genotypes—example 2

In the same experiment (see Chapter 17.4.2.1), viability coefficients were calculated for the gene locus

peroxidase-B (PER, EC 1.11.1.7). Figure 17.3 indicates a viability advantage for the homozygous genotype B$_5$B$_5$ under different experimental conditions (for details see Müller-Starck 1993). This example demonstrates that differences with respect to viabilities of genotypes are potential indicators of environmental stress.

A trend in favour of heterozygotes is evident in a study of genetic variation among seedling populations of *Picea abies* (L.) H. Karst. (Norway spruce) on two locations contaminated by heavy metal pollution and on corresponding control locations (Hosius *et al.*

233

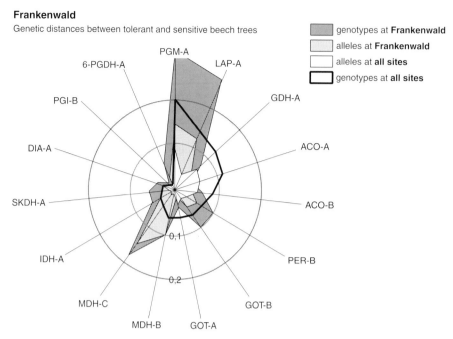

Frankenwald
Genetic distances between tolerant and sensitive beech trees

genotypes at **Frankenwald**
alleles at **Frankenwald**
alleles at **all sites**
genotypes at **all sites**

Figure 17.4   Profile of locus-specific genetic distances among tolerant and sensitive subsets in long-term air polluted stands of *Fagus sylvatica*. Data refer to the stand 'Frankenwald' in Bavaria which is particularly subjected to pollution governed by sulfur dioxide (after Ziehe *et al.* 1999, data taken from Müller-Starck 1993).

1996). In all provenance samples tested, viability advantages are evident for carriers of heterozygous genotypes at four isoenzyme coding gene loci.

17.4.2.3   *Locus-specific genetic distances between tolerant and sensitive subsets—example 3*

The comparison in pairs between tolerant and sensitive collectives in five adult stands of *Fagus sylvatica* under long-term exposure to environmental stress revealed significant genetic differences (Müller-Starck 1989). Pollution-specific trends were evident. In Figure 17.4, the genetic distances for each of the 15 gene loci studied are plotted, and are connected by lines (see Ziehe *et al.* 1999). Largest genetic distances are indicated for genotypes at the gene loci PGM-A, LAP-B and MDH-C. These distances can be compared with the average genetic distances at all five polluted *Fagus sylvatica* stands at different locations in Germany (see thick line). The allele frequencies (see dark shaded areas) reveal trends similar to the genotype frequencies. It appears that genetic distances between tolerant and sensitive collectives can be used for bioindication with respect to specific long-term field stresses.

17.4.3   Indicative potential of molecular markers

17.4.3.1   *Heat shock proteins—example 1*

Only a few genes have been identified in forest tree species (e.g. Kinlaw *et al.* 1996; Schubert *et al.* 1997a). The very large genome of coniferous tree species is considered as the main reason for the low number of investigations and the corresponding lack of information. The genes identified are known to occur also in angiosperms. They refer to a set of small heat shock proteins which can be pre-dated to the last common ancestor of seed plants, about 300 million years ago [Dong & Dunstan (1996) for *Picea glauca* (Moench) Voss (white spruce); Kaukinen *et al.* (1996) for *Pseudotsuga menziesii* (Mirb.) Franco (Douglas-fir) Schubert *et al.* (1997c) for *Picea abies*]. In addition to their strong responses to elevated temperatures, certain genes seem to be developmentally regulated during embryogenesis (see Dong & Dunstan 1996). In conifers, genes encoding small molecular weight heat shock proteins show only 28% to 66% identity at the deduced amino acid level with their angiosperm counterparts although certain protein motifs are known to be conserved during the evolution (Schubert *et al.* 1997c). Heat shock proteins have been

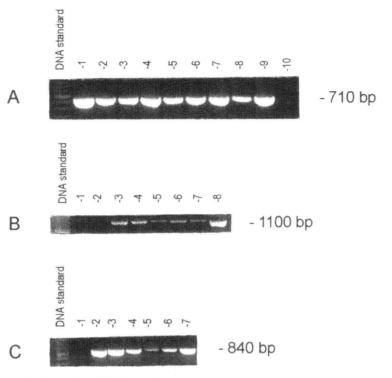

Figure 17.5 Polymerase chain reaction (PCR)-based detection of (a) *Phytophthora citricola* DNA, (b) *Phytophthora cambivora* DNA, and (c) *Phytophthora quercina* DNA extracted from oak seedlings artificially infected under controlled conditions with different *Phytophthora* isolates. The fragment sizes of the diagnostic amplicons in base pairs (bp) derived from DNA standard fragments are given at the right margin. The following tissue samples were analysed: (a) 1. symptomatic main root, 18 d after inoculation; 2. symptomatic main root, 19 d after inoculation; 3. symptomatic main root, 21 d after inoculation; 4. symptomatic main root, 25 d after inoculation; 5. symptomatic main root, 3 d after inoculation; 6. symptomatic main root, 5 d after inoculation; 7. symptomatic main root, 7 d after inoculation; 8. symptomatic main root, 11 d after inoculation; 9. *Phytophthora citricola* control DNA; and 10. PCR control reaction without any DNA. (b) 1. PCR control reaction without any DNA; 2. uninoculated control seedling; 3. symptomatic main root, 3 d after inoculation; 4. symptomatic main root, 4 d after inoculation; 5. main root revealing no symptoms, 4 d after inoculation; 6. symptomatic main root, 7 d after inoculation; 7. symptomatic main root, 8 d after inoculation; 8. *Phytophthora cambivora* control DNA. (c) 1. PCR control reaction without any DNA; 2. symptomatic main root, 3 d after inoculation; 3. symptomatic fine root, 6 d after inoculation; 4. symptomatic main root, 7 d after inoculation; 5. main root revealing no symptoms, 7 d after inoculation; 6. symptomatic fine root, 10 d after inoculation; 7. *Phytophthora quercina* control DNA (for the nucleotide sequences of the species-specific primer pairs and the PCR conditions used see Schubert *et al*. 1999) (from Schubert *et al*. 1999).

recognised for a long time as an important factor affecting survival of forest tree populations (e.g. Maguire 1955).

17.4.3.2   Stilbene synthase—example 2

Stilbene synthase (STS) genes are organised in large multi-gene families and have been identified only in a few unrelated plant species including *Arachis hypogaea* (groundnut) (Schröder *et al*. 1988), *Pinus sylvestris* L. (Scots pine) (Fliegmann *et al*. 1992), and *Vitis vinifera* L. (grapevine) (Sparvoli *et al*. 1994).

Their expression is known to be strongly induced by pathogen attack (Langcake 1981) and ultraviolet radiation (Fritzemeier & Kindl 1981). Moreover, it was shown that STS mRNA accumulated in *Pinus sylvestris* L. (Scots pine) seedlings treated with ozone (Zinser *et al*. 1998). Experiments with transgenic tobacco plants indicate that STS activity is mainly regulated at the transcriptional level and the ozone-responsive promoter region differs from the basal pathogen-responsive sequence (Schubert *et al*. 1997b). STS enzymes condense one molecule of *p*-coumaroyl-CoA or cinnamoyl-CoA with three molecules of malonyl-CoA to resveratrol or pinosylvin, respectively. Both stilbenes have antifungal activities and play an important role in plant defence (reviewed by Hart 1981).

17.4.3.3 Cinnamyl alcohol dehydrogenase — example 3

Cinnamyl alcohol dehydrogenase (CAD) catalyses the last step in the biosynthesis of the monomeric precursors of lignin. CAD enzymes are constitutively expressed in all lignifying plant tissues and several isoforms were characterised in different species (see Boudet *et al.* 1995). Elicitor-induced CAD activity has been reported from cell cultures of *Phaseolus vulgaris* L. (French bean) (Walter *et al.* 1988) and leaves of *Triticum aestivum* L. (wheat) (Mitchell *et al.* 1994). An increase of CAD activity was measured in seedlings of *Picea abies* following ozone fumigation (Galliano *et al.* 1993a, 1993b). Consequently, CAD deserves particular attention because it is both a physiological and a genetic indicator for lignification during development and plant stress responses. The increasing enzyme activity of CAD following exposure to ozone can be quantified and is correlated with an increase of corresponding mRNA. Genomic sequences encoding CAD have been cloned from *Picea abies* and their phylogenetic relationships with the corresponding angiosperm genes have been studied in great detail by Schubert *et al.* (1998).

17.4.3.4 Species-specific detection and quantification of pathogens (Phytophthora spp.)—example 4

There are more than 60 species in the genus *Phytophthora* and most of them are widespread destructive soil-borne root pathogens which infect crop plants, shrubs and trees (see Erwin & Ribeiro 1996). In view of the importance of *Phytophthora* spp. as the primary cause of root-rot diseases, oligonucleotide primers were developed for the polymerase chain reaction (PCR) based detection and quantification of the three *Phytophthora* species involved: *Phytophthora citricola*, *Phytophthora quercina* and *Phytophthora cambivora* (Schubert *et al.* 1999). These species have been frequently isolated from both naturally infected oak and *Fagus sylvatica* stands at different European locations (Jung *et al.* 1996).

Schubert *et al.* (1999) demonstrated that diagnostic PCR products are apparent three days after artificial inoculation with the three *Phytophthora* species concurrent with the occurrence of disease symptoms (Fig. 17.5). It was also possible to detect infection prior to the appearance of disease symptoms in *Phytophthora citricola*-infected seedlings of *Fagus sylvatica* as well as in seedlings of oaks which were inoculated with *Phytophthora cambivora* and *Phytophthora quercina*. Diagnostic amplicons (material verified by PCR amplification) were detected in nearly all DNA samples obtained from main roots, lateral roots and root tips of oak and *Fagus sylvatica* seedlings as well as hypocotyls and cotyledons in the case of *Fagus sylvatica*. No amplicons are derived from DNA samples of uninoculated controls.

These tools can be used efficiently to verify the intensity of infection in the field by *Phytophthora* species by means of quantitative PCR. Furthermore, such tests can identify tolerant and sensitive host individuals and also enable nursery material to be checked for certification of pathogen-free plant stock.

17.5 Concluding remarks

In biomonitoring, a variety of methods and organisms are used in order to describe the response of the biosphere to changes in environment. The main emphasis is put on the verification of effects caused by anthropogenically induced environmental stress. Most of the indicators reveal insights in the response of selected organisms to short-term exposure to particular stress factors. Such investigation will not help in the monitoring of long-term trends with respect to the response of ecosystems to complex stress conditions. The dilemma is evident that increasing complexity of ecosystems and of stress conditions coincides with decreasing information on the response of such systems.

Forest ecosystems reveal a high indicative potential for the verification of long-term environmental stress: they are long-lived, predominantly non-domesticated and subjected to a large variety of biotic and abiotic environmental stress components. Various problems need to be solved in order to proceed in a more effective way of biomonitoring in complex ecosystems. Genetic markers are considered to be efficient tools in such a biomonitoring. Like other indicators, most of the genetic markers do not fit into linear dose–effect relationships but this new class of markers is highly efficient in the monitoring of viability selection and other processes that respond directly to environmental stress.

It appears that genetic markers can cover a wide range of applications in bioindication including the processes following climate change or the species-specific detection and quantification of pathogens.

17.6 Acknowledgments

We wish to thank Eliane Röschter for her efficient help in the preparation of the manuscript. Parts of the cited molecular genetic studies of the authors were supported financially by the European Union (BIO 4 CT96-0706).

CRITERIA AND INDICATORS FOR THE CONSERVATION OF GENETIC DIVERSITY

Timothy J. Boyle

SUMMARY

Conservation of genetic diversity is usually one aspect of more complex management goals. It is associated with sustainable forest management, which is a concept that is difficult to measure or assess. The use of criteria and indicators, which outline conditions that should be met if forest management is to be deemed sustainable, offers a solution to this problem. The most important characteristic of an effective indicator is the practicality of assessment in a short period. As genes are inherently difficult to detect, this has usually resulted in the adoption of vague or very indirect indicators of genetic diversity. An approach is described which uses the assessment of processes which maintain genetic diversity as surrogate measures of genetic diversity itself. Using this conceptual framework, a number of indicators and verifiers (the variables actually measured) are proposed. Even given such an approach, problems related to the establishment of thresholds and choice of species to assess remain. Possible solutions to these problems are proposed, but empirical testing is required to verify the validity of such an approach.

18.1 Introduction

In some cases, the conservation of genetic diversity may be the sole or the major management goal. For example, in management of *Wollemia nobilis* W.G. Jones, K.D. Hill & J.M. Allan (Wollemi pine), with only 40 known individuals growing in the wild, conserving all known genetic variation is critical to the survival of the species. More often, however, conservation of genetic diversity will be one aspect of more complex management goals. The concept of sustainable yield has been a central theme of good forest management for centuries (e.g. Schlich 1905). Recently greater emphasis has been placed on the environmental, social and economic functions of forests, leading to concept of sustainability, or sustainable forest management.

Descriptions of what constitutes sustainable forest management are straightforward, and many such descriptions exist. For example, in tropical forests, sustainable management can be regarded as a process of managing forests for specified objectives of management with regard to the production of a continuous flow of forest products and services without undue reduction in its values and future productivity and without undesirable effects on the environment (International Tropical Timber Organization 1992b).

However, because of the multi-faceted nature of sustainability evident from this and other definitions, sustainable forest management is very difficult to measure or assess. As it is impossible to simultaneously optimise all functions of the forest, there must inevitably be trade-offs, involving the acceptance of suboptimal levels of some components of sustainability in order to benefit others. For example, optimising the economic yield of a forest area may require compromises in terms of the conservation value of the forest, and conversely, a strong emphasis on conservation may involve suboptimal economic returns from the forest. Both scenarios may be acceptable, and it is not possible to provide an unbiased judgement.

The use of criteria and indicators (C&I) offers a solution to the problem of assessment of sustainability. Rather than trying to quantify the acceptability of forest management, criteria and indicators outline conditions that should be met if forest management is to be deemed sustainable. Individual indicators may be quantitative or qualitative, but there is no attempt to provide an overall quantitative measure.

C&I have proved to be a popular assessment tool, but this popularity has, in itself, led to some confusion resulting from the profusion of jargon associated with the many proposed systems of C&I. However, C&I remain intuitively attractive precisely because they are frequently, if often subconsciously, used in everyday life. Phenomena such as beauty or safety cannot be measured quantitatively, so we use C&I to rank or assess the status of these phenomena. This comparison also highlights another aspect of sustainability in that we often consider beauty or safety to be subjective. By this we mean that the assessment depends on the specific attitudes of the assessor. So too, in assessing sustainability, we cannot ignore the specific goals and objectives of management—a forest area that may be considered sustainably managed for one set of objectives would be non-sustainable for a different set of objectives. However, while ranking of individual cases may be somewhat subjective, criteria that determine our perceptions of beauty or safety remain essentially constant.

As discussed by Brown *et al.* (2000), the conservation of genetic diversity is an essential component of sustainability for three reasons; that is, to:

(a) *maintain short-term viability of individuals and populations*—high levels of homozygosity within an individual or population reduce viability (i.e. the probability of survival) as experimentally demonstrated for populations of butterflies by Saccheri *et al.* (1998)

(b) *maintain the evolutionary potential of populations and species*—the ability of populations and species to evolve in response to changing environmental conditions depends on the existence of genetic variation

(c) *provide opportunities for use of genetic resources*—many plants and animals are exploited to derive useful products—a relatively small number of species are of such value that they are to some degree domesticated through active selection and breeding, but many more species, although not subject to planned programs of domestication, are genetically manipulated by communities living in or near forests.

Box 18.1   Principles, criteria, indicators and verifiers

Each level in the hierarchy is defined as follows (Prabhu *et al.* 1996).

Principle: *A fundamental truth or law as the basis of reasoning or action.* In the context of sustainable forest management, principles are seen as providing the primary framework for managing forests in a sustainable fashion. They provide the justification for criteria, indicators and verifiers. Consider that principles embody human wisdom, where wisdom is defined as: *a small increment in knowledge created by a person's (group's) deductive ability after attaining a sufficient level of understanding of a knowledge area.* Wisdom therefore depends on knowledge.

An example of a principle could be to 'maintain or enhance ecosystem integrity' or 'to ensure human well-being'.

Criterion: *A standard that a thing is judged by.* A criterion can therefore be seen as a 'second order' principle, one that adds meaning and operationability to a principle without itself being a direct measure of performance. Criteria are the intermediate points to which the information provided by indicators can be integrated and where an interpretable assessment can be formed. Principles form the final point of integration. In addition, criteria should be treated as reflections of *knowledge*. Knowledge is the accumulation of related information over time and can be viewed as a large-scale selective combination or union of related pieces of information.

An example of a criterion would be 'principal functions and processes of the forest ecosystem are maintained' or 'processes that maintain genetic variation are maintained'.

Indicator: *An indicator is any variable or component of the forest ecosystem or the relevant management systems used to infer attributes of the sustainability of the resource and its utilisation.* Indicators should convey a 'single meaningful message'. This single message is termed 'information' and represents an aggregate of one or more data elements with certain established relationships.

An example of an indicator could be 'directional change in allele or genotype frequencies'.

Verifier: *Data or information that enhances the specificity or the ease of assessment of an indicator.* At the fourth level of assessment tools, verifiers provide specific details that would indicate or reflect a desired condition of an indicator. They add meaning, precision and usually also site-specificity to an indicator. They may define the limits of a hypothetical zone from which recovery can still safely take place ('performance threshold/target'). However, they may also be defined as procedures needed to determine satisfaction of the conditions postulated in the indicator concerned ('means of verification').

An example of a verifier would be 'number of alleles'.

18.2 Criteria and indicators

C&I are tools that can be used to conceptualise, evaluate and implement sustainable forest management (Box 18.1). C&I themselves form part of a hierarchy of assessment tools. The four levels of the hierarchy are principles, criteria, indicators and verifiers. C&I may be applied at various scales, but they are most commonly used at the national (or ecoregional) scale and at a local scale which, in the context of forest sustainability, would normally be the forest management unit (FMU) level.

Both scales are important and need to be considered together. No single FMU can provide all possible forest functions. For example, the conservation of animals requiring extensive home ranges, such as large predators (tigers, *Panthera tigris*; wolves, *Canis lupus*) or large herbivores (elephants, *Elephas maximus*; rhinoceros, *Rhinoceros* spp.; moose, *Alces alces*), requires substantial areas of relatively undisturbed forest. Such conditions cannot be provided within a single FMU, so assessment over a larger scale is necessary for such criteria of sustainability. However, sustainable forest management depends largely on specific management actions undertaken within individual FMUs so assessment at this scale is essential for many criteria. A complete and comprehensive system of C&I for assessment of sustainability requires the integration and harmonisation of C&I at local and higher scales.

18.3 The basic problem of criteria and indicators for genetic diversity

A country such as Indonesia has 500 to 1000 FMUs. If each FMU were to be assessed every five years, this would require between 100 and 200 assessments of different FMUs each year. Thus, C&I assessment will clearly be costly. Consequently, it is envisaged that assessments of individual FMUs will be made by small teams of three to five people, over one to two weeks, because larger teams or longer periods would make the process too costly to be acceptable to those who bear the costs—industry, government or non-governmental organisations. These teams will need to assess C&I related to all aspects of sustainability: biophysical, social and economic. Therefore, the time available for the assessment of the conservation of genetic resources will be only a few (< 10) person days. These considerations dictate that the most important

characteristic of an effective indicator or verifier will be the practicality of assessment in a very short period. This need for practicality is a serious constraint for the assessment of 'conservation', which implies a need to consider temporal dynamics.

This constraint is particularly serious for the assessment of genetic diversity, as genes are both incredibly numerous and variable (millions within each individual or each species) and too small to detect without relatively sophisticated and expensive laboratory procedures. The problems in using C&I to assess genetic diversity can, therefore, be summarised as follows:

(a) selection of an appropriate conceptual framework for genetic C&I

(b) choice of species to assess

(c) methods to characterise or visualise genetic variation

(d) methods to detect temporal changes and trends in genetic parameters

(e) choice of thresholds or critical values to apply

(f) combining information from multiple indicators.

These problems are complex. Consequently, initiatives to develop C&I at a national scale have usually adopted vague, or very indirect, indicators of genetic diversity (Box 18.2). At the local scale, most systems of C&I have ignored genetic diversity, assuming that species conservation incorporates genetic conservation.

Box 18.2 National level genetic C&I proposed in various national/international initiatives

Helsinki process	Taraputo process
4.3 Changes in the proportions of stands managed for the conservation and utilisation of forest genetic resources (e.g. gene reserve forests, seed collection stands); differentiation between indigenous and introduced species	4c Measures for the conservation of genetic resources
4.4 Changes in the proportions of mixed stands of 2–3 tree species	
4.5 In relation to total area regenerated, proportions of annual area of natural regeneration	

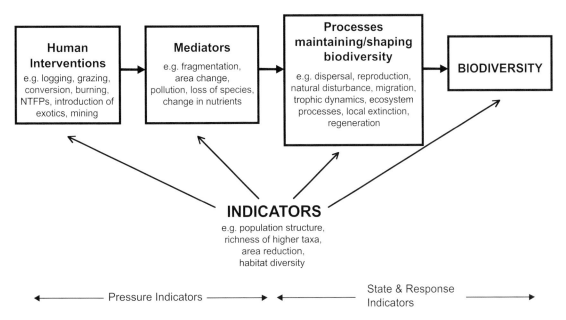

Figure 18.1 A conceptual model of the relationships between anthropogenic interventions under different forest management regimes, mediating processes, ecological processes which shape biodiversity and biodiversity. Indicators on the left-hand side of the figure are 'pressure' indicators, while those on the right are 'state' or 'response' indicators, which are better surrogates for biodiversity (from Stork *et al.* 1997).

18.4 Solutions

Two initiatives specifically aimed at the development of genetic C&I, tackled many of the problems listed in Chapter 18.3. Namkoong *et al.* (2000) attempted to develop a system of C&I at the FMU level, with particular reference to tropical forests, as part of a larger effort to define C&I for sustainability. In contrast, Brown *et al.* (2000) sought to develop indicators of genetic diversity for nationwide State of the Environment reporting within Australia. Despite the differences in goals and scale, these two groups took comparable approaches to the constraints posed by the listed problems.

18.4.1 A conceptual framework

C&I are often developed within a conceptual framework that recognises 'pressure', 'state' and 'response' indicators. The pressure–state–response indicators are equivalent to different components of a system that involves human interventions, mediating processes, processes that conserve diversity and the response of genetics or biodiversity itself (Fig. 18.1) (Namkoong *et al.* 2000; Stork *et al.* 1997). Brown *et al.* (2000) point out that pressure, state and process (equivalent to response) indicators have different

information contents. For example, assessing verifiers of the mating process is potentially very valuable, but is also rather difficult and expensive. Verifiers of population size may be used as surrogates of mating, but their information content is less. Conversely, individual genetic variation provides more information on the actual response to mating, but the effects of other processes are confounded in verifiers of individual variation. Therefore, the choice of appropriate indicators and verifiers is always going to be a trade-off between information content, scale of monitoring and cost.

Based on these considerations, Namkoong *et al.* (2000) concluded that a feasible approach to assessing conservation of genetic diversity on a local scale could be developed by examining the status of those genetic processes that serve to maintain genetic diversity and structure (see Chapter 2). Mutation is excluded from consideration since we are primarily concerned with relatively short-term processes of fewer than ten generations and with effects that are not likely to significantly change mutation rates. Therefore, the relevant processes are:

1 *random genetic drift* (i.e. the non-directional changes in genotypic frequencies among

243

generations due to random chance in small populations)

2   *selection* (i.e. relative differences among genotypes in viability or reproductive success)

3   *migration* (i.e. the exchange of genes between populations that differ in genotypic frequencies)

4   *mating* (i.e. the process mediating the recombination and assortment of genes between generations).

By considering the likely effect of various human activities in forests on the processes that maintain genetic diversity, and relating these processes to proposed genetic indicators, Namkoong *et al.* (2000) derived a table and figure relating human activities directly to indicators of genetic diversity, and this table is reproduced as Table 18.1 and Figure 18.2. The advantage of such an approach is that it allows forest managers or policy makers to identify which indicators need to be assessed, based only on their knowledge of activities occurring in the FMU. The one criterion and four indicators proposed by Namkoong *et al.* are given in Box 18.3.

The conceptual framework of Brown *et al.* (2000) was similar to that of Namkoong *et al.*, in that 'pressures', corresponding to human activities, were related to population characteristics, genetic processes and to effects on genetic diversity (Fig. 18.3). Brown *et al.* (2000) developed indicators corresponding to each step in Figure 18.3, and their proposed indicators are shown in Box 18.4.

18.4.2 Choice of species

It is clearly impossible to assess genetic diversity in all species. Therefore, a limited number of species need to be selected, and the selection should be made according to criteria relevant to the goals of the assessment. For example, Brown *et al.* (2000) proposed eight criteria for species selection, although they made no recommendations as to how many species should be used:

1   biological or ecological representativeness

2   taxonomic representativeness

3   sensitivity to particular pressures

4   practicality of sampling and analysis

5   existing knowledge base

6   amenability for laboratory rearing and captive breeding

7   cross-regional comparability

8   bacterial diversity

Namkoong *et al.* (2000) took a different approach, although there are many similarities in the criteria used. They argued that although a variety of species groups could be assessed, the focus should be on trees, both because they constitute the major structural component of forests, and because the knowledge base for trees tends to be better than for many other groups (cf. Brown *et al.* 2000, criterion 5). Namkoong *et al.* (2000) then considered which characteristics of tree species would determine sensitivity to different types of pressures (cf. Brown *et al.* 2000, criterion 3). For example, in the case of logging, the proposed criteria for species selection were that the species:

1   is ecologically valuable—it is a 'keystone' species**

2   flowers only after attaining a large size or advanced age**

3   has pollinators/seed dispersers that are highly specific**

4   is harvested, and has high value*

5   is rare*

6   has a spatial distribution making it susceptible to adverse impacts of logging*

7   has a reproductive niche that requires shade*

8   is dioecious*

Criteria marked with a single asterisk (*) are those for which information is usually available, while those marked with two asterisks (**) are criteria for which information may often be missing. Obviously, compromises need to be made among these criteria, as not all will be applicable to every species.

Namkoong *et al.* (2000) stress interregional comparison, rather than the cross-regional comparison proposed by Brown *et al.* (2000) in their criterion 7. Therefore, Namkoong *et al.* (2000) proposed that on a regional basis the criteria be used to identify the most sensitive species, and that to the extent possible, the same sensitive species be assessed in each FMU within a region. This would necessitate some compromise on the criterion of rarity, as truly

TABLE 18.1 Indicators relevant to different forest operations

	Indicators			
Modes of exploitation	Levels of genetic diversity are maintained	There is no directional change in genic/genotypic frequencies	There are no changes in gene flow/migration	There are no changes in the mating system
Logging—commercial species	X	X		X
Logging—non-commercial species	X	X		X
Grazing	X			
Fire	X		X	X
NTFP—reproductive[A]	X	X	X	X
NTFP—non-reproductive	X	X		
NTFP—whole individual	X	X		

[A] NTFP, non-timber forest product.

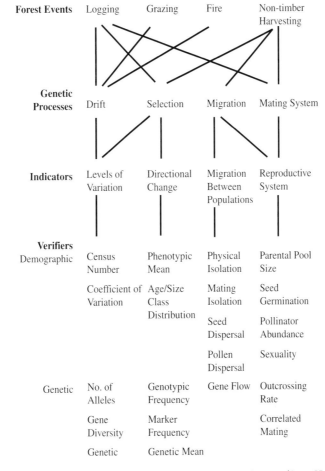

Figure 18.2 Relationship of forest events to genetic processes and their indicators (from Namkoong *et al.* 2000).

Box 18.3 Genetic C&I proposed by Namkoong *et al.* (2000)

Criterion: Conservation of the processes that maintain genetic variation

Indicator 1: Levels of variation

DEMOGRAPHIC VERIFIERS

1.D.1 Census number of sexually mature individuals

1.D.2 Census number of reproducing individuals

1.D.3 Coefficient of phenotypic variation

GENETIC VERIFIERS

1.G.1 Number of alleles

1.G.2 Gene diversity

1.G.3 Genetic variation

CRITICAL LEVELS

If census numbers in the target population are above critical absolute values (1.D.1 > 50 or 1.D.2 > 30) and the coefficient of phenotypic variation in the target population is higher or not significantly different from the reference population, then indicator 1 can be considered as indicating sustainability. Decisions based on demographic verifiers rely, therefore, on two conditions: one on the census number and one on the level of phenotypic variation. However, if either one of these conditions is not fulfilled, additional assessments with genetic verifiers have to be made before reaching a conclusion.

If demographic verifiers fail to indicate sustainability for indicator 1, genetic verifiers need to be assessed. The genetic verifiers proposed are more-or-less mutually exclusive. Recommendations are to select at least one that fits the resources and techniques available. If the selected verifier in the target population is no more than 25% less, or one standard error, less than in the reference population, then this indicator is acceptable.

Indicator 2: Directional change in allele or genotype frequencies

DEMOGRAPHIC VERIFIERS

2.D.1 Phenotypic shifts

2.D.2. Age/size class shifts

2.D.3. Environmental shifts

GENETIC VERIFIERS

2.G.1. Genotypic frequency shifts

2.G.2 Marker frequency shifts

2.G.3 Genetic mean shifts

CRITICAL LEVELS

Differences in the genetic verifiers between the reference or other comparison population and the affected population must be fairly strong for concern to be raised and hence if less than a 25% difference or a change of less than 1.5 standard errors are observed, sustainability can be assumed.

Indicator 3: Migration among populations

DEMOGRAPHIC VERIFIERS

3.D.1. Physical isolation

3.D.2. Mating isolation

3.D.3. Seed dispersal

3.D.4. Pollen dispersal

GENETIC VERIFIER

3.G.1 Gene flow

CRITICAL LEVELS

Changes in the verifiers of less than 50% in the demographic verifiers or of less than 10% in the genetic parameters would indicate sustainability for this indicator.

Indicator 4: Reproductive system

DEMOGRAPHIC VERIFIERS

4.D.1. Parental pool size

4.D.2. Seed germination

4.D.3. Pollinator abundance

4.D.4. Sexuality

GENETIC VERIFIERS

4.G.1. Outcrossing rate

4.G.2. Correlated mating

CRITICAL LEVELS

Changes in these verifiers of greater than 50% for the demographic verifiers or 10% of the genetic verifiers would be considered to be critical.

---

rare species would almost certainly not be present in all FMUs within a region. The assessment would proceed by sampling the most sensitive species first, and depending on the results of this assessment, no further species may be sampled (if the conclusion was 'sustainable'), or a bracketing process would be adopted (e.g. by jumping to the 5th most sensitive species) until sufficient precision in the overall assessment was obtained.

18.4.3 How to characterise/visualise the genetic variation

Bearing in mind the aspect of practicality in assessing genetic diversity within FMUs in only a very short time, Namkoong *et al.* (2000) concluded that, as a first step, each of their four indicators should be measured by means of 'demographic verifiers' (Box 18.3). These are essentially surrogates based on the same kind of considerations as those contained in indicator 1 of Brown *et al.* (2000) (population size, numbers, physical isolation). Only if no clear decision can be made based on these demographic verifiers (see Chapter 18.4.5) would the assessment move on to 'genetic verifiers' in order to gain precision. Genetic verifiers consist of the most appropriate molecular markers (equivalent to indicator 4 of Brown *et al.* 2000) which, in terms of cost-effectiveness, and given the state of technology in most tropical countries, would most likely be isozymes.

Because of the need for rapid assessment, Namkoong *et al.* (2000) concluded that no quantitative genetics data could be collected beyond simple estimates of

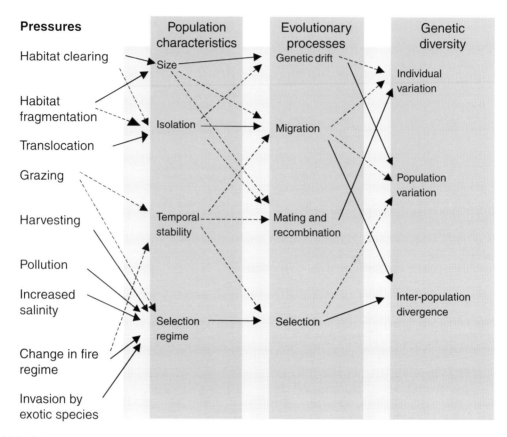

Figure 18.3 Relationships between pressures, evolutionary processes and biodiversity at the gene level (from Brown *et al*. 2000). The solid arrows represent 'strong relationships' and the dashed arrows 'weak relationships'.

phenotypic variation in easily observable traits. In contrast, indicator 5 of Brown *et al*. (2000) calls both for estimates of phenotypic variation and, mainly in the case of animals, estimation of additive genetic variation, requiring some knowledge of genetic relationships among individuals of a population. This difference is largely the result of the divergence in goals of the two groups. Namkoong *et al*. (2000) were concerned with assessing each FMU within a country, requiring much more data, while Brown *et al*. (2000) were interested in characterising variation within and among ecoregions of Australia.

18.4.4 How to detect temporal changes and trends in genetic parameters

The detection of trends in indicators is obviously relatively simple once a time series of the relevant parameters exist. For initial assessments a time series will not be available. In this case, two possible approaches may be used. The first, and simplest, is to select species for investigation (partly) on the basis of already having detailed studies of its population genetics. Depending on the detail and design of such studies, the historical status of genetic diversity can be inferred to some degree. In a tropical context, it is rather uncommon to encounter a species having been the subject of detailed population genetics studies. An alternative approach will therefore usually be necessary, and this involves the use of a 'pseudo-time series', in which localities within the FMU having different use histories are compared. Most obviously, the comparison of logged and unlogged areas can provide some indication of the consequences of logging. However, a critical assumption is that the two sites were originally identical in terms of levels and structure of genetic diversity. For this reason, localities that share similar environmental conditions and are within the same FMU should be used wherever possible, though compromises are obviously going to be required.

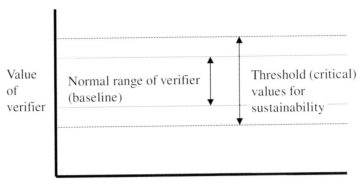

Figure 18.4   Normal and threshold values of a particular indicator. If the system is disturbed, and the new value of the indicator falls (or rises) beyond the threshold, non-sustainability applies. However, incorrect timing of assessment may provide a false result.

18.4.5   What thresholds or critical values to apply

As well as determining threshold values for verifiers/indicators, it may be necessary to establish baseline values. Both are problematic. Although values obtained from apparently undisturbed areas may give a first approximation of baseline values, such an interpretation is subject to some caveats. First, there is no truly undisturbed forest. Every forest area has been subjected to natural disturbance, and there are also relatively few areas that have not been disturbed by humans at some time. Without a detailed knowledge of at least the recent history of a stand, and an understanding of how events in that history will have affected indicator values, it cannot be assumed that values from an 'undisturbed' site are representative of long-term, regional averages. Second, there is no necessity that a value obtained from undisturbed sites is valid as a baseline for assessing sustainability. Such a value, if obtainable, may well be irrelevant to conditions of sustainable forest management.

The problems of thresholds are even more intractable. The imposition of any management regime will result in changes (increases or decreases) to indicator values. The relevant question is whether such changes result in a permanent or irreversible reduction in forest values, future productivity, or the quality of the physical and social environment. This is also related to the issue of temporal changes and timing of assessment. If an assessment is made while the system is still adjusting to a new value of an indicator, then a false conclusion may be reached (Fig. 18.4). Knowledge of the

relationship between genetic diversity and short-term viability, or long-term evolutionary potential, is almost completely lacking. Namkoong *et al.* (2000) provided some preliminary threshold values for their indicators (Box 18.3), but these are subjective and debatable. At present there is no established methodology for determining thresholds, and both Namkoong *et al.* (2000) and Brown *et al.* (2000) concluded that research on this issue is essential.

18.4.6   How to combine information from multiple indicators

The ultimate goal of assessment using C&I is to reach some conclusion about the sustainability of forest management. This requires a decision process. Stork *et al.* (1997), considering the broader issue of biodiversity C&I, proposed an approach which can be modified for application at the genetic level, as follows.

For each indicator, three types of decisions can be reached.

1   We have a definite conclusion on sustainability and we do not need to continue with other indicators. This conclusion may be either that management is not sustainable in terms of genetic diversity, or that it is.

2   We have reached a definite conclusion (either sustainable, or non-sustainable) for this indicator, but we want to continue to look at other indicators anyway.

Box 18.4 Genetic indicators proposed by Brown *et al.* (2000)

INDICATOR 1: Number of subspecific taxa

INDICATOR 2: Population size, numbers and physical isolation

INDICATOR 3: Environmental amplitude of populations

INDICATOR 4: Genetic diversity at marker loci within individuals and populations

INDICATOR 5: Quantitative genetic variation

INDICATOR 6: Interpopulation genetic structure

INDICATOR 7: Mating

---

3   We cannot reach a definite conclusion at this level by looking only at demographic verifiers, and we may need to consider using genetic verifiers.

All demographic verifiers are assessed first; that is, if it is not possible to reach a decision 1 conclusion for a particular indicator, the demographic verifiers of the next indicator are looked at. Only when no conclusions can be reached using demographic verifiers alone is it recommended that genetic verifiers are assessed. This can be illustrated in Table 18.2.

The problem occurs when results from two indicators give contrasting results or conclusions. One solution might be to recognise a hierarchy in importance of different indicators. For example, Namkoong *et al.* (2000) argue that if, for example, the critical values for indicator 2 in Box 18.3 (directional change) are exceeded, this may not indicate non-sustainability if values for indicators 3 (migration) and 4 (reproductive system) are very healthy. Such an approach recognises

the possibility of interaction among indicators. A more rigorous treatment of such interaction would involve the derivation of a joint risk function for simultaneous evaluation of critical levels of all indicators, but this approach requires much more empirical data.

18.5 Conclusions

Conservation of genetic diversity in forests subject to human use has to be nested in an ecosystem approach to forest management. However, even within such an approach, it will sometimes be possible that genetic diversity is eroding while the ecosystem as a whole shows no signs of stress. It will often be necessary, therefore, to assess genetic diversity, and C&I offer the only practical approach under normal operational conditions. The inherent difficulties of designing C&I for entities as complex as genes can be overcome by adopting a system based on processes that maintain

---

TABLE 18.2   **A decision-making process**

For each hypothetical example, the numbers refer to the type of decision possible, as discussed in the text.

Indicator	Example 1	Example 2	Example 3
Indicator 1	2	2	2
Indicator 2	2	2	3
Indicator 3	1	2	2
•	STOP	•	•
•		•	•
Indicator x		2	2
		STOP	STOP/or apply genetic verifiers

---

diversity. Even then, there remain substantial difficulties. Demographic verifiers offer a quick and easy, but indirect, method for assessing these processes, and when the need for direct information from genetic verifiers is indicated, the cost may become prohibitive.

Other problems that remain to be solved include the choice of appropriate species, determination of threshold levels, and the derivation of joint risk functions. While further research will inevitably be required to provide solutions to these problems, there exist low-diversity forest ecosystems with sufficient scientific understanding of most of the component tree species, for which application of genetic C&I based on current knowledge is feasible. Experience gained in such systems will serve to identify what kind of information can safely be transferred from one forest ecosystem to another, and which of the outstanding problems require ecosystem-specific research.

ECONOMICS AND CONSERVING FOREST GENETIC DIVERSITY

Jeffrey A. McNeely and
Frank Vorhies

SUMMARY

Although biodiversity conservation may be viewed by some policy makers or stakeholders as desirable due to potential future benefits, such benefits may not be obvious to those who are most directly utilising the forest. In some cases, market pressures and government policies drive local people to use forests unsustainably and thus to threaten forest genetic resources. Also, if conservation results in some stakeholders forgoing some immediate benefits that they would otherwise reasonably expect to enjoy, then some form of compensation for their opportunity costs might be necessary. Clearly, the conservation of biodiversity including genetic diversity involves an economic dimension. Thus, an understanding of basic economic principles is essential to any effort in conserving forest genetic resources.

19.1 Introduction: the basic economic concepts

Both ecological systems and economic systems are complex and dynamic and are based on such factors as production, scarcity, exchange, specialisation, competition, efficiency, growth, adaptation and evolution. The sciences of ecology and economics both examine the allocation of scarce resources and the consequences of different kinds of allocation processes. In both ecological systems and economic systems, exchange involves the flow of energy and resources through a hierarchy running from producers to consumers to decomposers ('recyclers') and back to primary producers. Thus, ecosystem managers and scientists should find it easy to understand the basic concepts of economic systems and be able to apply these to their work.

An understanding of economic systems and the appropriate application of economic tools may well be critically important for the conservation of ecosystems and the species and genetic resources within them. As Perrings (1996) has pointed out:

> A change in the relative price of resources, or the international trade regime, may be a far more direct threat and potent threat to the resilience of managed or impacted ecological systems. To understand and to analyse the biodiversity problem in these circumstances requires the integration of concepts and methods of both ecology and economics.

Economic arguments dominate today's policy debates and political decisions. The outcomes of these debates and decisions can have profound effects on forests and forest genetic diversity. In today's political economy, the world consists of economic agents—individuals, private companies and public organisations—who exchange goods and services based on monetary calculations of gain and loss in the context of the global market economy. Market transactions assign a monetary value (i.e. a price) to the utility of the goods or services exchanged. The prices paid for goods and services depends on the balance of the quantities supplied and the quantities demanded. This is the principle of 'supply and demand'. Scarce resources in high demand, for example, tend to have higher prices than abundant resources with low demand.

Because of both market failures and government failures, however, market-determined prices may not completely reflect social benefits and costs, including those associated with the conservation of genetic diversity. This can lead to a socially inefficient or socially undesirable allocation of resources. In the case of ecosystems such as forests, it can lead to degradation and even extinction. In particular, both market processes and government regulations often fail to deal with the public good aspects of genes and species, and ecosystems, especially in the context of increasing economic and political pressure to harvest commercial, private goods from these ecosystems. Public good aspects of ecosystems include ecological services such as freshwater catchments and carbon sequestration that are not easily managed through market processes; public good aspects of species include their bequest values; and public good aspects of genes include their evolutionary potential.

Although the values of particular genes may be realised through market transactions, markets as well as government regulations may fail to ensure that the public good aspects of forests are maintained as natural systems for storing or banking genes. Thus, a sustainable supply of commercially viable genetic resources from a forest may be undermined by weak or non-existent financial linkages between the commercialisation of forest genetic resources and the conservation of the forest ecosystem.

Economic theory would suggest that as genetic resources become rarer due to reduced supply brought about by an erosion of forest ecosystems, they would tend to become more valuable. Properly functioning markets and government regulations should lead to increased investment in conserving the forest as a storehouse of scarce and valuable genetic resources. However, if there are no financial links between the returns generated from private genetic resource use and the conservation of the ecosystems from which they come, then increased funding for conservation simply will not be forthcoming and biodiversity will be lost.

Economists also point out that economic agents apply a 'discount rate' to their decisions, preferring present to future returns. One dollar today is worth more than one dollar in a year. For example, if an economic agent is indifferent between $1 today and $1.25 in one year, then its discount rate is roughly 25%. Where an economic agent must make a decision about harvesting a biological resource with relatively slow growth rates—such as trees—and that agent has a high discount rate, there is a financial incentive to liquidate the resource (Clark 1990). When the intrinsic growth rate of a resource is significantly less than the discount rate, it may make financial sense to harvest the entire resource and invest the returns in higher-yield economic activities. Such a 'financially efficient' harvest would destroy any public good aspects of the resource, such as its ecological service as a natural bank of genetic resources.

Conservation efforts become exceedingly difficult when financial logic gives support to over-harvesting biological resources. However, if prices are expected to rise faster than the discount rate, the rational resource owner/manager would choose to conserve, not export. Thus, incorporation of the market values of genetic resources into the decision-making calculus could provide an additional incentive to conserve the ecosystem.

Lande *et al.* (1994) used models to demonstrate that even when the intrinsic growth rate of a resource is greater than the discount rate, various random effects can lead to rapid extinction of the population. They point out:

> Even a small discount rate drastically diminishes the maximum expected present value of a cumulative harvest before extinction, and little profit can be gained in comparison with immediate harvesting to extinction; this narrow profit margin will often be eroded by fixed costs of maintaining harvesting operations.

Lande *et al.* (1994) conclude that economic discounting probably should be avoided in the development of strategies for sustainable use of biological resources. However, it is unlikely that economic agents operating in the global market economy will be persuaded to ignore discount rates.

Economists also commonly use cost-benefit analyses in their identification and evaluation of problems and solutions, arguing that the production of a good is, by definition, economic only when the total benefits exceed the total costs (Pearce 1986). In principle, this should include the costs of dealing with, for example, loss of genetic diversity, soil pollution and atmospheric warming as well as the benefits of conserving biodiversity, reducing pollution and managing the climate system. But another common failure of markets and governments is the failure to account for these so-called 'externalities'. Failure to 'internalise' external environmental costs can lead to underpricing and overproduction, which may be one of the main reasons for the loss of biodiversity and genetic resources.

Externalisation of costs helps explain why the exploitation of biological resources may well be profitable in a commercial sense even if it is not socially optimal. For example, the way many forests are being harvested today is not socially efficient, since the costs—including, for example, the long-term cost to local communities, disruption of watersheds and loss of genetic resources—far exceed the short-term benefits (Repetto & Gillis 1988). Harvesting these forests is certainly profitable for the individuals who earn considerable financial benefits through the processes, but the institutional framework fails to force those individuals or agencies to pay the full costs of their exploitation. This usually occurs because governments have developed inadequate policies, property rights or regulatory systems and thus market responses do not account for environmental externalities (Daly & Cobb 1989).

Economists, therefore, distinguish between 'economic analysis', which uses social-welfare values of an activity and attempts to eliminate distortions in prices by means of correcting property rights or using shadow prices and other devices, and 'financial analysis', which is based on the perspective of a private economic agent, and uses existing prices and discount rates while ignoring external costs and benefits (Dixon & Sherman 1990). Because governments often fail to establish the necessary set of regulations and property rights to ensure that social benefits and costs are internalised, financial decisions in the marketplace are based on prices and interest rates which do not account for these social benefits and costs.

Economists also recognise that economic agents can receive benefits that they value but do not pay for. This is particularly the case for genetic resources where the market is unlikely to pay for the conservation of the forest or even for the intellectual property of the forest community. For example, a pharmaceutical firm may develop a new drug on the basis of a medicinal plant identified by a shaman in Brazil but provide no compensation to the shaman, to the forest or even to the government of Brazil.

As noted, critical to issues of internalising benefits and costs is consideration of property rights. Assigning property rights with respect to genetic resources is particularly challenging (Organization for Economic Cooperation and Development 1997, 1998). Who has or should have the rights over the access of genetic resources and how are the benefits and costs of this access to be distributed? Ironically, development policies and aid have often shifted rights and responsibilities for managing genetic resources from the people who live closest to resources and transferred these to distant central government agencies. Many essentially common property resources, such as forests and gene pools, have become the responsibility of governments rather than the grassroots communities that historically had managed them (Berkes 1989). Thus, development processes have tended to undermine the incentives of local communities to conserve natural systems and the genetic resources they contain.

Also ironically, the Convention on Biological Diversity may sometimes contribute to a perverse allocation of property rights. The Convention recognises the sovereignty of national governments over the genetic resources found within their borders, but nationalisation of genetic resources may not generate an array of incentives which will encourage local communities and the private sector to conserve these resources and the ecosystems in which they reside. Fortunately, however, few governments have yet been able to replace traditional forms of management with nationalised approaches. And with the strong international interest in the rights and responsibilities of local and indigenous communities with respect to the benefits generated from access to genetic resources, it is likely that most governments will develop more decentralised structures of ownership and control in order to provide incentives for the conservation and sustainable use of genetic resources (Organization for Economic Cooperation and Development 1997).

19.2 Economics and the loss of biodiversity

Economics should be studied by all those concerned with biodiversity because economic factors are driving the loss of biodiversity (e.g. see Barbier *et al.* 1990; Costanza 1991; Daly & Cobb 1989; Barbier *et al.* 1994). Three such factors are particularly worth mentioning (described in the following sections).

19.2.1 Subsidies to overexploitation: single product production processes and response to globalisation

Many governments subsidise the overexploitation of biological resources. For example, in 1991 the industrialised nations spent a total of US$322 billion on agricultural support and subsidies (Butler 1992); much of this has had significant negative effects on biodiversity, both in the USA and in other countries. In the USA, national forests are so heavily subsidised that for about 67% of them, below-cost sales of timber are consistently recorded (Repetto & Gillis 1988). Vast amounts of energy are being provided in the form of, for example, petroleum products, pesticides and transportation to spur economic growth. Most of this energy is a one-time-only and non-renewable windfall, depending on the productivity of species long dead and now converted to coal or petroleum; the exploitation of these energy sources is highly subsidised. These energy subsidies enable the true cost of exploiting the resources to be hidden, because the prices paid for timber, fish or crops (for example) do not adequately include the cost of replacing the oil consumed in harvesting, processing and marketing.

19.2.2 Single product production processes

Where forests once were used to provide a whole range of benefits, including wildlife, medicinal plants, watershed protection, construction materials, firewood and spiritual values, today the focus is on harvesting a single product, namely timber. This single-use focus tends to promote the harvesting of tropical forests with little attention to non-timber products, as well as promoting large commercial plantations of genetically similar species, often exotic, which are more financially attractive than natural forests (though not necessarily more economic).

19.2.3 Response to globalisation

Much of the loss of wild genetic resources has been due to the way we have changed the land by using it to provide commodities for export, thus responding to the global economy rather than a local system of supply and demand. The economic forces driving the loss of genetic diversity no longer respond to what local people will benefit or lose depending on the appropriateness of their practices. What were once basically locally self-sufficient and sustainable human systems have become part of much larger national and global systems of which the higher productivity is both welcome and undeniable, but the long-term sustainability is far from proven (Daly & Cobb 1989). Further, increased consumption, which is facilitated by this higher productivity, is also encouraging unsustainable land use practices, especially deforestation and the use of land for agriculture that would be more suitable for forests. When countries are all part of one system connected by powerful economic forces, it becomes very easy to over-exploit any part of the global system because it is assumed that other parts will compensate for such overexploitation (Dasmann 1975). The damage may not even be noticed by the nation or the international community until it is too late to do anything to avoid permanent degradation. Perhaps worse, globalisation may enhance the domination of the economically powerful, yet is based on a complex support structure which in itself can be very fragile—the global impacts of changes in oil prices, exchange rates, interest rates, natural disasters and local wars demonstrate the point.

19.3 Valuation and conservation of genetic resources

One element in building support for efforts to conserve forest genetic resources is to recognise the economic values of these resources. Some have argued that genetic resources are in one sense beyond value because they provide the biotic raw materials that underpin every major type of economic endeavour at its most fundamental level (Oldfield 1984). Ample economic justification can be marshalled, however, by those seeking to exploit genetic resources, so the same kinds of reasoning may need to be used to support alternative uses of the resources.

The species of an ecosystem, their mass, their arrangement and the genetic information they contain are the standing stock of an ecosystem—what might be considered biodiversity's free or public goods. The functions of an ecosystem—maintaining clean air and pure water, cycling nutrients and supporting a balance of creatures—are biodiversity's free or public services (Westman 1977). Such public goods and services are very difficult to put into monetary terms, but assigning monetary values to biodiversity may enable conservation decision-making to be put on a more economically defensible footing.

Valuation may simplify decisions for politicians, but it also implies that genetic diversity can be reduced to the simple metric of money. Giving a cash value to genetic diversity 'forces the great range of unique and distinct materials and processes that together sustain or even constitute life into an arbitrary and specious equivalence' (Rappaport 1993). Thus, valuation exercises must be carefully considered before being undertaken.

Further, not all problems affecting genetic diversity can be adequately characterised or described in quantitative, let alone monetary, terms. As Morowitz (1991) stated it, 'we are often left trying to balance the "good" of ethics with the "goods" of economics'. Assigning economic values to genetic diversity is difficult because—at least with present knowledge—species extinction cannot be reversed no matter how much money is spent. The preferences of future generations are impossible to predict, present benefits are difficult to balance against future costs, and commodity value and moral value are fundamentally different. Further, the genetic resources that are the rarest and the most narrowly distributed are both the ones most likely to be lost and the ones least likely to be missed by the biosphere. However, many of the rarest may be greatly missed by people—one dramatic example is the population of *Oryza nivara* (wild rice) S.D. Sharma & Shastry that is the only source of resistance to grassy stunt virus (Prescott-Allen & Prescott-Allen 1986).

In addition, values are not absolute, as different interests assign different values to the same resource: a forester may value a forest primarily for its timber, a geneticist for the genetic information it contains, a farmer for the clean water its streams provide, a tourism department for its attraction as a national park, a hunter for the trophy animals in it and conservationists for intrinsic reasons. The total value of the forest would be obtained by choosing the land use practices that maximise the sum of all the uses that can coexist.

Nevertheless, finding ways to assign economic value to biodiversity may be essential for conservation. Some economists contend that biodiversity is decreasing at least partly because so few genetic traits, species or ecosystems have market prices, the feedback signals that equilibrate market economics. Many economists maintain that prices that accurately reflect environmental costs would reduce consumption and keep the use of genetic resources in a closer balance with their sustainable availability. If the true value of genetic diversity could be included in the market system, they contend, markets could help conservation. The challenge is to develop the institutional structure that would enable such values to be incorporated into market processes, if one wants to play according to 'market economy' rules, although of course other resource allocation systems are also valid.

Today, most countries are finding it difficult to justify current, much less increased, expenditures on conserving genetic diversity, especially when these costs are accompanied by local and regional opportunity costs which entail politically costly land use conflicts with local people (Wells 1992). Although uncertainty is certainly a concern, even partial valuation in monetary terms of the benefits of conserving genetic resources can provide at least a lower limit to the full range of benefits and demonstrate to governments that conservation can yield a profit in terms that are meaningful to national development. Effective management of genetic resources includes issues of economic value. Food for the stomach comes before nourishment for the spirit, and the people in rural areas who must worry most about where their next meal is coming from often live in the midst of the greatest genetic diversity. In such conditions, it becomes imperative to seek economic support for conservation and this requires recognising the values of at least some of nature's goods and services.

Since policy makers need to quantify costs and benefits, a large literature has developed around the problems of assigning economic value to biological resources (summarised in Krutilla & Fisher 1985; Swanson & Barbier 1992; Pearce & Moran 1994). In addition to the relatively straightforward measurement of harvested goods which are sold in a market, such as timber, fish, medicinal plants and agricultural commodities, more subtle measures are available. Contingent valuation methods (CVM) use questionnaires and other interview techniques to measure the willingness to pay (WTP) for a resource or willingness to accept (WTA) compensation for forgoing the use of a resource. Indirect approaches derive preferences from observed market-based information—for example, the 'travel cost approach' can measure all travel-related expenditures to protected areas, to estimate the benefit arising from that experience. The 'replacement cost technique'

measures the cost of replacing or restoring a damaged biological resource to its original state. Several other techniques are also available. This multiplicity of approaches to assessing values is to be expected, because the benefits derived from a biological resource may be measured for one purpose by methods that may not be appropriate for other objectives, and the ways to measure one resource may not be the same for others. The value of a forest in terms of logs, for example, would be measured in quite a different way from the value of the forest for recreation, watershed protection or carbon sequestration.

Using such techniques, numerous studies have generated economic assessments on the values of biological resources. These include measures of direct use, basically through consumption; indirect use, for example through carbon storage or watershed protection; and non-use or option values based on an individual's WTP to safeguard the possibility of using a resource in the future, or for simply knowing that the resource exists or can be passed on to future generations (so-called existence or bequest values). However, the total economic value cannot be considered as simply the sum of all quantifiable values. Rather, trade-offs are involved: capturing the value of forest genetic resources in terms of logs may be very different from capturing the value of forest genetic resources in terms of their evolutionary information. Thus, determining the total valuation of the resources in any particular area of forest will require valuation using a wide range of assessment methods.

It is sometimes useful to distinguish between products that are continuously taken from nature, such as seeds or medicinal plants, and ones that are harvested once or infrequently to provide a small 'founder stock' which is then propagated or used as a genetic blueprint. With plants, the latter are often much the most important, as most plants can be propagated and cultivated—the value of these plants as genetic resources may be compared to an intellectual property right (Williams 1984; de Klemm 1985).

Some studies of direct use values seem to indicate that using tropical forests for their non-timber values is more economic than logging. For example, Peters *et al.* (1989) estimated that sustainable harvesting per hectare in the Peruvian Amazon would yield a sustainable benefit of $1987 per hectare, while clear-felling would bring in a one-time net revenue of only $1000. Sustainable harvesting of medicinal plants in

Belize would yield a net present value of $3327 per hectare, while plantation forestry with rotation felling yields only $3184. Travel cost evaluation of tourist trips to Costa Rica's protected areas for foreign visitors amounted to US$12.5 million per year, giving the protected areas a value per hectare which was over 12 times the market price of local non-protected area land. Nevertheless, a general case that non-timber values are more economical than logging has not been made.

Prescott-Allen and Prescott-Allen (1986) concluded that the productive use value of genetic resources demonstrates that they are indispensable to modern agriculture, that most of them come from a country other than where they are utilised, that the turnover of domestic genetic resources is rapid and that the use of new genetic resources is increasing (therefore requiring the lines of supply from other countries to be kept open and a great diversity of genetic resources to be maintained). The wild relatives of domestic plants will, therefore, be an essential component of ensuring food security for the next century (Hoyt 1988).

Although many of these direct values are substantial, indirect uses often yield even greater values. An annual carbon storage value of tropical forests has been estimated as $1300 to $5700 per hectare, while the total carbon storage value of the Brazilian Amazon has been calculated as $46 billion; and Western (1984) determined that each lion in Kenya's Amboseli National Park is worth US$27 000 per year in visitor attraction (the same lion would have a direct value of about US$1000 as a skin). Whether such estimates, however, have any relationship to what consumers are willing to pay is another matter.

Ecosystems preserve a reservoir of continually evolving genetic material—irrespective of whether the value of that material has yet been recognised—which enables the various species to adapt to changing conditions. The plants and animals conserved may spread into surrounding areas where they may be able to be cropped at some future date, or eventually contribute genetic material to domestic crops or livestock. Protecting ecosystems, therefore, can be seen as a means for nations, especially those in the species-rich tropics, to keep at least part of their biological resources intact for the future benefit of their citizens. As a result, society as a whole may be willing to pay to retain the option of having future access to a given level of genetic diversity. As the demand for genetic

resources grows while the supply continues to dwindle (if current trends continue), their value is likely to increase. Therefore, some economists suggest that conventional cost-benefit relationships need to incorporate mechanisms to deal with the probability of higher future values and the irretrievability of lost opportunities to preserve genetic material.

While monetary figures such as these are often merely minimal estimates of the total economic value of ecosystems and the resources they contain, they do provide policy-makers with ammunition they can use in the kinds of battles they face in political forums. Biologists also need to be aware of the danger that placing values on species or protected areas may open them up to market forces, and some policy-makers might conclude that a price tag on a resource means that it is for sale.

In conclusion, wild species and the genetic variation within them make contributions to forestry, agriculture, medicine and industry worth many millions and maybe even billions of dollars per year. Perhaps even more important are the essential ecological processes that are carried out by nature, including recycling nutrients, protecting soil, protecting nurseries and breeding grounds and regulating climate. Conserving these processes cannot be divorced from conserving the genetic diversity of the individual species that make up natural ecosystems. Thus, genetic variability important to human welfare is not limited to that found in domestic or wild plants that are relatives of plantation crops, or animals or plants that are harvested for food, fuel or medicine. They also include the genetic variability found in species such as earthworms, bees and termites that may make even more important contributions to society in terms of helping to maintain healthy and productive ecosystems.

19.4 The use of incentives, charges and other market instruments

Whatever methodology is used to assess the value of genetic resources, valuation is one contribution of economics to biodiversity conservation. It informs planners, resource managers and local people about how important genetic diversity may be to local communities and to national development objectives, indicates how important an area is for the genetic resources it contains, reveals common interests in conservation among various sectors and facilitates comparison of costs and benefits of different management proposals.

Another contribution of economics is to determine how biological resources can be conserved. This is where various kinds of socioeconomic measures, such as incentives and disincentives, can be important in ensuring that the benefits suggested above are delivered to the community, and that the community, in turn, is enabled to protect the biological resources upon which its continued prosperity depends. Hence, the Convention on Biological Diversity calls for the use of measures that are economically and socially sound to promote the conservation and sustainable use of genetic resources.

Regulations such as controls on hunting or the establishment of protected areas have been the mainstay of environmental policies and resource protection in virtually all countries, usually being the preferred method of control by governments (Kröller 1992). Regulations have proven their value, although often at a high cost in terms of litigation and bureaucratic interventions. Biodiversity, however, is too complex to be managed solely by such a command and control regulatory framework (Dudley 1992). To make further progress, the regulatory framework now needs to be complemented by market-based and community-based measures, producing new mixes of incentives and regulations. These new mixtures aim to influence human behaviour as it affects biodiversity (Repetto 1987; McNeely 1988, 1993; Katzman & Cale 1990; McNeely & Dobias 1991).

Most economic incentives have been used to promote resource exploitation—indeed, many of these might more accurately be termed 'perverse incentives' (Repetto 1988). Subsidies that deplete genetic diversity have been important to modern societies, but conservation biologists might use economic arguments to design approaches to forestry that are not coupled to resource exploitation which is inimical to biodiversity. For example, support could be provided in terms of incentives for soil, water and wetland conservation, promotion of diverse crops and livestock breeds and other measures that would conserve genetic diversity. Further, governments could begin to grant conservation concessions, just as they grant mining and logging concessions.

Incentives—such as subsidies, tax differentials and other fiscal mechanisms—can be positively used to divert land,

capital and labour towards conservation. They can ensure more equitable distribution of the costs and benefits of conserving biological resources, compensate local people for losses suffered through regulations controlling exploitation, and reward the local people who make sacrifices for the benefit of the larger public. Incentives can be attractive to policy makers when they help conserve biological resources at a lower economic cost than the economic benefits received.

Bach and Gram (1996) argue that economic incentives at production level are far more efficient than trade measures in promoting sustainable forest management practices. However, high profits do not by themselves provide an incentive for improved forest management, but should induce policy makers in government to increase the overall taxation of tropical forest operations and apply the resulting revenue to sustainably managed concessions. Bach and Gram argue that 'if sustainable-management practices are to be implemented rapidly, it is necessary to leave concessionaires using those practices better off'.

Biologists should also be aware of the many other market instruments that could have a significant positive effect on biodiversity. Economic instruments that might be mobilised to support the conservation of forest genetic resources include management agreements, covenants, compensation agreements, tax concessions, ownership and use rights. Repetto et al. (1992), for example, analysed a wide range of environmental charges, recreation fees for use of national forests and other public lands, product charges on ozone-depleting substances and agricultural chemicals, and the reduction of subsidies for mineral extraction and other commodities produced on public lands. Their sample of potential environmental charges would reduce a wide range of damaging activities while raising over US$40 billion in revenue. Recreation fees in national forests, for example, could yield US$5 billion per year. These findings refute the argument that environmental quality can be obtained only at the cost of lost jobs and income. Instead, providing a better framework of market incentives by restructuring revenue systems can both improve environmental quality and make economies more competitive by taxing people on their consumption of resources rather than on their salaries, property and profits. The use of such economic instruments is sometimes more effective than regulations, although in most cases some combination of regulations and incentives (the carrot and the stick)

Box 19.1 Non-consumptive benefits of conserving forest resources

The benefits accruing to society in return for investments in conserving genetic resources will vary considerably among areas and resources. Most benefits will fall into one of the following categories:

1 photosynthetic fixation of solar energy, transferring this energy through green plants into natural food chains, and thereby providing the support system for species which are harvested

2 ecosystem functions involving reproduction, including pollination, gene flow, cross-fertilisation; maintenance of environmental forces and species that influence the acquisition of useful genetic traits in economic species; and maintenance of evolutionary processes, leading to constant dynamic tension among competitors in ecosystems

3 maintaining water cycles, including recharging groundwater, protecting watersheds, and buffering extreme water conditions (e.g. flood, drought)

4 regulation of climate, at both macro-climatic and micro-climatic levels (including influences on temperature, precipitation, air turbulence)

5 production of soil and protection of soil from erosion, including protecting coasts from erosion by the sea

6 storage and cycling of essential nutrients (e.g. carbon, nitrogen, oxygen) and maintenance of the oxygen–carbon dioxide balance

7 absorption of breakdown of pollutants, including the decomposition of organic wastes, pesticides and air and water pollutants

8 provision of recreational aesthetic, sociocultural, scientific, educational, spiritual and historical values of natural environments.

---

will be required. Thus market dimensions can be added to existing regulatory and licence structures that are designed to ensure the sustainable use of forest genetic resources.

19.5 Compensating people in rural areas for opportunity costs

A critical issue in conserving genetic resources is the question of distribution of costs and benefits. When local people are prohibited from using forest genetic resources that they had long considered 'theirs', they are being forced to pay an opportunity cost. In other cases, farmers are outraged over government forest policies that enable outside concessionaires to deplete the forests that had long been the source of their irrigation water, construction materials, medicinal plants and game animals. Very often, the individuals who are earning excessive rents from resource

exploitation do so because they are able to externalise the environmental costs. When logging activities affect downstream agricultural systems, for example by reducing the quality or quantity of irrigation water by reducing the storage capacity of a reservoir through increased siltation rates, the affected farmers pay. Timber concessionaires typically externalise such environmental costs, if indeed they are even conscious of them.

The opportunity costs of conserving genetic diversity are paid disproportionately by the people who live closest to the greatest biological diversity. In one of conservation's greatest ironies, individuals who live amid the greatest biological wealth tend to be the poorest. In another of conservation's ironies, when a central government establishes a national park because of its outstanding value to the nation and to humanity, it simultaneously prohibits access to any marketable value for the local people who once benefited from the goods and services that the area provided, and for which the national park was established to protect.

Thus, the opportunity costs of modern conservation programs that restrict access to resources are falling disproportionately upon the very communities that development projects are designed to assist. If conservation programs are to be socially accepted, then new and more appropriate means of apportioning opportunity costs, or providing compensation for them, need to be sought.

The moral justification for a system of protected areas or any other program to conserve genetic diversity will be strengthened by increasing the flow of benefits that it provides to people. Indeed, the more people benefit directly from the proper management of wild genetic resources, the greater the incentive for them to protect the genetic resources and the lower the cost to government of doing so. If a government decides that an area is important enough to be conserved, and this involves opportunity costs to local communities, then it would seem only appropriate that these local communities be compensated for their opportunity costs through a series of economic incentives.

19.6 Conclusions

The Convention on Biological Diversity provides global political legitimacy to issues of conservation, sustainable use and equitable sharing of benefits arising from the use of genetic resources. It is apparent that public support is crucial to any successful conservation program—such support will need to be based on a sound ethical footing, reliable information and economic benefits. Conservation biologists will need to build on science to demonstrate the benefits of conserving genetic diversity to farmers, fishers and foresters, balance the attention given to loss of genetic diversity with concern for sustainable use of harvestable species, and build a broader constituency among business, the public and academics. An effective overall strategy for mobilising political and economic support for conserving genetic diversity will:

(a) give management responsibility and tenure rights to the people most directly involved

(b) ensure that prices fully reflect environmental costs

(c) provide incentive measures to encourage individual behaviour that is in the long-term benefit of the larger society

(d) provide the best available science to support decisions

(e) seek a diversity of local solutions to local problems.

In short, conservation biologists need to contribute to approaches for managing genetic resources that are ecologically sound, economically feasible and politically palatable.

RETHINKING STAKEHOLDER INVOLVEMENT IN BIODIVERSITY CONSERVATION PROJECTS

Thomas Enters

SUMMARY

According to the conventional wisdom of biodiversity conservation, the economic activities of rural people were viewed as threats to the undisturbed functioning of natural ecosystems and their genetic diversity. Thus, representative samples of ecosystems were put under strict protection.

The poor success rate of such measures has lead to the exploration of alternative conservation approaches. Most notable among these are community involvement and the devolution of resource management responsibilities. These approaches are also problematic as they are based on numerous—sometimes untested—assumptions regarding the interest, commitment and skills of local people, the notion of stable and homogenous communities and the superiority of community-based natural resource tenure and knowledge systems.

The success, in both critical social and biological terms, of biodiversity conservation efforts hinges on abandoning naive assumptions. It is important to challenge the 'received wisdom' about livelihood strategies of 'forest-dependent' people, stakeholder involvement, community cohesion and the interest of local people in forest conservation and management. It is just as crucial to reconsider the new 'conservation-through-development' orthodoxies that have influenced project designs and policies in recent decades and to stress that many issues surrounding forests and their sustainable use are very location specific.

20.1 Introduction

Historically, people and their economic activities have been viewed as threats to the undisturbed functioning of natural ecosystems. In the classical approach to conservation, people played no part and were excluded from protected areas, which were designed according to biological concepts and inventories. As it became evident that the social costs of conservation projects were sometimes rather high, and that their success rate was disturbingly disappointing, the classical approach to biodiversity conservation was replaced by integrated conservation and development projects (ICDP). In this latest approach to conservation, stakeholders and particularly local people are viewed, at least in theory, as active partners.

While the success rate of ICDPs is also not very encouraging, the general belief is that only the involvement of local people in biodiversity management leads to the desired result of maintaining tropical ecosystems and their life-supporting functions. Such is the enthusiasm for the populist approach that it has become heresy to question it and to refute the widely accepted images of intact forest-managing rural communities.

In this chapter I will dare to commit a bit of heresy by casting some doubts on the assumptions that ICDPs are based on. Perrings *et al.* (1995) remind us of the massive ignorance and uncertainty about the extent and significance of change in the level of tropical forest diversity. Similar deficiencies remain with regard to forest use and management by various stakeholders. Lack of knowledge is compounded by the distance between popular and scientific debate, as well as diverging perspectives of forest dwellers. Biodiversity means different things to different people and I do not claim to resolve this dilemma. Some readers might argue

that extreme examples were chosen. It is not my intention to distort the general situation excessively but to make my points. This is only possible or is made more powerfully by selecting the extreme, uncommon or extraordinary. It also serves the purpose of rethinking stakeholder involvement. The discussions focus on local people in tropical countries, particularly in Asia. This is not to say that problems are not similar in other ecoregions and that similar questions should be raised regarding the conservation of non-forest environments.

20.2 Setting the scene

The preservation of natural ecosystems has long been on the agenda of institutions concerned with biodiversity. Representative samples of ecoregions have been set aside and put under strict protection. This western vision of an untouched wilderness has permeated global policies and politics in resource management for decades and has resulted in the classic approach to meeting biodiversity conservation needs, which is still at the heart of the conservation agendas of many countries (Gilmour 1995). Basically the enactment of the classical approach requires the user to (Biot *et al.* 1995):

(a) identify biodiversity loss as serious, indicating that conservation is urgently needed

(b) design a project in which, in case exclusion is not an option, the cooperation of local communities is sought

(c) implement plans through a combination of encouragement, persuasion and subtle threats, sometimes by more coercive powers. (For more information on the use of coercion in conservation projects, see Peluso 1993; Enters 1996; Lee 1996).

Box 20.1 Provoking resentment of protected areas

(from Kothari 1997, pp. 49–50)

By law or by administrative dictate, customary rights and activities have been curtailed in many PAs [protected areas in India]; in some, people have been summarily displaced. Almost never have adequate alternatives or rehabilitation been provided. As a result, village communities now hate the PA concept; it is associated with restrictions on access to resources, harassment by forestry officials, exposure of crops and livestock to ravage by wild animals, and invasion by noisy tourists—now given free rein in the same areas where villagers have been banned.

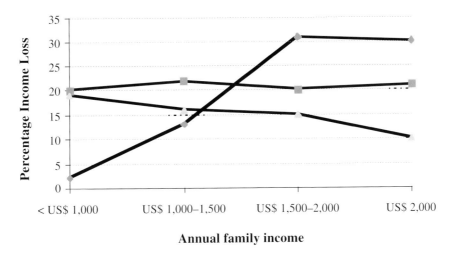

Figure 20.1   Effect of restricting forest-based activities on different income groups; diamond, cardamon production; square, shifting cultivators; triangles, non-timber forest products (after Gunatilake 1995).

Key points are that local people are viewed as 'the target population' or 'beneficiaries' and that they are frequently excluded from the areas considered important for biodiversity conservation. Unfortunately, as numerous examples show (e.g. Braatz 1992; Pinedo-Vasquez & Padoch 1993; Colchester 1994; Fairhead & Leach 1994), conservation projects and programs that fail to consider the interest of local residents undermine existing indigenous management systems and restrict local authorities in their decision making on forest management only intensify the loss of biological diversity. Furthermore, they raise highly contentious debates of national and local interests (McNeely 1997) and can lead to open protests and conflicts (Nepal & Weber 1993; Pimbert & Pretty 1995). Thus, it is not surprising that the classical approach produced disappointing results.

The recognition that solutions to biological problems lie in social, cultural and economic systems stimulated the development of a new paradigm that views people as part of the solution and not as the problem. Thus, co-operation has been replaced with forms of participation. According to the new thinking the classical approach failed because of numerous major problems of which the lack of fit between recommendations and local livelihood strategies and the exclusion of local people in project design, planning, implementation, monitoring and evaluation are most relevant to the following discussion.

Further impetus to the earlier critique of the classical approach has come from studies of traditional knowledge and practices. A new consensus appears to have developed according to which local people do not pose a threat to biodiversity but rather they were the victims of its loss. Hence, attention has shifted from blaming shifting cultivators for deforestation to identifying the constraints that are the primary cause of unsustainable land use practices and overexploitation of forest resources.

In the classical approach it is assumed that the supplier of biodiversity is the national government that has sovereignty and nominal control over the areas required for biodiversity conservation (Panayotou & Glover 1994). In reality, however, effective control rests with any forest user (or stakeholder) and particularly with the resident population who live in and around the forests. In economic terms they are the providers of biodiversity and it is their everyday activities that ultimately decide the fate of much terrestrial biodiversity. As a result, use restrictions and the establishment of protected areas represent opportunity costs (Child 1994), which can be substantial (Fig. 20.1). Furthermore, it is neither politically feasible nor ethically justifiable to deny the use of natural resources to the people who depend upon their availability without providing alternative means of making a living (Wells 1995).

This has lead to the point where proposing any activity which is not profoundly grounded in the involvement of local people is considered unacceptable. The top-down approach has been replaced by bottom-up initiatives that link development and conservation in mutually supportive ways. Partnership in biodiversity conservation is actively pursued and each activity within the framework of a conservation project involves local people as partners.

Most projects today emphasise community-based participation in making decisions and tangible incentive systems that favour forest conservation over exploitation (Wells & Brandon 1993). Yet this so-called populist or participatory approach does not appear to be more effective than the classical approach. A closer examination reveals that one reason for the poor performance is that genuine participation is still the exception rather than the rule. Most organisations pay only lip service to the goal of involving local communities. In reality, they go about their business as usual. Projects tend to be short-term in nature and over-reliant on expatriate expertise. In addition, most projects are deficient in clear criteria through which to determine whether local people benefit from conservation and/or whether conservation objectives are reached more effectively by involving local people.

A more profound problem, however, is that unproved and optimistic assumptions remain unchallenged, leading to unclear project objectives (Wells 1994/95). The major problem is that the concept of partnership in conservation is based on the following untested assumptions:

(a) farmers and forest dwellers depend on forest products and services

(b) local populations are interested in forest resource use and conservation

(c) contemporary rural communities are homogeneous and stable

(d) community-based tenurial systems are more suitable for forest conservation.

The following discussion rethinks these four issues in light of opportunities for involving local communities in biodiversity conservation projects and protected area management.

The focus of the discussion is on forest and/or forest margin dwellers (indigenous or non-indigenous) and their livelihood strategies in tropical forests. Some, although not all, communities still depend largely on forest resources for subsistence and income generation. Lowland communities, however, have been predominantly delinked from forest resources and their livelihood strategies are, from the perspective of community-based forest conservation, less important.

20.3 Forest dependence by farmers and forest dwellers

While policy makers have tended to overestimate the employment benefits associated with timber harvests, the significance of employment and income generation for local economies was underestimated and remains to a large extent obscure even today. To many forest managers non-timber forest products (NTFPs) are still what they used to be called until recently: minor forest products. The example of India, where NTFPs contribute about 50% of forest revenue and 70% of income through export (Campbell 1992, cited in Sekar *et al.* 1996), indicates that this view is highly distorted. In Indonesia, the rattan industry alone provides employment for 200 000 people (Haury & Saragih 1995). More than 320 000 people are involved in NTFP production in Vietnam (Tien 1994) and in Bangladesh NTFPs provide employment for nearly 300 000 people (Basit 1995). These figures are impressive in themselves but obscure the magnitude of the contribution of forest-based activities to total income of many rural households, which case studies in Sri Lanka (Gunatilake 1995), Indonesia (De Foresta & Michon 1995) and India (Hegde *et al.* 1996; Kant 1997) have shown to be between 50% and 75%.

Some tropical forest dwellers use sophisticated and complex forest management techniques to increase the market value of their forests while maintaining other values such as high plant diversity, a multiplicity of outputs and flexibility of production (Padoch & Pinedo-Vasquez 1996). Many communities preserve designated areas from clearing specifically as sources for construction materials and services that only 'old-growth' forests can provide (Mayer 1996). However, forest dependence and the interest in particular forest products and services change over time. Faichampa (1990) has shown how the views on forest resource use change with a transformation of the local economy in Northern Thailand. The first migrants to forest areas particularly valued the fertile forest soils for dryland farming. As a result, considerable areas were clearcut for agricultural purposes. With the introduction of wet rice farming, the watershed functions became important and communities set aside forests in the upper watershed to secure a continuous water supply for irrigation. Today wet rice farming has lost its significance and consequently forests are valued either for their aesthetic or recreational functions.

Finally, dependence on forest resources is also a relative term. For example, commercially oriented forest product collection in parts of East Kalimantan is not necessarily critical to the livelihood of the Dayaks. Momberg *et al.* (1995), cited in Vayda (1996), explain that longer *gaharu* collecting expeditions are mounted 'to pay off debts incurred by purchasing motor boats, engines, radios, watches and whole wardrobes from traders and part-time expedition sponsors'.

20.4 Interest in forest resource use and conservation

Despite the continuing forest dependence of many rural people and industries, most products are not managed sustainably. Instead overexploitation of naturally occurring products is common, thus casting some doubt on the possibility of promoting the use of NTFPs in ICDPs. The fragility of extractive economies in general and the unsustainable use of NTFPs in particular have been pointed out by Homma (1992), Hall and Bawa (1993), Gupta (1994), Antolin (1995), Ros-Tonen *et al.* (1995) and Parnwell and Taylor (1996). While the diminishing natural resource can be partially explained by forest conversion and destructive logging operations, the reasons for overexploitation in remaining natural forests are very complex.

Several social and environmental constraints have held over-harvesting of NTFPs and timber for local purposes in check (Peluso 1991). Where population numbers are low, accessibility restricted and subsistence use predominates, most products are still used sustainably and traditional restrictions and regulations are heeded. Today, however, even remote areas are accessible, resulting in the breakdown of traditional controls and subsequently a very aggressive collection behaviour for commercially important products. Empirical evidence suggests that very few wild resources can sustain commercial exploitation, and that trade will either result in local extinction or initiate domestication (Wilkie & Godoy 1996). As Peters (1996a, p. 27) explains, 'although the fact is seldom mentioned in much of the literature on the

subject, a large number of non-timber forest products are actually harvested destructively'. This at times seriously reduces the abundance of particular species. Examples include ironwood (Peluso 1992), wild honey (Kaplan & Kopischke 1992; R. Hegde, pers. comm. 1997), sandalwood and fruits (Balachander 1995) as well as mushrooms, rattan, bird nests and *gaharu* (Peters 1994).

The traditional collectors suffer most from the demand on available supplies. They are usually not organised. Furthermore, the existence of monopsonies in marketing (a market situation in which one buyer exerts a disproportionate influence on the market) leads to inefficiency in marketing and very low returns on labour to collectors. Market expansion for many products has led at the same time to greater competition among collectors and traders. As Basha (1996) pointed out for bamboo, the traditional NTFP sector is less harmful to the resource than the newer commercial sector.

While other studies generally confirm this finding, the issue is far more complex, particularly because first there is no clear distinction between subsistence and commercial use. With the increasing monetisation of local economies, forest products have become more important for income generation (Balachander 1995). Second, the forest products sector is very dynamic. As a result products recently classified as belonging to the traditional sector belong, today, to a very organised commercial sector.

A major complicating issue in identifying viable conservation strategies and in assessing the interest of local communities in forest conservation is that images of successful intact resource-managing communities are used as a basis to advocate stronger legal rights and government recognition for community-based systems. Forward-looking in intention, policy advocacy work of this kind sometimes eludes the description of local communities, past and present, with description of an ideal type (Li 1996). In addition, doomsday descriptions of environmental disaster are presented as fact and alert policy makers, although what is claimed to be a fact is sometimes nothing more than a perception, a belief or an ideology ingrained in people's minds either out of convenience or ignorance (Ives & Messerli 1989), while the real environment has changed less over the last 20 years than the optic through which it is viewed. Also, many surveys on which research results and project implementation are

---

Box 20.2   The continuous problem of survey biases

(from Mayer 1996, p. 194)

...few women could answer questions about forest use and products, and few were willing to express opinions about the forest management scenarios described in the questions. It was also decided to direct questions only to household heads over 25 years old. Younger men could provide only limited information about conditions before 1982 in households they now head. Thus the survey is biased toward the perspectives and blind spots of middle-aged men, at the expense of those of women, younger men, or other groups. In addition, in the few villages where a substantial portion of the population support themselves primarily by work other than agriculture and agroforestry, the farmers and longer-resident ethnic groups are over-represented. This bias also reflects the tendency to interview older and longer-term residents, rather than younger people and newcomers.

---

based are biased toward people making their living primarily in subsistence agriculture and forest-oriented work (Box 20.2).

The result of biases and advocacy is a normative image of community and household behaviour in which, particularly, indigenous people are prescribed to have 'a different world view, consisting of a custodial and non-materialist attitude to land and natural resources, and want to pursue a separate development to that proffered by the dominant society' (Verlaat 1995, p. 47). Examples from the tropical forests around the world, however, indicate that local communities at the forest margin and in the forests themselves are increasingly affected by rapid marketisation and modernisation processes that characterise especially the economies of South-East Asia (Rigg 1997).

Even in the past, many traditional societies were not conservationist *per se*, but rather manipulators of the natural forests (Sekhran 1996). This is perhaps no more evident than on the island of Borneo where many Dayaks have manipulated old-growth and secondary forests for several hundred years in order to raise their productivity (Peters 1996b). Such practices have been termed 'forest agriculture' and Colfer *et al.* (1993) has described the Dayaks as managers of the forests rather than 'marauders'. Even though they exploit their forest fruit gardens actively such forests have a species diversity and a vegetation structure resembling natural forests (de Jong 1995). Their effect on the forest is minor, as long as population densities remain low and no other stakeholders appear. However, this is rarely the case and Colfer and Soedjito (1996) explain that an area that until the early 1960s was a lowland primary dipterocarp forest is today covered by a significantly

different forest that is affected by fires, extensive logging and small-scale agroforestry.

Local communities' interest in biodiversity conservation depends, at least in part, on how much they are still part of the ecosystem and how much their behaviour directly affects their own survival. It appears that cultural mechanisms that have been developed as adaptations to the forest environment over hundreds of years are easily cast aside when trade and new technologies free people from traditional ecological constraints (McNeely *et al.* 1995). That means that traditional resource use patterns are only sustainable as long as population densities are low, land is abundant, technologies are simple and involvement in the market economy is limited. As Alvard (1993, p. 356) points out, under such conditions people can be exceedingly wasteful of resources 'yet not have a large enough impact to cause a significant negative impact'. Accordingly, it might as well be that they do not use their indigenous knowledge to maintain an ecological balance—they just happen to be biodiversity custodians by default.

But how many people still live under these conditions? Equally important is the question of how many people *want* to live under these circumstances. Human relationships with the landscapes they are inhabiting must be conceptualised in time and space—they are dynamic relationships, not static activities that can be categorised as interactions independent of time (Zube & Busch 1990). That's why it is particularly important to gain insights of the perspectives not only of the household heads but rather of their children (i.e. the next generation). There is considerable evidence that the younger generations do not want to step into the footsteps of their forefathers, and non-farm employment opportunities are favoured over agriculture (Parnwell & Taylor 1996; Rigg 1997). In Papua New Guinea, for example, younger people are alienated from traditional culture and oblivious to past natural resource management practices. Their harbouring of 'high' development expectations is viewed as a serious threat to biodiversity conservation (Sekhran 1996). Local people, whether the hill tribes of Northern Thailand, tribal people of India, Dayaks of Kalimantan, the Penan of Sarawak or the native population of Amazonia must be expected to desire the same material benefits that people in more-developed economies enjoy, including adequate nutrition, shelter, health care and education (Alvard 1993).

Sustainable development and biodiversity conservation ultimately require that renewable resource consumption is stationary—that the product of number of people and the resource amount which each consumes not increase (Daly 1980, cited in Robinson 1993). Technological advances are probably sufficient to satisfy basic needs even under the scenario of increasing populations. However, they are not sufficient to cater for the increase in wants that puts a much heavier strain on biodiversity.

To assume that there are always ways to improve local incomes without depleting biodiversity is, at best, naive (Wells 1994/95). Even if it is possible, it is just as naive to assume that people are interested in biodiversity conservation and that they prefer to hang on to traditional practices and knowledge. Instead, many local people even in remote locations of the landscape yearn to acquire material possessions and wealth (Langub 1996) and welcome the material goods they can obtain, although they knowingly destroy the resource that they depend on (Rajasekaran & Warren 1994). At the same time, they are reformulating their images of themselves, and seek the benefits of a fuller citizenship and demand access to roads, education and health facilities, which in their view are symbols of modernity (Li 1995a).

National governments are responding to these demands by bringing development and modernity to rural areas. While the number of indigenous people who have been able to attend school is still small, it changes the outlook of peoples' lives. One effect of western education is the weakening of the old systems of social stratification (Ngo 1996) as well as the changes of historical community structures, assuming they existed in the first place.

20.5 The reality in contemporary local communities

The term 'local community' is often used loosely. As Gilmour and Fisher (1991) argue, it is a loose synonym for a group of people and of little use in implementing community forestry projects. They suggest to work instead with 'interest groups'; that is 'a group of people who have similar sets of interests in respect of a particular situation' (p. 69). They acknowledge that shared resources and livelihood guarantees are characteristics of small groups, such as small tribes, neighbourhoods or extended families, but seldom of whole village communities that tend rather to be heterogeneous, factional and stratified. This heterogeneity is very dynamic and constantly changing, which explains why community regulations for forest use fail sometimes to continue effectively. Many community-based conservation projects frequently lack this dynamic perspective.

Since conservation means different things to different people (Elliot 1996) and interest groups (Kremen *et al.* 1994) it is crucial to avoid a situation where a project targets the wrong people due to ignorance about local power structures and the distribution of interest (Ingles 1996). However, who is to choose who is right and who is wrong?

The premise that local communities should be given a central role in reaching conservation objectives has the inherent dilemma of defining the 'local community', their 'indigenous knowledge' and 'traditional culture'. The attempt to catalogue tradition and locate an authoritative source able to present 'the community' or 'culture' leads to simplifications inevitably ridden with power (see also Box 20.3), as articulate representatives rendered more powerful by State support overlook ambiguities in the meaning of indigenous terms and practices (Li 1996). Discourse and practice in support of community participation, organisation or control remain problematic in some countries such as Indonesia (Barber 1989, cited in Li 1996). In Thailand, however, it is the hierarchical nature of the society that provides much of the explanation for disappointing performance of participatory projects and grass roots development (Rigg 1991).

Box 20.3 Inventing traditional communities and imagining communities

(from Li 1995)

In an effort to consolidate land rights, the state looks for simplified notions of tradition; in particular, it is interested in identifying the individual "owners" holding "traditional" rights to land. The 1947 Agrarian Law which recognizes the land rights of "traditional" communities (so long as these are compatible with the national interest) requires that the identity of this "community" and its "tradition" first be defined and pinned down. It is now, in these modern conditions and in the context of the commoditization of land, that the Lauje are being "traditionalized" or "tribalized" for the first time. Historically, the Lauje were a scattered and individualistic group, without a strong sense of ethnic identity, and with little need for a formalized Adat articulated in terms of "Lauje Tradition". They must now, however, begin to make land claims based on being a "traditional community". Without a centralized leadership structure, there is no agreement on who should speak for the group in articulating these questions of identity, community and tradition. Meanwhile, as some become quite effective in making and defending claims, colonizing and monopolizing land and having their names entered into government records, others are struggling to hold on to land and livelihoods.

(from Blaikie & Jeanrenaud 1996)

...the biodiversity issue in agrarian societies in the South revolves around competition for scarce resources, strategies for gaining access and struggles which sometimes involve direct physical confrontation as well as the creation, use and manipulation of legal means. There is a comforting and misleading notion of "community" which is used in many conservation documents. It has become a social construction which policy makers and foreign donors need and upon which they base assumptions about local management of resources. Anderson (1983) talks of "imagined communities" which meet policy objectives. In reality, "communities" are often highly differentiated—along lines of gender, age, wealth, for example—and therefore their members may have very different perceptions and definitions of biodiversity. Also, the implications of biodiversity loss—as well as the costs of conserving biodiversity—must be differentiated according to wealth, gender and age. There is a need to "deconstruct" the notion of community.

In addition, many local communities are not asking for less State involvement and more isolation, but rather a better State, one which is more responsive to their needs and offers them access to services and facilities.

The most daunting task is that different interest groups subsumed in the category 'community' interact with the local environment and its resources in different ways. This interaction is constantly changing and depends as much on the type of prevailing agroecosystem (Fig. 20.2) as it does on the local economy and influences of external forces.

Many rural communities do not view society and nature as indistinguishable and regard themselves as controlled by the natural environment. Rapidly evolving social systems and community values make it difficult—perhaps even dangerous—to generalise but it appears that increasingly rural communities perceive

that natural resources may be dominated and sacrificed for personal gain. Social systems and values change even more rapidly during a conservation project which may also affect tree and land tenure regimes. This is particularly to be expected where improved accessibility and marketability leads to a commercialisation of former subsistence activities.

20.6 Community-based tenurial systems and their suitability for forest conservation

Community-based tenurial systems are rarely acknowledged by national governments or logging operators in any meaningful way. While it is admitted that transferring authority over forests to local communities is not a panacea to the problem of resource degradation, it is usually viewed as a prerequisite for biodiversity conservation. Increased

Coevolution and the Development of Agroecosystems

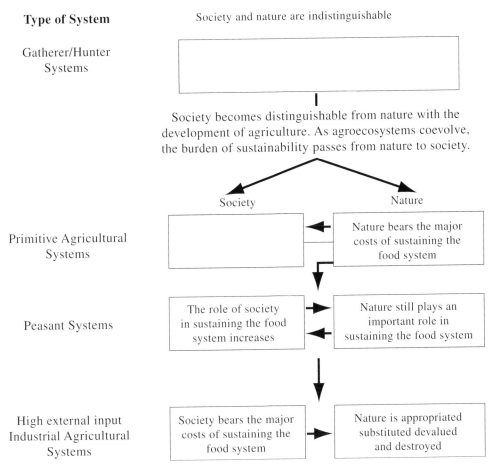

Figure 20.2 Agricultural evolution (after Redclift & Woodgate 1994).

tenure security is a necessary condition for farmers to adopt sustainable farming practices (Cook & Grut 1989; Lutz & Young 1992). The Penan of Sarawak, for example, cannot participate meaningfully in agricultural projects until they are certain of their land rights (Langub 1996). What is true for agriculture is also true for forest management. People are only willing to invest their scarce resources in conservation activities if they know that ultimately they will reap the benefits.

However, tenure security is not a sufficient condition for sustainable forest management. In Papua New Guinea, for example, many communities have been campaigning for many years, for various reasons, including the desire for rent capture and consolidating power, to attract extractive development, mining and

logging, to their communally held areas (McCallum & Sekhran 1996). The benefits that miners and loggers can offer substantially outweigh the ones of conservation (Box 20.4). Freedom to use their natural resources as they please results as a consequence in forest degradation, not unlike the situation in other regions where the natural forest belongs to the State.

In this context, forest-edge communities are opting to sell harvest rights to timber companies as a means of obtaining cash and social services. This provides 'development', as they perceive it, and addresses their concerns regarding economic exclusion. The fact that it does little to establish a framework for durable development rarely enters the decision-making calculus. Conservation, because it yields future, diffuse and often intangible benefits, many of which have no

---

Box 20.4 Communal forests for wholesale

(from McCallum & Sekhran 1996)
Many communities have a world view that is inclined towards the short-term, planning for their perceived immediate needs rather than for their long-term welfare necessities. The medium to long-term social, economic and environmental consequences of their current land use practices are poorly understood and often ignored.

---

direct monetary value, tends to be undervalued in this context and seen to conflict with community aspirations. This ethos tends to be reinforced by many developers and Government agencies.

Also, resource and property rights are changing in tune with the transformation of the agrarian societies and livelihood strategies. For example, in Kalimantan many of the forests managed by Dayak families but communally shared for many products are rapidly privatised (Peluso & Padoch 1996). In particular where roads change the relative isolation of formerly remote areas, markets for land develop and it is not uncommon to find parcels of forests, that *de jure* belong to the State, but *de facto* are traded in the market. The interest in communally held resources or common property is thus diminishing.

20.7 Conclusions

I acknowledge that the discussion above is narrow in scope. Beside rural people, many other stakeholders influence biodiversity and affect the success or failure of any conservation project. Other stakeholders have not received any attention in the discussion, because first, at least in theory, their activities are easier to regulate, and second their dependence on natural forests is not as crucial for their livelihoods, meaning that they will be able to adjust more easily to a new situation or the imposition of restrictions. Also, the discussion focused on the problems of involving people in 'a project'. There is more to biodiversity conservation, however, than demarcating a protected area on the ground and implementing 'a project'. A project-based approach has inherent limitations which are often overlooked. Many factors leading to deforestation and the loss of terrestrial biodiversity in tropical forest can neither be influenced by local communities nor by project personnel during even a long-term project.

'Putting people first' has been borrowed from general development practitioners (Wells 1995) and it cannot be emphasised enough that participation is crucial. People need to be involved at all stages of decision-making processes. But its meaning is not restricted to empowering people to mobilise their own capacities and gaining control over their livelihood strategies. Many never had that control and may be at a complete loss. The questions—who is participating and for what reasons—are crucial as the discussion of heterogeneous communities has highlighted.

Identifying the human consumers of natural resources also requires considering other socioeconomic and political groups. The needs and interests of other stakeholders frequently contradict those of the direct users. There is a need to recognise the stratified nature of many rural societies. While they may have been traditionally egalitarian and non-hierarchical, internal divisions are emerging because of marketisation, modernisation and the commodification of the natural resources.

As Neumann (1996) stresses, it is crucial that project design recognises that villagers are often politically fractured and socially differentiated. Fractures in the local community may arise because of gender, class, age or ethnic group. Also, divisions within the communities shift, perhaps at times nearly disappearing, as communities present a unified front to a perceived threat from outside, and sometimes multiply in internal struggles over land and resources. It can be assumed that the community the way it existed before the onset of any project is different from the community at the end. Social differentiation is probably more profound and the villagers aspiration will have changed. Decision-making bodies and processes within the village can change dramatically, calling for an extremely flexible approach in dealing with the community and its interest groups.

An important lesson is that biodiversity conservation projects need to ignore or consciously abandon those areas where communities have already made choices that will likely cause long-term conflict with conservation imperatives. The solution to forest loss is not in finding additional economic incentives for the rural poor, but in generating more attractive alternatives elsewhere. There will always be the potential for conflicts of interest between rural people's ability to earn a living and the conservation of areas of high ecological value. Conservation and sustainable development projects can aim to mitigate such conflicts of interest by promoting alternative income sources and education programs. However, some conflicts will persist and the need for protecting forest areas through policing and enforcement will often be inescapable (Wells 1994/95). Sustainable use is a powerful approach to conservation, but it is not the only one, and the conservation of much forest biodiversity also requires a preservationist approach.

Finally, the effects of integrated conservation and development projects have been disappointing because the linkages between socioeconomic development for local residents and the needed behavioural response to reduce the pressure on the remaining forest resources are not well established. In large part, the inability to establish these linkages results, first, from a lack of understanding of livelihood strategies at the forest margin and their relationship with the forest resource. Second, our limited knowledge of household behaviour prevents us from predicting the effects of many interventions. The assumption that the living standards of forest residents can be improved and biodiversity conservation objectives can be reached simultaneously remains untested. Therefore, Ferraro and Kramer (1995, p. 35) conclude that 'if you cannot identify a very precise conceptual link between a proposed intervention and household decision-making, do not proceed with the intervention'. While it is probably impossible to follow this rule of thumb during the project design stage, it should guide project implementors during project implementation.

This is perhaps the most important conclusion. It should motivate us to re-examine our perceptions of what rural livelihoods and communities at the forest margin are all about. It should make us understand that communities not only are heterogeneous entities but also that substantial differences among villages, districts, economies and forests exist. It should spur us to develop a typology of rural settings not unlike the agroecosystems one suggested by Redclift and Woodgate (1994) and to examine the potential nexus between the interest of resource users and the objectives of biodiversity conservation. This nexus and the ability to conserve forests depends on numerous variables. Population density, the arrival of technological innovations (e.g. chainsaws), improved access to infrastructure including education and markets have been discussed above. They are good indications of the extent of market penetration and modernisation within a community. The more advanced a community is, the more stratified will be its members which will make the introduction of community-based biodiversity conservation projects a considerable challenge. Education and modernisation lead to a fast disappearance of local knowledge. It is futile to work with the oldest community members in an attempt to fix tradition when it is the younger people whose involvement in sustainable activities ultimately counts more.

CHAPTER 21

POLITICS, POLICIES AND THE CONSERVATION OF FOREST GENETIC DIVERSITY

Peter J. Kanowski

SUMMARY

The conservation of forest genetic diversity is a dominant theme in environmental politics in the late 20th century. Debate has focused on several issues central to the Convention on Biological Diversity: sovereignty over and access to genetic resources, the sharing of benefits and costs, the dynamic between conservation and use, and the relationship between economic efficiency and social and environmental objectives. Multilateral intergovernmental resolution of these issues is proving difficult, but various innovative approaches have emerged in particular situations. These initiatives are characterised by their diversity and flexibility, and by the establishment of partnerships between interested parties.

Policies for the conservation of forest genetic diversity, therefore, need to be based on the recognition and accommodation of the diversity of interests and rights. They need also foster coordination, integration and innovation at the policy level, and acknowledge institutional capacity. Realising these goals is perhaps more challenging than developing the technical elements of genetic conservation policies, in which we are considerably assisted by scientific advances reviewed elsewhere in this book. Key technical issues with a significant policy dimension are the respective roles of *in situ* and *ex situ* conservation, the identification of priority areas for conservation, the integration of on-reserve and off-reserve management, and assessment of the effects of forest management on genetic diversity. These issues are being addressed principally through the development of integrated conservation strategies, reappraisal of the role and management of protected areas, and the development and evaluation of sustainable forest management regimes.

21.1 Introduction

Concern for the conservation and wise use of biological diversity has been a dominant theme of environmental politics and policies in the latter part of the 20th century. This environmental theme, exemplified in global politics by the negotiation of the Convention on Biological Diversity (CBD), has been paralleled in commerce by an emphasis on free trade and stronger property rights, exemplified by the World Trade Agreement (WTA). There are both tensions and possible synergies between these two themes and the interests that underlie them—simplistically, those of environmental conservation and economic development. These tensions and synergies are also evident between the international regimes which have developed in respect of each theme and in their relationships with issues of national sovereignty and local livelihoods. The principal actors in this policy debate have been multilateral agreements and agencies, large—often transnational—corporate enterprises, multilateral non-government organisations and national governments and their agencies. Together, these actors and their interests define much of the recent and contemporary debate around the conservation of forest genetic diversity.

The politics of the conservation and use of forest genetic diversity are embedded in this broader context, as are policies and strategies to achieve conservation goals. In the simplest terms, strategies for genetic conservation have generally been based around both *in situ* and *ex situ* approaches; *in situ* approaches have typically included both 'strict' conservation in reserve systems and 'conservation through use' in managed forests and agroecosystems. While national strategies for the conservation of forest genetic diversity share these common elements, they often differ substantially in their expression—reflecting differing political situations and priorities, different development strategies and institutional structures, diverse social and economic circumstances, varying histories of land allocation and use and differing interpretations of incomplete scientific knowledge.

Because the conservation of forest genetic diversity cannot be dissociated from the control and use of forest resources more generally, policies to achieve genetic conservation goals remain strongly contested—at international, national and local levels. This chapter recounts important elements of the history of policies for conservation of forest genetic

diversity, considers the current policy environment and explores how the cause of the conservation and wise use of forest genetic diversity might be advanced.

21.1.1 History

The history of concern about and action for the conservation of forest genetic diversity could be partitioned arbitrarily into four overlapping eras:

(a) prehistory

(b) reservation for protection

(c) laissez-faire cooperation

(d) environmental politics.

21.1.1.1 Prehistory

In the sense used by Arndt (1987) to describe the evolution of ideas about economic development, 'prehistory' describes an era prior to the emergence of an organised body of thought and action. This concept of prehistory is also relevant to forest genetic diversity, as research in anthropology, history and genetics has begun to reveal the substantial, sometimes profound, effects of earlier civilisations and societies on forms, patterns and levels of forest genetic diversity. Ledig (1992) classifies these effects in terms of the conversion and fragmentation of forest ecosystems; their alteration through the harvesting of forest products, use of fire or more general environmental change; and the translocation and domestication of plant and animal species. Although the effects of these processes on forest genetic diversity have generally been adverse (e.g. in the local or total extinction of species, or the erosion of gene pools) there are also important counter examples that demonstrate how human interventions have maintained or enhanced genetic diversity. Relatively well-documented cases include the home and forest gardens of Asia, the apêtê forest patches of the Brazilian cerrado (Posey 1985), the semi-deciduous forest islands among savanna in Guinea (Fairhead & Leach 1998) and the *Leucaena* agroforestry systems of south-central Mexico (Hughes 1998).

The era of prehistory is important in contemporary politics and policies about forest genetic diversity for two principal reasons: its implications for current forms, patterns and levels of forest genetic diversity, and the recognition it suggests for indigenous and local peoples' knowledge of, and rights to, genetic diversity.

Box 21.1   IUCN Protected Area Categories

[from IUCN (1994) in World Conservation Union/WWF (1998)]

IUCN category & description

I(a)   Strict Nature Reserve

I(b)   Wilderness Area

Protected areas managed for science or wilderness protection

II   National Park

Protected areas managed mainly for ecosystem conservation and recreation

III   Natural Monument/Natural Landmark

Protected areas managed mainly for conservation of specific natural features

IV   Habitat/Species Management Area

Protected areas managed mainly for landscape/seascape conservation and recreation

V   Protected Landscape/Seascape

Protected areas managed mainly for landscape/seascape conservation and recreation

VI   Managed Resource Protected Area

Areas managed mainly for the sustainable use of natural ecosystems

---

21.1.1.2   Reservation for protection

By the late 19th century, the unprecedented rate and scale of the effects of European settlement on the forests of the New World, notably North America and Australia, generated sufficient public and political concern that the pleas of advocates of sustainable use and of preservation (e.g. Pinchot and Muir, respectively, in the USA) were heeded. The consequence of societies' 'enhanced sensibilities' (Western & Wright 1994) were the reservation of forests, either for sustained production or preservation of the natural environment.

The New World origins of what we now call national parks, as 'wilderness' areas protected from human intervention, have had a profound legacy for our conception and management of protected areas (Western & Wright 1994). The era of reservation for protection established many existing protected areas, and the philosophy it represents remains dominant in most national conservation strategies. The World Conservation Union (IUCN) now recognises six categories of protected area (Box 21.1); around 8% of the world's forests are designated as protected areas

consistent with these categories (World Conservation Monitoring Centre 1998). The Worldwide Fund for Nature (WWF) and IUCN are campaigning for 10% of each nation's forests to be included in protected area systems (World Wide Fund for Nature/IUCN 1996); 22 countries have agreed to this goal, and some (e.g. Australia) are committed to exceeding it—the Australian target is 15% of pre-1750 forest ecosystems (Australia & New Zealand Environment & Conservation Council/Ministerial Council on Forestry, Fisheries & Aquaculture 1997). [Countries that have agreed to this goal are Argentina, Armenia, Australia, Austria, Bolivia, Brazil, Canada, Chile, PR China, Colombia, Greece, Lithuania, Malawi, Mozambique, New Zealand, Nicaragua, Romania, Russian Republic of Sakhs, Slovak Republic, Tunisia, Vietnam, Uzbekistan (World Conservation Union/ WWF 1998)].

Unfortunately, the value of most existing national protected area systems to the conservation of forest genetic diversity is limited by a number of factors (Ledig 1988; Frankel *et al.* 1995; Kramer *et al.* 1997; Kanowski *et al.* 1999). Most represent a biased sample of ecosystems and communities within ecosystems;

relatively few are of sufficient size and sufficiently connected to other forests to achieve long-term conservation goals. For a variety of social, economic and political reasons, many areas are protected more on paper than in practice. Nor does the traditional model of protected areas fit well with some traditional systems of land tenure, such as those of Melanesia (e.g. Tacconi & Bennett 1997; Filer & Sekhran 1998) or southern Africa (e.g. Nhira *et al*. 1998), or where landscapes are now more cultural than natural, such as in Western Europe. Implications for the conservation of genetic diversity are discussed further in Chapter 21.2.4.

21.1.1.3 Laissez-faire cooperation

Although its genesis is in the 16th to 19th century period of botanical and colonial exploration, and the associated establishment of botanical gardens and international transfer of seeds or plants of species of known or presumed economic importance (e.g. Grove 1995), the major era of laissez-faire cooperation in forest genetic diversity can conveniently be ascribed to the first seven decades of the 20th century. In the first part of the era, bilateral or colonial relationships were dominant, but these were superseded in the latter part by several multilateral frameworks. In all cases, these programs were predominantly intergovernmental or publicly sponsored, relatively informal and free of regulation other than that related to quarantine, and focused on a limited number of species of economic importance (i.e. on genetic resources rather than genetic diversity in the broader sense). As the National Research Council (1991) observed, their main objectives were 'the acquisition and distribution of high-quality seeds and tree improvement, not genetic conservation'.

The first multilateral framework for cooperation in forest genetic resources was provided by the establishment of the International Union of Forestry Research Organisations (IUFRO) in 1892. IUFRO operates principally as a series of subject-specific Working Parties; those concerned with forest genetics began facilitating the transfer and exchange of forest tree germplasm for research and *ex situ* conservation in the early years of the 20th century (Burley & Adlard 1992). More formal multilateral cooperation was initiated in 1968 with the establishment of the Food and Agriculture Organization of the United Nations (FAO) Panel of Experts on Forest Gene Resources, with a global mandate to review relevant activities, develop priorities and recommend action (Food and Agriculture Organization of the United Nations 1997b). The Panel's recommendations were influential in coordinating the work of existing national institutions, such as France's CIRAD Forêt and the United Kingdom's Oxford Forestry Institute, and in catalysing the establishment and activities of new national centres, such as CSIRO's Australian Tree Seed Centre and the Danish International Development Agency's Tree Seed Centre. These and other national institutions subsequently assumed leading roles in the exploration, distribution and *ex situ* conservation of forest genetic resources; together, they formed an informal international network, loosely but effectively coordinated by FAO's Panel and IUFRO's Working Parties, and funded by their governments or other development assistance sponsors to work on a particular suite of species. Their activities were reviewed by the National Research Council (1991).

These institutions and many development assistance agencies were also instrumental in fostering the establishment of corresponding national or regional centres in many 'developing' countries; examples of the latter include the Association of South East Asian Nations' (ASEAN) Tree Seed Centre in Thailand, the seed bank of ESNACIFOR in Honduras, and the SADAC Regional Seedbank in Zimbabwe. The latter part of this era also saw the emergence of new forms of partnership, involving private industry as well as government: the Central America and Mexico Coniferous Resources Cooperative (CAMCORE), established in 1980, remains the exemplar, with activities far broader than its name suggests. The 1970s also saw the establishment of two centres concerned with forest genetic resources in the multilateral network of the Consultative Group for International Agricultural Research (CGIAR); the International Plant Genetic Resources Institute (IPGRI), and the International Centre for Research in Agroforestry (ICRAF). Although broader in principle, IPGRI's priorities have remained focused on agricultural crop species and fruit or food-product trees, while ICRAF's have been limited to some of the tree species important in agroforestry systems. A third CGIAR centre focused on forests, the Centre for International Forestry Research (CIFOR), was established in 1990, and its priorities reflect more those of the next era. The activities of these centres in relation to forest genetic resources were reviewed by Palmberg-Lerche (1994).

The emphasis in this era of laissez-faire policy for forest genetic diversity was the facilitation of access, on a minimal-cost basis, to genetic resources of potential economic value. Concern about the loss or degradation of forests also promoted an emphasis on *ex situ* conservation of economically important species. Forest genetic resources were seen primarily as a common heritage, and their free exchange to be the best means of realising their benefits. In the latter part of this era, the notion of free access began to be challenged by some nations as the commercial value of their forest genetic resources became apparent. Some were quite successful in developing policies to capture some part of this value, and the issues of access and benefit became increasingly troubled for all parties.

21.1.1.4   Environmental politics

The era of environmental politics for forest genetic diversity is defined by increasing public concern and political activity about forests, and which began to emerge in the 1970s, intensified in the 1980s with a focus on tropical forest loss and degradation, reached an international political crescendo in 1992 at the United Nations (UN) Conference on Environment and Development (the Rio de Janiero 'Earth Summit'), and which continues in various manifestations as the international community and individual nations deal with the legacies of the Earth Summit for all forests— tropical, temperate and boreal. During this era, policy around forest genetic diversity became much more complex—both in recognition of the complexity of issues, and as a consequence of the strength and diversity of competing interests, associated with forest genetic diversity.

Some international initiatives or events both reflect and help describe the evolution of thought, politics and action in this era, and provide some guide to the complexity. One group is those concerned with the environment and with forests in general; a second is that concerned more specifically with genetic diversity. Principal among the former are: the 1972 UN Conference on the Human Environment; the 1980 World Conservation Strategy and its 1990 successor *Caring for the Earth*; the 1983 International Tropical Timber Agreement; the 1985 Tropical Forestry Action Plan; the 1987 report of the World Commission for Environment and Development, *Our Common Future*; the 1992 Earth Summit, and Agenda 21 which emerged from it; the subsequent Intergovernmental Panel and its successor Forum on Forests; and the emergence of various processes (e.g. Helsinki,

Montreal, Tarapoto) to promote sustainable forest management, and to develop criteria and indicators— including those related to genetic diversity (e.g. Namkoong *et al.* 1996; Stork *et al.* 1997)—against which to assess it. Together, these agreements and activities have progressively defined an international forests regime summarised by, among others, Tarasofsky (1995), Thomson (1996) and Grayson and Maynard (1997).

The second group of initiatives, concerned specifically with genetic diversity, focused initially on the genetic resources of crop plants and subsequently on biological diversity more generally. Issues about forest genetic diversity have been peripheral to the former, and subsumed and submerged within the latter; nevertheless, these initiatives are important because of the policy debates they reflect and contexts they set. Since 1983, intergovernmental discussions about plant genetic resources have focused the Commission on Genetic Resources for Food and Agriculture, comprising the 174 members States of FAO. The Commission generated the first comprehensive international agreement on plant genetic resources, the non-binding International Undertaking (IU), in 1983.

In March 1997 the IU had 111 adherents, with notable exceptions being Brazil, Canada, China, Japan, Malaysia and the United States of America (International Institute for Sustainable Development 1997).

The most recent discussions in the Commission (International Institute for Sustainable Development 1997) have highlighted difficulties in resolving the controversial topics of access to genetic resources and to 'Farmers' Rights', the concept describing recognition of 'the contributions of farmers to conserving, improving, and making available plant genetic resources' (Food and Agriculture Organization 1989). The Commission's 1996 adoption of a *Global Plan of Action for the Conservation and Sustainable Utilisation of Genetic Resources for Food and Agriculture* specifically excluded forest genetic resources, referring them to the Intergovernmental Panel on Forests. However, the Panel and its successor Forum have not yet been willing to engage with issues of forest genetic diversity, in large part because of uncertainty and debate about its role in relation to that of the Convention on Biological Diversity.

The agreement at the 1992 Earth Summit of the legally binding Convention on Biological Diversity (CBD) is of seminal but as yet unrealised importance to forest genetic diversity. At May 1998 the CBD had been ratified by 172 nations; the USA remains the only significant absentee (Secretariat to CBD 1998).

The objectives of CBD are expressed as three central elements (United Nations Environment Programme 1994):

(a) the conservation of biological diversity

(b) the sustainable use of the components of biological diversity

(c) the fair and equitable sharing of benefits arising out of the utilisation of genetic resources.

The key issues discussed by the CBD (Glowka *et al.* 1994) remain the focal points for debate about forest genetic diversity: the tension between national sovereignty and the 'common concern of humankind', the dynamic between conservation and sustainable use, and the questions of access to and payment for both genetic resources and (bio)technologies. Sovereignty, access and payment, with their implications for property rights and trade, have become issues of friction not just within the framework of the CBD but with the World Trade Agreement and its provisions on Trade Related Aspects of Intellectual Property (TRIPs). The fundamental tension, as *Nature* (1998) discusses, is that between equity in the distribution of benefits arising from the use of genetic resources and economic efficiency in exploiting them. Although there have been some innovative responses to these challenges, such as the Merck-INBio bioprospecting agreement (Reid *et al.* 1993) and others supported by the International Cooperative Biodiversity Group (*Nature* 1998), a more general resolution appears to remain elusive.

Agreement of the CBD reflected acknowledgment by governments of many concerns about biological diversity which non-government organisations had articulated in the preceding decade; however, the failure of parties at the Earth Summit or subsequently to agree to an international Forests Convention reflects continuing political divisions between, especially, North and South about how to address concerns over forest conservation and management, including those related to forest genetic diversity. These political divisions about forests have continued to overwhelm attempts to address issues about forest genetic diversity within the framework of the CBD; consequently, these issues remained after the 4th

Conference of the Parties to the CBD in May 1998, embroiled in the larger international debates about forests and about trade.

21.1.1.5 Synopsis

Three themes in global politics dominate debate and policy about forest genetic diversity at the end of the 20th century. The first embodies the tensions between competing interests in the exploitation of genetic resources and the distribution of any consequent benefits, and is evident in both the Commission on Genetic Resources for Food and Agriculture and the Conference of the Parties to the CBD and in their interaction with the World Trade Organization. The second is the continuing debate about the conservation and wise use of all the world's forests, being conducted in the International Forum on Forests, around other initiatives such as those related to forest certification and product labelling, and within nations. The third, and probably the ascendant, is the free trade agenda of the World Trade Agreement and its provisions, which will have the most effect on issues associated with access to, and recompense from, exploitation of genetic resources.

Whilst the international regime relevant to the conservation of forest genetic diversity will continue to evolve, the policies by which individual nations interpret and implement genetic conservation objectives will remain critical. The next section of this chapter considers the principal elements of such policies.

21.2 Policies for forest genetic conservation: issues and elements

Forest genetic diversity is complex, heterogeneous and dynamic; it is shaped by interactions between the physical environment, the biology of forest systems and populations, and the influences of people and societies. Policies for its conservation need to recognise these forces and their interdependencies.

This context suggests some key issues and elements to be involved in policies if they are to achieve genetic conservation objectives:

(a) recognition and accommodation of the diversity of interests and rights

(b) policy integration, coordination and innovation

(c) institutional capacity and cooperation

(d) respective roles of *in situ* and *ex situ* conservation

(e) development of integrated conservation strategies

(f) respective roles of both protected areas and off-reserve forests.

Each of these is discussed below.

21.2.1 Recognition and accommodation of the diversity of interests and rights

Discussions around the IU on Plant Genetic Resources, between the Conference of the Parties to the CBD, and in the context of the WTA, have been severely hampered by the difficulties of defining broadly acceptable means of recognising and accommodating the diversity of interests in, and rights to, genetic diversity. The spectrum of immediate interests encompasses (among others) indigenous peoples and local communities, farmers and forest owners, governments and their agencies and enterprises seeking to commercialise or otherwise profit from forest genetic diversity. There are also, as the CBD recognises, the broader interests of the global community and of future generations. These issues have been reviewed in the context of forest genetic diversity by, for example, ten Kate (1995) and Kanowski *et al.* (1997).

As with other arenas of public policy, establishing a policy framework that achieves this objective, and thus acts as an enduring foundation for more specific policy development, is particularly challenging—the negotiations which preceded and have followed the international agreements relevant to genetic diversity, or associated with corresponding national initiatives, amply demonstrate the difficulties of reconciling diverse and often competing interests. These processes can be instructive, however, and there is now access to experience and literature specifically concerned with how these challenges for forests may be met: examples range from the international (e.g. Thomson 1996) and national (e.g. Carew-Reid *et al.* 1994) levels to those negotiated with and between local communities in recognition of their rights (e.g. Reid *et al.* 1993; Western *et al.* 1994; Posey & Dutfield 1996; Hughes 1998). Flexibility, diversity and negotiation are important elements of these approaches, and comanagement and partnership are important common themes that emerge from them.

21.2.2 Policy integration, coordination and innovation

Competing or inconsistent policies frequently arise as a means of satisfying different interest groups, thus emphasising the importance of developing the coherent and consistent policy framework discussed above. Notwithstanding nor diminishing its limitations, the CBD—as a generally-agreed legally binding framework with objectives broadly consistent with those declared by national governments for forests under their jurisdiction—offers one framework for the coordination of disparate policies relevant to the conservation of forest genetic diversity. The development of national biodiversity strategies consistent with the objectives of the CBD illustrates how it might enact such a coordinating role. Other forums, such as those provided by other intergovernmental processes and by the FAO, can also complement the activities of the CBD.

The separation of authority and responsibilities between government ministries and agencies concerned with management of forests for conservation and for production is a particular challenge to policy integration and coordination; experience with agencies integrating these functions suggests, unsurprisingly, that agency restructuring alone is not a solution, and generates its own set of challenges. Another common limitation of major consequence for achievement of conservation objectives is the vexed relationship between public policy and forests under private ownership or control. Although a range of incentives and regulations have been employed to promote conservation objectives on private land, the effectiveness of these measures varies widely (Tasmanian Public Land Use Commission 1997). Indeed, some have acted as perverse incentives for the conversion or unsustainable management of forests (e.g. Kotey *et al.* 1998). National strategies which engaged rather than avoided the complexities of achieving conservation goals on private land would advance considerably the cause of the conservation of forest genetic diversity.

Engagement between public and private sectors is also an imperative because of the prevailing political ideology of the late 20th century, in which the traditional role of the State is diminished. The relative decline of State authority over forests has been paralleled by an increasing role for the private sector and for the community at large. In many countries 'civil society' has asserted its interests in public forest management; those same interests, particularly in respect of environmental values, are slowly being extended across tenure boundaries to forest in private ownership. One of the principal manifestations of these expressions of interest in forest management is the development of consultative or collaborative processes through which stakeholders exercise influence over forest management decisions. Corporations are also finding themselves subject to pressure, internally from shareholder associations, and externally from community expectation, and many are responding with the establishment of environmental management systems and reporting mechanisms. All of these developments should assist the realisation of genetic conservation objectives.

There are already encouraging examples of initiatives within the private sector, and of partnerships between the private, government and non-governmental sectors, which illustrate the potential of private ownership and enterprise to contribute to genetic conservation objectives. Work exploring the role of the private sector in forestry (e.g. Bass & Hearne 1997), and in achieving conservation objectives (e.g. Tasmanian Public Land Use Commission 1997), offer good foundations for developing successful partnerships between the public, private and community sectors. In recognition of these changes, the traditional emphasis in conservation-directed policies on regulatory frameworks and sharply defined institutional roles is likely to evolve to encompass more flexible and innovative institutional arrangements, and stronger cross-sectoral partnerships supported by incentive structures and market-mediated mechanisms.

21.2.3 Institutional capacity and cooperation

Issues of institutional capacity and cooperation also emerge from the discussion above and from that in Chapter 21.1.1.3. Institutional capacity has been limited by the low status historically accorded the environment and the conservation of forest genetic diversity relative to that of other priorities, such as agricultural expansion; this is reflected both within national governments and in their positions at intergovernmental forums. Consequently, much of the onus for strategic development, coordination and action for the conservation of forest genetic diversity has fallen to relatively informal multilateral

mechanisms, such as FAO's Panel of Experts or IUFRO Working Parties, to those national institutions which have been able to assume international responsibilities, and more recently to non-governmental organisations and the few international or regional centres with a focus on forests. A major constraint is that few of these are politically influential, and well-resourced or securely-resourced. It is less a lack of institutional structures or collaborative frameworks, than the adequacy of support for those that already exist, which most limits effective cooperative action. The lack of support includes limited resources for research, education, and the exchange of information to promote the conservation of forest genetic diversity.

Funding mechanisms emerging from new collaborative partnerships with the private sector (e.g. *Nature* 1998) offer one way forward; these include innovative use of the capital markets (e.g. Mansley 1996), as evidenced in debt-for-nature swaps (World Wide Fund for Nature/IUCN 1996), or in the emergence of carbon trading regimes (e.g. Moura-Costa & Stuart 1998). New or renewed multilateral mechanisms, such as the role being developed for the Global Environmental Facility in supporting implementation of the provisions of the CBD (International Institute for Sustainable Development 1998), represent another important source of funds to support achievement of genetic conservation objectives.

21.2.4 The respective roles of *in situ* and *ex situ* conservation

The complexity of forest ecosystems, the dominant role of tree species in them, the environmental and economic value of forests and trees and the poor conservation status of most tree populations *ex situ* has led to the characterisation of forest trees as a paradigm of *in situ* conservation (Frankel *et al.* 1995). Effective *in situ* conservation demands that both ecosystem function and process, and intraspecific population genetic processes, are maintained in a network of sites that are comprehensive and representative in terms of all levels of genetic organisation.

The *ex situ* conservation status of forest species is generally correlated with the extent of their domestication, and is therefore either poor or non-existent for most. Only a trivial proportion of forest species (e.g. around 100 tree species) are conserved

---

Box 21.2 Examples of mechanisms promoting conservation on private land

(from Kanowski et al. 1999; drawing principally from Binning & Young 1997, McNeely 2000), Tasmanian Public Land Use Commission 1997 and Western & Wright 1994)

- **Voluntary agreements** between forest owners and governments or non-government conservation organisations. These take many forms, including conservation management agreements, stewardship agreements, conservation easements, contracts and leases. Most are established under relevant legislation; some are bound to the title of the land and others are for a fixed period; some draw on incentive mechanisms, either financial or non-financial. Typically, they impose constraints or make requirements for active management. Non-government organisations—ranging from large internationally-active organisations such as The Nature Conservancy or Birdlife International to local conservation trusts—have been particularly effective in facilitating conservation agreements with landowners.

- **Community-based conservation**, through a variety of mechanisms developed in partnership with local communities in order to benefit them as well as to achieve conservation goals. Examples include Costa Rica's BOSCOSA, Zimbabwe's CAMPFIRE, and the United Kingdom's Farm Scheme.

- **Business partnerships** to support conservation can take many forms. They may be based around the direct commercial values of forests, for example through joint ecotourism or pharmaceutical product ventures, or voluntary cooperative programs in support of conservation, such as those involving energy enterprises operating adjacent to Indonesia's Kutai National Park.

---

adequately *ex situ*. These species are almost exclusively those whose genetic resources have been assembled for domestication programs, with which almost all substantive *ex situ* forest conservation activities are associated (National Research Council 1991). For forest species, the value of *ex situ* seed storage is further limited by the relatively large number of species, many of economic importance, whose seed is not amenable to storage. Although some progress has been made with other storage technologies, few are operationally feasible for trees (Haines 1994). Although research to develop these technologies has merit, their technical limitations and cost will continue to preclude their use, other than when exceptional circumstances or strong economic imperatives prevail.

The situation of trees in terms of *ex situ* conservation is also true for most other forest species—most of which are not yet described by science. Consequently, *in situ* conservation strategies—in the broad sense discussed below—will remain fundamentally important for achieving the conservation of forest genetic diversity.

21.2.5 *In situ* conservation strategies

Most *in situ* conservation strategies have been developed around the foundation of a reserve system of protected areas—the ultimate expression and focus of *in situ* conservation (e.g. Western & Wright 1995; World Commission on Protected Areas/IUCN 1997). Acknowledgment that protected areas do not function as islands isolated from their broader environments—ecological, economic or social—has fostered the development of a parallel suite of policies that seek to reinforce the effectiveness of the reserve system through complementary off-reserve management. These policies are now focused on the concepts of sustainable forest management, and its assessment against criteria and indicators (Braatz 1997; Wijewardana *et al.* 1997).

Perceptions of the purpose of both protected areas and forests outside reserves, and of their roles in achieving conservation objectives, continue to evolve—and are the subject of considerable debate (e.g. Western & Wright 1994; Brandon 1995; Wood 1995; Hale & Lamb 1997; Kramer *et al.* 1997; World Conservation Union/WWF 1998). Experience at both policy and operational levels has led, for example, to a reassessment of assumptions about protected area function and management (e.g. Western & Wright 1995; World Commission on Protected Areas/IUCN 1997), and the reinterpretation of IUCN's Protected Area categories (World Conservation Union/WWF 1998). A common theme is that achieving conservation objectives requires more than the declaration of conservation reserves, and demands the integration of on-reserve and off-reserve management through the development and implementation of integrated conservation strategies.

In response both to the limitations of the traditional model of protected areas, and as a consequence of the changing role of governments, there has also been increasing diversification in the means by which protected areas are established, financed and managed; examples include the involvement of NGOs such as The Nature Conservancy or WWF, the establishment of community-based partnerships such as those described by Western *et al.* (1994), and the emergence of various forms of voluntary conservation agreements between private forest owners and governments or NGOs (Binning & Young 1997). A major challenge to contemporary conservation efforts is to develop policies which build on the role and contributions of protected areas, while recognising their limits.

21.2.5.1 *Identifying priority areas for in situ conservation*

Although the means by which conservation objectives should be achieved is the subject of considerable debate, there is broad agreement that efforts should be focused on areas of highest priority. Recognition of the need to identify priorities for the conservation of biological diversity has generated considerable methodological development in planning for *in situ* conservation. Contemporary conservation planning methodologies seek to maximise the achievement of forest conservation goals, through the identification of priority areas for conservation, while minimising the opportunity costs associated with realising the goals. In practice, this requires comparison of the values of different forest areas, so that the benefits and costs of different forest conservation options can be assessed and used to inform decisions. Suites of conservation planning tools (e.g. BioRap, Margules & Redhead 1995; C-Plan, Pressey & Logan 1997) are now

available for this purpose; they are based on common principles (Kanowski *et al.* 1999) as follows:

(a) making the best use of both biological and environmental data, which will inevitably be incomplete

(b) incorporating the social and cultural values of stakeholders by including them in the planning process

(c) minimising bias and maximising efficiency in the achievement of conservation goals

(d) incorporating economic costs into the process for achieving conservation goals

(e) providing the information necessary for informed negotiation about the best way to achieve conservation goals, and to strike a balance with other goals.

The best geographical basis for conservation planning is a 'bioregion', which is defined by ecosystem rather than cadastral boundaries (e.g. Saunier & Maganck 1995; Breckwoldt 1996). A bioregion can be defined as (Bridgewater *et al.* 1996):

> a land and water territory whose limits are defined not by political boundaries, but by the geographical limits of human communities and ecological systems.

The scale of a bioregion will reflect the level at which planning is being conducted and the pattern of ecological variation. Australia, for example, has been partitioned into 80 bioregions for conservation planning (Thackway & Cresswell 1995). Within each bioregion, planning seeks to identify a set of priority areas which are (Australia and New Zealand Environment & Conservation Council/Ministerial Council on Forestry, Fisheries & Aquaculture 1997):

(a) *comprehensive*—all forest ecosystems are represented

(b) *adequate*—the viability and integrity of populations, species and communities are maintained, and risks of loss minimised by replication across the landscape

(c) *representative*—diversity within ecosystems is sampled by replication within the same ecosystem across the landscape

(d) *efficient*—the network of priority areas meets the three criteria above according to any efficiency

Box 21.3 Applying conservation planning methods: an example from Papua New Guinea

(from C.R. Margules, CSIRO Australia, in Kanowski et al. 1999)

With the support of the World Bank and assistance from CSIRO and Australian agencies, the Papua New Guinea Department of Environment and Conservation embarked on a conservation planning process using the BioRap (Margules & Readhead 1995) suite of conservation planning tools. The objectives of the planning process are to:

1   identify a set of areas which together represent the biodiversity of the country

2   in doing so, minimise potential conflict with forestry and agriculture, avoid areas of high land use intensity and high population density, incorporate existing protected areas, and give preference to areas of high conservation value nominated by an expert panel.

The first step was to identify and then map biodiversity surrogates (parameters which act as proxies for biological diversity) and store these in a data base. A total of 1239 biodiversity surrogates were selected, comprising:

1   608 'environmental domains', reflecting different combinations of climate, landform and geology

2   a further 621 surrogates derived from 208 vegetation types and 10 bioclimatic regions

3   10 species groups representing 87 species from specialist assessment of museum and herbarium records.

Other information used in the planning process includes:

1   Opportunity costs
•   estimates of timber volume

•   agricultural potential

2   Commitments
•   existing protected areas

•   rare and threatened species

•   areas of endemism

3   Areas to be excluded
•   areas with high land use intensity

•   areas with high population density

4   Preferred areas
•   areas with low land use intensity

•   areas with low population density

•   areas identified by experts as having high conservation value.

This process identified a set of conservation priority areas, which together:

•   represent all biodiversity surrogates

- incorporate existing protected areas

- minimise forgone forestry and agricultural opportunities

- sample all areas of bird endemism and all rare and threatened species

- avoid areas of high land use intensity.

criteria specified (two such criteria might be the minimisation of area and of opportunity cost).

This approach to conservation planning begins with stakeholders agreeing on a set of regional goals for forest conservation; for example, that 10% of each ecosystem should be protected, by whatever mechanisms are most appropriate. The region is then divided into smaller spatial units, for example, grid cells, catchments or some other mapped unit. The contribution each of these units makes to the various goals is then measured or estimated, and—in consultation with stakeholders—priorities for the use of each are assigned according to these contributions (e.g. Box 21.3).

These priority areas for *in situ* conservation will occur across the landscape and, most likely, on a range of tenures: some may be in existing protected areas, whereas others will be on forests outside protected areas—under public, communal, private or other ownership or control and subject to various forest management regimes. Their identification in the conservation planning process allows informed consideration and negotiation about how their genetic diversity might best be conserved, for example, through the establishment of reserves, or the modification or maintenance of particular forest management regimes.

Four common elements are evident from experience with these conservation planning processes (Kanowski *et al.* 1999):

(a) they work best when they make use of all sources of data, including traditional and local knowledge

(b) recent technological advances—especially in Geographic Information Systems—are revolutionising our capacity to capture data describing forest values and represent it in easily interpretable ways

(c) they provide a powerful means of informing all interested parties about priorities and options for achieving conservation goals

(d) supportive policy frameworks and institutional processes are essential if the values held by all interested parties are to be recognised and incorporated in the conservation planning process.

21.2.5.2 Integration of on-reserve and off-reserve management

Recognition of the importance of forests and trees outside reserves—sometimes defined as *circa situ* to emphasise its distinction from conservation within reserves (Kanowski & Boshier 1997; Hughes 1998)—in contributing to conservation objectives has focused attention both on the sustainable management of off-reserve forests and the integration of on-reserve and off-reserve management. The concept of a spectrum or continuum of contributions from different parts of the forest landscape has been used by, among others, Pressey and Logan (1997) and Kanowski *et al.* (1999) to describe the process of achieving conservation objectives on a bioregional scale.

Consequently, Kanowski *et al.* (1997, 1998) have argued that the principles of landscape ecology and of adaptive management (e.g. Margules & Lindenmayer 1996; Kohm & Franklin 1997; Ludwig *et al.* 1997), which recognise the importance of the biogeographical context and the limits to knowledge, define essential elements of off-reserve forest management. These principles include:

(a) the maintenance or restoration of connectivity between protected areas

(b) the maintenance of heterogeneity across the forest landscape

(c) the maintenance of structural complexity and floristic diversity within forest stands

(d) the use of an array of management strategies implemented at different spatial scales

(e) the state of processes that generate and maintain genetic structure and diversity (e.g. Stork *et al.* 1997).

The translation of these principles to operational forest management regimes poses many challenges, but one to which researchers and forest managers are beginning to respond (e.g. Boyle & Boontawee 1995; University of British Columbia & Universiti Pertanian Malaysia 1996; Hale & Lamb 1997; Bachman *et al.* 1998). For example, Pressey & Logan (1997) demonstrated how conservation planning tools could be adapted to assist in the development of off-reserve forest management regimes; Halladay & Gilmour (1995) reviewed the role of traditional agroecosystems in contributing to conservation objectives; suites of indicators (e.g. Namkoong *et al.* 1996; Stork *et al.* 1997; Montreal Implementation Group 1998; Saunders *et al.* 1998) are being proposed and tested, under the auspices of a number of international processes (Wijewardana *et al.* 1997), to assess the impacts of forest management on forest genetic diversity.

21.3 Conclusions

The dominant political theme associated with the conservation of forest genetic diversity is that which permeates the CBD, debates about the management of protected areas and of forests outside those areas, and the dialogue in various forums about the relationship between trade and the environment: that of the tensions, and potential synergies, between the conservation and use of biological diversity. In addition to agreement of the CBD and associated negotiations between its Conference of Parties, policy responses have included the development of national strategies for biodiversity conservation, concerted attempts to protect a greater proportion of remaining natural forests *in situ*, the integration into conservation strategies of regimes for sustainable forest management, and the proposal of suites of

criteria and indicators against which to assess the conservation of forest genetic diversity.

Recent advances in our understanding of ecology and genetics are dramatically improving our capacity to identify the key scientific elements of effective genetic conservation strategies—helping to resolve issues such as optimum reserve location and design, the most appropriate silvicultural regimes and the respective and complementary contributions of various *in situ* and *ex situ* measures. Although these advances have highlighted the limitations of our knowledge, they are also informing us about the most effective use of human and financial resources to achieve genetic conservation objectives.

Scientific advances have been paralleled by the evolution of our thinking about how best to achieve these conservation objectives in the context of diverse social, economic and political circumstances. Locally appropriate responses, consistent with the international regimes defined by the CBD and World Trade Organisation, are emerging as a result. New forms of partnership between many of the actors with interests in forests, which recognise the diversity of roles and contributions, are especially important in delivering conservation outcomes. Perhaps our greatest policy challenge in the conservation of forest genetic diversity is creating the space for these locally appropriate responses to develop in a world in which both commerce and regulation are increasingly globalised.

21.4 Acknowledgment

Reviews by David Boshier, Ken Eldridge and Christel Palmberg-Lerche assisted greatly in improving this Chapter. Errors and omissions remain my responsibility.

This chapter is dedicated to Professor Gene Namroong; a version of it was presented at the symposium in his honour, *Unifying Perspectives of Evolution, Conservation and Breeding*, in July 1999.

Forest Conservation Genetics: Limitations and Future Directions

David H. Boshier and
Andrew G. Young

Summary

Maintenance of tree populations with sufficient genetic variation to adapt to future conditions, whether in reserves or in production systems, is essential. Given the limited resources available for conservation, genetic criteria will, however, generally form only one component within strategies that reflect the integral nature of factors such as demography, ecology and genetics, as components of natural systems. The circumstances under which genetic aspects may be limiting are discussed, while recognising the limits to the use of genetic information and the need for integrated conservation strategies. Limitations of current approaches and knowledge are also considered along with the direction of more recent and future work such as: genetic adaptation of trees within more complex forest environments, the integration of metapopulation and demographic models and modelling of the effects of climate change. The need for more research is highlighted, with the generation of baseline information, as well as the use of a range of model species from which species may be grouped into management guilds being identified as priorities.

22.1 Introduction

In drawing up effective conservation strategies, principles of conservation genetics need to be interpreted and applied to specific ecosystems, species and socioeconomic conditions. The predominance of national parks or other protected lands in ecologically marginal areas such as at higher elevations, or on poor soil, or their selection on the basis of other conservation goals such as the protection of charismatic megafauna, means that only a limited and biased proportion of a tree species' gene pool may be conserved (Ledig 1988) (Fig. 22.1 and see Fig. 10.2). To avoid these problems, forest managers are now advocating and developing management strategies that integrate the conservation of forest genetic resources within production systems (e.g. Riggs 1990; Kuusipalo & Kangas 1994; Namkoong 1994) and emphasise the roles of remnant forest patches and trees on farms (see Chapter 20). However, given the limitation of resources for conservation as a whole, genetic criteria will generally form only one component of an overall conservation or management strategy. We therefore conclude this book by considering under what circumstances genetic aspects may be limiting and, conversely, reviewing the limits to the use of genetic information. This chapter also discusses the limitations of knowledge and application, and where such work may lead in the future.

22.2 Limits of genetics

The importance or even relevance of genetics to conservation issues has formed fertile ground for discussion. Various authors have expounded the view that ecological, demographic and stochastic issues are of greater importance than genetic ones for endangered species (e.g. Lande 1988; Caro & Laurenson 1994). Certainly the responses to, and the consequences of, changes and loss of genetic diversity are on a longer time scale to most demographic or ecological factors and genetic information alone may only rarely translate automatically into definitive conservation strategies (Avise 1996).

However, as mentioned in Chapter 5, the dangers of inbreeding in many tree species are clear, with reduced fertility, slower growth rates in progeny, limited environmental tolerance and increased susceptibility to pests or diseases. Extreme losses of genetic diversity may also lead to problems, such as for *Trochetiopsis*

erythroxylon (Forst. f.) Marais (St Helena redwood), which through overexploitation for timber was reduced to one individual and now exhibits a shrub rather than tree form (Rowe & Cronk 1995). Small populations of some angiosperm tree species may also be particularly susceptible to loss of incompatibility alleles (see Chapter 5), which may then directly threaten a population's viability through most pollinations being incompatible—leading to a reduction or failure in seed production. Similarly genetic threats to rare native tree populations may arise by pollen contamination from a more common congener [e.g. *Cercocarpus traskiae* Eastw. (Catalina Island mahogany), Rieseberg & Swensen 1996; see also Chapter 13] or urban tree plantings (e.g. *Pinus radiata* D. Don; Eldridge 1995, 1998).

There are different reproductive strategies among forest tree species and identifying the potential for management to critically alter or endanger these processes is key. Reducing the possibility or effect of inbreeding and maintaining diversity in naturally outcrossing tree species will be important, while maintenance of breeding system flexibility will be a priority for species that naturally combine outcrossing and inbreeding. To determine the importance of observed levels of genetic diversity and inbreeding requires the establishment of baseline information from which deviations can be measured. Studies are often compromised, however, by the ability to decide on a realistic baseline population from which change can be observed and to understand the significance of any observed changes. In addition, populations may rarely be genetically endangered except in extreme circumstances, and loss of genetic diversity may often be a symptom rather than the cause of endangerment. Any increased levels of inbreeding may be unimportant from an evolutionary viewpoint, with selfed individuals selected against at various stages of regeneration. Increased levels of inbreeding may, however, be critical in terms of the levels of diversity sampled for *ex situ* conservation, tree breeding or plantation programmes.

Ultimately the largest cause of biodiversity loss is habitat destruction, which unless halted renders identification of genetic problems ultimately irrelevant (Holsinger 1996). To portray genetic perspectives as inherently in opposition to other conservation outlooks is, however, to view the field too narrowly and disregard the integral nature of factors, such as demography, ecology and genetics, as components of

Figure 22.1   The predominance of protected areas in hilly regions leads to the conservation of a biased proportion of a tree species' gene pool. Rice fields established following elimination of forest between Lomas Barbudal Biological Reserve (near side) and Paloverde National Park (far side) in Costa Rica. (Photograph: David Boshier).

natural systems. Such a narrow view tends to implicitly associate genetics with nothing more than heterozygosity assessment, which in itself not only can say little about population viability (Avise 1996), but also fails to recognise that just as genetic diversity exists at various levels (see Chapter 3), so knowledge of it and its dynamics under different conditions, both natural (see Chapters 4 & 5), or artificial (see Chapters 14 & 15), offers a multi-faceted insight to conservation biology. Resource limitations similarly dictate the need for integrated approaches at a variety of levels, for example:

(a) incorporation of genetic criteria into more general management procedures

(b) extrapolation of appropriate strategies for most taxa from the results of studies of relatively few—the 'model species' approach

(c) identification of under what circumstances genetic aspects will or may become limiting for what type of species

(d) use of genetic data to inform on demographic processes.

A good example of this last role, which is becoming increasingly important, is in the study of the long-term demographic trends of threatened plant species. Demographic processes are crucial to reductions in population size and hence demographic properties (e.g. mean and variance of population growth rate) of small populations are of particular interest and have formed a basis for analysing population viability of rare plant species (e.g. Menges 1990, 1992). Demographic parameters (e.g. population size, mating system) vary, however, from generation to generation, and consequently, direct demographic observations over the short term may not provide a representative view of the long-term situation. A current small population size may be unimportant to a population that has always experienced low numbers, whereas it may be critical to one that has only recently become small. Although long-term demographic studies can provide good insight into population trends, costs and immediate threats to populations dictate a need for rapid, informed solutions. As demographic properties of populations influence their genetic properties, information from genetic markers can go beyond marker diversity and provide estimates of demographic parameters of interest to conservation biologists.

For example, tests have been developed which discriminate recently bottle-necked populations from

291

stable populations, using allele frequency data to determine whether a population exhibits a significant number of loci with heterozygosity excess (BOTTLENECK program; Cornuet & Luikart 1996; Luikart & Cornuet 1998). Populations that have experienced a recent reduction in their effective population size show a correlated reduction in allele numbers and heterozygosities at polymorphic loci. Allelic diversity is, however, reduced more quickly than heterozygosity, resulting in a larger observed heterozygosity than that expected from the observed allele number. Thus, in a population where the effective size has remained constant in the past, there is approximately an equal probability that a locus shows an heterozygosity excess or an heterozygosity deficit (detailed discussion in Maruyama & Fuerst 1985; Cornuet & Luikart 1996). Genealogical analysis of DNA sequence data can also be used to quantify important demographic parameters (e.g. effective population size, mating system, migration rate). Owing to the historical nature of sequence change information, the analysis reflects the long-term demographic properties of the population and can therefore be used to infer long-term demography (Milligan & Strand 1996).

Comparisons between such long-term estimates and current demographic observations can be used to elucidate if, and how, current trends deviate from historical ones, information essential to the evaluation and management of rare plant species. Given the scarcity of resources available for conservation, the integration of population genetics and demography, with the aim of quantifying both properties of populations, and historical trends in those properties, has much to offer conservation biology.

22.3 Neutral and adaptive genetic variation

This book has emphasised the use of neutral genetic markers to inform on the dynamics of genetic processes within natural and human-influenced tree populations, as a guide to conserving genetic diversity and population integrity. In many situations, effective conservation requires both ecological restoration and 'conservation through use' of trees in agroecosystems (see Chapter 20), with the use of local seed sources often stressed, particularly with regard to the adaptation of tree populations (Linhart 1995). Where information is very limited, discussion of suitable seed sources may acquire a nationalistic air, rather than

being based on hard evidence for the scale of adaptation and hence the definition of 'local'. Understanding the relationships between a species' genetic diversity, the heterogeneity of its habitat and scale of adaptation, is therefore important for its conservation and future adaptability to changing environmental conditions. The maintenance of genetic variation, enabling adaptability to a range of environmental conditions over both space and time should be an important part of any conservation efforts involving trees, rather than the absolute genetic state, which by its very nature is in constant flux. The scale over which species show adaptation to their environment inevitably depends on the degree of heterogeneity of the specific habitat characteristics which affect a species, and the amount of gene flow. Gene flow may counteract even fairly strong levels of selection, preventing formation of locally adapted populations, although very strong environmental variation (hence selection pressure), may produce adaptive differences over short distances, despite continued high levels of gene flow (e.g. *Eucalyptus urnigera* Hook. f. over a 450 m altitudinal transect on Mount Wellington, Tasmania, Barber & Jackson 1957; Barber 1966).

Commercial tree species, unlike agronomic crops, mostly have little history of domestication, long rotation times, and are often planted in 'natural' settings where genetic diversity is important for adaptability, as well as increased productivity. As such, forestry trials can be informative about patterns of variation in response to habitat heterogeneity (Millar & Westfall 1992). Provenance and progeny trials have provided information about levels and patterns of quantitative genetic variation and the extent of genotype–environment interaction for a wide variety of temperate and tropical tree species. In both temperate and tropical trees most morphological genetic variation occurs within rather than between provenances, with evidence for adaptive variation over reasonably short distances in many temperate tree species, owing to features such as aspect and altitude (e.g. Adams & Campbell 1981; Zobel & Talbert 1984). In most tropical trees there is little evidence from provenance trials for or against adaptive variation over very short distances, ranking reversals (adaptation) or significant genotype by environment interaction occurring only with large environmental site changes (e.g. dry/wet zones, alkali/acidic soils).

In looking principally at production variables of interest to foresters, and not at fitness-related traits, provenance trials have, however, inherent limitations in providing information on adaptive variation. These limitations become increasingly important as interest grows in ecological restoration, where both planting and natural regeneration are involved. In particular they are poor at informing about survival at the seed/seedling phase, and competition among species, while the initial seed collections often avoid poorly formed or diseased trees, such that there is inherent bias in the sampling of populations. In commercial forestry, the low chance of encountering a rare, extreme environment (e.g. a 1 in 20 year frost) may not limit production under short rotations, whereas such events may be influential in adaptation of natural populations. With much research funded on a short-term basis, and given their long life cycles, the potential to overlook important long-term adaptation in trees is great. The type of species studied is also often limited to fast-growing pioneer or long-lived pioneers, such that there is a lack of information on shade-tolerant, slow-growing, climax tropical tree species.

Reciprocal transplant experiments, which test directly for localised adaptation to environmental heterogeneity by testing the fitness of 'home' and 'away' genotypes within the sites from which the genotypes originate (Primack & Kang 1989) may be of increasing usefulness in studying adaptive variation in trees. In subjecting the plants to more natural conditions than in most provenance trials, it is possible to study responses to natural processes within the environment at each site, including competition with the native flora. Although most reciprocal transplant experiments of plants have identified localised adaptation at the finest scale examined (e.g. Antonovics & Bradshaw 1970; Hickey & McNeilly 1975; Waser & Price 1985; Kindell *et al.* 1996), few truly reciprocal transplant experiments have been undertaken on trees (e.g. Sork *et al.* 1993; Rice *et al.* 1997) and currently no generalisations can be made. Reciprocal transplant experiments also face problems of interpretation in that given the long life of trees and the rapid changes to habitats experienced under deforestation, the environment may also have altered, such that the source site no longer represents the conditions under which the population evolved.

As seed composition is affected by patterns of pollen flow, the detection of localised adaptation may also be buffered by the extent of pollen flow from differing environments, which may be considerable when adjacent populations differ over a short distance, or where pollen flow changes as a result of human disturbance. Increased levels of pollen flow between populations (see Chapters 10 & 16), also raise the possibility of outbreeding depression from the break up of coadapted allelic complexes, or dilution of adapted alleles, although current evidence suggests that such negative effects may only occur at extreme distances (Hardner *et al.* 1998; Boshier & Billingham 2000).

Highly species-diverse tropical forests, where competition between plant species is intense, present more complex conditions for the evaluation of adaptation compared to the relatively monospecific temperate conifer forests. Sampling problems associated with the wide dispersion of individuals, environmental heterogeneity, high genetic variability of trees, and the need for sample sizes sufficient to detect statistically significant differences, make testing for genetic adaptation over short distances (e.g. < 1 km) problematic. Given the evidence for extensive gene flow in trees (see Chapter 5), however, high selection pressures will be required to produce adaptive differences over such short distances.

Despite the practical difficulties and expense associated there is a clear need for more studies that examine genetic adaptation of trees within more complex forest environments. Future studies need to look at among others:

(a) adaptation over a greater ecological and reproductive range of species

(b) adaptation in tropical trees over a mosaic of spatial scales, focussing particularly on where abrupt habitat changes occur within the pollen dispersal range

(c) adaptation at seed and seedling phase, particularly under conditions of natural regeneration and interspecific competition (Boshier & Billingham 2000).

22.4 Integration of information

In some species there may be large genetic differences among populations across the geographical range at the molecular level (e.g. because of genetic drift), but few significant adaptive differences. In other cases there may be adaptive differences among populations, although no molecular differences. For example, over its north to south range (about 1000 km), populations of Scots pine in Finland show high levels of heterozygosity (using a range of molecular markers) and low differentiation (2%; Hedrick & Savolainen 1996). In contrast, common garden and transplant experiments show the northern and southern populations to be highly differentiated with respect to several traits that confer adaptation to the severe conditions (e.g. date of first budset—34% north–south differentiation; Karhu et al. 1996), with the probability of a seed crop less than 5% in the north, as the growing season is generally too short for seed maturation (Henttonen et al. 1986). Furthermore, experiments in Sweden showed that survival is reduced by 7% per degree increase in latitude and 16% per 100 m increase in altitude (Persson & Stahl 1990).

These differences are because molecular markers are primarily influenced by gene flow and mating and so reflect patterns of coancestry. In contrast, adaptive differences are determined by differences in selection regime. Thus, different types of genetic information are providing data about the spatial dynamics of different genetic processes. Just as there is a need to integrate management for a variety of purposes, so there is a need to integrate the diverse sources of genetic information to help in making informed management decisions.

A good example of the integration of different kinds of genetic information is the field, greenhouse and laboratory studies undertaken in north-western United States of America (USA), for a range of conifer species. These showed that a significant proportion (typically 25–45%) of the genetic variation within populations was accounted for by climatic variables (e.g. mean annual rainfall and temperature) or location (e.g. latitude, altitude, slope aspect), reflecting a range of environmental factors specific to each location. Consequently, problems in transferring seed for planting elsewhere within a species distribution were correlated more with environmental changes than the geographical distance moved (Adams & Campbell 1981). Much work went into the integration of information (models developed using regression, principal components and classification analysis) to assess the relative risks of moving seed from one location to another and the subsequent development of seed zones (e.g. Campbell & Sugano 1993; Sorensen 1994). Other examples of such integration of information have taken a variety of forms, with

varying levels of sophistication (e.g. Genetic Management Resource Units/Genetic Conservation Areas in north-western USA, Millar & Westfall 1992; Millar & Westfall 1996; seed zones for *Tectona grandis* L.f. in Thailand, Graudel et al. 1997).

22.5 Model species to modelling

Integration of information may be, as above, at the genetic level for a specific species. Generally, however, genetic information is only one input into management options. To fully understand the relative merits of different management strategies within a site, there is a need to integrate genetic, ecological, silvicultural and sociological information to reflect the complex interactions of natural systems and issues of sustainable management. The lack of information, the inevitable limitation of resources and the need for more immediate and general action in many situations will necessitate pragmatic 'best guess' approaches to conservation strategies. Such limitations also dictate the need for studies that permit extrapolation of appropriate strategies for most taxa from results on relatively few (Moran & Hopper 1987; National Research Council 1991). Simulation models such as ECO-GENE (Degen et al. 1996), which have been developed to study the effect of silvicultural treatments in temperate forests, combine population genetic processes with forest growth models to simulate the effect of management types and intensities on within-species genetic diversity and species composition.

Adaptation and validation of such models to more complex tropical forest conditions will require extensive reproductive and regeneration ecology and genetic data on selected model species. Species selection criteria should take into account mating system, spatial distribution and ecological 'guild'. Many timber species fall into the ecological groupings of pioneer or long-lived, shade-intolerant trees. The choice of model species should not, however, be specifically limited to commercial species, or the ability to model the effects of management on a range of species types will be compromised. Many pioneer or long-lived, shade-intolerant trees, by their nature are likely to be genetically fairly resilient to disturbance. The species most likely to be genetically susceptible to management are probably the rare, slow-growing, shade-intolerant species, with only occasional seed production, reaching reproductive age late and having specialised pollination syndromes.

Although information exists on the reproductive ecology, genetics and mating patterns of several tropical species, there are relatively few studies of the effect of management practices on these processes (e.g. see Chapters 15 & 16). Such studies to date are equivocal as to whether effects are negative, positive or neutral, but generally show altered patterns of gene flow. The indirect measure of gene flow (Nm, see Chapters 3 & 6), reflects only the past status of populations, and will be slow to reflect any changes owing to management. Its use as the only basis for estimating the susceptibility of populations to management is, therefore, extremely limited. Use of existing data from unmanaged or undisturbed forests is therefore unlikely to produce a valid model, making it difficult to identify critical scenarios and make accurate predictions. The validation of models and their subsequent ability to accurately model the effects of management on genetic diversity depends on the generation of new data, with direct estimates of pollen flow within populations, under logging conditions.

Identification, by a validated model, of the factors that leave species genetically susceptible to management offers the possibility to group species into management guilds (e.g. combination of ecological grouping, spatial distribution, aspects of reproductive biology) and the extension of guidelines to a much greater range of species. The ability to extrapolate from the modelling results to make more general management recommendations for groups of species will depend on the existence of good basic biological information that enables species to be so classified.

The potential of models to develop and incorporate genetic indicators for assessing sustainable management into management guidelines will also depend on the full integration with management operations, so that genetic indicators are developed, with subsequent testing and feedback, that are useable day-to-day. In this context, indigenous knowledge also has a potentially important role:

(a)   as a source of technical information

(b)   providing insights into indigenous management practices (e.g. levels of fruit harvesting, selective management for female trees in dioecious species), and their impact on genetic resources.

For many populations, forest loss and fragmentation has created metapopulations out of formerly continuous populations with associated extinction and recolonisation dynamics within landscape mosaics, where 'ecological distance' between populations and direct physical distance may have divergent implications for gene flow. The integration of landscape models, which use spatially explicit information about the mosaic of habitat types, with metapopulation models that describe a set of connected populations occurring within a landscape, offer a means to examine the influence of different landscapes and habitats on genetic processes and structure of populations (e.g. Antonovics et al. 1997; Steinberg & Jordan 1997; Wiens 1997; Sork et al. 1998). Connecting genetic and demographic models at landscape scales requires the adoption of more relevant scales of study than those over which migration is currently measured, and ones that are sensitive to recent changes in gene flow (both pollen and seed). Direct parentage analysis based methods estimate current rates of gene flow but have generally been applied over modest spatial scales, while recent studies suggest that under fragmentation pollen flow distances may increase by factors greater than 10 (e.g. White & Boshier 2000). Current analytical models have not, however, been used to their maximum capabilities to resolve such long-distance pollen dispersal events (Nason 1998). In addition, while pollen and seed movement may influence genetic structure differentially, from the perspective of plant population demographic processes (i.e. colonisation) in metapopulation and landscape models, seed dispersal data may be as important as pollen dispersal data. Differences in the inheritance patterns of molecular markers permit direct comparisons between the relative levels of gene flow owing to pollen and seed dispersal.

Consequences for longer term adaptation have been addressed by Mátyás (1994, 1996) and Schmidtling (1994, 1997), who advocated the use of old provenance trials to study the possible effect of climate change on growth of tree species. Some research has predicted productivity increases in boreal forests as a response to global warming, although the models used assume genetic adaptation of the trees to the new climate (e.g. Kauppi & Posch 1988). Given the rate at which such climatic change may occur, the ability of trees to respond by migration or selection may be limited. Growth of Picea abies (L.) H. Karst. (Norway spruce), Pinus taeda L. (loblolly pine), and three other USA pines in long-running provenance tests were interpreted using regression models to relate growth to

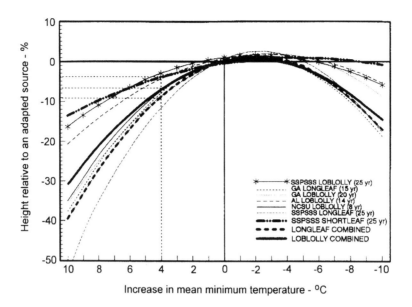

Figure 22.2 Height deviations from 'local' sources versus change in mean minimum temperature (source to planting), for three southern pine species (Reproduced by permission. From *Ecological Issues and Environmental Impact Assessment* copyright © 1997, Gulf Publishing Company, Houston, Texas, 800-231-6275. All rights reserved.).

temperature variables (Mátyás 1994; Schmidtling 1997) (Fig. 22.2). Data from different trials were combined by expressing growth as a percentage deviation from the local seed source and expressing temperature at the source as a deviation from that at the planting site. With a 4°C mean temperature increase, the models predicted growth losses of about 5%–12% in height and 8%–20% volume, relative to that expected from a genetically adapted source. Therefore, although growth may increase with an increase in temperature, the results showed that the increase would be considerably less than expected unless a genetically adapted source has time to evolve, migrate or is planted by forest managers.

22.6 Conclusions

There can be little doubt that effective long-term management of forest resources is a complex task. However, it is also clear that, given the extent of human intervention, we must ensure that we leave tree populations with sufficient genetic diversity so that they can adapt to whatever future conditions may exist, thus allowing them to continue to provide essential ecosystem services such as water and nutrient cycling, and critical resources such as timber

and fruit. Exactly how we go about integrating different types of genetic, ecological and socioeconomic information, so as to be able to set thresholds for resource extraction that will ensure such sustainability and generate workable forest management policy, is less clear.

Detailed genetic studies can provide insights into both past and present population dynamics. They can also identify patterns of adaptation and genetic diversity. In the presence of all this knowledge and the absence of financial and time constraints, optimal decisions regarding genetic resource management for forests could be made. In reality, the genetic information will only rarely exist for the species of interest, and the time to get it will not be available. Furthermore, conservation decisions are not made in a vacuum. Forests are one of the most pressured resources on the planet and current patterns of utilisation will also determine what can be conserved and what can not. This will often depend on the economic and social importance of forest to local people.

Given these constraints, any sensible way forward for the development of forest genetic conservation strategies must involve at least four elements.

(a) Continued accumulation of accurate baseline biological data on a broad range of forest plant species.

(b) Detailed genetic and demographic studies of a few 'model species' that are ecologically and evolutionarily representative of the range of life histories that exist. Data regarding the response of these species to forest changes can then be extrapolated to other species with which they share characteristics such as pollination or dispersal syndromes.

(c) Recognition of the need to integrate conservation and resource extraction objectives, within a local socioeconomic context. Forest conservation genetics is a resource management issue like any other, except it is arguably more important than most because genetic variation is the basic material of future evolution and often a fundamental human resource.

(d) Percolation of genetic, ecological and socioeconomic realities upwards into policy development and deployment.

22.7 Acknowledgments

Work by David Boshier on this publication is an output from research partly funded by the United Kingdom Department for International Development (DFID) for the benefit of developing countries. The views expressed are not necessarily those of DFID. Projects R5729, R6516 Forestry Research Programme.

Aagaard, J.E., Vollmer, S.S., Sorensen, F.C. & Strauss, S.H. 1995. Mitochondrial DNA products among RAPD profiles are frequent and strongly differentiated between races of Douglas-fir. *Molecular Ecology*, 4, 441–7.

Abernethy, K. 1994. The establishment of a hybrid zone between red and sika deer (genus *Cervus*). *Molecular Ecology*, 3, 551–62.

Ackerman, J.D., Mesler, M.R., Lu, K.L. & Montalvo, A.M. 1982. Food foraging behavior of male Euglossini (Hymenoptera: Apidae): vagabonds or trapliners? *Biotropica*, 14, 241–8.

Adalgisa, F. 1985. Castanheira anã, um gigante da Amazônia. *Globo Rural*, 1, 36–41.

Adams, T. & Campbell, R.K. 1981. Genetic adaptation and seed source specificity. In *Reforestation of Skeletal Soils: Proceedings of a Workshop*. Eds S.D. Hobbs & O.T. Helgerson. 17–19 November 1981, Medford, OR. pp. 78–85. Forest Research Laboratory, Oregon State University, Corvallis, OR, USA.

Adams, W.T. 1981. Population genetics and gene conservation in Pacific Northwest conifers. In *Evolution Today*, Proceedings of the Second International Conference of Systematic and Evolutionary Biology. Eds G.G.E. Scudder & J.L. Reveal. pp. 401–15. Hunt Institute for Botanical Documentation, Carnegie-Mellon University, Pittsburgh, PA, USA.

Adams, W.T. 1992. Gene dispersal within forest tree populations. *New Forests*, 6, 217–40.

Adams, W.T. & Birkes, D.S. 1989. Mating patterns in seed orchards. In *Proceedings of the 20th Southern Forest Tree Improvement Conference*, pp. 75–86. 26–30 June 1989, Charleston, South Carolina, copies available from The National Technical Information Service, Springfield, VA, USA.

Adams, W.T. & Birkes, D.S. 1991. Estimating mating patterns in forest tree populations. In *Biochemical Markers in the Population Genetics of Forest Trees*. Eds S. Fineschi, M.E. Malvolti, F.H.H. Cannata & S.P.B. Hattemer. pp. 157–72. Academic Publishing, bv, The Hague.

Adams, W.T., Birkes, D.S. & Erickson, V.J. 1992a. Using genetic markers to measure gene flow and pollen dispersal in forest tree seed orchards. In *Ecology and Evolution of Plant Reproduction*. Ed. R. Wyatt. pp. 37–61. Chapman and Hall, New York.

Adams, W.T., Hipkins, V.D., Burczyk, J. & Randall, W.K. 1997. Pollen contamination trends in a maturing Douglas-fir seed orchard. *Canadian Journal of Forest Research*, 27, 131–4.

Adams, W.T., Neale, D.B., Doerksen, A.H. & Smith, D.B. 1991. Inheritance and linkage of isozyme variants from seed and vegetative bud tissues in coastal Douglas-fir (*Pseudotsuga menziesii var. menziesii* (Mirb.) Franco). *Silvae Genetica*, 39, 153–67.

Adams, W.T., Strauss, S.H., Copes, D.L. & Griffin, A.R. 1992b. *Population Genetics of Forest Trees*. Kluwer Academic Publishers, Dordrecht.

Ager, A.A. & Guries, R.P. 1982. Barriers to interspecific hybridization in *Ulmus americana*. *Euphytica*, 31, 909–20.

Aitken, S.N. & Adams, W.T. 1995. Impacts of genetic selection for height growth on annual developmental cycle traits in coastal Douglas-fir. In *Evolution of Breeding Strategies for Conifers from the Pacific Northwest*. P. AD-1 Proceedings, IUFRO Conference, 31 July–4 August 1995, Limoges, France.

Akerman, S., Tammisola, J., Regina, M., Kauppinen, V. & Lapinjoki, S. 1996. Segregation of AFLP markers in *Betula pendula* (Roth). *Forest Genetics*, 3, 117–23.

Allard, R.W. 1970. Population structure and sampling methods. In *Genetic Resources in Plants—Their Exploration and Conservation*. Eds O.H. Frankel & E. Bennett. pp. 97–107. Blackwell Scientific Publications, Oxford.

Allard, R.W. & Workman, P.L. 1963. Population studies in predominantly self-pollinated species. IV. Seasonal fluctuations in estimated values of genetic parameters in lima bean populations. *Evolution*, 17, 470–80.

Allendorf, F.W. & Leary, R.F. 1988. Conservation and distribution of genetic variation in a polytypic species, the cutthroat trout. *Conservation Biology*, 2, 170–84.

Alvard, M.S. 1993. Testing the 'ecologically noble savage' hypothesis: interspecific prey choice by Piro hunters of Amazonia Peru. *Human Ecology*, 21, 355–87.

Amir, H.M.S., Mona, Z., Ghazali, M.H. & Rozita, A. 1989. Nutrient dynamics of Tekam Forest Reserve, Peninsular Malaysia, under different logging phases. *Journal of Tropical Forest Science*, 2, 71–80.

Anderson, E. 1948. Hybridization of the habitat. *Evolution*, 2, 1–9.

Andersson, B. & Danell, O. 1997. Is *Pinus sylvestris* resistance to pine twist rust associated with fitness costs or benefits? *Evolution*, 51, 1808–14.

Anonymous 1976. *Twentieth Annual Report on Cooperative Tree Improvement and Hardwood Research Program*. North Carolina State University, Raleigh, NC, USA.

Antolin, A.T. 1995. Rattan: a source of employment for upland communities of Northeastern Luzon. In *Beyond Timber: Social, Economic and Cultural Dimensions of Non-Wood Forest Products in Asia and the Pacific*. Eds P.B. Durst & A. Bishop. pp. 55–60. Proceedings of a Regional Expert Consultation held in Bangkok, 28 November–2 December 1994. Food and Agriculture Organization of the United Nations, Bangkok.

Antonovics J. & Bradshaw, A.D. 1970. Evolution on closely adjacent plant populations. VIII. Clinal patterns at a mine boundary. *Heredity*, 25, 349–62.

Antonovics, J., Thrall, P.H. & Jarosz, A.M. 1997. Genetics and the spatial ecology of species interactions: the *Silene-Ustilago* system. In *Spatial Ecology: The Role of Space in Population Dynamics and Interspecific Interactions*. Eds D. Tilman & P. Kareiva. pp. 158–80. Princeton University Press, Princeton, NJ, USA.

Appanah, S. & Chan, H.T. 1981. Thrips, the pollinators of some dipterocarps. *Malaysian Forester*, 44, 234–52.

Apsit, V.J., Nakamura, R.R. & Wheeler, N.C. 1989. Differential male reproductive success in Douglas-fir. *Theoretical and Applied Genetics*, 77, 681–4.

Arndt, H.W. 1987. *Economic Development*. University of Chicago Press, Chicago.

Arndt, U., Fomin, A. & Lorenz, S. 1992. Key reactions in forest desease used as effects criteria for biomonitoring. In *Ecological Indicators*. Vol. 2. Eds D.H. McKenzie et al. pp. 829–40. Elsevier Applied Science Publishers, London.

Arndt, U., Fomen, A. & Lorenz, S. 1995 (Eds). *Bioindikation—Neue Entwicklungen, Nomenklatur, Synökologische Aspekte*. (Bioindication—New Trends, Nomenclature, Synecological Aspects). Verlag Günter Heimbach, Ostfildern.

Arndt, U., Nobel, W. & Schweizer, B. 1987. *Bioindikatoren—Möglichkeiten, Grenzen und neue Erkenntnisse*. (Bioindicators—Chances, Limitations and Trends). Ulmer Verlag, Stuttgart.

Arnold, M.L. 1997. *Natural Hybridization and Evolution*. Oxford University Press, Oxford.

Arnold, M.L., Hamrick, J.L., & Bennett, B.D. 1993. Interspecific pollen competition and reproductive isolation in *Iris*. *Journal of Heredity*, 84, 13–16.

Arriola, P.E. & Ellstrand, N.C. 1996. Crop-to-weed gene flow in the genus *Sorghum* (Poaceae): spontaneous interspecific hybridization between Johnsongrass, *Sorghum halepense*, and crop sorghum, *S. bicolor*. *American Journal of Botany*, 83, 1153–60.

Arriola, P.E. & Ellstrand, N.C. 1997. Fitness of interspecific hybrids in the genus *Sorghum*: persistence of crop genes in wild populations. *Ecological Applications*, 7, 512–18.

Ashton, P.S. 1969. Speciation among tropical forest trees: some deductions in the light of recent evidence. *Biological Journal of the Linnean Society of London*, 1, 155–96.

Askew, G.R. 1986. Implications of non-synchronous flowering in clonal conifer seed orchards. In *IUFRO Conference: A Joint Meeting of Working Parties on Breeding Theory, Progeny Testing and Seed Orchards*. Eds A.V. Hatcher & R.J. Weir (coordinators). pp. 182–91. The North Carolina State University–Industry Cooperative Tree Improvement Program, Williamsburg, VA, USA.

Askew, G.R. & Blush, T.D. 1990. Short note, An index of phenological overlap in flowering for clonal conifer seed orchards. *Silvae Genetica*, 39, 168–71.

Augspurger, C.K. 1983. Phenology, flowering synchrony, and fruit set of six neotropical shrubs. *Biotropica*, 15, 257–67.

Australia and New Zealand Environment & Conservation Council (ANZECC)/Ministerial Council on Forestry, Fisheries & Aquaculture (MCFFA) 1997. *Nationally Agreed Criteria for the Establishment of Comprehensive, Adequate and Representative Reserve System for Forests in Australia*. Commonwealth of Australia, Canberra.

Australian Department of Primary Industries & Energy 1997. *Montreal Process Working Group: Montreal Process Criteria and Indicators*. http://www.dpie.gov.au/agfor/forests/montreal/c-i.html

Avise, J.C. 1996. Introduction: the scope of conservation genetics. In *Conservation Genetics: Case Studies from Nature*. Eds J.C. Avise & J.L. Hamrick. pp. 1–9. Chapman and Hall, New York.

Bach, C.F. & Gram, S. 1996. The tropical timber triangle. *Ambio*, 25, 166–70.

Bachman, P., Köhl, M. & Päivinen, R. 1998. *Assessment of Biodiversity for Improved Forest Planning*. Forestry Sciences 51, Kluwer Academic Publishers, Dordrecht.

Bacilieri, R., Ducousso, A., Petit, R.J. & Kremer, A. 1996. Mating system and asymmetric hybridization in a mixed stand of European oaks. *Evolution*, 50, 900–8.

Baharuddin, K. & Rahim, A.N. 1994. Suspended sediment yield resulting from selective logging practices in a small watershed in Peninsular Malaysia. *Journal of Tropical Forest Science*, 7, 286–95.

Bahl, A., Loitsch, S.M. & Kahl, G. 1993. Air pollution and plant gene expression. In *Plant Responses to the Environment*. Ed. P.M. Gresshoff. pp. 71–96. CRC Press, Boca Raton, FL, USA.

Bakhtiyarova, R.M., Starova, N.V. & Yanbaev, Y.A. 1995. Genetic changes in populations of Scots pine growing under industrial air pollution conditions. *Silvae Genetica*, 44, 157–60.

Balachander, G. 1995. Fire, grazing, and extraction of non-timber forest products in the Nilgiris Biosphere Reserve, Southern India: implications for forest policy, sustainable use and local economies. Dissertation submitted to the Graduate School, New Brunswick, Rutgers, NJ, USA.

Balduman, L. 1995. Variation in Adaptive Traits of Coastal Douglas-fir: Genetic and Environmental Components. MSc thesis, Oregon State University, USA.

Ballal, S.R., Foré, S.A. & Guttman, S.I. 1994. Apparent gene flow and genetic structure of *Acer saccharum* subpopulations in forest fragments. *Canadian Journal of Botany*, 72, 1311–15.

Band, M. & Ron, M. 1997. Heterozygote deficiency caused by a null allele at the bovine ARO23 microsatellite. *Animal Biotechnology*, 8, 187–90.

Baradat, Ph., Adams, W.T. & Müller-Starck, G. (Eds) 1995a. *Population Genetics and Gene Conservation of Forest Trees*. SPB Academic Publishing bv, The Hague.

Baradat, Ph., Maillart, M., Marpeau, A., Slak, M.F., Yani, A. & Pastiszka, P. 1995b. Utility of terpenes to assess population structure and mating patterns in conifers. In *Population Genetics and Genetic Conservation of Forest Trees*. Eds Ph. Baradat, W.T. Adams & G. Müller-Starck. pp. 5–27. SPB Academic Publishing, Amsterdam.

Baranger, A., Chevre, A.M., Eber, F. & Renard, M. 1995. Effect of oilseed rape genotype on the spontaneous hybridization rate with a weedy species: an assessment of transgene dispersal. *Theoretical and Applied Genetics*, 91, 956–63.

Barber, H.N. 1966. Selection in natural populations. *Heredity*, 20, 551–72.

Barber, H.N. & Jackson, W.D. 1957. Natural selection in action in *Eucalyptus*. *Nature*, 179, 1267–9.

Barbier, E.B., Burgess, J.C. & Folke, C. 1994. *Paradise Lost? The Ecological Economics of Biodiversity*. Earthscan Publications, London.

Barbier, E., Burgess, J., Swanson, T. & Pearce, D.W. 1990. *Elephants, Economics and Ivory*. Earthscan Publications, London.

Barrett, C., Lefort, F. & Douglas, G.C. 1997. Genetic characterization of oak seedlings, epicormic, crown and micropropagated shoots from mature trees by RAPD and microsatellite PCR. *Scientia Horticulturae*, 70, 319–30.

Barrett, S.C.H. & Bush, E.J. 1991. Population processes in plants and the evolution of resistance to gaseous air pollutants. In *Ecological Genetics and Air Pollution*. Eds G.E. Taylor Jr., L.F. Pitelka & M.T. Clegg. pp. 137–65. Springer, Berlin.

Barrett, S.C.H. & Charlesworth, D. 1991. Effects of change in the level of inbreeding on the genetic load. *Nature*, 352, 522–4.

Barrett, S.C.H. & Kohn, J.R. 1991. Genetic and evolutionary consequences of small population sizes in plants: implications for conservation. In *Genetics and Conservation of Rare Plants*. Eds D.A. Falk & K.E. Holsinger. pp. 3–30. Oxford University Press, New York.

Barton, N.H. & Hewitt, G.M. 1985. Analysis of hybrid zones. *Annual Review of Ecology and Systematics*, 16, 113–48.

Barton, N.H. & Slatkin, M. 1986. A quasi-equilibrium theory of the distribution of rare alleles in a subdivided population. *Heredity*, 56, 409–15.

Basha, S.C. 1996. Ochlandra (bamboo reed) a vanishing asset of forests in Kerala, South India. In *Bamboo in Asia and the Pacific*. Proceedings of the Fourth International Bamboo Workshop held in Chiang Mai, Thailand, 27–30 November 1991. Technical Document GCP/RAS/ 134/ASB, FORSPA Publication 6. pp. 18–26. Forestry Research and Support Programme for Asia and the Pacific, Bangkok.

Basit, M.A. 1995. Non-wood forest products from the mangrove forests of Bangladesh. In *Beyond Timber: Social, Economic and Cultural Dimensions of Non-Wood Forest Products in Asia and the Pacific*. Eds P.B. Durst & A. Bishop. pp. 193–200. Proceedings of a Regional Expert Consultation held in Bangkok, 28 November –2 December 1994. Food and Agriculture Organization of the United Nations, Bangkok.

Bass, S. & Hearne, R.R. 1997. *Private Sector Forestry: A Review of Instruments for Ensuring Sustainability*. International Institute for Environment & Development, London.

Bauer, S., Galliano, H., Pfeiffer, F., Meßner, B., Sandermann, H. Jr. & Ernst, D. 1993. Isolation and characterization of a cDNA clone encoding a novel-short-chain alcohol dehydrogenase from Norway spruce (*Picea abies* L. Karst.). *Plant Physiology*, 103, 1479–80.

Bawa, K.S. 1974. Breeding systems of tree species of a lowland tropical community. *Evolution*, 28, 85–92.

Bawa, K.S. 1977. Reproductive biology of *Cupania guatemalensis* Radlk. (Sapindaceae). *Evolution*, 31, 52–63.

Bawa, K.S. 1979. Breeding systems of trees in a tropical wet forest. *New Zealand Journal of Botany*, 17, 521–4.

Bawa, K.S. 1983. Patterns of flowering in tropical plants. In *Handbook of Experimental Pollination Biology*. Eds C.E. Jones & R.J. Little. pp. 394–410. Scientific and Academic Editions, New York.

Bawa, K.S. & Ashton, P.S. 1991. Conservation of rare trees in tropical rainforests, a genetic perspective. In *Genetics and Conservation of Rare and Endangered Plants*. Eds D.A. Falk & K.E. Holsinger. pp. 62–71. University Press, Oxford.

Bawa, K.S. & Buckley, D.P. 1988. Seed, ovule ratios, selective abortion and mating systems in Leguminosae. In *Advances in Legume Biology*. Eds C.H. Stirton & J.L. Zarucchi. pp. 243–62. Missouri Botanical Gardens Monographs of Systematic Botany, Missouri Botanical Garden, St Louis, MO, USA.

Bawa, K.S. & Opler, P.A. 1975. Dioecism in tropical forest trees. *Evolution*, 29, 167–79.

Bawa, K.S., Ashton, P.S. & Salleh, M.N. 1990. Reproductive ecology of tropical forest plants, management issues. In *Reproductive Ecology of Tropical Forest Plants*. Eds K.S. Bawa & M. Hadley. pp. 3–13. Man and the Biosphere Series, vol. 7. Unesco/Parthenon Publishing, Paris/Carnforth.

Bawa, K.S., Perry, D.R. & Beach, J.H. 1985. Reproductive biology of tropical lowland rainforest trees 1. Sexual systems and incompatibility mechanisms. *American Journal of Botany*, 72, 331–45.

Bazzigher G. 1982. The Swiss Endothia-resistance breeding program. In *Resistance to Disease and Pests in Forest Trees*. Eds H.M. Heybroek, B.R. Stephan & K. von Weissenberg. pp. 396–403. Centre for Agricultural Publishing and Documentation, Wageningen, Netherlands.

Beckman, J.S. & Weber, J.L. 1992. Survey of human and rat microsatellites. *Genomics*, 12, 627–31.

Beismann, H., Barker, J.H.A., Karp, A. & Speck, T. 1997. AFLP analysis sheds light on distribution of two *Salix* species and their hybrid along a natural gradient. *Molecular Ecology*, 6, 989–93.

Benet, H., Guries, R.P., Boury, S. & Smalley, E.B. 1995. Identification of RAPD markers linked to a black leaf spot resistance gene in Chinese elm. *Theoretical and Applied Genetics*, 90, 1068–73.

Bengtsson, B.O., Weibull, P. & Ghatnekar, L. 1995. The loss of alleles by sampling: a study of the common outbreeding grass *Festuca ovina* over three geographic scales. *Hereditas*, 122, 221–38.

Bennett, M.D. & Leitch, I.J. 1998. Nuclear DNA amounts in angiosperms—583 new estimates. *Annals of Botany*, 80, 169–96.

Bergmann, F. & Gregorius, H.-R. 1993. Ecogeographical distribution and thermostability of isocitrate dehydrogenase (IDH) alloenzymes in European silver fir. *Biochemical Systematics and Ecology*, 21, 597–605.

Bergmann, F. & Hosius, B. 1996. Effects of heavy-metal polluted soils on the genetic structure of Norway spruce seedling populations. *Water, Air, and Soil Pollution*, 89, 363–73.

Bergmann, F. & Ruetz, W. 1991. Isozyme genetic variation and heterozygosity in random tree samples and selected orchard clones from the same *Picea abies* populations. *Forest Ecology and Management*, 46, 39–47.

Bergmann, F. & Scholz, F. 1987. The impact of air pollution on the genetic structure of Norway spruce. *Silvae Genetica*, 36, 80–3.

Bergmann, F. & Scholz, F. 1989. Selection effects of air pollution in Norway spruce. In *Genetic Effects of Air Pollutants in Forest Tree Populations*. Eds F. Scholz, H.-R. Gregorius & D. Rudin. pp. 143–60. Springer, Berlin.

Bergmann, F., Gregorius, H.-R. & Larsen, J.B. 1990. Levels of genetic variation in European silver fir (*Abies alba*)—are they related to the species' decline? *Genetica*, 82, 1–10.

Berkes, V. 1989. *Common Property Resources: Ecology and Community-based Sustainable Development*. Belhaven Press, London.

Bernatzky, R. & Mulcahy, D.L. 1992. Marker-aided selection in a backcross breeding program for resistance to chestnut blight in the American chestnut. *Canadian Journal of Forest Research*, 22, 1031–5.

Berrang, P., Karnosky, D.F. & Bennett, J.P. 1991. Natural selection for ozone tolerance in *Populus tremuloides*: an evaluation of nationwide trends. *Canadian Journal of Forest Research*, 21, 1091–7.

Białobok, S., Karolewski, P. & Oleksyn, J. 1980a. Sensitivity of Scots pine needles from mother trees and their progenies to the action of SO_2, O_3, a mixture of these gases, NO_2 and HF. *Arboretum Kórnickie*, 25, 289–303.

Białobok, S., Oleksyn, J. & Karolewski, P. 1980b. Zróznicowanie wrazliwosci na dzialanie dwutlenku siarki 6 polskich proweniencji swierka pospolitego (*Picea abies* (L.) Karst.). [Differentiation in sensitivity to the action of sulfur dioxide of Norwegian spruce (*Picea abies* (L.) Karst.) of 6 Polish provenances]. *Arboretum Kórnickie*, 25, 305–10.

Billington, H.L. 1991. Effect of population size on genetic variation in a dioecious conifer. *Conservation Biology*, 5, 115–19.

Bing, D.J., Downey, R.K. & Rakow, G.F.W. 1996. Hybridizations among *Brassica napus*, *B. rapa* and *B. juncea* and their tow weedy relatives *B. nigra* and *Sinapis arvensis* under open pollination conditions in the field. *Plant Breeding*, 115, 470–3.

Binning, C. & Young, M. 1997. *Motivating People: Using Management Agreements To Conserve Remnant Vegetation*. Paper 1/97. Land & Water Resources Research & Development Corporation, Canberra.

Biot, Y., Blaikie, P.M., Jackson, C. & Palmer-Jones, R. 1995. *Rethinking Research on Land Degradation in Developing Countries*. World Bank Discussion Paper 289. The World Bank, Washington, DC.

Birks, J.S. & Kanowski, P.J. 1988. Interpretation of the composition of coniferous resin. *Silvae Genetica*, 37, 29–39.

Birks, J.S. & Kanowski, P.J. 1995. Resin compostional data: issues and analysis. In *Population Genetics and Genetic Conservation of Forest Trees*. Eds Ph. Baradat, W.T. Adams & G. Müller-Starck. pp. 29–40. SPB Academic Publishing, Amsterdam.

Birky, C.W. 1995. Uniparental inheritance of mitochondrial and chloroplast genes: mechanisms and evolution. *Proceedings of the National Academy of Sciences, USA*, 92, 11331–8.

Blaikie, P. & Jeanrenaud, S. 1996. *Biodiversity and Human Welfare*. Discussion Paper 72. United Nations Research Institute for Social Development, Geneva.

Bodenes, C., Labbe, T., Pradere, S. & Kremer, A. 1997. General vs. local differentiation between two closely related white oak species. *Molecular Ecology*, 6, 713–24.

Boes, T.K., Brandle, J.R. & Lovett, W.R. 1991. Characterisation of the flowering phenology and seed yield in a *Pinus sylvestris* clonal seed orchard in Nebraska. *Canadian Journal of Forest Research*, 21, 1721–9.

Boot, R.G.A. & Gullison, R.E. 1995. Approaches to developing sustainable extraction systems for tropical forest products. *Ecological Applications*, 5, 896–903.

Boscherini, G., Morgante, M., Rossi, P. & Vendramin, G.G. 1994. Allozyme and chloroplast DNA variation in Italian and Greek populations of *Pinus leucodermis*. *Heredity*, 73, 284–90.

Boshier, D.H. 1992. A study of the reproductive biology of *Cordia alliodora* (R. and P.) Oken. D. Phil. Thesis, University of Oxford, Oxford.

Boshier, D.H. 1995. Incompatibility in *Cordia alliodora* (Boraginaceae), a neotropical tree. *Canadian Journal of Botany*, 73, 445–56.

Boshier, D.H. & Billingham, M.R. 2000. Genetic variation and adaptation in tree populations. In *Ecological Consequences of Habitat Heterogeneity*. Ed. M.J. Hutchings. Blackwell Science, UK (in press).

Boshier, D.H. & Lamb, A.T. 1997. *Cordia alliodora, genetics and tree improvement*. Tropical Forestry Papers No. 36. Oxford Forestry Institute, Oxford.

Boshier, D.H., Chase, M.R. & Bawa, K.S. 1995a. Population genetics of *Cordia alliodora* (Boraginaceae), a neotropical tree. 3. Gene flow, neighborhood, and population substructure. *American Journal of Botany*, 82, 484–90.

Boshier, D.H., Chase, M.R. & Bawa, K.S. 1995b. Population genetics of *Cordia alliodora* (Boraginaceae), a neotropical tree. 2. Mating system. *American Journal of Botany*, 82, 476–83.

Bossart, J.L. & Pashley-Prowell, D. 1998. Genetic estimates of population structure and gene flow limitations, lessons and new directions. *Trends in Ecology and Evolution*, 13, 202–6.

Botstein, D., White, R.L., Skolnick, M. & Davis, R.W. 1980. Construction of a genetic linkage map in man using restriction fragment length polymorphisms. *American Journal of Human Genetics*, 32, 314–31.

Boudet, A.M., Lapierre, C. & Grima-Pettenati, J. 1995. Biochemistry and molecular biology of lignification. *New Phytologist*, 129, 203–36.

Bousquet, J., Cheliak, W.M. & Lalonde, M. 1987. Allozyme variability in natural populations of green alder (*Alnus crispa*) in Quebec. *Genome*, 21, 550–70.

Boyle, T.J.B. & Boontawee, B. 1995. *Measuring and Monitoring Biodiversity in Tropical and Temperate Forests*. Centre for International Forestry Research, Bogor, Indonesia.

Boyle, T.J.B. & Yeh. F.C. 1988. Within-population genetic variation and its implications for selection and breeding. In *Tree Improvement —Progressing Together: Proceedings of the 21st Meeting of the Canadian Tree Improvement Association*. Part 2, N.S. Truro, Eds E.K. Morgenstern & T.J.B. Boyle. pp. 20–42. Forestry Canada, Ottawa.

Boyle, T.J.B., Liengsiri, C. & Piewluang, C. 1991. Genetic studies in a tropical pine—*Pinus kesiya* III. The mating system in four populations from Northern Thailand. *Journal of Tropical Forest Science*, 4, 37–44.

Braatz, S. 1992. Conserving Biological Diversity. A Strategy for Protected Areas in the Asia-Pacific Region. World Bank Technical Paper Number 193. Asia Technical Department Series, The World Bank, Washington, DC.

Braatz, S.M. 1997. State of the world's forests 1997. *Nature & Resources*, 33 (3–4), 18–25.

Bradshaw, A.D. & McNeilly, T. 1981. *Evolution and Pollution*. Edward Arnold Publishing, London.

Bradshaw, H.D., Villar, M., Watson, B.D., Otto, K.G., Stewart, S. & Stettler, R.F. 1994. Molecular genetics of growth and development in *Populus*. III. A genetic linkage map of a hybrid poplar composed of RFLP, STS, and RAPD markers. *Theoretical and Applied Genetics*, 89, 167–78.

Bramlett, D.L., Askew, G.R., Blush, T.D., Bridgwater, F.E., & Jett, J.B. (Eds). 1993. *Advances in Pollen Management*. United States Department of Agriculture, Forest Service, Agriculture Handbook 698, Washington, DC.

Brandon, K. 1995. People, parks, forests or fields. *Land Use Policy*, 12, 137–44.

Brasier, C.M. 1996. *Phytophthora cinnamomi* and oak decline in southern Europe. Environmental constraints including climate change. *Annales des Sciences Forestiers*, 53, 347–58.

Breckwoldt, R. (Ed) 1996. *Approaches to Bioregional Planning*. Biodiversity Series Paper No. 10, Department of Environment, Sport and Territories, Australia.

Brewer, G.J. & Sing, C.F. 1970. *An Introduction to Isozyme Techniques*. Academic Press, New York.

Bridgwater, F.E., Blush, T.D. & Wheeler, N.C. 1993. Supplemental mass pollination. In *Advances in Pollen Management*. Eds D.L. Bramlett, G.R. Askew, T.D. Blush, F.E. Bridgwater & J.B. Jett. pp. 69–77. United States Department of Agriculture, Forest Service, Agricultural Handbook 698, Washington, DC.

Bridgewater, P.B., Cresswell, I. & Thackway, R . 1996. A bioregional framework for planning a national system of protected areas. In *Approaches to Bioregional Planning*. Biodiversity Series. Ed. R. Breckwoldt. Paper No. 10, pp. 67–72. Department of Environment, Sport and Territories, Australia.

Briggs, D. & Walters, S.M. 1997. *Plant Variation and Evolution*. Cambridge University Press, Cambridge.

Brooks, A.B. 1937. *Castanea dentata*. *Castanea*, 2, 61–7.

Brotschol, J.V., Roberds, J.H. & Namkoong, G. 1986. Allozyme variation among North Carolina population of *Liriodendron tulipifera* L. *Silvae Genetica*, 35, 131–8.

Browder, J.O. 1992. Social and economic constraints on the development of market-oriented extractive reserves in Amazon rain forests. In *Non-Timber Forest Products from Tropical Forests: Evaluation of a Conservation and Development Strategy*. Eds D.C. Nepstad & S. Schwartzman. pp. 33–42. Advances in Economic Botany 9. New York Botanical Garden, New York.

Brown, A.H.D. 1978. Isozymes, plant population genetic structure and genetic conservation. *Theoretical and Applied Genetics*, 52, 145–57.

Brown, A.H.D. 1979. Enzyme polymorphisms in plant populations. *Theoretical Population Biology*, Vol. 15, pp. 1–42.

Brown, A.H.D. 1989. The case for core collections. In *The Use of Plant Genetic Resources*. Eds A.H.D. Brown, O.H. Frankel, D.R. Marshall & J.T. Williams. pp. 136–56. Cambridge University Press, Cambridge.

Brown, A.H.D. 1990. Genetic characterization of plant mating systems. In *Plant Population Genetics, Breeding, and Genetic Resources*. Eds A.H.D. Brown, M.T. Clegg, A.L. Kahler & B.S. Weir. pp. 145–62. Sinauer Associates, Sunderland, MA, USA.

Brown, A.H.D. & Allard, R.W. 1970. Estimation of the mating system in open-pollinated maize populations using isozyme polymorphisms. *Genetics*, 66, 133–45.

Brown, A.H.D. & Briggs, J.D. 1991. Sampling strategies for genetic variation in ex situ collections of endangered plant species. In *Genetics and Conservation of Rare Plants*. Eds D.A. Falk & K.E. Holsinger. pp. 99–119. Oxford University Press, New York.

Brown, A.H.D. & Feldman, M.W. 1981. Population structure of multilocus associations. *Proceedings of the National Academy of Sciences, USA*, 78, 5913–16.

Brown, A.H.D. & Marshall, D.R. 1995. A basic sampling strategy: theory and practice. In *Collecting Plant Genetic Diversity: Technical Guidelines*.. Eds L. Guarino, V. Ramantha Rao & R. Read. pp. 75–92. CAB International, Wallingford, UK.

Brown, A.H.D. & Moran, G.F. 1981. Isozymes and genetic resources of forest trees. In *Isozymes of North American Forest Trees and Forest Insects*. Ed. M.T. Conkle. pp. 1–10. USDA, Berkeley, CA, USA.

Brown, A.H.D. & Schoen, D.J. 1994. Optimal sampling strategies for core collections of plant genetic resources. In *Conservation Genetics*. Eds V. Loeschcke, J. Tomiuk & S.K. Jain. pp. 357–70. Birkhauser Verlag, Basel.

Brown, A.H.D., Barrett, S.C.H. & Moran, G.F. 1985. Mating system estimation in forest trees, models, methods and meanings. In *Population Genetics in Forestry*. Ed. H.R. Gregorius. Springer-Verlag, Berlin.

Brown, A.H.D., Feldman, M.W. & Nevo, E. 1980. Multilocus structure of natural populations of *Hordeum spontaneum*. *Genetics*, 96, 523–36.

Brown, A.H.D., Matheson, A.C. & Eldridge, K.G. 1975. Estimation of the mating system of *Eucalyptus obliqua* L'Herit. by using allozyme polymorphisms. *Australian Journal of Botany*, 23, 931–49.

Brown, A.H.D., Marshall, D.R. & Weir, B.S. 1981. Current status of the charge state model for protein polymorphism. In *The Genetic Studies of Drosophila Populations. Proceedings of the Kioloa Conference*. Eds J.B. Gibson & J.G. Oakeshott. pp. 15–43. Australian National University Press, Canberra.

Brown, A.H.D., Young, A.G., Burdon, J.J., Christidis, L., Clarke, G., Coates, D. & Sherwin, W. 2000. Genetic indicators for State of the Environment Reporting. Department of Environment, Sports and Territories Technical Report, Canberra.

Brown, I.R. 1971. Flowering and seed production in grafted clones of Scots pine. *Silvae Genetica*, 20, 121–32.

Brown, T.A. 1999. *Genomes*. BIOS Scientific Publishers, Oxford.

Bruce, R.C. & Boyce, S.G. 1984. Measurements of diversity on the Nantahala National Forest. In Natural Diversity in Forest Ecosystems: Proceedings of the Workshop. Eds J.L. Cooley & J.H. Cooley. pp. 71–85. Institute of Ecology, University of Georgia, Athens, GA, USA.

Brune, A. 1990. Reproductive biology and tropical plantation forestry. In *Reproductive Ecology of Tropical Forest Plants*. Eds K.S. Bawa & M. Hadley. pp. 3–13. Man and the Biosphere Series, vol. 7. Unesco/Parthenon Publishing, Paris and Carnforth.

Brunet, J. & Charlesworth, D. 1995. Floral sex allocation in sequentially blooming plants. *Evolution*, 49, 70–9.

Buchert, G.P., Rajora, O.P., Hood, J.V. & Dancik, B.P. 1997. Effects of harvesting on genetic diversity in old-growth eastern white pine in Ontario, Canada. *Conservation Biology*, 11, 747–58.

Bullock, S.H. & Bawa, K.S. 1981. Sexual dimorphism and the annual flowering pattern in *Jacaratia dolichaula* (D. Smith) Woodson (Caricaceae) in a Costa Rican rain forest. *Ecology*, 62, 1494–1504.

Bullock, S.H. & Bawa, K.S. 1994. Wind pollination in neotropical dioecious trees. *Biotropica*, 26, 172–9.

Burczyk, J. 1992. System kojarzenia a fenologia i intensywnosc kwitnienia na wybranej plantacji nasiennej sosny zwyczajnej (*Pinus sylvestris* L.). [Mating system, floral phenology and flowering intensity in a Scots pine (*Pinus sylvestris* L.) seed orchard.] Doctoral dissertation. University of A. Mickiewicz, Poznan, Poland [In Polish].

Burczyk, J. & Chalupka, W. 1997. Flowering and cone production variability and its effect on parental balance in a Scots pine clonal orchard. *Annals Science Forestier*, 54, 129–44.

Burczyk, J. & Prat, D. 1997. Male reproductive success in *Pseudotsuga menziesii* (Mirb.) Franco, the effects of spatial structure and flowering characteristics. *Heredity*, 79, 638–47.

Burczyk, J., Adams, T.W. & Birkes, D.S. 1993. NEIGHBOR. A computer program for estimating mating patterns in conifer populations from genetic marker data. Release 1. Oregon State University, OR, USA.

Burczyk, J., Adams, T.W. & Shimuzu, J.Y. 1996. Mating patterns and pollen dispersal in a natural knobcone pine (*Pinus attenuata*) stand. *Heredity*, 77, 251–60.

Burczyk, J., Kosinski, G. & Lewandowski, A. 1991. Mating pattern and empty seed formation in relation to crown level of *Larix decidua* (Mill.) clones. *Silva Fennica*, 25, 201–5.

Burdon, J.J. 1982. The effect of fungal pathogens on plant communities. pp. 99–112. In *The Plant Community as a Working Mechanism*. Ed. E.I. Newman. Blackwell Scientific Publications, Oxford.

Burdon, J.J. 1987. *Diseases and Plant Population Biology*. p. 208. Cambridge University Press, Cambridge.

Burdon, J.J. 1994. The role of parasites in plant populations and communities. In *Biodiversity and Ecosystem Function*. Eds E. Schulze & H.A. Mooney. pp. 165–79. Springer-Verlag, Berlin.

Burley, J. & Adlard, P.G. 1992. Plantation silvicultural research and genetic improvement. In *Proceedings of the International Union of Forestry Research Organisations Centennial Meeting*. Berlin, September 1992. International Union of Forestry Research Organisations, Vienna.

Bush, R.M & Smouse, P. 1992. Evidence for adaptive significance of allozymes in forest trees. *New Forests*, 6, 179–96.

Butcher, P.A., Doran, J.C. & Slee, M.U. 1994. Intraspecific variation in leaf oils of *Melaleuca alternifolia*. *Biochemical Systematics and Ecology*, 22, 419–30.

Butcher, P.A., Moran, G.F. & Perkins, H. 1998. RFLP diversity in the nuclear genome of *Acacia mangium*. *Heredity*, 81, 205–13.

Butler, D. 1992. Why exorbitant farming subsidies have to end. *World Link*, 5, 6–10.

Byrne, M., Marquez-Garcia, M.I., Uren, T., Smith, D.S. & Moran, G.F. 1996. Conservation and genetic diversity of microsatellite loci in the genus *Eucalyptus*. *Australian Journal of Botany*, 44, 331–41.

Byrne, M., Moran, G.F., Stukely, M., Emebiri, L.C. & Williams, E.R. 1997. Identification of QTL for resistance to *Phytophthora cinnamomi* in *Eucalyptus marginata*. p. 111. Proceedings of the Australian Plant Pathology Society, Eleventh Biennial Conference, Perth, Australia.

Byrne, M., Parrish, T.L. & Moran, G.F. 1998. Nuclear RFLP diversity in *Eucalyptus nitens*. *Heredity*, 81, 225–33.

Caballero, A. 1994. Developments in the prediction of effective population size. *Heredity*, 73, 657–79.

Caballero, A. & Keightley, P. 1996. Genomic mutation rates for lifetime reproductive output and lifespan in *Caenorhabditis elegans*. *Proceedings of the National Academy of Sciences, USA*, 94, 3823–7.

California Department of Fish and Game, Natural Diversity Data Base. 1997a. *RareFind Printout on Threats Involving Hybridization*, USA.

California Department of Fish and Game, Natural Diversity Data Base. April 1997b. Special Plants List. Quarterly publication, USA.

Campbell, R.K. 1979. Genecology of Douglas-fir in a watershed in the Oregon Cascades. *Ecology*, 60, 1036–50.

Campbell, R.K. 1987. Biogeographical distribution limits of Douglas-fir in Southwest Oregon. *Forest Ecology and Management*, 18, 1–34.

Campbell, R.K. & Sorensen, F.C. 1984. Genetic implications of nursery practices. In *Forest Nursery Manual: Production of Bareroot Seedlings*. Eds M.L. Duryea & T.D. Landis. pp. 183–91. Matinus Nijhoff/Dr W. Junk, Boston.

Campbell, R.K. & Sugano, A.I. 1993. Genetic variation and seed zones of Douglas-fir in the Siskiyou National Forest. Research Paper, PNW-RP-461. United States Department of Agriculture, Forest Service, Pacific Northwest Research Station, Portland, OR, USA.

Carew-Reid, J., Prescott-Allen, R., Bass, S. & Dalal-Clayton, B. 1994. *Strategies for National Sustainable Development*. Earthscan Publications, London.

Caro, T.M. & Laurenson, M.K. 1994. Ecological and genetic factors in conservation: a cautionary tale. *Science*, 263, 485–6.

Carson, H.L. 1990. Increased genetic variance after a population bottleneck. *Trends in Ecology and Evolution*, 7, 228–30.

Cervera, M.T., Gusmao, J., Steenackers, M., Avan, G., Mvan, M., Boerjan, W., Van Gysel, A. & Van Montagu, M. 1996a. Application of AFLP^{TM}-based molecular markers to breeding of *Populus* spp. *Plant Growth and Regulation*, 20, 47–52.

Cervera, M.T., Gusmao, J., Steenackers, M., Peleman, J., Storme, V., Vanden Broeck, A., Van Montagu, M. & Boerjan, W. 1996b. Identification of AFLP molecular markers for resistance against *Melampsora larici-populina* in *Populus*. *Theoretical and Applied Genetics*, 93, 733–7.

Chaisurisri, K. & El-Kassaby, Y.A. 1993. Estimation of clonal contribution to cone and seed crops in a *Picea sitchensis* seed orchard. *Annales des Sciences Forestieres*, 50, 461–7.

Chaisurisri, K. & El-Kassaby, Y.A. 1994. Genetic diversity in a seed production population vs. natural populations of *Picea sitchensis*. *Biodiversity and Conservation*, 3, 512–23.

Chaisurisri, K., Edwards, D.G.W. & El-Kassaby, Y.A. 1992. Genetic control of seed size and germination in Sitka spruce. *Silvae Genetica*, 41, 348–55.

Chaisurisri, K., Edwards, D.G.W. & El-Kassaby, Y.A. 1993. Accelerating aging in Sitka spruce seed. *Silvae Genetica*, 42, 303–8.

Chakraborty, R. 1981. The distribution of the number of heterozygous loci in an individual in natural populations. *Genetics*, 98, 461–6.

Chalmers, K.J., Newton, A.C., Waugh, R., Wilson, J. & Powell, W. 1994. Evaluation of the extent of genetic variation in mahoganies (Meliaceae) using RAPD markers. *Theoretical and Applied Genetics*, 89, 504–8.

Chamberlain, J.R. 1998. *Calliandra calothyrsus*, reproductive biology in relation to its role as an important multipurpose tree. In *Reproductive Biology in Systematics, Conservation and Economic Botany*. Eds S.J. Owens & P.J. Rudall. pp. 439–48. Royal Botanic Gardens, Kew.

Chan, H.T. 1981. Reproductive biology of some Malaysian dipterocarps. III. Breeding systems. *Malaysian Forester*, 44, 28–36.

Chan, H.T. & Appanah, S. 1980. Reproductive biology of some Malaysian dipterocarps. I. Flowering biology. *Malaysian Forester*, 43, 132–43.

Chappelka, A.H., Kush, J.S., Meldahl, R.S. & Lockaby B.G. 1990. An ozone-low temperature interaction in loblolly pine (*Pinus taeda* L.). *New Phytologist*, 114, 721–6.

Charlesworth, D. & Charlesworth, B. 1987. Inbreeding depression and its evolutionary consequences. *Annual Review of Ecology and Systematics*, 18, 237–68.

Chase, M.R., Boshier, D.H. & Bawa, K.S. 1995. Population genetics of *Cordia alliodora* (Boraginaceae), a neotropical tree. 1. Genetic variation in natural populations. *American Journal of Botany*, 82, 468–75.

Chase, M., Kesseli, R. & Bawa, K. 1996a. Microsatellite markers for population and conservation genetics of tropical trees. *American Journal of Botany*, 83, 51–7.

Chase, M.R., Moller, C., Kesseli, R., Bawa, K.S. 1996b. Distant gene flow in tropical trees. *Nature (London)*, 383, 398–9.

Chat, J., Chalak, L. & Petit, R.J. 1999. Strict paternal inheritance of chloroplast DNA and maternal inheritance of mitochondrial DNA in intraspecific crosses of kiwifruit. *Theoretical and Applied Genetics*, 99, 314–22.

Cheliak, W.M. & Pitel, J.A. 1984. Genetic control of allozyme variants in mature needle tissue of white spruce trees. *Journal of Heredity*, 75, 34–40.

Cheliak, W.M., Dancik, B.P., Morgan, K., Yeh, F.C.H. & Strobeck, K.M.C. 1985. Temporal variation of the mating system in a natural population of jack pine. *Genetics*, 109, 569–84.

Child, G. 1994. Strengthening protected-area management: a focus for the 1990s, a platform for the future. *Biodiversity and Conservation*, 3, 459–63.

Clark, C.W. 1990. *Mathematical Bioeconomics*. Wiley, New York.

Clegg, M.T. 1980. Measuring plant mating systems. *Bioscience*, 30, 814–18.

Coates, D.J. 1988. Genetic diversity and populations genetic structure in the rare Chittering grass wattle, *Acacia anomala* Court. *Australian Journal of Botany*, 36, 273–86.

Coates, D.J. & Sokolowski, R.E. 1989. Geographic patterns of genetic diversity in karri (*Eucalyptus diversicolor* F. Muell.). *Australian Journal of Botany*, 37, 145–56.

Coates, D.J., & Sokolowski, R.E.S. 1992. The mating system and patterns of genetic variation in *Banksia cuneata* A.S. George (Proteacae). *Heredity*, 69, 11–20.

Colchester, M. 1994. *Salvaging Nature—Indigenous Peoples, Protected Areas and Biodiversity Conservation*. Discussion Paper 55. United Nations Research Institute for Social Development, Geneva.

Colfer, C.J.P. & Soedjito, H. 1996. Food, forests, and fields in a Bornean rain forest: toward appropriate agroforestry development. In *Borneo in Transition*. Eds C. Padoch & N. Lee Peluso. pp. 162–86. Oxford University Press, Kuala Lumpur.

Colfer, C.J.P., Pierce, C.J. with Dudley, R.G. 1993. *Shifting Cultivators in Indonesia: Marauders or Managers of the Forest?* FAO Community Case Study Series, No. 6. The Food and Agriculture Organization of the United Nations, Rome.

Compton, S.G., Ross, S.J. & Thornton, I.W.B. 1994. Pollinator limitation of fig tree reproduction on the island of Anak Krakatau (Indonesia). *Biotropica*, 26, 180–6.

Condon, M.A. & Gilbert, L.E. 1988. Sex expression of *Gurania* and *Psiguria* (Cucurbitaceae), neotropical vines that change sex. *American Journal of Botany*, 75, 875–84.

Conkle, M.T., Hodgskiss, P.D., Nunnaly, L.B. & Hunter, S.C. 1982. *Starch Gel Electrophoresis of Conifer Seeds: A Laboratory Manual*. General Technical Report PSW-64. United States Department of Agriculture, Berkeley, CA, USA.

Cook, C.C. & Grut, M. 1989. *Agroforestry in Sub-Saharan Africa. A Farmer's Perspective*. World Bank Technical Paper Number 112. The World Bank, Washington, DC.

Cook, S.A., Copsey, A.D. and Dickman, A.W. 1989. Response of *Abies* to fire and *Phellinus*. In *The Evolutionary Ecology of Plants*. Eds J. Bock & Y.B. Linhart. pp. 363–92. Westview, Boulder, CO, USA.

Copes, D.L. & Sniezko, R.A. 1991. The influence of floral bud phenology on the potential mating system of a wind-pollinated Douglas-fir seed orchard. *Canadian Journal of Forest Research*, 21, 813–20.

Corner, E.J.H. 1954. The evolution of tropical forests. In *Evolution as a Process*. Eds J. Huxley, A.C. Hardy & E.C. Ford. Allen & Unwin, London.

Cornuet J.M. & Luikart G. 1996. Description and power analysis of two tests for detecting recent population bottlenecks from allele frequency data. *Genetics*, 144, 2001–14.

Costanza, R. (Ed.) 1991. *Ecological Economics: The Science and Management of Sustainability*. Columbia University Press, New York.

Cox, R.M. 1989. Natural variation in sensitivity of reproductive processes in some boreal forest trees to acidity. In *Genetic Effects of Air Pollutants in Forest Tree Populations*. Eds F. Scholz, H.-R. Gregorius & D. Rudin. pp. 77–88. Springer, Berlin.

Cronk, Q.C.B. 1986. The decline of the St. Helena ebony *Trochetiopsis melanoxylon*. *Biological Conservation*, 35, 159–72.

Cronk, Q.C.B. 1995. A new species and hybrid in the St. Helena endemic genus *Trochetiopsis*. *Edinburgh Journal of Botany*, 52, 205–13.

Crossa, J. 1989. Methodologies for estimating the sample size required for genetic conservation of outbreeding crops. *Theoretical & Applied Genetics*, 77, 153–61.

Crow, J.F. & Denniston, C. 1988. Inbreeding and variance effective population numbers. *Evolution*, 42, 482–95.

Crow, J.F. & Kimura, M. 1970. *An Introduction to Population Genetic Theory*. Harper and Row, New York.

Cruden, R.W. 1977. Pollen-ovule ratios, a conservative indicator of breeding systems in flowering plants. *Evolution*, 31, 32–46.

Cruzan, M.B. & Arnold, M.L. 1993. Ecological and genetic associations in an *Iris* hybrid zone. *Evolution*, 47, 1432–45.

Cruzan, M.B. & Arnold, M.L. 1994. Assortative mating and natural selection in an *Iris* hybrid zone. *Evolution*, 48, 1946–58.

Cylinder, P. 1997. Monterey pine forest conservation strategy. *Fremontia*, 25, 21–6.

Czabator, F.J. 1962. Germination value: an index combining speed and completeness of pine seed germination. *Forest Science*, 8, 21–31.

Dafni, A. 1992. *Pollination Ecology—A Practical Approach*. The Practical Approach Series. IRL Press at Oxford University Press, Oxford.

Daly, H.E. & Cobb, J.B. Jr 1989. *For the Common Good: Redirecting the Economy Towards Community, the Environment and a Sustainable Future*. Beacon Press, Boston.

Dancik, B.P. and Yeh, F.C. 1983. Allozyme variability and evolution of lodgepole pine (*Pinus contorta* var. *latifolia*) and jack pine (*P. banksiana*) in Alberta. *Canadian Journal of Genetics and Cytology*, 25, 57–64.

Darwin, C. 1876. *The Effects of Cross and Self Fertilisation in the Vegetable Kingdom*. John Murray, London.

Dasmann, R. 1975. National parks, nature conservation, and 'future primitive'. *Ecologist*, 65, 164–7.

Daubenmire, R. 1972. Phenology and other characteristics of tropical semi-deciduous forest in north-western Costa Rica. *Journal of Ecology*, 60, 147–70.

Davis, D.D. & Gerhold, F.D. 1976. Selection of trees for tolerance of air pollutants. In *Better Trees for Metropolitan Landscapes*. Eds F.S. Santamour, H.D. Gerhold & S. Little. pp. 61–6. General Technical Report, Northeastern Forest Experiment Station, USDA Forest Service. No. NE-22, USA.

Davis, M.B. 1983. Holocene vegetational history of the eastern United States. In *Late-Quaternary Environments of the United States*. Vol. 2. Ed. H.E. Wright. pp. 166–81. The Holocene University of Minnesota Press, Minneapolis, MN, USA.

Dawson, I.K., Waugh, R., Simons, A.J. & Powell, W. 1997. Simple sequence repeats provide a direct estimate of pollen-mediated gene dispersal in the tropical tree *Gliricidia sepium*. *Molecular Ecology*, 6, 179–83.

De Foresta, H. & Michon, G. 1995. *Several Ecology and Economic Aspects of Damar Garden in Krui–West Lampung*. Paper presented at a one day seminar on 'Damar Agroforests in Krui–Lampung as a Model of Community Forest'. Krui, Lampung.

Degen, B. 1997. Abschlußbericht 'Ökologische Genetik'. (Final Report Ecological Genetics). In *Auswertungen der Waldschadensforschungsergebnisse (1982–1992) zur Aufklärung Komplexer Ursache-Wirkungsbeziehungen mit Hilfe System Analytischer Methoden*. pp. 407–501. Berichte 6/97, Umweltbundesamt, Berlin.

Degen, B., Gregorius, H.-R. & Scholz, F. 1996. ECO-GENE, a model for simulation studies on the spatial and temporal dynamics of genetic structures of tree populations. *Silvae Genetica*, 45, 323–9.

DeHayes, D.H. & Hawley, G.J. 1992. Genetic implications in the decline of red spruce. *Water, Air, and Soil Pollution*, 62, 233–48.

de Jong, W. 1995. Recreating the forest: successful examples of ethnoconservation among dayak groups in central Kalimantan. In *Management of Tropical Forests: Towards an Integrated Perspective*. Ed. Ø. Sandbukt. pp. 295–304. Center for Development and the Environment, University of Oslo, Oslo.

De Klemm, C. 1985. Preserving genetic diversity: a legal review. *Landscape Planning*, 12, 221–8.

Delatour, C. 1996. View of pathological problems in broadleaved forest trees in Europe. In *Forest Trees and Palms. Diseases and Control*. Eds S.P. Raychaudhuri & K. Maramorosch. pp. 73–80. Science Publishers, NH, USA.

Delouche, J.C. & Baskin, C.C. 1973. Accelerated aging techniques for predicting the relative storability of seed lots. *Seed Science and Technology*, 1, 427–52.

DeMauro, M.M. 1993. Relationship of breeding system to rarity in the Lakeside Daisy (*Hymenoxys acaulis* var. *glabra*). *Conservation Biology*, 7, 542–50.

Denti, D. & Schoen, D.J. 1988. Self-fertilization rates in *Picea glauca*: effects of pollen and seed production. *Journal of Heredity*, 79, 284–8.

Devey, M.E., Delfino-Mix, A., Kinloch Jr, B.B. & Neale, D.B. 1995. Random amplified polymorphic DNA markers tightly linked to a gene for resistance to white pine blister rust in sugar pine. *Proceedings of the National Academy of Sciences, USA*, 92, 2066–70.

Devey, M.E., Fiddler, T.A., Lui, B.H., Knapp, S.J. & Neale, D.B. 1994. An RFLP linkage map for loblolly pine based on a three-generation pedigree. *Theoretical and Applied Genetics*, 88, 273–8.

Devlin, B. & Ellstrand, N.C. 1990. The development and application of a refined method for estimating gene flow from angiosperm paternity analysis. *Evolution*, 44, 248–59.

Di-Giovanni, F. & Kevan, P.G. 1991. Factors affecting pollen dynamics and its importance to pollen contamination: a review. *Canadian Journal of Forest Research*, 21, 1155–70.

Di-Giovanni, F., Kevan, P.G. & Arnold, J. 1996. Lower planetary boundary layer profiles of atmospheric conifer pollen above a seed orchard in northern Ontario, Canada. *Forest Ecology and Management*, 83, 87–97.

Diamond, J.M. 1989. Overview of recent extinctions. In *Conservation for the Twenty–First Century*. Eds D. Western & M. Pearl. pp. 37–41. Oxford University Press, New York.

Dickman, C.R., Pressey, R.L., Lim, L. & Parnaby, H.A. 1993. Mammals of particular conservation concern in the Western Division of NSW. *Biological Conservation* 69, 219–48.

Didham, R.K., Ghazoul, J., Stork, N.E. & Davis, A.J. 1996. Insects in fragmented forests: a functional approach. *Trends in Ecology and Evolution*, 11, 255–60.

Diggle, P.K. 1993. Developmental plasticity, genetic variation, and the evolution of andromonoecy in *Solanum hirtum* (Solanaceae). *American Journal of Botany*, 80, 967–73.

Dixon, J.A. & Sherman, P.B. 1990. *Economics of Protected Areas: A New Look at Benefits and Costs*. Island Press, Washington, DC.

Dobzhansky, T. 1937a. Genetic nature of species differences. *American Naturalist*, 71, 404–20.

Dobzhansky, T. 1937b. *Genetics and the Origin of Species*. Columbia University Press, New York.

Dobzhansky, T. 1940. Speciation as a stage in evolutionary divergence. *American Naturalist*, 74, 312–21.

Dong, J.-Z. & Dunstan, D.I. 1996. Characterization of three heat-shock-protein genes and their developmental regulation during somatic embryogenesis in white spruce (*Picea glauca* (Mönch) Voss). *Planta*, 200, 85–91.

Dong, J. & Wagner, D.B. 1993. Taxonomic and population differentiation of mitochondrial diversity in *Pinus banksiana* and *Pinus contorta*. *Theoretical and Applied Genetics*, 86, 573–8.

Dong, J. & Wagner, D.B. 1994. Paternally inherited chloroplast polymorphism in *Pinus*: estimation of diversity & within population subdivision, & tests of disequilibrium with a maternally inherited mitochondrial polymorphism. *Genetics*, 136, 1187–94.

Doran, J.C. 2000. Genetic improvement of eucalypts, with particular reference to oil-bearing species. In *Medicinal & Aromatic Plants—Industrial Profiles—Eucalyptus*. Ed. J.J.W. Coppen. Harwood Academic Publishers, Netherlands (in press).

Dow, B.D. & Ashley, M.V. 1996. Microsatellite analysis of seed dispersal and parentage of saplings in bur oak, *Quercus macrocarpa*. *Molecular Ecology*, 5, 615–27.

Dow, B.D., Ashley, M.V. & Howe, H.F. 1995. Characterization of highly variable (GA/CT)$_n$ microsatellites in the bur oak, *Quercus macrocarpa*. *Theoretical and Applied Genetics*, 91, 137–41.

Dudley, J.P. 1992. Rejoinder to Rohlf and O'Connell: biodiversity as a regulatory criterion. *Conservation Biology*, 6, 587–9.

Dumolin-Lapegue, S., Bodenes, C. & Petit, R.J. 1996. Detection of rare polymorphisms in mitochondrial DNA of oaks with PCR-RFLP combined to SSCP analysis. *Forest Genetics*, 3, 227–30.

Dumolin-Lapegue, S., Demesure, B., Fineschi, S. & Le Corre, V. 1997. Phylogeographic structure of white oaks throughout the European continent. *Genetics*, 146, 1997.

Dyson, W.G. & Freeman, G.H. 1968. Seed orchard designs for sites with a constant prevailing wind. *Silvae Genetica*, 17, 12–15.

Echt, C.S., Maymarquardt, P., Hseih, M. & Zahorchak, R. 1996. Characterization of microsatellite markers in eastern white pine. *Genome*, 39, 1102–8.

Eckberg, I., Eriksson, G. & Dormling, I. 1979. Photoperiodic reactions in conifer species. *Holarctic Ecology*, 2, 255–63.

Edwards, D.G.W. & El-Kassaby, Y.A. 1995. Douglas-fir genotypic response to seed stratification germination parameters. *Seed Science and Technology*, 23, 771–8.

Edwards, D.G.W. and El-Kassaby, Y.A. 1996. The biology and management of forest seeds: genetic perspectives. *The Forestry Chronicle*, 72, 481–4.

Edwards, K.J., Barker, J.H.A., Daly, A., Jones, C. & Karp, A. 1996. Microsatellite libraries enriched for several microsatellite sequences in plants. *BioTechniques*, 20, 758–60.

Ehrenberg, C. E. & Simak, M. 1956. Flowering and pollination in Scots pine (*Pinus sylvestris* L.) Medd. *Statens Skogsforskningsinst*, 46, 1–27.

Ehrlich, P.R. & Raven, P.H. 1969. Differentiation of populations. *Science*, 165, 1228–32.

Eldridge, K.G. 1995. Pitch canker: assessing the risk to Australia. *Institute of Foresters of Australia Newsletter*, 36 (6), 9–13.

Eldridge, K.G. 1998. *Californian Radiata Pine Seed in Store: What to Do with It?* CSIRO Forestry and Forest Products, Client Report No. 389. Canberra.

Eldridge, K.G. & Griffin, A.R.1983. Selfing effects in *Eucalyptus regnans*. *Silvae Genetica*, 32, 216–21.

Eldridge, K.G., Davidson, J., Harwood, C.E. & Van Wyk, G. 1993. *Eucalypt domestication and breeding*. Clarendon Press, Oxford.

El-Kassaby, Y.A. 1989. Genetics of seed orchards: expectations and realities. In Proceedings of the 20th Southern Forest Tree Improvement Conference, June, 1989. The Southern Forest Tree Improvement Committee in cooperation with WestVaco Corporation and Clemson University. pp. 87–109. Charleston, SC, USA.

El-Kassaby, Y.A. 1991. Genetic variation within and among conifer populations: review and evaluation of methods. In *Biochemical Markers in the Population Genetics of*

Forest Trees. Eds S. Fineschi, M.E. Malvolti, F. Cannata & H.H. Hattemer. pp. 61–76. SPB Academic Publishing, bv, The Hague.

El-Kassaby, Y.A. 1992. Domestication and genetic diversity—should we be concerned? *The Forestry Chronicle*, 68, 687–700.

El-Kassaby, Y.A. 1995. Evaluation of tree–improvement delivery system: factors affecting genetic potential. *Tree Physiology*, 15, 545–50.

El-Kassaby, Y.A. & Askew, G.R. 1991. The relation between reproductive phenology and output in determining the gametic pool profile in a Douglas-fir seed orchard. *Forest Sciences*, 37, 827–35.

El-Kassaby, Y.A. & Barclay, H.J. 1992. Cost of reproduction in Douglas-fir. *Canadian Journal of Botany*, 70, 1429–32.

El-Kassaby, Y.A. & Cook, C. 1994. Female reproductive energy and reproductive success in a Douglas-fir seed orchard and its impact on genetic diversity. *Silvae Genetica*, 43, 243–6.

El-Kassaby, Y.A. & Davidson, R. 1990. Impact of crop management practices on the seed crop genetic quality in a Douglas-fir seed orchard. *Silvae Genetica*, 39, 230–7.

El-Kassaby, Y.A. & Davidson, R. 1991. Impact of pollination environment manipulation on the apparent outcrossing rate in a Douglas-fir seed orchard. *Heredity*, 66, 55–9.

El-Kassaby, Y.A. & Reynolds, S. 1990. Reproductive phenology, parental balance, and supplemental mass pollination in a Sitka-spruce seed orchard. *Forest Ecology and Management*, 31, 45–54.

El-Kassaby, Y.A. & Ritland, K. 1986a. Low levels of pollen contamination in a Douglas-fir seed orchard as detected by allozyme markers. *Silvae Genetica*, 35, 224–9.

El-Kassaby, Y.A. & Ritland, K. 1986b. The relation of outcrossing and contamination to reproductive phenology and supplemental mass pollination in a Douglas-fir seed orchard. *Silvae Genetica*, 35, 240–4.

El-Kassaby, Y.A. & Ritland, K. 1992. Frequency-dependent male reproductive success in a polycross of Douglas-fir. *Theoretical and Applied Genetics*, 83, 752–8.

El-Kassaby, Y.A. & Ritland, K. 1995a. Genetic variation in low elevation Douglas-fir of British Columbia and its relevance to gene conservation. *Biodiversity and Conservation*, 5, 779–94.

El-Kassaby, Y.A. & Ritland, K. 1995b. Impact of selection and breeding on the genetic diversity in Douglas-fir. *Biodiversity and Conservation*, 5, 795–813.

El-Kassaby, Y.A. & Thomson, A.J. 1990. Continued reliance on bulked seed orchard crops: is it reasonable? In Joint Meeting of Western Forest Genetics Association and IUFRO Working Parties S2-02-05, 06, 12 and 14, Douglas-fir, Contorta pine, Sitka spruce and Anbies Breeding and Genetic Resources. Sponsored by Weyerhaeuser Company. Olympia, Washington (Aug. 1990) 4, 56–65.

El-Kassaby, Y.A. & Thomson, A.J. 1996. Parental rank changes associated with seed biology and nursery practices in Douglas-fir. *Forest Science*, 42, 228–35.

El-Kassaby, Y.A. & Yanchuk, A.D. 1994. Genetic diversity, differentiation, and inbreeding in pacific yew from British Columbia. *Journal of Heredity*, 85, 112–17.

El-Kassaby, Y.A., Barnes, S., Cook, C. & MacLeod, D.A. 1993a. Supplemental-mass pollination success rate in a mature Douglas-fir seed orchard. *Canadian Journal of Forest Research*, 23, 1096–9.

El-Kassaby, Y.A., Chaisurisri, K., Edwards, D.G.W. & Taylor, D.W. 1993b. Genetic control of germination parameters of Douglas-fir, Sitka spruce, *Thuja plicata*, and yellow cedar and its impact on container nursery production. In *Dormancy and Barriers to Germination*, Proceedings of the International Symposium. IUFRO Project Group P2.04–00 (Seed Problems) Ed. and compiler D.G.W. Edwards. pp. 37–42. Victoria, BC, Canada.

El-Kassaby, Y.A., Davidson, R. & Webber, J.W. 1986. Genetics of seed orchards: a Douglas-fir case study. In *IUFRO Conference, A Joint Meeting of Working Parties on Breeding Theory, Progeny Testing and Seed Orchards*. Eds A.V. Hatcher & R.J. Weir (coordinators). pp. 440–50. The North Carolina State University–Industry Cooperative Tree Improvement Program. Williamsburg, VA, USA.

El-Kassaby, Y.A., Edwards, D.G.W. & Taylor, D.W. 1992. Genetic control of germination parameters in Douglas-fir and its importance for domestication. *Silvae Genetica*, 41, 48–54.

El-Kassaby, Y.A., Fashler, A.M.K. & Crown, M. 1989a. Variation in fruitfulness in a Douglas fir seed orchard and its effect on management decisions. *Silvae Genetica*, 38, 113–21.

El-Kassaby, Y.A., Fashler, A.M.K. & Sziklai, O. 1984. Reproductive phenology and its impact on genetically improved seed production in a Douglas fir seed orchard. *Silvae Genetica*, 33, 120–5.

El-Kassaby, Y.A., Ritland, K., Fashler, A.M.K. & Devitt, W.J.B. 1988. The role of reproductive phenology upon the mating system of a Douglas fir seed orchard. *Silvae Genetica*, 37, 76–82.

El-Kassaby, Y.A., Rudin, D. & Yazdani, R. 1989b. Levels of outcrossing and contamination in two *Pinus sylvestris* L. seed orchards in northern Sweden. *Scandinavian Journal of Forest Research*, 4, 41–9.

El-Kassaby, Y.A., Russell, J. & Ritland, K. 1993c. Mixed-mating in an experimental population of western redcedar, *Thuja plicata*. *Journal of Heredity*, 85, 227–31.

Ellenberg, H., Beva, H.E., Düll, R., Wirth, V., Werner, W. & Paulissen, D. 1991. *Zeigerwerte von Pflanzen in Mitteleuropa*. (Indicator Values of Plants in Central Europe). Verlag Goltze, Göttingen.

Elliot, C. 1996. Paradigms of forest conservation. *Unasylva*, 187, 3–9.

Ellstrand, N.C. 1988. Pollen as a vehicle for the escape of engineered genes? *Trends in Biotechnology*, 6, S30–1.

Ellstrand, N.C. 1992a. Gene flow by pollen: implications for plant conservation genetics. *Oikos*, 63, 77–86.

Ellstrand, N.C. 1992b. Gene flow among seed plant populations. *New Forests*, 6, 241–56.

Ellstrand, N.C. & Elam, D.R. 1993. Population genetic consequences of small population size: implications for plant conservation. *Annual Review of Ecology and Systematics*, 24, 217–42.

Ellstrand, N.C. & Marshall, D.L. 1985. Interpopulation gene flow by pollen in wild radish, *Raphanus sativus*. *American Naturalist*, 126, 606–16.

Ellstrand, N.C., Whitkus, R. & Rieseberg, L.H. 1996. Distribution of spontaneous plant hybrids. *Proceedings of the National Academy of Sciences USA*, 93, 5090–3.

Ellsworth, D.L., Rittenhouse, K.D. & Honeycutt, R.L. 1993. Artifactual variation in randomly amplified polymorphic DNA banding patterns. *BioTechniques*, 14, 214–18.

El-Mousadik, A. & Petit, R.J. 1996. Chloroplast DNA phylogeography of the argan tree of Morocco. *Molecular Ecology*, 5, 547–55.

Emlen, J.M. 1975. Niches and genes: some further thoughts. *American Naturalist*, 109, 472–6.

Endler, J.L. 1986. *Natural Selection in the Wild*. Princeton University Press, Princeton, NJ, USA.

Ennos, R.A. 1994. Estimating the relative rates of pollen and seed migration among plant populations. *Heredity*, 72, 250–9.

Enters, T. 1996. The token line: adoption and non-adoption of soil conservation practices in the Highlands of Northern Thailand. In *Soil Conservation Extension*. Eds S. Sombatpanit, M.A. Zöbisch, D.W. Sanders & M.G. Cook. pp. 417–27. Science Publishers, Lebanon.

Environment Canada 1991. The State of Canada's Environment. Environment Canada, Ottawa.

Erickson, V.J. 1987. The influence of distance and floral phenology on pollen gene flow and mating system patterns in a coastal Douglas-fir seed orchard. MS Thesis, Oregon State University, Corvallis, OR, USA.

Erickson, V.J. & Adams, W.T. 1989. Mating success in a coastal Douglas-fir seed orchard as affected by distance and floral phenology. *Canadian Journal of Forest Research*, 19, 1248–55.

Eriksson, G. 1982. Ecological genetics of conifers in Sweden. *Silva Fennica*, 16, 149–56.

Eriksson, G., Namkoong, G. & Roberds, J.H. 1993. Dynamic gene conservation for uncertain futures. *Forest Ecology and Management*, 62, 15–37.

Eriksson, U.R., Yazdani, R., Wilhelmsson, L. & Danell, Ö. 1994. Success rate of supplemental mass pollination in a seed orchard of *Pinus sylvestris* L. *Scandinavian Journal of Forest Research*, 9, 60–7.

Erwin, D.C. & Ribeiro, O.K. 1996 (Eds). Phytophthora *Diseases World-wide*. American Phytopathological Society Press, St Paul, MN, USA.

Exum, E.M. 1992. Tree in a coma. *American Forests*, 98, 20–5, 59.

Fady, B., Arbez, M. & Marpeau, A. 1992. Geographic variability of terpene composition. In *Abies cephalonica* Loudon and *Abies* species around the Aegean: hypotheses for their possible phylogeny from the Miocene. *Trends in Ecology and Evolution (Berlin)*, 6, 162–71.

Faichampa, K. 1990. Community forestry in Thailand: a case study from the North. In *Deforestation and Poverty: Can Commercial and Social Forestry Break the Vicious Circle?* Eds S. Tongpan, T. Panayotou, S. Jetanavanich, K. Faichampa & C. Mehl. pp. 138–71. The 1990 TDRI Year-End Conference Research Report No. 2. Thailand Development Research Institute, Bangkok.

Fairhead, J. & Leach, M. 1993. Contested Forests: Modern Conservation and Historical Land Use of Guinea's Ziama Reserve. Working Paper 7. Connaissance et Organisation Locales Agro-ecologiques (COLA), Kissidougou (Guinea).

Fairhead, J. & Leach, M. 1994. Contested forests: modern conservation and historical land use of Guinea's Ziama Reserve. *African Affairs*, 93 (373), 481–512.

Fairhead, J. & Leach, M. 1998. *Reframing Deforestation—Global Analysis and Local Realities: Studies in West Africa*. Global Environmental Change series. Routledge, London.

Falconer, D.S. 1981. *Introduction to Quantitative Genetics*. 2nd edn. Longman, London.

Falconer, D.S. 1989. *Introduction to Quantitative Genetics*. 3rd edn. Longmann Scientific and Technical, Harlow, Essex, UK.

Farris, M.A. & Mitton, J.B. 1984. Population density, outcrossing rate, and heterozygote superiority in ponderosa pine. *Evolution*, 38, 1151–4.

Fashler, A.M.K. & El-Kassaby, Y.A. 1987. The effect of water spray cooling treatment on reproductive phenology in a Douglas-fir seed orchard. *Silvae Genetica*, 36, 245–9.

Feder, W.A. 1968. Reduction in tobacco pollen germination and tube elongation induced by low levels of ozone. *Science*, 160, 1122.

Federov, A.A. 1966. The structure of the tropical rain forest and speciation in the humid tropics. *Journal of Ecology*, 54, 1–11.

Feret, P.P., Diebel. K.E. & Sharik, T.L. 1990. Effects of simulated acid rain on reproductive attributes of red spruce (*Picea rubens* Sarg.). *Environmental and Experimental Botany*, 30, 309–12.

Fergus, C. 1991. The Florida panther verges on extinction. *Science*, 251, 193–7.

Ferraro, P.J. & Kramer, R.K. 1995. *A Framework for Affecting Household Behavior to Promote Biodiversity Conservation*. Prepared for United States Agency for International Development, Bureau of Africa by Environmental and Natural Resources Policy and

Training (EPAT) Project. Winrock International Environmental Alliance, Arlington, VA, USA.

Filer, C. & Sekhran, N. 1998. *Loggers, Donors and Resource Owners. Policy that Works for Forests and People—Papua New Guinea.* International Institute for Environment & Development, London.

Fineschi, S., Malvolti, M.E., Cannata, F. and Hattemer, H.H. 1991. *Biochemical Markers in the Population Genetics of Forest Trees.* SPB Academic Publishing bv, The Hague.

Finkeldey, R. & Gregorius, H.-R. 1994. Genetic resources: selection criteria and design. In *Conservation and Manipulation of Genetic Resources in Forestry.* Eds Z.-S. Kim & H.H. Hattemer. pp. 322–47. Kwang Moon Kag Publishing, Seoul.

Fins, L. Friedmann, S.T. & Brotschol, J.V. (Eds) 1992. *Handbook of Quantitative Forest Genetics.* Kluwer Academic Publishers, Dordrecht.

Fisher, R.A. 1930. *The Genetic Theory of Natural Selection.* Clarendon Press, Oxford.

Fjellstrom, R.G. & Parfitt, D.E. 1994. Walnut (*Juglans* spp.) genetic diversity determined by restriction fragment length polymorphisms. *Genome*, 37, 690–700.

Fliegmann, J., Schröder, G., Schanz, S., Britsch, L. & Schröder, J. 1992. Molecular analysis of chalcone and dihydropinosylvin synthase from Scots pine (*Pinus sylvestris*), and differential regulation of these and related enzyme activities in stressed plants. *Plant Molecular Biology*, 18, 489–503.

Food and Agriculture Organization of the United Nations (FAO) 1989. *Plant Genetic Resources: Their Conservation* in situ *for Human Use.* FAO, Rome.

Food and Agriculture Organization of the United Nations (FAO) 1993. *Forest Resources Assessment, 1990: Tropical Countries.* FAO Forestry Paper 112, FAO, Rome.

Food and Agriculture Organization of the United Nations (FAO) 1997a. *State of the World's Forest Resources.* FAO, Rome.

Food and Agriculture Organization of the United Nations (FAO) 1997b. *Conservation and Sustainable Utilisation of Forest Genetic Resources.* Item 8, 13th Session, Rome, Italy, 10–13 March 1997, Briefing Note, Committee on Forestry, FAO, Rome.

Foré, S.A., Hickey, R.J., Vankat, J.L., Guttman, S.I. & Schaefer, R.L. 1992. Genetic structure after forest fragmentation: a landscape ecology perspective on *Acer saccharum. Canadian Journal of Botany*, 70, 1659–68.

Forest Resources Division, FAO, 1995. Collecting woody perennials. In *Collecting Plant Genetic Diversity:Technical Guidelines.* Eds L. Guarino, V. Ramantha Rao & R. Read. pp. 485–509. CAB International, Wallingford, UK.

Fowells, H.A. & Schubert, G.H. 1956. *Seed Crops of Forest Trees in the Pine Region of California.* US Department of Agriculture, Technical Bulletin No. 1150, USA.

Fox, J.E.D. 1968. Logging damage and the influence of climber cutting prior to logging in the lowland dipterocarp forest of Sabah. *Malaysian Forester*, XXXI (4), 326–47.

Fox, J., Kanter, R., Yarnasarn, S., Elkasingh, M. & Jones, R. 1994a. Farmer decision making and spatial variables in northern Thailand. *Environmental Management*, 18, 391–9.

Fox, J., Krummel, J., Yarnasarn, S., Ekasingh, M. & Podger, N. 1994b. Land use and landscape dynamics in northern Thailand: assessing change in three upland watersheds since 1954. In *Spatial Information and Ethnoecology: Case Studies from Indonesia, Nepal, and Thailand.* Ed. J. Fox. pp. 27–42. East-West Centre Working Paper, Environment Series No. 38. East-West Centre, Honolulu.

Fox, J., Yonzon, P. & Podger, N. 1994c. Maps, yaks, and red pandas: using GIS to model conflicts between biodiversity and human needs. In *Spatial Information and Ethnoecology: Case Studies from Indonesia, Nepal, and Thailand.* Ed. J. Fox. pp. 15–26. East-West Centre Working Paper, Environment Series No. 38. East-West Centre, Honolulu.

Frankel, O.H. 1983. The place of management in conservation. In *Genetics and Conservation: A Reference for Managing Wild Animals and Plant Populations.* Eds C.M. Schonewald-Cox, S.M. Chambers, B. MacBryde & L. Thomas. pp. 1–14. Benjamin/Cummings, Menlo Park, CA, USA.

Frankel, O.H., Brown, A.H.D. & Burdon, J.J. 1995. *The Conservation of Plant Biodiversity.* Cambridge University Press, Cambridge.

Frankie, G.W. 1976. Pollination of widely dispersed trees by animals in Central America with emphasis on bee-pollination systems. In *Tropical Trees, Variation, Breeding and Conservation.* Eds J. Burley & B.T. Styles. pp. 151–9. Linnaen Society Symposium Series 2. Academic Press, London.

Frankie, G.W. & Baker, H.G. 1974. The importance of pollinator behavior in the reproductive biology of tropical trees. *Anuario Instituto Biologico Universidad de México*, 45, Serie Botanica (1), 1–10.

Frankie, G.W. & Haber, W. 1983. Why bees move among mass-flowering neotropical trees. In *Handbook of Experimental Pollination Biology.* Eds C.E. Jones & R.J. Little. pp. 360–72. Scientific & Academic Editions, New York.

Frankie, G.W., Baker, H.G. & Opler, P.A. 1974. Comparative phenological studies of trees in tropical wet and dry forests in the lowlands of Costa Rica. *Journal of Ecology*, 62, 881–919.

Frankie, G.W., Newstrom, L., Vinson, S.B. & Barthell, J.F. 1993. Nesting-habitat preferences of selected *Centris* bee species in Costa Rican dry forest. *Biotropica*, 25, 322–33.

Frankie, G.W., Opler, P.A. & Bawa, K.S. 1976. Foraging behaviour of solitary bees: implications for outcrossing of a neotropical forest tree species. *Journal of Ecology*, 64, 1049–57.

Friedman, S.T. & Adams, W.T. 1985. Estimation of gene flow into two seed orchards of loblolly pine (*Pinus taeda* L.). *Theoretical and Applied Genetics*, 69, 609–15.

Fritzemeier, K.-H. and Kindl, H. 1981. Coordinate induction by UV light of stilbene synthase, phenylalanine ammonia-lyase and cinnamate 4-hydroxylase in leaves of Vitaceae. *Planta*, 151, 48–52.

Führer, E. 1990. Forest decline in Central Europe: additional aspects of its cause. *Forest Ecology and Management*, 37, 249–57.

Gaiotto, F.A., Bramucci, M. & Grattapaglia, D. 1997. Estimation of outcrossing rate in a breeding population of *Eucalyptus urophylla* with dominant RAPD and AFLP markers. *Theoretical and Applied Genetics*, 95, 842–9.

Galliano, H., Cabané, M., Eckerskorn, C., Lottspeich, F., Sandermann, H. Jr. & Ernst, D. 1993a. Molecular cloning, sequence analysis and elcitor-/ozone—induced accumulation of cinnamyl alcohol dehydrogenase from Norway spruce (*Picea abies* L.). *Plant Molecular Biology*, 23, 145–56.

Galliano, H., Heller, W. & Sandermann Jr., H. 1993b. Ozone induction and purification of spruce cinnamyl alcohol dehydrogenase. *Phytochemistry*, 32, 557–63.

Garsed, S.G. & Rutter, A.J. 1982. Relative performance of conifer populations in various tests for sensitivity to SO_2, and the implications for selecting trees for planting in polluted areas. *New Phytologist*, 92, 349–67.

Gartside, D.W. & McNeilly, T. 1974. The potential for evolution of heavy metal tolerances in plants. II. Copper tolerance in normal populations of different plant species. *Heredity*, 32, 335–48.

Geburek, Th. & Knowles, P. 1992. Ecological-genetic investigations in environmentally stressed mature sugar maple (*Acer saccharum* Marsh.) populations. *Water, Air, and Soil Pollution*, 62, 261–8.

Geburek, Th. & Scholz, F. 1989. Response of *Picea abies* (L.) Karst. provenances to aluminum in hydroponics. In *Genetic Effects of Air Pollutants in Forest Tree Populations*. Eds F. Scholz, H.-R. Gregorius & D. Rudin. pp. 55–65. Springer, Berlin.

Geburek, Th. & Scholz, F. 1992. Response of *Picea abies* (L.) Karst. provenances to sulphur dioxide and aluminum: a pilot study. *Water, Air, and Soil Pollution*, 62, 227–32.

Geburek, Th., Scholz, F., Knabe, W. & Vornweg, A. 1987. Genetic studies by isozyme gene loci on tolerance and sensitivity in an air polluted *Pinus sylvestris* field trial. *Silvae Genetica*, 36, 49–53.

Gentry, A.H. 1974. Flowering phenology and diversity in tropical Bignoniaceae. *Biotropica*, 6, 64–8.

Genys, J.B. & Heggestad, H.E. 1978. Susceptibility of different species, clones and strains of pines to acute injury caused by ozone and sulphur dioxide. *Plant Disease Reporter*, 62, 687–91.

Genys, J.B. & Heggestad, H.E. 1983. Relative sensitivity of various types of eastern white pine, *Pinus strobus*, to

sulphur dioxide. *Canadian Journal of Forest Research*, 13, 1262–5.

Ghazoul, J., Liston, K.A. & Boyle, T.J.B. 1998. Disturbance induced density-dependent seed set in *Shorea siamensis* (Dipterocapaceae), a tropical forest tree. *Journal of Ecology*, 86, 462–73.

Gibbs, P.E. 1988. Self-incompatibility mechanisms in flowering plants: some complications and clarifications. *Lagascalia*, 15(Extra), 17–28.

Gibbs, P.E. & Bianchi, M. 1993. Post-pollination events in species of *Chorisia* (Bombacaceae) and *Tabebuia* (Bignoniaceae) with late-acting self-incompatibility. *Botanica Acta*, 106, 64–71.

Giertych, M. 1984. Istebna spruce in the light of international provenance trials. *Sylwan*, 128, 27–42.

Gilbert, L.E. 1975. Ecological consequences of a co-evolved mutualism between butterflies and plants. In *Coevolution of Animals and Plants*. Eds L.E. Gilbert & P.H. Raven. pp. 210–40. University of Texas Press, Austin, TX, USA.

Gillies, A.M.C., Cornelius, J.P., Newton, A.C., Navarro, C., Hernandez, M. & Wilson, J. 1997. Genetic variation in Costa Rican populations of the tropical timber species *Cedrela odorata* L., assessed using RAPDs. *Molecular Ecology*, 6, 1133–45.

Gilmour, D. 1995. Conservation and development: seeking the linkages. In *Management of Tropical Forests: Towards an Integrated Perspective*. Ed. Ø. Sandbukt. pp. 255–67. Center for Development and the Environment, University of Oslo, Oslo.

Gilmour, D.A. & Fisher, R.J. 1991. *Villagers, Forests and Foresters*. Sahayogi Press, Kathmandu, Nepal.

Glaubitz, J.C. 1995. Applications of Molecular Markers to Forest Genetics: Gene Expression, Genetic Linkage Mapping, and Genetic Diversity. PhD dissertation, University of British Columbia, Vancouver.

Gleeson, S.K. 1982. Heterodichogamy in walnuts: inheritance and stable ratios. *Evolution*, 36, 892–902.

Glowka, L., Burhenne-Guilmin, F., Synge, H., McNeely, J.A. & Gündling, L. 1994. *A Guide to the Convention on Biological Diversity*. Environmental Policy and Law Paper No. 30. IUCN (World Conservation Union), Gland.

Gockel, J., Harr, B., Schlotterer, C., Arnold, W., Gerlach, G. & Tautz, D. 1997. Isolation and characterization of microsatellite loci from *Apodermus flavicolis* (Rodentia, Muridae) and *Clethrionomys glareolus* (Rodentia, Cricetidae). *Molecular Ecology*, 6, 597–9.

Golding, G.B. & Strobeck, C. 1983. Increased number of alleles found in hybrid populations due to intragenic recombination. *Evolution*, 37, 17–29.

Gontcharenko, G.G., Padutov, V.E. & Silin, A.E. 1993a. Allozyme variation in natural populations of Eurasian pines. I. Population structure, genetic variation and differentiation in *Pinus pumila* (Pall.) Regel from Chukotsk and Sakhalin. *Silvae Genetica*, 42, 237–46.

Gontcharenko, G.G., Padutov, V.E. & Silin, A.E. 1993b. Allozyme variation in natural populations of Eurasian pines. II. Genetic variation, diversity, differentiation, and gene flow in *Pinus sibirica* Du Tour in some lowland and mountain populations. *Silvae Genetica*, 42, 247–53.

Gontcharenko, G.G., Silin, A.E. & Padutov, V.E. 1994. Allozyme variation in natural populations of Eurasian pines. III. Population structure, diversity, differentiation and gene flow in central and isolated populations of *Pinus sylvestris* L. in eastern Europe and Siberia. *Silvae Genetica*, 43, 119–32.

Gooding, G. 1998. Genetic diversity in Willamette Valley ponderosa pine. MSc thesis, Oregon State University, OR, USA.

Goodnight, C.J. 1988. Epistasis and the effect of founder events on the additive genetic variance. *Evolution* 42, 441–54.

Gordon, T.R., Wikler, K.R., Storer, A.J. & Wood, D.L. 1997. Pitch canker and its potential impacts on monterey pine forests in California. *Fremontia*, 25, 5–9.

Govindaraju, D.R. 1989. Estimates of gene flow in forest trees. *Biological Journal of the Linnean Society*, 37, 345–57.

Grant, P.R. & Grant, B.R. 1992. Hybridization of bird species. *Science*, 256, 193–7.

Grant, V. 1981. *Plant Speciation*. Columbia University Press, New York.

Grattapaglia, D. & Sederoff, R. 1994. Genetic linkage maps of *Eucalyptus grandis* and *Eucalyptus urophylla* using a pseudo-testcross mapping strategy and RAPD markers. *Genetics*, 137, 1121–37.

Graudal, L., Kjaer, E., Thomsen, A. & Larsen, A.B. 1997. Planning national programmes for conservation of forest genetic resources. Technical Note 48. Danida Forest Seed Centre, Humlebaek, Denmark.

Grayson, A.J. & Maynard, W.B. 1997. *The World's Forests—Rio +5: International Initiatives Towards Sustainable Management*. Commonwealth Forestry Association, Oxford.

Greenwood, M. & Rucker, T. 1985. Estimating pollen contamination in loblolly pine seed orchards by pollen trapping. In Proceedings of the 18th Southern Forest Tree Improvement Conference, pp. 179–86. 21–23 May 1985, Long Beach, Mississippi, copies available from The National Technical Information Service, Springfield, VA, USA.

Gregorius, H.-R. 1980. The probability of losing an allele when diploid genotypes are sampled. *Biometrics*, 36, 643–52.

Gregorius, H.-R. 1995. Measurement of genetic diversity with special reference to the adaptive potential of populations. In *Measuring and Monitoring Biodiversity in Tropical and Temperate Forests*. Eds T.J.B. Boyle & B. Boontawee. pp. 145–75. Center for International Forestry Research, Bogor, Indonesia.

Gregorius, H.-R. & Namkoong, G. 1986. Joint analysis of genotypic and environmental functions. *Theoretical and Applied Genetics*, 72, 413–22.

Gregorius, H.-R. & Namkoong, G. 1987. Resolving the dilemma of interaction, separability and additivity. *Mathematical Bioscience*, 85, 51–69.

Gregorius, H.-R. & Roberds, J.H. 1986. Measurement of genetical differentiation among subpopulations. *Theoretical and Applied Genetics*, 71, 826–34.

Gregorius, H.-R., Hattemer, T.T., Bergmann, F. & Müller-Starck, G. 1985. Unweltbelastung und Anpassungsfähigkeit von Baumpopulationen. *Silvae Genetica*, 34, 230–41.

Griffin, A.R. 1982. Clonal variation in radiata pine seed orchards. I. some flowering, cone and seed production traits. *Australian Journal of Forest Research*, 12, 295–302.

Griffin, A.R. 1991. Effects of inbreeding on growth of forest trees and implications for management of seed supplies for plantation programmes. In *Reproductive Ecology of Tropical Forest Plants*. Eds K.S. Bawa & M. Hadley. pp. 355–74. Man and the Biosphere Series, vol. 7. Unesco/Parthenon Publishing, Paris & Carnforth.

Griffin, A.R. & Lindgren, D. 1985. Effect of inbreeding on production of filled seed in *Pinus radiata*—experimental results and a model of gene action. *Theoretical and Applied Genetics*, 71, 334–43.

Griffin, C.R., Shallenberger, R.J. & Fefer, S.I. 1989. Hawaii's endangered waterbirds: a resource management challenge. In *Proceedings of the Freshwater Wetlands and Wildlife Symposium*. Eds R.R. Sharitz & I.W. Gibbons. pp. 155–69. Savannah River Ecology Laboratory, Aiken, SC, USA.

Griffiths, A.J.F., Miller, J.H. & Suzuki, D.T. 1996. *An Introduction to Genetic Analysis*. Freeman, New York.

Großkopf, E., Wegener-Strake, A., Sandermann, H. Jr. & Ernst, D. 1994. Ozone-induced metabolic changes in Scots pine: mRNA isolation and analysis of in vitro translated proteins. *Canadian Journal of Forest Research*, 24, 2030–3.

Grove, R.H. 1995. *Green Imperialism*. Cambridge University Press, Cambridge.

Grumbine, R.E. 1992. *Ghost Bears*. Island Press, Washington, DC.

Gunatilake, H.M. 1995. An economic impact assessment of the proposed conservation program on peripheral communities in the Knuckles Forest Range of Sri Lanka. *Journal of Sustainable Forestry*, 3, 1–14.

Gupta, B.N. 1994. India. In *Non Wood Forest Products in Asia*. Eds P.B. Durst, W. Ulrich & M. Kashio. pp. 19–48. RAPA Publication 1994/28. Food and Agriculture Organization of the United Nations, Bangkok.

Guries, R.P. & Ledig, F.T. 1982. Genetic diversity and population structure in pitch pine (*Pinus rigida* Mill.). *Evolution* 36, 387–402.

Ha, C.O., Sands, V.E., Soepadmo, E. & Jong, K. 1988. Reproductive patterns of selected understorey trees in the Malaysian rain forest: the apomictic species. *Botanical Journal of the Linnean Society*, 97, 317–31.

Haines, R.J. 1994. *Biotechnology in Forest Tree Improvement*. Forestry Paper 118. Food and Agriculture Organization of the United Nations, Rome.

Hale, P. & Lamb, D. (Eds). 1997. *Conservation Outside Nature Reserves*. Centre for Conservation Biology, University of Queensland, Brisbane.

Hall, P. & Bawa, K. 1993. Methods to assess the impact of extraction of non-timber forest products on plant populations. *Economic Botany*, 47, 234–47.

Hall, P., Orrell, L.C. & Bawa, K. 1994. Genetic diversity and the mating system in a tropical tree, *Carapa guianensis* (Meliaceae). *American Journal of Botany*, 81, 1104–11.

Hall, P., Walker, S. & Bawa, K. 1996. Effect of forest fragmentation on genetic diversity and mating system in a tropical tree, *Pithecellobium elegans*. *Conservation Biology*, 10, 757–68.

Halladay, P. & Gilmour, D.A. 1995. *Conserving Biodiversity Outside Protected Areas—The Role of Traditional Agro-ecosystems*. IUCN (World Conservation Union), Gland.

Hamrick, J.L. 1987. Gene flow and distribution of genetic variation in plant populations. In *Differentiation Patterns in Higher Plants*. Ed. K. Urbanska. pp. 53–76. Academic Press, New York.

Hamrick, J.L. 1994a. Distribution of genetic diversity in tropical tree populations: implications for the conservation of genetic resources. In Breeding Tropical Trees. Eds C. Lambeth & W. Dvorak. pp. 74–82. IUFRO, Cartagena-Cali, Columbia.

Hamrick, J.L. 1994b. Genetic diversity and conservation in tropical forests. In *Proceedings: International Symposium on Genetic Conservation and Production of Tropical Forest Tree Seed*. Eds R.M. Drysdale, S.E.T. John & A.C. Yappa. ASEAN Forest Tree Seed Centre Project, Muak-Lek, Saraburi, Thailand.

Hamrick, J.L & Godt, M.J.W. 1989. Allozyme diversity in plant species. In *Plant Population Genetics, Breeding and Genetic Resources*. Eds A.H.D. Brown, M.T. Clegg, A.L. Kahler & B.S. Weir. pp. 43–63. Sinauer Associates, Sunderland, MA, USA.

Hamrick, J.L. & Godt, M.J.W. 1996a. Effects of life history traits on genetic diversity in plant species. *Proceedings of the Royal Academy of Sciences B*, 351, 1291–8.

Hamrick, J.L. & Godt, M.J.W. 1996b. Conservation genetics of endemic plant species. In *Conservation Genetics: Case Histories from Nature*. Eds J.C. Avise & J.L. Hamrick. pp. 281–304. Chapman and Hall, New York.

Hamrick, J.L. & Loveless, M.D. 1989. Genetic structure of tropical populations: associations with reproductive biology. In *The Evolutionary Ecology of Plants*. Eds J.H. Brock & Y.B. Linhart. Westview Press, Boulder, CO, USA.

Hamrick, J.L. & Murawski, D.A. 1990. The breeding structure of tropical tree populations. *Plant Species Biology*, 5, 157–65.

Hamrick, J.L. & Nason, J.D. 1996. Consequences of dispersal in plants. In *Population Dynamics in Ecological Space and Time*. Eds O.E. Rhodes, R.K. Chesser & M.H. Smith. pp. 203–36. University of Chicago Press, Chicago.

Hamrick, J.L., Godt, M.J.W. & Sherman-Broyles, S. 1992. Factors influencing levels of genetic diversity in woody plant species. *New Forests*, 6, 95–124.

Hamrick, J.L., Linhart, Y.B. & Mitton, J.B. 1979. Relationship between life history characteristics and electrophoretically detectable genetic variation in plants. *Annual Review of Ecology and Systematics*, 10, 175–200.

Hamrick, J.L., Murawski, D.A. & Nason, J.D. 1993. The influence of seed dispersal mechanisms on the genetic structure of tropical tree populations. *Vegetatio*, 107/108, 281–97.

Hamrick, J.L., Schnabel, A. & Wells, P.V. 1994. Distributions of genetic diversity within and among populations of Great Basin conifers. In *Natural History of the Colorado Plateau and Great Basin*. Eds K.T. Harper, L.L. St Clair, K.H. Thorne & W.W. Hess. pp. 147–61. University of Colorado Press, Niwot, CO, USA.

Hanover, J.W. 1992. Applications of terpene analysis in forest genetics. *New Forests*, 6, 159–78.

Hanski, I. & Gilpin, M.E. (Eds) 1997. *Metapopulation Biology. Ecology, Genetics, and Evolution*. Academic Press, New York.

Hardner, C.M., Potts, B.M. & Gore, P.L. 1998. The relationship between cross success and spatial proximity of *Eucalyptus globulus* ssp. *globulus* parents. *Evolution* 52, 614–18.

Hardner, C.M., Vaillancourt, R.E. & Potts, B.M. 1996. Stand density influences outcrossing rate and growth of open-pollinated families of *Eucalyptus globulus*. *Silvae Genetica*, 45, 266–8.

Hardy, G.H. 1908. Mendelian proportions in a mixed population. *Science*, 28, 49–50.

Harju, A. & Muona, O. 1989. Background pollination in *Pinus sylvestris* seed orchards. *Scandinavian Journal of Forest Research*, 4, 513–20.

Harju, A.M. & Nikkanen, T. 1996. Reproductive success of orchard and nonorchard pollens during different stages of pollen shedding in a Scots pine seed orchard. *Canadian Journal of Forest Research*, 26, 1096–1102.

Harris, L.D. 1984. *The Fragmented Forest: Island Biogeography Theory and the Preservation of Biotic Diversity*. The University of Chicago Press, Chicago.

Harris, S.A., Fagg, C.W. & Barnes, R.D. 1997. Isozyme variation in *Faidherbia albida* (Leguminosae; Mimosoideae). *Plant Systematics and Evolution*, 207, 119–32.

Harrison, R.G. 1986. Pattern and process in a narrow hybrid zone. *Heredity*, 56, 337–49.

Harry, D.E., Temesgen, B. & Neale, D.B. 1998. Codominant PCR-based markers for *Pinus taeda* developed from mapped cDNA clones. *Theoretical and Applied Genetics*, 97, 327–36.

Hart, C. 1973. *British Trees in Colour*. Book Club Associates, London.

Hart, J.H. 1981. Role of phytostilbenes in decay and disease resistance. *Annual Review of Phytopathology*, 19, 437–58.

Hartl, D.L. 1980. *Principles of Population Genetics*. Sinauer Associates, Sunderland, MA, USA.

Hartl, D.L. & Clark, A.G. 1989. *Principles of Population Genetics*. 2nd edn. Sinauer Associates, Sunderland, MA, USA.

Harwood, C.E., Moran, G.F. & Bell, J.C. 1997. Genetic differentiation in natural populations of *Grevillea robusta*. *Australian Journal of Botany*, 45, 669–78.

Haury, D. & Saragih, B. 1995. Processing and Marketing Rattan. Promotion of Sustainable Forest Management Systems in East Kalimantan (SFMP), Samarinda, Indonesia.

Hayashi, K. 1992. PCR-SSCP: a method for detection of mutations. *GATA*, 9, 73–9.

Hedgcock, G.G. 1912. Winter-killing and smelter-injury in the forest of Montana. *Torreya*, 12, 25–30.

Hedrick, P.W. 1984. *Population Biology: The Evolution and Ecology of Populations*. Jones Bartlett Publishers, Boston.

Hedrick, P.W. 1994. Purging inbreeding depression and the probability of extinction: full-sib mating. *Heredity*, 73, 363–72.

Hedrick, P.W. & Gilpin, M.E. 1997. Genetic effective size of a metapopulation. In *Metapopulation Biology. Ecology, Genetics and Evolution*. Eds I. Hanski & M.E. Gilpin. pp. 165–81. Academic Press, New York.

Hedrick, P.W. & Savolainen, O. 1996. Molecular and adaptive variation: a perspective for endangered plants. In *Southwestern Rare and Endangered Plants: Proceedings of the Second Conference*. Eds J. Maschinski, H.D. Hammond, & L. Holter. General Technical Report RM-GTR-283. pp. 92–102. United States Department of Agriculture, Forest Service, Rocky Mountain and Range Experiment Station. Fort Collins, CO, USA.

Hegde, R., Suryaprakash, S., Achot, L. & Bawa, K.S. 1996. Extraction of non-timber forest products in the forests of Biligiri Rangan Hills, India. 1. Contribution to rural income. *Economic Botany*, 50, 243–51.

Heinrich, B. 1975. Energetics of pollination. *Annual Review of Ecology and Systematics*, 6, 139–70.

Heithaus, E.R., Fleming, T.H. & Opler, P.A. 1975. Foraging patterns and resource utilisation in seven species of bats in a seasonal tropical forest. *Ecology*, 56, 841–54.

Henttonen, H., Kanninen, M., Nygren, M. & Ojansuu, R. 1986. The maturation of *Pinus sylvestris* seeds in relation to temperature climate in northern Finland. *Scandinavian Journal of Forestry Research*, 1, 243–9.

Hertel, H. & Zander, M. 1991. Genetische Unterschiede zwischen gesunden und geschädigten Buchen eines belasteten Bestandes. pp. 227–9. Berichte des Forschungszentrums Waldökosysteme, Reihe B, Bd. 22, Göttingen, Germany.

Hewitt, G.M. 1996. Some consequences of ice ages, and their role in divergence and speciation. *Biological Journal of the Linnean Society*, 58, 247–76.

Heydon, M.J. & Bulloh, P. 1996. The impact of selective logging on sympatric civet species in Borneo. *Oryx*, 30, 31–6.

Hickey, D.A. & McNeilly, T. 1975. Competition between metal tolerant and normal plant populations; a field experiment on normal soil. *Evolution*, 29, 458–64.

Hodges, S.A., Burke, J.M. & Arnold, M.L. 1996. Natural formation of *Iris* hybrids: experimental evidence on the establishment of hybrid zones. *Evolution*, 50, 2504–9.

Hodgson, L.M. 1976. Some aspects of flowering and reproductive biology in *Eucalyptus grandis* at J.D.M. Keet Forest Research Station. 2. The fruit, seed, seedlings, self-fertility, selfing and inbreeding effects. *South African Forestry Journal*, 98, 32–43.

Holsinger, K.E. 1996. The scope and the limits of conservation genetics. *Evolution*, 50, 2558–61.

Holsinger, K.S. 1993. The evolutionary dynamics of fragmented plant populations. In *Biotic Interactions and Global Climate Change*. Eds P.M. Kareiva, J.G. Kingsolver & R.B. Huey. pp. 198–216. Sinauer Associates, Sunderland, MA, USA.

Homma, A.K.O. 1992. The dynamics of extraction in Amazonia: a historical perspective. *Advances in Economic Botany*, 9, 23–32.

Hosius, B., Bergmann, F. & Hattemer, H.H. 1996. Physiologische und genetische Anpassung von Fichtensämlingen verschiedener Provenienz an schwermetallkontaminierten Böden. Forstarchiv, 67, 108–14.

House, A.P.N. & Bell, J.C. 1994. Isozyme variation and mating system in *Eucalyptus urophylla* S.T.Blake. *Silvae Genetica*, 43, 167–76.

Houston, D.B. & Stairs, G.R. 1973. Genetic control of sulphur dioxide and ozone tolerance of eastern white pine. *Forest Science*, 19, 267–71.

Hoyt, E. 1988. *Conserving the Wild Relatives of Crops*. International Board for Plant Genetic Resources, IUCN (World Conservation Union), Rome.

Huang, H., Dane, F. & Kubisiak, T.L. 1998. Allozyme and RAPD analysis of the genetic diversity and geographic variation in wild populations of the American chestnut (Fagaceae). *American Journal of Botany*, 85, 1013–21.

Hubbard, A.L., McOrist, S., Jones, T.W., Boid, R., Scott, R. & Easterbee, N. 1992. Is survival of European wildcats *Felis silvestris* in Britain threatened by interbreeding with domestic cats? *Biological Conservation*, 61, 203–8.

Hubbs, C.L. 1955. Hybridization between fish species in nature. *Systematic Zoology*, 4, 1–20.

Hughes, C.E. 1998. *Leucaena —A Genetic Resources Handbook*. Tropical Forestry Papers No. 37. Oxford Forestry Institute, Oxford.

Hughes, R.N. & Cox, R.M. 1994. Acidic fog and temperature effects on stigmatic receptivity in two birch species. *Journal of Environmental Quality*, 23, 686–92.

Hunt, D.M. 1997. Forest Reservation in New Zealand. Ecology Division, Department of Scientific and Industrial Research, NZ.

Hunt, R.S. & Van Sickle, G.A. 1984. Variation in susceptibility to sweet fern rust among *Pinus contorta* and *P. banksiana. Canadian Journal of Forest Research*, 14, 672–5.

Huntley, B. & Birks, H.J.B. 1983. *An Atlas of Past and Present Pollen Maps for Europe: 0–13,000 Years Ago*. Cambridge University Press, Cambridge.

Huttunen, S. 1978. The effect of air pollution on provenances of Scots pine and Norway spruce in northern Finland. *Silva Fennica*, 12, 1–16.

Huttunen, S. & Törmälehto, H. 1982. Air pollution resistance of some Finnish *Pinus sylvestris* L. provenances. *Aquilo Seria Botanica*, 18, 1–9.

Ingles, A.W. 1996. *The Role of Participatory Learning and Action Approaches in Integrated Conservation and Development Projects*. Project Methodology Paper 96/1. NTFP Project. Department of Forestry (Lao PDR) and The World Conservation Union, Vientiane.

Ingram, C. 1988. *The Evolutionary Basis of Ecological Amplitude*. PhD dissertation, University of Liverpool, UK.

Innes, J.L. 1994. The occurrence of flowering and fruiting on individual trees over 3 years and their effects on subsequent crown condition. *Trends in Ecology and Evolution*, 8, 139–50.

Intachat, J., Holloway, J.D. & Speight, M.R. 1997. The effects of different forest mangement practices on geometroid moth populations and their diversity in Peninsular Malaysia. *Journal of Tropical Forest Science*, 9, 411–30.

International Institute for Sustainable Development (IISD) 1997. *Earth Negotiations Bulletin* 9 (76), 8 December 1997. http://www.iisd.ca/linkages/iu.html.

International Institute for Sustainable Development (IISD) 1998. *Earth Negotiations Bulletin* 9 (96), 18 May 1998. http://www.iisd.ca/linkages/biodiv.html.

International Institute for Sustainable Development (IISD) 1999. Ministerial conference on the protection of forests in Europe liaison unit. In *Helsinki: European Criteria and Most Suitable Quantitative Indicators for Sustainable Forest Management*. http://www.iisd.ca/linkages/forestry/indicat.html.

International Tropical Timber Organization 1992a. ITTO guidelines for the sustainable management of natural tropical forests. *ITTO Policy Development Series No. 1*, Yokohama, Japan.

International Tropical Timber Organization 1992b. Criteria for the measurement of sustainable tropical forest management. *ITTO Policy Development Series No. 3*, Yokohama, Japan.

Isabel, N., Beaulieu, J. & Bousquet, J. 1995. Complete congruence between gene diversity estimates derived from genotypic data at enzyme and RAPD loci in black spruce. *Proceedings of the National Academy of Sciences, USA*, 92, 6369–73.

Isagi, Y. & Suhandono, S. 1997. PCR primers amplifying microsatellite loci of *Quercus myrsinifolia* Blume and their conservation between oak species. *Molecular Ecology*, 6, 897–9.

Ives, J.D. & Messerli, B. 1989. *The Himalayan Dilemma. Reconciling Development and Conservation*. Routledge, London.

Jalkanen, R. 1982. *Lophodermella sulcigena* in clones and progenies of scots pine in Finland. In *Resistance to Disease and Pests in Forest Trees*. Eds H.M. Heybroek, B.R. Stephan & K. von Weissenberg. pp. 441–7. Centre for Agricultural Publishing and Documentation, Wageningen, Netherlands.

James, S.H. & Kennington, W.J. 1993. Selection against homozygotes and resource allocation in the mating system of *Eucalyptus camaldulensis* Dehnh. *Australian Journal of Botany*, 41, 381–91.

Janzen, D.H. 1967. Synchronization of sexual reproduction of trees within the dry season in Central America. *Evolution*, 21, 620–37.

Janzen, D.H. 1971. Euglossine bees as long-distance pollinators of tropical plants. *Science*, 171, 203–5.

Janzen, D.H. 1989. Natural history of a wind-pollinated Central American dry forest legume tree (*Ateleia herbert-smithii* Pittier). In *Advances in Legume Biology*. Eds C.H. Stirton & J.L. Zarucchi. pp. 293–376. Monographs in Systematic Botany from the Missouri Botanical Garden, no. 29. Missouri Botanical Garden, St Louis, MO, USA.

Jarne, P. & Lagoda, P.J.L. 1996. Microsatellites, from molecules to populations and back. *Trends in Ecology and Evolution*, 11, 424–9.

John, B. 1990. *Meiosis*. Cambridge University Press, Cambridge.

Johnsen, O. 1989a. Phenotypic changes in progenies of northern clones of *Picea abies* (L.) Karst. grown in southern seed orchard. I. Frost hardiness in a phytotron experiment. *Scandinavian Journal of Forestry Research*, 4, 317–30.

Johnsen, O. 1989b. Phenotypic changes in progenies of north clones of *Picea abies* (L.) Karst. grown in southern seed orchard.II. Seasonal growth rhythm and height in field trials. *Scandinavian Journal of Forestry Research*, 4, 331–41.

Johnsen, O. 1989c. Phenotypic changes in progenies of northern clones of *Picea abies* (L.) Karst. grown in southern seed orchard. III. Climatic damage and growth rhythm in a progeny trial. *Scandinavian Journal of Forestry Research*, 4, 343–50.

Jones, C.J., Edwards, K.J., Castaglione, S., Winfield, M.O., Sala, F., van de Wiel, C., Bredemeijer, G., Vosman, B., Matthes, M., Daly, A., Brettschneider, R., Bettini, P., Buiatti, M., Maestri, E., Malcevschi, A., Marmiroli, N., Aert, R., Volckaert, G., Rueda, J., Linacero, R., Vazquez, A. & Karp, A. 1997. Reproducibility testing of RAPD, AFLP, and SSR markers in plants by a network of European laboratories. *Molecular Breeding*, 3, 381–90.

Jonsson, T. & Lindgren, P. 1990. *Logging Technology for Tropical Forests—For or Against?* The Forest Operations Institute 'Skogsarbeten', Sweden.

Jung, T., Blaschke; H. & Neumann, P. 1996. Isolation, identification and pathogenicity of *Phytophthora* species from declining oak stands. *European Journal of Forest Pathology*, 26, 253–72.

Kageyama, P.Y. 1990. Genetic structure of tropical tree species of Brazil. In *Reproductive Ecology of Tropical Forest Plants*. Eds K.S. Bawa & M. Hadley. pp. 375–87. Man and the Biosphere Series, vol. 7. Unesco/Parthenon Publishing, Paris/Carnforth.

Kainulainen, P., Holopainen, J.K. & Oksanen, J. 1995. Effects of SO_2 on the concentration of carbohydrates and secondary compounds in Scots pine (*Pinus sylvestris* L.) and Norway spruce (*Picea abies* (L.) Karst.) seedlings. *New Phytologist*, 130, 231–8.

Kalchenko, V.A., Rubanovich, A.V., Fedotov, I.S. & Arkhipov, N.P. 1993. Genetic effects induced by the Chernobyl accident in the gametes of Scots pine (*Pinus sylvestris* L.). *Genetica Moskva*, 29, 1205–12.

Kanashiro, M. 1986. Reproductive biology of *Cordia goeldiana*; a neotropical heterostylous species. MSc thesis, North Carolina State University, NC, USA.

Kanazin, V., Marak, L.F. & Shoemaker, R.C. 1996. Resistance gene analogues are conserved and clustered in soybean. *Proceedings of the Academy of Sciences, USA*, 93, 11746–50.

Kannenberg, L.W. 1983. Utilization of genetic diversity in crop breeding. In *Plant Gene Resources: A Conservation Imperative*. Eds C.W. Yeatman, D. Kafton & G. Wilkes. pp. 93–110. AAAS Selected Symposium 87. Westview Press, Boulder, CO, USA.

Kanowski, P.J. & Boshier, D.H. 1997. Conserving the genetic resources of trees *in situ*. In *Plant Conservation: The in situ Approach*. Eds N. Maxted, B.V. Ford-Lloyd and J.G. Hawkes. pp. 207–19. Chapman and Hall, London.

Kanowski, P.J., Cork, S.J., Lamb, D. & Dudley, N. 1998. Assessing the success of off-reserve forest management in contributing to biodiversity conservation. In *Fostering Stakeholder Input to Advance Development of Scientifically-Based Indicators for Sustainable Forest Management*. Proceedings of the International Union of Forest Research Organisations/Food and Agriculture Organization of the United Nations/Centre for International Forestry Research Conference, 24–28 August 1998, Melbourne.

Kanowski, P.J., Gilmour, D.A., Margules, C.R. & Potter, C.S. 1999. *International Forest Conservation: Protected Areas and Beyond*. Discussion Paper to Intergovernmental Forum on Forests. Commonwealth of Australia, Canberra.

Kanowski, P.J., Margules, C.R. & Lindenmayer, D.B. 1997. *Forests and Biological Diversity*. Discussion Paper, Secretariat to the Convention on Biological Diversity, Montreal.

Kant, S. 1997. Integration of biodiversity conservation in tropical forest and economic development of local communities. *Journal of Sustainable Forestry*, 4, 33–61.

Kaplan, H. & Kopischke, K. 1992. Resource use, traditional technology, and change among native peoples of Lowland South America. In *Conservation of Neotropical Forests*. Eds K.H. Redford & C. Padoch. pp. 83–107. Columbia University Press, New York.

Karhu, A., Hurme, P., Karjalainen, M., Karvonen, P., Kärkkäinen, K., Neale, D. & Savolainen, O. 1996. Do molecular markers reflect patterns of differentiation in adaptive traits of conifers? *Theoretical and Applied Genetics*, 93, 215–21.

Kärkkäinen, K., Koski, V. & Savolainen, O. 1996. Geographical variation in the inbreeding depression of Scots pine. *Evolution*, 50, 111–19.

Karnosky, D.F. 1977. Evidence for genetic control of response to sulphur dioxide and ozone in *Populus tremuloides. Canadian Journal of Forest Research*, 7, 437–40.

Karnosky, D.F. 1989. Air pollution induced population changes in North American forest. In *Pollution and Forest Decline*. Eds J.B. Bucher & I. Bucher-Wallin. pp. 315–17. Proceedings of the 14th International Meeting for Specialists in Air Pollution Effects on Forest Ecosystems. Interlaken, Switzerland.

Karnosky, D.F. & Steiner, K.C. 1981. Provenance and family variation in response of *Fraxinus americana* and *F. pennsylvanica* to ozone and sulphur dioxide. *Phytopathology*, 71, 804–7.

Karolewski, P. 1989. Free proline content of 18 European Provenances of Scotch Pine *(Pinus sylvestris* L.) and their susceptibility to the action of SO_2, NO_2, and HF. *Folia Drendologica*, 16, 365–82.

Karolewski, P. & Białobok, S. 1979. Wpływ dwutlenku siarki, ozonu, mieszaniny tych gazów i fluowodoru na uszkozenie igieł modrzewia europejskiego. *Arboretum Kórnickie*, 24, 297–305.

Karron, J.D., Thumser, N.N., Tucker, R. & Hessenauer, A.J. 1995. The influence of population density on outcrossing rates in *Mimulus ringens. Heredity*, 75, 175–80.

Katzman, M. & Cale, W. 1990. Tropical forest preservation using economic incentives: a proposal of conservation easements. *BioScience*, 40, 827–32.

Kaukinen, K.H., Tranbarger, T.J. & Misra, S. 1996. Post-germination-induced and hormonally dependent expression of low-molecular-weight heat shock protein genes in Douglas fir. *Plant Molecular Biology*, 30, 1115–28.

Kauppi, P. & Posch, M. 1988. A case study of the effects of CO_2-induced Fclimatic warming on forest growth and the forest sector: A. Productivity reactions of northern boreal forests. In *Impact of Climatic Variations on Agriculture. Vol. 1—Assessment in Cool Temperature Regions*. Eds M.L. Parry, T.R. Carter & N.T. Konijn. pp. 183–94. Kluwer Academic Publishers, Dordrecht.

Kaur, A., Ha, C.O., Jong, K., Sands, V.E., Chan, H.T., Soepadmo, E. & Ashton, P.A. 1978. Apomixis may be widespread among trees of the climax rain forest. *Nature*, 271, 440–2.

Kaur, A., Jong, K., Sands, V.E., Chan, H.T. & Soepadmo, E. 1986. Cytoembryology of some Malaysian dipterocarps, with some evidence of apomixis. *Botanical Journal of the Linnean Society*, 92, 75–88.

Kavanagh, K.L., Yoder, B.J., Aitken S.N., Gartner, B.L. & Knowe, S. 1998. Root and shoot vulnerability to xylem cavitation in four populations of Douglas-fir seedlings. *Tree Physiology*, 19, 31–7.

Keane, P. 1997. Diseases in natural plant communities. In *Plant Pathogens and Plant Diseases*. Eds J.F. Brown & H.J. Ogle. pp. 518–32. Rockvale Publications, Armidale, Australia.

Kelly, A. & Coates, D.J. 1995. Population dynamics, reproductive biology and conservation of *Banksia brownii* and *Banksia verticillata*. Report to Australian Nature Conservation Agency, Canberra. Department of Conservation and Land Management, Western Australia.

Kennington, W.J., Waycott, M. & James, S.H. 1996. DNA fingerprinting supports notions of clonality in a rare mallee, *Eucalyptus argutifolia*. *Molecular Ecology*, 5, 693–6.

Key, K.H.L. 1968. The concept of stasipatric speciation. *Systematic Zoology*, 17, 14–22.

Khoshoo, T.N. 1959. Polyploidy in gymnosperms. *Evolution*, 13, 24–39.

Kiem, P., Paige, K.N., Whitlam, T.G. & Lark, K.G. 1989. Genetic analysis of an interspecific hybrid swarm of *Populus*: occurrence of unidirectional introgression. *Genetics*, 123, 557–65.

Kijas, J.M.H., Fowler, J.C.S., Garbett, C.A. & Thomas, M.R. 1994. Enrichment of microsatellites from the citrus genome using biotinylated oligonucleotide sequences bound to streptavidin-coated magnetic particles. *BioTechniques*, 16, 656–62.

Kijas, J.M.H., Thomas, M.R., Fowler, J.C.S. & Roose, M.L. 1997. Integration of trinucleotide microsatellites into a linkage map of *Citrus*. *Theoretical and Applied Genetics*, 94, 701–6.

Kim, Z.-S. & Hattemer, H.H. 1994. *Conservation and Manipulation of Genetic Resources in Forestry*. Kwang Moon Kag Publishing Company, Seoul, Republic of Korea.

Kimura, M. & Crow, J.F. 1964. The number of alleles that can be maintained in a finite population. *Genetics*, 49, 725–38.

Kimura, M. & Ohta, T. 1975. Distribution of allelic frequencies in a finite population under stepwise production of neutral alleles. *Proceedings of the National Academy of Sciences USA*, 72, 2761–4.

Kindell, C.E., Winn, A.A. & Miller, T.E. 1996. The effects of surrounding vegetation and transplant age on the detection of local adaptation in the perennial grass *Aristida stricta*. *Journal of Ecology*, 84, 745–54.

King, R.A. & Ferris, C. 1998. Chloroplast DNA phylogeography of *Alnus glutinosa* (L.) Gaerton. *Molecular Ecology*, 7, 1151–61.

Kinlaw, C.S., Ho, T., Gertula, S.M, Gladstone, E. & Harry, D.E. 1996. Gene discovery in Loblolly pine trough cDNA sequencing. In *Somatic Cell Genetics and Molecular Genetics of Trees*. Eds M.R. Ahuja, W. Boerjan & D.B. Neale. pp. 175–82. Kluwer Academic Publishers, Dordrecht.

Kinloch Jr, B.B. 1982. Mechanisms and inheritance of rust resistance in conifers. In *Resistance to Disease and Pests in Forest Trees*. Eds H.M. Heybroek, B.R. Stephan & K. von Weissenberg. pp. 119–29. Centre for Agricultural Publishing and Documentation, Wageningen, Netherlands.

Kinloch Jr, B.B. 1991. Distribution and frequency of a gene for resistance to white pine blister rust in natural populations of sugar pine. *Canadian Journal of Botany*, 70, 1319–23.

Kinloch Jr, B.B. & Littlefield, J.L. 1977. White pine blister rust: hypersensitive resistance in sugar pine. *Canadian Journal of Botany*, 55, 1148–55.

Kinloch, B.B., Westfall, R.D. & Forrest, G.I. 1986. Caledonian Scots pine: origins and genetic structure. *New Phytologist*, 104, 703–29.

Kitamura, K., Mohamad Yusof, A.R., Ochiai, O. & Yoshimaru, H. 1994. Estimation of outcrossing rate on *Dryobalanops aromatica* Gaertn. F. in primary and secondary forests in Brunei, Borneo, Southeast Asia. *Plant Species Biology*, 9, 37–41.

Kitching, J.J. & Margules, C.R. 1995. Assessing priority areas for biodiversity and protected area networks. In *Measuring and Monitoring Biodiversity in Tropical and Temperate Forests*. Eds T.J.B. Boyle and B. Boontawee. pp. 355–64. Center for International Forestry Research, Kuala Lumpur.

Kitzmiller, J.H. 1990. Managing genetic diversity in a tree improvement program. *Forest Ecology and Management*, 35, 131–49.

Klug, W.S. & Cummings, M.K. 1999. *Concepts of Genetics*. Prentice Hall, Prentice Hall, Upper Saddle River, NJ, USA.

Knabe, W., Urfer, W. & Venne, H. 1990. Die Variabilität der Immissionsresistenz von Fichtenherkünften—ein Beitrag zum IUFRO-Fichtenprovenienzversuch 1964/1968. *Silvae Genetica* 39, 8–17.

Knowles, P., Furnier, G.R., Aleksiuk, M.A. & Perry, D.J. 1987. Significant levels of self-fertilization in natural populations of tamarack. *Canadian Journal of Botany*, 65, 1087–91.

Kohm, K.A. & Franklin, J.F. (Eds) 1997. *Creating a Forestry for the 21st Century*. Island Press, Washington, DC.

Komorek, B., Shearer, B., Smith, B. & Fairman, R. 1994. The control of *Phytophthora* in native plant communities. In *Report to the Australian Nature Conservation Agency*, pp. 1–14. Department of Conservation and Land Management, Perth.

Konnert, M. 1992. Genetische Untersuchungen in geschädigten Weißtannenbeständen (*Abies alba*) Südwestdeutschlands. Mitteilungen der Forstlichen Versuchs- und Forschungsanstalt Baden-Württemberg, Heft 167, 119 p.

Koski, V. 1980. On the variation of flowering and seed crops in mature stands of *Pinus sylvestris* L. *Silva Fennica*, 14, 71–5.

Koski, V. 1996. Management guidelines for *in situ* gene conservation of wind pollinated temperate conifers. *Forest Genetic Resources*, 24, 2–7.

Kotey, J.F., Owusu, J.G.K., Yeboah, R., Amanor, K. & Antwi, L. 1998. *Falling into Place. Policy that Works for Forests and People—Ghana*. International Institute for Environment & Development, London.

Kothari, A. 1997. India explores joint management of protected areas. *Ecodecision*, Winter, 49–52.

Kozlowski, T.T. & Constantinidou, H.A. 1986a. Environmental pollution and tree growth. Part I. Sources and types of pollutants and plant responses. *Forestry Abstracts*, 47, 5–51.

Kozlowski, T.T. & Constantinidou, H.A. 1986b. Environmental pollution and tree growth. Part II. Factors affecting responses to pollution and alleviation of pollution effects. *Forestry Abstracts*, 47, 105–32.

Kramer, R., van Schaik, C. & Johnson, J. (Eds) 1997. *Last Stand: Protected Areas and the Defense of Tropical Biodiversity*. Oxford University Press, Oxford.

Kremen, C., Merenlender, A.M. & Murphy, D.D. 1994. Ecological monitoring: a vital need for integrated conservation and development programs in the tropics. *Conservation Biology*, 8, 388–97.

Kremer, A., Savill, P.S. & Steiner, K.C. 1993. (Eds.) Genetics of oaks. *Annales des Sciences Forestieres*, 50, Suppl. 1, 233–44.

Kress, L.W., Skelly, J.M. & Hinkelmann, K.H. 1982. Relative sensitivity of 18 full-sib families of *Pinus taeda* to O_3. *Canadian Journal of Forest Research*, 12, 203–9.

Kriebel, H.B. & Leben, C. 1981. The impact of SO_2 air pollution on the gene pool of eastern white pine. In *17th IUFRO World Congress, Proceedings Division 2*, pp. 185–9, Japanese IUFRO Congress Committee c/o Forestry and Forest Products Research Institute P.O. Box 16, Tsukuba Norin Kenkyudanchi-nai, Ibaraki 305, Japan.

Kröller, E. 1992. Report on the Development Cooperation Directorate/ Development Centre workshop on the use of economic instruments for environmental management in developing countries. OECD, Paris.

Krug, E. 1990. Reduced fertilization capacity of SO_2-fumigated *Picea omorika*-pollen. *European Journal of Forest Pathology*, 20, 122–6.

Krusche, D. & Geburek, Th. 1991. Conservation of forest gene resources as related to sample size. *Forest Ecology and Management*, 40, 145–50.

Krutilla, J.V. & Fisher, A.C. 1975. *The Economics of Natural Environments: Studies in the Valuation of Commodity and Amenities Resources*. Resources for the Future/Johns Hopkins University Press, Baltimore, MD.

Kubis, S., Schmidt, T. & Heslop-Harrison, J.S. 1998. Repetitive DNA elements as a major component of plant genomes. *Annals of Botany*, 82 (Supplement A), 45–55.

Kuittinen, H., Muona, O., Kärkkäinen, K. & Borzan, Z. 1991. Serbian spruce, a narrow endemic, contains much genetic variation. *Canadian Journal of Forest Research*, 21, 363–7.

Kuusipalo, J. & Kangas, J. 1994. Managing biodiversity in a forestry environment. *Conservation Biology*, 8, 450–60.

Kvarnheden, A. & Engstrom, P. 1992. Genetically stable, individual specific differences in hypervariable DNA in Norway spruce, detected by hybridization to a phage M13 probe. *Canadian Journal of Forest Research*, 22, 117–23.

Lacy, R.C. 1987. Loss of genetic diversity from managed populations: interacting effects of drift, mutation, immigration, selection and population subdivision. *Conservation Biology*, 1, 143–58.

Lagercrantz, U. & Ryman, N. 1990. Genetic structure of Norway spruce (*Picea abies*): concordance of morphological and allozymic variation. *Evolution*, 44, 38–53.

Lagudah, E.S., Moullet, O. & Appels, R. 1997. Map-based cloning of a gene sequence encoding a nucleotide-binding domain and a leucine-rich region at the Cre3 nematode resistance locus of wheat. *Genome*, 40, 659–65.

Landbo, L. & Jorgensen, R.B. 1997. Seed germination in weedy *Brassica campestris* and its hybrids with *B. napus*: implications for risk assessment of transgenic oilseed rape. *Euphytica*, 97, 209–16.

Lande, R. 1988. Genetics and demography in biological conservation. *Science*, 241, 1455–60.

Lande, R. 1992. Neutral theory of quantitative genetic variance in an island model with local extinction and colonisation. *Evolution*, 46, 381–9.

Lande, R. 1995. Mutation and conservation. *Conservation Biology*, 9, 782–91.

Lande, R. & Barrowclough, G.F. 1987. Effective population size, genetic variation, and their use in population management. In *Viable Populations for Conservation*. Ed. M.E. Soulé. pp. 87–123. Cambridge University Press, Cambridge.

Lande, R., Engen, S. & Saether, B.-E. 1994. Optimal harvesting, economic discounting and extinction risk in fluctuating populations. *Nature*, 372, 88–90.

Lang, K.J, Neumann, P. & Schütt, P. 1971. Der Einfluß von Samenherkunft und Düngung auf die SO_2-Härte von *Pinus contorta*-Sämlingen. *Flora*, 160, 1–9.

Langcake, P. 1981. Disease resistance of *Vitis* spp. and the production of the stress metabolites resveratrol, e-viniferin, a-viniferin and pterostilbene. *Physiological Plant Pathology*, 18, 213–26.

Langevin, S.A., Clay, K. & Grace, J.B. 1990. The incidence and effects of hybridization between cultivated rice and its related weed red rice (*Oryza sativa* L.). *Evolution*, 44, 1000–8.

Langub, J. 1996. Penan response to change and development. In *Borneo in Transition*. Eds C. Padoch & N. Lee Peluso. pp. 103–20. Oxford University Press, Kuala Lumpur.

Larsen, J.B. 1986. Das Tannensterben: Eine neue Hypothese zur Klärung des Hintergrundes dieser rätselhaften Komplexkrankheit der Weißtanne (*Abies alba* Mill.). *Forstwissenschaftliches Centralblatt*, 105, 381–96.

Larsen, J.B. 1991. Breeding for physiological adaptability in order to counteract an expected increase in environmental heterogeneity. *Forest Tree Improvement*, No. 23, 5–9.

Larsen, J.B. & Friedrich, J. 1988. Wachstumsreaktionen verschiedener Provenienzen der Weißtanne (*Abies alba* Mill.) nach winterlicher SO₂-Begasung. *European Journal of Forest Pathology*, 18, 190–9.

Larsen, J.B., Quian, X.M., Scholz, F. & Wagner, I. 1988. Ecophysiological reactions of different provenances of European silver fir (*Abies alba* Mill.) to SO₂ exposure during winter. *European Journal of Forest Pathology*, 18, 44–50.

Latta, R.G. & Mitton, J.B. 1997. A comparison of population differentiation across four classes of gene marker in limber pine (*Pinus flexilis* James). *Genetics*, 146, 1153–63.

Lawrence, M.J. & Marshall, D.F. 1997. Plant population genetics. In *Plant Genetic Conservation: The In Situ Approach*. Eds N. Maxted, B.V. Ford-Lloyd & J.G. Hawkes. pp. 99–113. Chapman and Hall, London.

Lawrence, M.J., Marshall, D.F. & Davies, P. 1995a. Genetic of genetic conservation. I. Sample size when collecting germplasm. *Euphytica*, 84, 89–99.

Lawrence, M.J., Marshall, D.F. & Davies, P. 1995b. Genetic of genetic conservation. II. Sample size when collecting seed of cross-pollinating species and the information that can be obtained from the evaluation of material held in genebanks. *Euphytica*, 84, 101–7.

Ledig, F.T. 1986. Conservation strategies for forest gene resources. *Forest Ecology and Management*, 14, 77–90.

Ledig, F.T. 1988. The conservation of diversity in forest trees: why and how should genes be conserved? *BioScience*, 38, 471–9.

Ledig, F.T. 1992. Human impacts on genetic diversity in forest ecosystems. *Oikos*, 63, 87–108.

Ledig, F.T. & Conkle, M.T. 1983. Gene diversity and genetic structure in a narrow endemic, Torrey pine (*Pinus torreyana* Parry ex Carr.). *Evolution*, 37, 79–85.

Ledig, F.T. & Kitzmiller, J.H. 1992. Genetic strategies for reforestation in the face of global climate change. *Forest Ecology and Management*, 50, 153–69.

Ledig, F.T., Jacob-Cervantes, V., Hodgkiss, P.D. & Eguiluz-Piedra, T. 1997. Recent evolution and divergence among populations of a rare Mexican endemic, Chihuahua spruce, following Holocene climatic warming. *Evolution*, 51, 1808–14.

Lee, H.Y. 1967. Studies in *Swietenia* (Meliaceae), Observations on the sexuality of the flowers. *Journal of the Arnold Arboretum*, 48, 101–4.

Lee, J. 1996. Participation and Pressure in the Mist Kingdom of Sumba, a Local NGO's Approach for Tree Planting. PhD dissertation, University of Adelaide, Adelaide.

Lefol, E., Fleury, A. & Darmency, H. 1996. Gene dispersal from transgenic crops II. Hybridization between oilseed rape and hoary mustard. *Sexual Plant Reproduction*, 9, 189–96.

Leitch, I.J. & Bennett, M.D. 1997. Polyploidy in angiosperms. *Trends in Plant Science*, 2, 470–6.

Lerceteau, E., Robert, T., Petiard, V. & Crouzillat, D. 1997. Evaluation of the extent of genetic variability among *Theobroma cacao* accessions using RAPD and RFLP markers. *Theoretical and Applied Genetics*, 95, 10–19.

Lesica, P. & Allendorf, F.W. 1995. When are peripheral populations valuable for conservation? *Conservation Biology*, 9, 753–60.

Lever, C. 1987. *Naturalized Birds of the World*. John Wiley and Sons, New York.

Levin, D.A. & Kerster, H.W. 1974. Gene flow in seed plants. In *Evolutionary Biology 7*. Eds T. Dobzhansky, M.T. Hecht & W.C. Steere. pp. 139–220. Plenum Press, New York.

Levin, D.A. 1981. Dispersal versus gene flow in plants. *Annuals of Missouri Botanical Garden*, 68, 233–53.

Levin, D.A., Francisco–Ortega, J.K. & Jansen, R.K. 1996. Hybridization and the extinction of rare plant species. *Conservation Biology*, 10, 10–16.

Lewontin, R.C. 1964. The interaction of selection and linkage. I. General considerations; heterotic models. *Genetics*, 49, 49–67.

Lewontin, R.C. 1972. *The Genetic Basis of Evolutionary Change*. Columbia University Press, New York.

Lewontin, R.C. 1988. On measures of gametic disequilibrium. *Genetics*, 120, 849–52.

Li, C.C. 1955. *Population Genetics*. The University of Chicago Press, Chicago.

Li, P. & Adams, W.T. 1989. Range-wide patterns of allozyme variation in Douglas-fir (*Pseudotsuga menziesii*). *Canadian Journal of Forest Research*, 19, 149–61.

Li, T. Murray, 1995a. Agrarian Transformation in the Indonesian Uplands. In *Agrarian Transformation in the Indonesian Uplands*. Eds T. Murray Li and L. Uhryniuk. pp. 1–11. Conference Proceedings. EMDI Environmental Reports 48. Ministry of State for Environment, Jakarta.

Li, T. Murray, 1995b. Contested terrain, contested terms: changing agrarian relations in Upland Sulawesi. In *Agrarian Transformation in the Indonesian Uplands*. Eds T. Murray Li & L. Uhrynuik. pp. 22–4. Conference Proceedings. EMDI Environmental Reports 48. Ministry of State for Environment, Jakarta.

Li, T. Murray, 1996. Images of community: discourse and strategy in property relations. *Development and Change*, 27, 501–27.

Libby, W.J. 1982. What is the safe number of clones per plantation? In *Resistance to Disease and Pests in Forest Trees*. Eds H.M. Heybroek, B.R. Stephan & K. von Weissenberg. pp. 342–60. Centre for Agricultural Publishing and Documentation, Wageningen, Netherlands.

Libby, W.J., Slettler, R.F. & Seitz, F.W. 1969. Forest genetics and forest tree breeding. *Annual Review of Genetics*, 3, 469–94.

Lieberman, D. 1982. Seasonality and phenology in a dry tropical forest in Ghana. *Journal of Ecology*, 70, 791–806.

Liengsiri, C., Piewluang, C. & Boyle, T.J.B. 1990. *Starch Gel Electrophoresis of Tropical Trees—A Manual*. ASEAN-Canada Forest Tree SeedCentre, Muak Lek, Thailand.

Lindgren, D. 1991. Can shields stop aliens from upper space? Movement of contaminating pollen in a Scots pine seed orchard. In *Pollen Contamination in Seed Orchards*. Ed. D. Lindgren. pp. 34–42. Proceedings of the Meeting of the Nordic Group for Tree Breeding 1991, Report 10, Swedish University of Agricultural Sciences, Department of Forest Genetics and Plant Physiology, Umeå, Sweden.

Lindgren, D. 1994. Prediction and optimization of genetic gain with regard to genotype x environment interaction. *Studia Forestalia Suecica*, 166, 15–24.

Linhart, Y.B. 1973. Ecological and behavioural determinants of pollen dispersal in hummingbird pollinated *Heliconia*. *American Naturalist*, 107, 511–23.

Linhart, Y.B. 1989. Interactions between genetic and ecological patchiness in forest trees and their dependent species. In *The Evolutionary Ecology of Plants*. Eds H. Bock & Y.B. Linhart. pp. 393–430. Westview Press, Boulder, CO, USA.

Linhart, Y. 1995. Restoration, revegetation, and the importance of genetic and evolutionary perspectives. In *Proceedings: Wildland Shrub and Arid Land Restoration Symposium*. Compiled by B.A. Roundy, E.D. McArthur, J.S. Haley & D.K. Mann, October 19–21 1993, Las Vegas. General Technical Report INT-GTR-315, pp. 271–87. United States Department of Agriculture, Forest Service, Intermountain Research Station, Ogden, UT, USA.

Lloyd, D.G. 1979. Evolution towards dioecy in heterostylous populations. *Plant Systematics and Evolution*, 131, 71–80.

Löchelt, S. 1994. Bestimmung genetischer Merkmale von Fichten (*Picea abies* [L.] Karst.) mit unterschiedlich ausgeprägten Schadsymptomen auf baden-württembergischen Dauerbeobachtungsflächen. *Allgemeine Forst- und Jagdzeitung*, 165, 21–7.

Loveless, M.D. 1992. Isozyme variation in tropical trees, patterns of genetic organization. *New Forests*, 6, 67–94.

Loveless, M.D., Hamrick, J.L. & Foster, R.B. 1998. Population structure and mating system in *Tachigali versicolor*, a monocarpic neotropical tree. *Heredity*, 81, 134–43.

Ludwig, J.A., Tongway, D.J., Freudenberger, D., Noble, J. & Hodgkinson, K. (Eds) 1997. *Landscape Ecology, Function and Management: Principles from Australia's Rangelands*. CSIRO Australia, Melbourne.

Lui, Z. & Furnier, G.R. 1993. Comparison of allozyme, RFLP, and RAPD markers for revealing genetic variation within and between trembling aspen and bigtooth aspen. *Theoretical and Applied Genetics*, 87, 97–105.

Luikart, G. & Cornuet, J.M. 1998. Empirical evaluation of a test for identifying recently bottlenecked populations from allele frequency data. *Conservation Biology*, 12, 228–37.

Lutz, E. & Young, M. 1992. Integration of environmental concerns into agricultural policies of industrial and developing countries. *World Development*, 20, 241–53.

Lynch, M. 1996. A quantitative-genetic perspective on conservation issues. In *Conservation Genetics: Case Histories From Nature*. Eds J.C Avise & J.L. Hamrick. pp. 471–501. Chapman and Hall, New York.

Lynch, M. & Gabriel, W. 1990. Mutation load and the survival of small populations. *Evolution*, 44, 1725–37.

Lynch, M. & Lande, R. 1993. Evolution and extinction in response to environmental change. In *Biotic Interactions and Global Climate Change*. Eds P.M. Kareiva, J.G. Kingsolver & R.B. Huey. pp. 234–50. Sinauer Associates, Sunderland, MA, USA.

Lynch, M. & Milligan, B.G. 1994. Analysis of population genetic structure with RAPD markers. *Molecular Ecology*, 3, 91–9.

Lynch, M., Conery, J. & Burger, R. 1995. Mutation accumulation and the extinction of small populations. *American Naturalist*, 146, 489–518.

Macnair, M.R. 1993. The genetics of metal tolerance in vascular plants. Tanskley Review no. 49, *New Phytologist*, 124, 541–59.

Macnair M.R. 1997. The evolution of plants in metal-contaminated environments. In *Environmental Stress, Adaptation and Evolution*. Eds R. Bijlsma & V. Loeschke. pp. 3–24. Birkhäuser, Basel.

Maguire, B. 1976. Apomixis in the genus *Clusia* (Clusiaceae)—a preliminary report. *Taxon*, 25, 241–4.

Maguire, W.P. 1955. Radiation, surface temperature, and seedling survival. *Forest Sciences*, 1, 277–85.

Makkonen-Spieker, K. 1985. Auswirkungen des Alumniums auf junge Fichten (*Picea abies* Karst.) verschiedener Provenienzen. *Forstwissenschaftliches Centralblatt*, 104, 341–53.

Manchon, N., Lefranc, M., Bilger, I., Mazer, S.J. & Sarr, A. 1997. Allozyme variation in *Ulmus* species from France: analysis of differentiation. *Heredity*, 78, 12–20.

Manderscheid, R., Jäger, H.J. & Kress, L.W. 1992. Dose-response relationships of ozone effects on foliar nitrogen metabolism of *Pinus taeda* (L.): rise in protein content following increase in amino-N turnover. *New Phytologist*, 121, 623–35.

Manion, P.D. 1981. *Tree Disease Concepts*. Prentice-Hall, New Jersey, USA.

Manion, P.D. & Griffin, D.H. 1986. Sixty-five years of research on hypoxylon canker of aspen. *Plant Disease*, 70, 803–8.

Mansley, M. 1996. Achieving sustainable forestry: the role of capital markets. In *Making Forest Policy Work 1996*. Ed. K.L. Harris. pp. 129–36. Oxford Forestry Institute, Oxford.

Margules, C.R. & Lindenmayer, D.B. 1996. Landscape level concepts and indicators for the conservation of forest biodiversity and sustainable forest management. Paper to: *Economic, Social and Political Issues in Certification of Forest Management*. 12–16 May 1996. pp. 65–83. University of British Columbia & Universiti Pertanian Malaysia. University of British Columbia, Vancouver.

Margules, C.R. & Redhead, T.D. 1995. *BioRap—Guidelines for Using the BioRap Methodology and Tools*. CSIRO Australia, Melbourne.

Margules, C.R., Nicholls, A.O. & Pressey, R.L. 1988. Selecting networks of reserves to maximize biological diversity. *Biological Conservation*, 43, 663–76.

Margules, C.R., Pressey, R.L. & Nicholls, A.O. 1991. Selecting nature reserves. In *Nature Conservation: Cost Effective Biological Surveys and Data Analysis*. Eds C.R. Margules & M.P. Austin. pp. 90–7. CSIRO, Melbourne.

Margulis, L. 1970. *Origin of Eukaryotic Cells*. Yale University Press, New Haven, CT, USA.

Marques, C.M., Araujo, J.A., Ferreira, J.G., Whetten, R., O'Malley, D.M., Lui, B.H. & Sederoff, R. 1998. AFLP genetic maps of *Eucalyptus globulus* and *E. tereticornis*. *Theoretical and Applied Genetics*, 96, 727–37.

Marshall, D.R. 1989. Crop genetic resources: current and emerging issues. In *Plant Population Genetics, Breeding and Genetic Resources*. Eds A.H.D. Brown, M.T. Clegg, A.L. Kahler & B.S. Weir. pp. 367–88. Sinauer Associates, Sunderland, MA, USA.

Marshall, D.R. & Brown, A.H.D. 1975. Optimum sampling strategies in genetic conservation. In *Crop Genetic Resources for Today and Tomorrow*. Eds O.H. Frankel & J.G. Hawkes. pp. 21–40. Cambridge University Press, Cambridge.

Marshall, D.R. & Brown, A.H.D. 1998. Sampling wild legume populations. In *Genetic Resources of Mediterranean Pasture and Forage Legumes*. Proceedings of an International Workshop. pp. 78–89. Kluwer Academic Publishers, Dordrecht.

Martin, M.A., Shiozawa, D.K., Loudenslager, E.J. & Jensen, J.N. 1985. Electrophoretic study of cutthroat trout populations in Utah. *Great Basin Naturalist*, 45, 677–87.

Martin, W. 1999. A briefly argued case that mitochondria and plastids are descendants of endosymbionts, but that the nuclear compartment is not. *Proceedings of the Royal Society of London, B*, 266, 1387–95.

Martinsen, G.D. & Whitham, T.G. 1994. More birds nest in hybrid cottonwood trees. *Wilson Bulletin*, 106, 474–81.

Martinson, S.R. 1996. Association among geographic, isozyme and growth variables for sugar pine in South-West Oregon and through out species range. PhD dissertation, North Carolina State University, Raleigh, NC.

Maruyama T. & Fuerst P.A. 1985. Population bottlenecks and non equilibrium models in population genetics. II. Number of alleles in a small population that was formed by a recent bottleneck. *Genetics*, 111, 675–89.

Masters, C.J. 1974. The controlled pollination techniques and analyses of intraspecific hybrids for black walnut (*Juglans nigra* L.). PhD dissertation, Purdue University, W. Lafayette, IN, USA.

Masterson, J. 1994. Stomatal size in fossil plants: evidence for polyploidy in majority of angiosperms. *Science*, 264, 421–3.

Matheson, A.C., Bell, J.C. & Barnes, R.D. 1989. Breeding systems and genetic structure in some Central American pine populations. *Silvae Genetica*, 38, 107–13.

Mattila, A., Pakkanen, A., Vakkari, P. & Raisio, J. 1994. Genetic variation in English oak (*Quercus robur*) in Finland. *Silva Fennica*, 28, 251–6.

Mátyás, C. 1994. Modeling climate change effects with provenance test data. *Tree Physiology*, 14, 797–804.

Mátyás, C. 1996. Climatic adaptation of trees: rediscovering provenance tests. *Euphytica*, 92, 45–54.

Mayer, J. 1996. Impacts of the East Kalimantan forest fires of 1982–1983 on village life, forest use, and land use. In *Borneo in Transition*. Eds C. Padoch & N. Lee Peluso. pp. 187–218. Oxford University Press, Kuala Lumpur.

Mayr, E. 1942. *Systematics and the Origin of Species*. Columbia University Press, New York.

Mayr, E. 1963. *Animal Species and Evolution*. Belknap Press of Harvard University Press, Cambridge, MA, USA.

Mazourek, J.C. & Gray, P.N. 1994. The Florida duck or the mallard. *Florida Wildlife*, 48, 29–31.

McCallum, R. & Sekhran, N. 1996. Lessons Learned Through ICAD Experimentation in Papua New Guinea: The Lak Experience. Paper presented at the Heads of Forestry Meeting, Port Vila, Vanuatu, 23–28 September.

McClenaghan, Jr., L.R. & Beauchamp, A.C. 1986. Low genic differentiation among isolated populations of the California fan palm (*Washingtonia filifera*). *Evolution*, 40, 315–22.

McCormick, L.H. & Steiner, K.C. 1978. Variation in aluminum tolerance among six genera of trees. *Forest Science*, 24, 565–8.

McCracken, A.R. & Dawson, W.M. 1997. Growing clonal mixtures of willow to reduce effect of *Melampsora epitea* var. *epitea*. *European Journal of Forest Pathology*, 27, 319–29.

McGranahan, M., Bell, J.C., Moran, G.F. & Slee, M. 1997. High genetic divergence between geographic regions in the highly outcrossing species *Acacia aulacocarpa* (Cunn. ex Benth.). *Forest Genetics*, 4, 1–13.

McNeely, J.A. 1988. *Economics and Biological Diversity: Developing and Using Economic Incentives to Conserve Biological Diversity*. IUCN (World Conservation Union), Gland.

McNeely, J.A. 1993. Economic Incentives for conserving biodiversity: lessons from Africa. *Ambio*, 22, 144–50.

McNeely, J.M. 1997. New trends in protecting and managing biodiversity. *Ecodecision*, Winter, 20–3.

McNeely, J.A. 2000. *Mobilizing broader support for Asia's biodiversity: how civil society can contribute to protected area management*. pp. 248. Asian Development Bank, Manila (in press).

McNeely, J.A. & Dobias, R.J. 1991. Economic incentives for conserving biological diversity in Thailand. *Ambio*, 20, 86–90.

McNeely, J.A. & Wachtel, P.S. 1988. *Soul of the Tiger*. Oxford University Press, Oxford.

McNeely, J.A., Gadgil, M., Leveque, C., Padoch, C. & Redford, K. 1995. Human influences on biodiversity. In *Global Biodiversity Assessment*. UNEP. Cambridge University Press, Cambridge.

Mejnartowicz, L. 1983. Changes in genetic structures of Scots pine (*Pinus sylvestris* L.) population affected by industrial emission of fluoride and sulphur dioxide. *Genetica Polonica*, 24, 41–50.

Menges, E.S. 1990. Population viability analysis for an endangered plant. *Conservation Biology*, 4, 52–62.

Menges, E.S. 1992. Stochastic modeling of extinction in plant populations. In *Conservation Biology, The Theory and Practice of Nature Conservation, Preservation and Management*. Eds P.L. Fiedler & S.K. Jain. pp. 253–75. Chapman and Hall, New York.

Merkel, H.W. 1905. A deadly fungus on the American chestnut. New York Zoological Society, 10th Annual Report, pp. 97–103.

Michelmore, R. 1995. Molecular approaches to manipulation of disease resistance genes. *Annual Review of Phytopathology*, 15, 393–427.

Michelmore, R. 1996. Flood warning—resistance genes unleashed. *Nature Genetics*, 14, 376–8.

Mikola, J. 1982. Bud-set phenology as an indicator of climatic adaptation of Scots pine in Finland. *Silva Fennica*, 16, 178–84.

Millar, C.I. 1993. Impact of the Eocene on the evolution of *Pinus* L. *Annals of the Missouri Botanical Garden*, 80, 471–98.

Millar, C.I. & Libby, W.J. 1991. Strategies for conserving clinal, ecotypic, and disjunct population diversity in widespread species. In *Genetics and Conservation of Rare Plants*. Eds D.A. Falk & K.E. Holsinger. pp. 149–70. Oxford University Press, New York.

Millar, C. & Westfall, R. 1992. Allozyme markers in forest genetic conservation. *New Forests*, 6, 347–71.

Millar, C.I. & Westfall, R.D. 1996. Integrated management and monitoring of Genetic Conservation Areas on National Forests in California. In *The Status of Temperate North American Forest Genetic Resources*. Eds D.L. Rogers & F.T. Ledig. Report No. 16. University of California Genetic Resources Conservation Program, Davis, CA, USA.

Milligan, B.G. & Strand, A.E. 1996. Genetics and conservation biology: assessing historical trends in the demography of populations. In *Southwestern Rare and Endangered Plants: Proceedings of the Second Conference*. General Technical Report RM-GTR-283. Eds J. Maschinski, H.D. Hammond & L. Holter. pp. 125–37. United States Department of Agriculture, Forest Service, Rocky Mountain and Range Experiment Station, Fort Collins, CO, USA.

Mitchell, H.J., Hall, J.L. & Barber, M.S. 1994. Elicitor-induced cinnamyl alcohol deydrogenase activity in lignifying wheat (*Triticum aestivum* L.) leaves. *Plant Physiology*, 104, 551–6.

Mitton, J.B. 1983. Conifers. In *Isozymes in Plant Genetics and Breeding*. Eds S.D. Tanksley & T.J. Orton. pp. 443–72. Elsevier, Amsterdam.

Mitton, J.B. 1992. The dynamic mating system of conifers. *New Forests*, 6, 197–216.

Mitton, J.B. & Grant, M.C. 1984. Associations among protein heterozygosity, growth rate, and developmental homeostasis. *Annual Revue of Ecology and Systematics*, 15, 479–99.

Mitton, J.B. & Pierce, B.A. 1980. The distribution of individual heterozygosity in natural populations. *Genetics*, 95, 1043–54.

Montreal Implementation Group (Australia). 1998. *A Framework of Regional (Sub-national) Level Criteria and Indicators of Sustainable Forest Management in Australia*. Commonwealth of Australia, Canberra.

Moore, W.S. 1977. An evaluation of narrow hybrid zones in vertebrates. *The Quarterly Review of Biology*, 52, 263–77.

Mopper, S., Mitton, J.B., Whitham, Th.G., Cobb, N.S. & Christensen, K.M. 1991. Genetic differentiation and heterozygosity in pinyon pine associated with resistance to herbivory and environmental stress. *Evolution*, 45, 989–99.

Moran, G.F. 1992. Patterns of genetic diversity in Australian tree species. *New Forests*, 6, 49–66.

Moran, G.F. & Bell, J.C. 1983. *Eucalyptus*. In *Isozymes in Plant Genetics and Breeding*. Eds S.D. Tanksley & T.J. Orton. pp. 423–41. Elsevier, Amsterdam.

Moran, G.F. & Bell, J.C. 1987. The origin and genetic diversity of *Pinus radiata* in Australia. *Theoretical and Applied Genetics*, 73, 616–22.

Moran, G.F. & Brown, A.H.D. 1980. Temporal heterogeneity of outcrossing rates in Alpine ash (*Eucalyptus delegatensis* R.T. Bak.). *Theoretical and Applied Genetics*, 57, 113–20.

Moran, G.F. & Hopper, S.D. 1983. Genetic diversity and the insular population structure of the rare granite rock species, *Eucalyptus caesia* Benth. *Australian Journal of Botany*, 31, 161–72.

Moran, G.F. & Hopper, S.D. 1987. Conservation of the genetic resources of rare and widespread eucalypts in remnant vegetation. In *Nature Conservation: The Role of Remnants of Native Vegetation*. Eds D.A. Saunders, G.W. Arnold, A.A. Burnbidge & A.J.M. Hopkins. pp. 151–62. Surrey Beatty & Sons, CSIRO/CALM, WA, Chipping Norton, NSW, Australia.

Moran, G.F., Bell, J.C. & Eldridge, K.G. 1988. The genetic structure and the conservation of the five natural populations of *Pinus radiata*. *Canadian Journal of Forest Research*, 18, 506–14.

Moran, G.F., Bell, J.C. & Griffin, A.R. 1989a. Reduction in levels of inbreeding in a seed orchard of *Eucalyptus regnans* F. Muell. compared with natural populations. *Silvae Genetica*, 38, 32–6.

Moran, G.F., Bell, J.C. & Hilliker, A.J. 1983. Greater meiotic recombination in male vs female gametes in *Pinus radiata*. *Journal of Heredity*, 74, 62.

Moran, G.F., Bell, J.C. & Turnbull, J.W. 1989b. A cline in genetic diversity in *Casuarina cunninghamiana*. *Australian Journal of Botany*, 37, 169–80.

Moran, G.F., Muona, O. & Bell, J.C. 1989c. *Acacia mangium*: a tropical forest tree of the coastal lowlands with low genetic diversity. *Evolution*, 43, 231–5.

Moran, G.F., Muona, O. & Bell, J.C. 1989d. Breeding systems and genetic diversity in *Acacia auriculiformis* and *A. crassicarpa*. *Biotropica*, 21, 250–6.

Morgante, M. & Olivieri, A.M. 1993. PCR-amplified microsatellites as markers in plant genetics. *The Plant Journal*, 3, 175–82.

Moritz, C. 1994. Defining 'Evolutionarily Significant Units' for conservation. *Trends in Evolution and Ecology*, 9, 373–5.

Morowitz, H.J. 1991. Balancing species preservation and economic considerations. *Science*, 253, 752–4.

Mortensen, L.M. 1990. Effects of ozone on growth and dry matter partitioning in different provenances of Norway spruce (*Picea abies* (L.) Karst.). *Norwegian Journal of Agricultural Sciences*, 4, 61–6.

Mosseler, A. 1992. Life history and genetic diversity in red pine, implications for gene conservation in forestry. *Forestry Chronicle*, 68, 701–8.

Mosseler, A., Egger, K.N. & Hughes, G.A. 1992. Low levels of genetic diversity in red pine confirmed by random amplified polymorphic DNA markers. *Canadian Journal of Forest Research*, 22, 1332–7.

Mosseler, A., Innes, D.J. & Roberts, B.A. 1991. Lack of allozymic variation in disjunct Newfounland populations of red pine (*Pinus resinosa*). *Canadian Journal of Forest Research*, 21, 525–8.

Moura-Costa, P. & Stuart, M.D. 1998. Forestry-based greenhouse gas mitigation: the story of market evolution. *Commonwealth Forestry Review*, 77, 191–202.

Mukai, Y., Suyama, Y., Tsumura, Y., Kawahara, T., Yoshimaru, H., Kondo, T., Tomaru, N., Kuramoto, N. & Murai, M. 1995. A linkage map for sugi (*Crytomeria japonica*) based on RFLP, RAPD, and isozyme loci. *Theoretical and Applied Genetics*, 90, 835–40.

Müller, M., Köhler, B., Grill, D., Guttenberger, H. & Lütz, C. 1994. The effects of various soils, different provenances and air pollution on root tip chromosomes in Norway spruce. *Trends in Ecology and Evolution*, 9, 73–9.

Müller-Starck, G. 1982. Tracing external pollen contribution to the offspring of a Scots pine seed orchard. In Proceedings of the IUFRO joint meeting on breeding strategies including multiclonal varieties, p. 176. Lower Saxony Forest Research Institute, Department of Forest Tree Breeding, Staufenberg-Escherode, Germany.

Müller-Starck, G. 1985. Genetic differences between "tolerant" and "sensitive" beeches (*Fagus sylvatica* L.) in an environmentally stressed adult forest stand. *Silvae Genetica*, 34, 241–7.

Müller-Starck, G. 1989. Genetic implications of environmental stress in adult forest stands of *Fagus sylvatica* L. In *Genetic Aspects of Air Pollutants in Forest Tree Populations*. Eds F. Scholz, H.-R. Gregorius & D. Rudin. pp. 127–42. Springer Verlag, Berlin.

Müller-Starck, G. 1993. *Auswirkungen von Umweltbelastungen auf genetische Strukturen von Waldbeständen am Beispiel der Buche* (*Fagus sylvatica* L.). [Impacts of Environmental Stress on Genetic Structures of Forest Stands as Exemplified by European Beech (*Fagus sylvatica* L.)]. Sauerländer's Verlag, Frankfurt a. M., Germany.

Müller-Starck, G. 1994. Die Bedeutung der genetischen Variation für die Anpassung gegenüber Umweltstreß. (The significance of genetic variation for adaptation to environmental stress). *Schweiz. Zeitschrift f. Forstwesen*, 145/12, 977–98.

Müller-Starck, G. 1995. Protection of genetic variability in forest trees. *Forest Genetics*, 2, 121–4.

Müller-Starck, G. & Gregorius, H.-R. 1986. Monitoring genetic variation in forest tree populations. In *Proceedings of the 18th World IUFRO Congress*, pp. 589–99. Ljubljana, Div. 2, Vol. II. Ljubljana, Yugoslavia.

Müller-Starck, G. & Ziehe, M. 1991. Genetic variation in populations of *Fagus sylvatica* L., *Quercus robur* L., and *Q. petrea* Liebl. in Germany. In *Genetic Variation in European Populations of Forest Trees*. Eds G. Müller-Starck & M. Ziehe. pp. 125–40. J.D. Sauerländer's, Frankfurt a.M., Germany.

Müller-Starck, G., Baradet, P. & Bergmann, F. 1992. Genetic variation within European tree species. *New Forests*, 6, 23–47.

Mullis, K.B. & Faloona, F.A. 1987. Specific synthesis of DNA *in vitro* by a polymerase catalysed chain reaction. *Methods in Enzymology*, 155, 335–50.

Muona, O. 1989. Population genetics in forest tree improvement. In *Plant Population Genetics, Breeding, and Genetic Resources*. Eds A.H.D. Brown,

M.T. Clegg, A.L. Kahler & B.S. Weir. pp. 282–98. Sinauer Associates, Sunderland, MA, USA.

Muona, O. & Harju, A. 1989. Effective population sizes, genetic variability and mating system in natural stands and seed orchards of *Pinus sylvestris. Silvae Genetica*, 38, 221–8.

Muona, O., Moran, G. & Bell, J.C. 1991. Hierarchical patterns of correlated mating in *Acacia melanoxylon. Genetics*, 127, 619–26.

Muona, O., Paule, L. Szmidt, A.E. & Kärkkäinen, K. 1990. Mating system analysis in a central and northern European population of *Picea abies. Scandinavian Journal of Forest Research*, 5, 97–102.

Murali, K.S., Uma Shanker, Uma Shaanker, R., Ganeshaiah, K.N. & Bawa, K.S. 1996. Extraction of non-timber forest products in the forests of Biligiri Rangan Hills, India. 2. Impact of NTFP extraction on regeneration, population structure, and species composition. *Economic Botany*, 50, 252–69.

Muralidharan, K. & Wakeland, E.K. 1993. Concentration of primer and template qualitatively affects products in random-amplified polymorphic DNA PCR. *BioTechniques*, 14, 362–4.

Murawski, D.A. 1995. Reproductive biology and genetics of tropical trees from a canopy perspective. In *Forest Canopies*. Eds M.D. Lowman & N.M. Nadkarni. pp. 457–93. Academic Press, San Diego.

Murawski, D.A. & Bawa, K.S. 1994. Genetic structure and mating system of *Stemonoporous oblongifolius* (Dipterocarpaceae) in Sri Lanka. *American Journal of Botany*, 81, 155–60.

Murawski, D.A. & Hamrick, J.L. 1991. The effect of the density of flowering individuals on the mating systems of nine tropical tree species. *Heredity*, 67, 167–74.

Murawski, D.A. & Hamrick, J.L. 1992. The mating system of *Cavanillesia platanifolia* under extremes of flowering tree density: a test of predictions. *Biotropica*, 24, 99–101.

Murawski, D.A., Dayanandan, B. & Bawa, K.S. 1994a. Outcrossing rates of two endemic *Shorea* species from Sri Lankan tropical rain forests. *Biotropica*, 26, 23–9.

Murawski, D.A., Gunatilleke, I.A.U.N. & Bawa, K.S. 1994b. The effects of selective logging on inbreeding in *Shorea megistophylla* (Dipterocarpaceae) from Sri Lanka. *Conservation Biology*, 8, 997–1002.

Murawski, D.A., Hamrick, J.L., Hubbell, S.P. & Foster, R.B. 1990. Mating systems of two Bombacaceous trees of a neotropical moist forest. *Oecologia*, 82, 501–6.

Murcia, C. 1995. Edge effects in fragmented forests: implications for conservation. *Trends in Ecology and Evolution*, 10, 58–62.

Murray, B.G. 1998. Nuclear DNA amount in gymnosperms. *Annals of Botany*, 82 (Supplement A), 3–15.

Murray, V. 1989. Improved double-stranded DNA sequencing using the linear polymerase chain reaction. *Nucleic Acids Research*, 17, 8889.

Myers, R.M., Maniatis, T. & Lerman, L.S. 1987. Detection and localization of single base changes by denaturing gradient gel electrophoresis. *Methods in Enzymology*, 155, 501–27.

Nagasaka, K. & Szmidt, A.E. 1984. Multilocus analysis of external pollen contamination of a Scots pine (*Pinus sylvestris* L.) seed orchard. No. VII. In *Genetic Studies of Scots Pine (*Pinus sylvestris *L.) Domestication by Means of Isozyme Analysis*. Ed. A.E. Szmidt. pp. 1–10. Swedish University of Agricultural Studies, Umeå, Sweden.

Nagasaka, K. & Szmidt, A.E. 1985. Multilocus analysis of external pollen contamination of Scots pine (*Pinus sylvestris* L.) seed orchard. In *Population Genetics in Forestry*. Ed. H.R. Gregorius. pp. 134–8. Lecture Notes in Biomathematics 60, Springer Verlag, Berlin.

Nair, V.M.G. 1996. Oak wilt: an internationally dangerous tree disease. pp. 181–98. In *Forest Trees and Palms. Disease and Control*. Eds S.P. Raychaudhuri & K. Maramorosch. Science Publishers, NH, USA.

Nakamura, R.R. & Wheeler, N.C. 1992. Pollen competition and parental success in Douglas-fir. *Evolution*, 46, 846–51.

Namkoong, G. 1984a. A control concept of gene conservation. *Silvae Genetica*, 33, 160–3.

Namkoong, G. 1984b. Strategies for gene conservation in forest tree breeding. In *Plant Gene Resources: A Conservation Imperative*. Eds C.W. Yeatman, D. Krafton & G. Wilkes. AAA Selected Symposium 87. Westview Press, Boulder, CO, USA.

Namkoong, G. 1988. Sampling for germplasm collections. *HortScience*, 23, 79–81.

Namkoong, G. 1989. Population genetics and dynamics of conservation. In *Biotic Diversity and Germplasm Preservation*. Eds L. Knutson & A.K. Stoner. pp. 161–81. Global Imperatives. Kluwer Academic Publishers, Dordrecht.

Namkoong, G. 1991a. Maintaining genetic diversity in breeding for resistance in forest trees. *Annual Review of Phytopathology*, 29, 325–42.

Namkoong, G. 1991b. Biodiversity—issues in genetics, forestry and ethics. *Forestry Chronicle*, 68, 438–43.

Namkoong, G. 1994. *An evolutionary concept of breeding*. Marcus Wallenberg Prize Lecture. Stockholm, Sweden, 22 September 1994.

Namkoong, G., Barnes, R.D. & Burley, J. 1980. A philosophy of breeding strategy for tropical forest trees. *Tropical Forestry Papers*, No. 16., Commonwealth Forestry Institute, Oxford University, Oxford.

Namkoong, G., Boyle, T.J.B., El-Kassaby, Y., Eriksson, G., Gregorius, H.-R., Joly, H., Kremer, A., Savolainen, O., Wickneswari, R., Young, A.G., Zeh-Nlo, M. & Prabhu, P. 2000. *Testing Criteria and Indicators for Assessing the Sustainability of Forest Management: Genetic Criteria and Indicators* (in press).

Namkoong, G., Boyle T.J.B., Gregorious, H.-R., Joly, H., Savolainen, O., Wickneswari, R. & Young, A. 1996. *Testing Criteria and Indicators for Assessing the Sustainability of Forest Management: Genetic Criteria and Indicators*. Working Paper No. 10, Centre for International Forestry Research, Bogor, Indonesia.

Nason, J. 1998. Scaling-up: enlarging the spatial scale of parentage analysis. In *Proceedings from a Workshop on Gene Flow in Fragmented, Managed, and Continuous Populations*. Eds V.L. Sork, D. Campbell, R. Dyer, J. Fernandez, J. Nason, R. Petit, P. Smouse & E. Steinberg. Research Paper No. 3. pp. 20–1. National Center for Ecological Analysis and Synthesis, Santa Barbara, CA, USA.

Nason, J.D. & Hamrick, J.L. 1997. Reproductive and genetic consequences of forest fragmentation: two case studies of neotropical canopy trees. *Journal of Heredity*, 88, 264–76.

Nason, J.D., Herre, E.A. & Hamrick, J.L. 1996. Paternity analysis of the breeding structure of strangler fig populations: evidence for substantial long-distance wasp dispersal. *Journal of Biogeography*, 23, 501–12.

Nason, J.D., Herre, E.A. & Hamrick, J.L. 1998. The breeding structure of a tropical keystone plant resource. *Nature*, 391, 685–7.

National Research Council 1991. *Managing Global Genetic Resources: Forest Trees*. Committee on Managing Global Genetic Resources, Agricultural Imperatives, Subcommittee on Managing Plant Genetic Resources, Forest Genetic Resources Work Group, Board on Agriculture, National Research Council. National Academy of Sciences, National Academy Press, Washington, DC.

Nature 1998. The complex realities of sharing genetic resources. *Nature* 392, 525.

Neale, D.B. 1984. Population genetic structure of the Douglas fir shelterwood regeneration system in southwest Oregon. PhD dissertation. Oregon State University, Corvallis, Oregon.

Neale, D.B. & Adams, W.T. 1985. Allozyme and mating-system variation in balsam fir (*Abies balsamea*) across a continuous elevational transect. *Canadian Journal of Botany*, 63, 2448–53.

Nei, M. 1972. Genetic distance between populations. *American Naturalist*, 106, 283–92.

Nei, M. 1973. Analysis of gene diversity in subdivided populations. *Proceedings of the National Academy of Sciences, U SA*, 70, 3321–3.

Nei, M. 1974. A new measure of genetic distance. In *Genetic Distance*. Eds J.F. Crow & C. Denniston. pp. 63–76. Plenum Press, New York.

Nei, M. 1975. *Molecular Population Genetics and Evolution*. p. 288. North-Holland Publishing Company, Amsterdam.

Nei, M. 1987. *Molecular Evolutionary Genetics*. Columbia University Press, New York.

Nei, M., Maruyama, T. & Chakraborty, R. 1975. The bottleneck effect and genetic variability in populations. *Evolution*, 29, 1–10.

Nei, M., Tajima, F. & Tateno, Y. 1983. Accuracy of estimated phylogenetic trees from molecular data. II.

Gene frequency data. *Journal of Molecular Evolution*, 19, 153–70.

Nelson, C.D., Kubisiak, T.L., Stine, M. & Nance, W.L. 1994. A genetic linkage map of longleaf pine (*Pinus palustris* Mill.) based on random amplified polymorphic DNAs. *Journal of Heredity*, 85, 433–9.

Nepal, S.K. & Weber, K.E. 1993. *Struggle for Existence: Park–People Conflict in the Royal Chitwan National Park, Nepal*. HSD Monograph 28. Asian Institute of Technology, Bangkok.

Nepstad, D.C., Brown, I.F., Luz, L., Alechandra, A. & Viana, V. 1992. Biotic impoverishment of Amazonian forests by tappers, loggers and cattle ranchers. In *Non-timber Forest Products from Tropical Forests: Evaluation of a Conservation and Development Strategy*. Eds D.C. Nepstad & S. Schwartzman. pp. 1–14. Advances in Economic Botany 9. New York Botanical Garden, New York.

Nesbitt, K.A., Potts, B.M., Vaillancourt, R.E., West, A.K. & Reid, J.B. 1995. Partitioning and distribution of RAPD variation in a forest tree species, *Eucalyptus globulus* (Myrtaceae). *Heredity*, 74, 628–37.

Neumann, R.P. 1996. Forest products research in relation to conservation policies in Africa. In *Current Issues in Non-Timber Forest Products Research*. Eds M. Ruiz Pérez & J.E.M. Arnold. pp. 161–76. Center for International Forestry Research, Bogor, Indonesia.

Nevo, E., Lave, B. & Ben-Shlom, R. 1983. Selection of allelic isozyme polymorphisms in marine organisms: pattern, theory and application. pp. 69–92. In *Isozymes: Current Topics in Biological an Medical Research*. Eds M.C. Rattazi, J.G. Scandalios & G.S. Whill. Alan R. Liss, New York.

Newcombe, G. & Bradshaw Jr, H.D. 1996. Quantitative trait loci conferring resistance in hybrid poplar to *Septoria populicola*, the cause of leaf spot. *Canadian Journal of Forest Research*, 26, 1943–50.

Newstrom, L.E., Frankie, G.W., Baker, H.G. & Colwell, R.K. 1993. Diversity of flowering patterns at La Selva. In *La Selva, Ecology and Natural History of a Lowland Tropical Rainforest*. Eds L.A. McDade, K.S. Bawa, G.S. Hartshorn & H.A. Hespenheide. pp. 142–60. University of Chicago Press, Chicago.

Newstrom, L.E., Frankie, G.W. & Baker, H.G. 1994. A new classification for plant phenology based on flowering patterns in lowland tropical rain forest trees at La Selva, Costa Rica. *Biotropica*, 26, 141–59.

Newton, R.J., Funkhouser, E.A., Fong, F. & Tauer, C.G. 1991. Molecular and physiological genetics of drought tolerance in forest species. *Forest Ecology and Management*, 43, 225–50.

Ngo, T.H.G. Mering, 1996. A new perspective on property rights: examples from the Kayan of Kalimantan. In *Borneo in Transition*. Eds C. Padoch & N. Lee Peluso. pp. 137–49. Oxford University Press, Kuala Lumpur.

Nhira, C., Baker, S., Gondo, P., Mangono, J.J. & Marunda, C. 1998. *Contesting Inequality in Access to Forests. Policy That Works for Forests and People—Zimbabwe*.

International Institute for Environment & Development, London.

Nicholson, D.I. 1958. An analysis of logging damage in tropical rain forest, North Borneo. *Malaysian Forester*, XXI (4), 235–45.

Niebling, C.R. & Conkle, M.T. 1990. Diversity of Washoe pine and comparisons with allozymes of ponderosa pine races. *Canadian Journal of Forest Research*, 20, 298–308.

O'Brien, S.J., Roelke, M.E., Yuhki, N., Richards, K.W., Johnson, W.E., Franklin, W.L., Anderson, A.E., Bass, O.L., Jr., Belden, R.C. & Martenson, J.S. 1990. Genetic introgression within the Florida panther *Felis concolor coryi*. *National Geographic Research*, 6, 485–94.

Oldfield, M. 1984. *The Value of Conserving Genetic Resources*. US Department of Interior, National Park Service, Washington, DC.

Oleksyn, J. 1987. Air pollution effects on 15 European and Siberian Scots pine (*Pinus sylvestris* L.) provenances growing in a 75-year-old experiment. *Arboretum Kórnickie*, 32, 151–62.

Oleksyn, J. 1988. Height growth of different European Scots pine *Pinus sylvestris* L. provenances in a heavily polluted and a control environment. *Environmental Pollution*, 55, 289–99.

Oleksyn, J., Chalupka, W., Tjoelker, M.G. & Reich, P.B. 1992. Geographic origin of *Pinus sylvestris* populations influences the effect of air pollution on flowering and growth. *Water, Air, and Soil Pollution*, 62, 201–12.

Oleksyn, J., Fritts, H.C. & Hughes, M.K. 1993. Tree-ring analysis of different *Pinus sylvestris* provenances, *Quercus robur*, *Larix decidua* and *L. decidua* x *L. kaempferi* affected by air pollution. *Arboretum Kórnickie*, 38, 87–111.

Oleksyn, J., Karolewski, P. & Rachwał, L. 1988. Susceptibility of European *Pinus sylvestris* L. provenances to SO$_2$, NO$_2$, SO$_2$ + NO$_2$ and HF under laboratory and field conditions. *Acta Societatis Botanicorum Poloniae*, 57, 107–15.

Oleksyn, J., Prus-Głowacki, W., Giertych, M. & Reich, P.B. 1994. Relation between genetic diversity and pollution impact in a 1912 experiment with east European *Pinus sylvestris* provenances. *Canadian Journal of Forest Research*, 24, 2390–4.

O'Malley, D.M. & Bawa, K.S. 1987. Mating system of a tropical rain forest tree species. *American Journal of Botany*, 74, 1143–9.

O'Malley, D.M., Buckley, D.P., Prance, G.T. & Bawa, K.S. 1988. Genetics of Brazil nut (*Bertholletia excelsa* Humb. and Bonpl., Lecythidaceae) 2. Mating system. *Theoretical and Applied Genetics*, 76, 929–32.

Omi, S.K. & Adams, W.T. 1986. Variation in seed set and proportions of outcrossed progeny with clones, crown position, and top pruning in a Douglas-fir seed orchard. *Canadian Journal of Forest Research*, 16, 502–7.

Opler, P.A., Baker, H.G. & Frankie, G.W. 1975. Reproductive biology of some Costa Rican *Cordia* species (Boraginaceae). *Biotropica*, 7, 234–47.

Opler, P.A., Frankie, G.W. & Baker, H.G. 1976. Rainfall as a factor in the release, timing, and synchronization of anthesis by tropical trees and shrubs. *Journal of Biogeography*, 3, 231–6.

Opler, P.A., Frankie, G.W. & Baker, H.G. 1980. Comparative phenological studies of treelet and shrub species in tropical wet and dry forests in the lowlands of Costa Rica. *Journal of Ecology*, 68, 167–88.

Organization for Economic Cooperation and Development 1997. *Issues in the Sharing of Benefits Arising out of the Utilization of Genetic Resources*. OECD, Paris.

Organization for Economic Cooperation and Development 1998. *The Economics of Benefit Sharing: Introduction and Practical Experiences*. OECD, Paris.

Orgel, L.E. & Crick, F.H.C. 1980. Selfish DNA: the ultimate parasite. *Nature*, 284, 604–7.

Ouberg, N.J., van Treuren, R. & van Damme, J.M.M. 1991. The significance of genetic erosion in the process of extinction II. Morphological variation and fitness components in populations of varying size of *Salvia pratensis* L. and *Scabiosa columbaria* L. *Oecologia*, 86, 359–67.

Padoch, C. & Pinedo-Vasquez, M. 1996. Smallholder forest management: looking beyond non-timber forest products. In *Current Issues in Non-Timber Forest Products Research*. Eds M. Ruiz Pérez & J.E.M. Arnold. pp. 103–17. Center for International Forestry Research, Bogor, Indonesia.

Pakkanen, A. & Pulkkinen, P. 1991. Pollen production and background pollination levels in Scots pine seed orchards of northern Finnish origin. In *Pollen Contamination in Seed Orchards*. Ed. D. Lindgren. pp. 14–21. Proceedings of the Meeting of the Nordic Group for Tree Breeding 1991, Report 10, Swedish University of Agricultural Sciences, Department of Forest Genetics and Plant Physiology, Umeå, Sweden.

Pakkanen, A., Pulkkinen, P. & Vakkari, P. 1991. Pollen contamination in the years 1988–1989 in some old Scots pine seed orchards of northern Finnish origin. *Reports from the Foundation of Forest Tree Breeding, Helsinki, Finland*, 3, 3–8.

Palmberg, C. 1975. Geographic variation and early growth in South-eastern semiarid Australia of *Pinus halepensis* Mill. and *Pinus brutia* Ten. species complex. *Silvae Genetica*, 24, 150–60.

Palmberg-Lerche, C. 1994. International programmes for the conservation of forest genetic resources. In *Proceedings of the International Symposium on Genetic Conservation and Production of Tropical Tree Seed*. Eds R.M. Drysdale, S.E.T. John & A.C. Yappa. pp. 78–101. ASEAN–Canada Forest Tree Seed Project, Muak-Lek, Thailand.

Pan, F.J. 1985. Systematics and genetics of the *Leucaena diversifolia* (Schlecht.) Benth. Complex. PhD dissertation. University of Hawaii at Manoa, Hawaii, USA.

Panayatou, T. & Ashton, P.S. 1992. *Not by Timber Alone: Economics and Ecology for Sustaining Tropical Forests*. Island Press, Washington, DC.

Panayotou, T. & Glover, D. 1994. Economic and financial incentives for biodiversity conservation and development. Paper commissioned by IUCN for presentation at the Regional Conference on Biodiversity Conservation, 6–8 June 1994, Asian Development Bank, Manila.

Parani, M., Lakshmi, M., Elango, S., Ram, N., Anuratha, C.S. & Parida, A. 1997. Molecular phylogeny of mangroves. 2. Intra- and inter-specific variation in *Avicennia* revealed by RAPD and RFLP markers. *Genome*, 40, 487–95.

Park, Y.S. & Fowler, D.P. 1982. Effects of inbreeding and natural variances in a natural population of tamarack (*Larix laricina* (Du Roi) K. Koch) in eastern Canada. *Silvae Genetica*, 31, 21–6.

Parnwell, M.J.G. & Taylor, D.M. 1996. Environmental degradation, non-timber forest products and Iban communities in Sarawak. In *Environmental Change in South-East Asia*. Eds M.J.G. Parnwell & R.L. Bryant. pp. 269–300. Routledge, London.

Parsons P.A. 1987. Evolutionary rates under environmental stress. *Evolutionary Biology*, 21, 311–47.

Patterson, W.A. & Olson, J.J. 1983. Effects of heavy metals on radicle growth of selected woody species on filter paper, mineral and organic soil substrates. *Canadian Journal of Forest Research*, 13, 233–8.

Paule, L. 1991. Clone identity and contamination in a Scots pine seed orchard. In *Pollen Contamination in Seed Orchards*. Proceedings of the Meeting of the Nordic Group for Tree Breeding 1991, Report 10, Ed. D. Lindgren. pp. 22–32. Swedish University of Agricultural Sciences, Department of Forest Genetics and Plant Physiology, Umeå, Sweden.

Paule, L. & Gomory, D. 1992. Genetic processes in seed orchards as illustrated by European larch (*Larix decidua* Mill.) and Scots pine (*Pinus sylvestris* L.). In *Proceedings of the International Conference on Forest–Wood–Ecology, Section Phytotechnique and Forest Management in Present Ecological Conditions*. Ed. S. Korpel. pp. 109–14. Technical University, Zvolen, Slovakia.

Paule, L., Lindgren, D. & Yazdani, R. 1991. Pollen contamination in Norway spruce seed orchards investigated by allozymes. In *Pollen Contamination in Seed Orchards*. Proceedings of the Meeting of the Nordic Group for Tree Breeding 1991. Report 10, Ed. D. Lindgren. p. 52. Swedish University of Agricultural Sciences, Department of Forest Genetics and Plant Physiology, Umeå, Sweden.

Pearce, D.W. 1986. *Cost-Benefit Analysis*. Macmillan, Basingstoke, UK.

Pearce, D. & Moran, D. 1994. *The Economic Value of Biodiversity*. Earthscan Publications and IUCN (World Conservation Union), London.

Peluso, N.L. 1991. Case Study One, rattan industries in East Kalimantan, Indonesia. In *Case Studies in Forest-based Small Scale Enterprises in Asia*. Ed. J.Y. Campbell. pp. 5–28. Community Forestry Case Study 4. Food and Agriculture Organization of the United Nations, Bangkok.

Peluso, N.L. 1992. The ironwood problem: (mis)management and development of an extractive rainforest product. *Conservation Biology*, 6, 210–19.

Peluso, N.L. 1993. Coercing conservation? *Global Environmental Change*, 3, 199–217.

Peluso, N.L. & Padoch, C. 1996. Changing resource rights in managed forests of West Kalimantan. In *Borneo in Transition*. Eds C. Padoch & N. Lee Peluso. pp. 121–36. Oxford University Press, Kuala Lumpur.

Pennacchini, F. & Ducci, F. 1991. Prova di resistenza ad inquinanti di 6 provenienze italiane di abete bianco. *Annali dell'Istituto Sperimentale per la Selvicoltura Arezzo*, 22, 73–93.

Penner, G.A., Bush, A., Wise, R., Kim, W., Domier, L., Kasha, K., Laroche, A., Scoles, G., Molnar, S.J. and, Fedak, G. 1993. Reproducibility of random amplified polymorphic DNA (RAPD) analysis among laboratories. *PCR Methods and Applications*, 2, 341–5.

Perez de la Vega, M. & Allard, R.W. 1984. Mating system and genetic polymorphism in populations of *Secale cereale* and *S. vavilovii*. *Canadian Journal of Genetics and Cytology*, 26, 308–17.

Perring, F.H. & Farrell, L. 1983. *British Red Data Books, 1. Vascular Plants*. Royal Society for Nature Conservation, Lincoln, UK.

Perrings, C. 1996. Economics, ecology and the Global Biodiversity Assessment. *Trends in Evolution and Ecology*, 11, 290.

Perrings, C., Mäler, K.G., Folke, C., Holling, C.S. & Bengt-Owe, J. 1995. *Biodiversity Loss*. Cambridge University Press, Cambridge.

Perry, D.J. & Dancik, P.J. 1986. Mating system dynamics of lodgepole pine in Alberta, Canada. *Silvae Genetica*, 35, 190–5.

Persson, B. & Stahl, E.G. 1990. Survival and yield of *Pinus sylvestris* L. as related to provenance transfer and spacing at high altitudes in northern Sweden. *Scandinavian Journal of Forestry Research*, 5, 381–95.

Peters, C.M. 1994. *Sustainable Harvest of Non-Timber Plant Resources in Tropical Moist Forest: An Ecological Primer*. Biodiversity Support Program, Washington, DC.

Peters, C.M. 1996a. Observations on the sustainable exploitation of non-tropical forest products. In *Current Issues in Non-Timber Forest Products Research*. Eds M. Ruiz Pérez & J.E.M. Arnold. pp. 19–39. Center for International Forestry Research, Bogor, Indonesia.

Peters, C.M. 1996b. Illipe nuts (*Shorea* spp.) in West Kalimantan: use, ecology, and management potential of an important forest resource. In *Borneo in Transition*. Eds C. Padoch & N. Lee Peluso. pp. 230–44. Oxford University Press, Kuala Lumpur.

Peters, C.M., Gentry, A.H. & Mendelsohn, R.O. 1989. Valuation of an Amazonian rainforest. *Nature*, 33, 655–6.

Petit, R.J. & Kremer, A. 1993. Geographic structure of chloroplast DNA polymorphisms in European oaks. *Theoretical and Applied Genetics*, 87, 122–8.

Petit, R.J., Kremer, A. & Wagner, D.B. 1993. Geographic structure of chloroplast DNA polymorphisms in European oaks. *Theoretical and Applied Genetics*, 87, 122–8.

Petit, R.J., Pineau, E., Demesure B., Bacilieri, R., Ducousso, A. & Kremer, A. 1997. Chloroplast DNA footprints of postglacial recolonization by oaks. *Proceedings of the National Academy of Sciences, USA*, 94, 9996–10001.

Pfeiffer, A., Olivieri, A.M. & Morgante, M. 1997. Identification and characterization of microsatellites in Norway spruce (*Picea abies* K). *Genome*, 40, 411–19.

Pielou, E.C. 1975. *Ecological Diversity*. John Wiley and Sons, New York.

Piesch, R.F. 1987. Tree improvement comes of age in the Pacific Northwest— implications for the nursery manager. *Tree Planter's Notes*, 38, 3–8.

Pimbert, M.P. & Pretty, J.N. 1995. Parks, People and Professionals. Discussion Paper 57. United Nations Research Institute for Social Development, Geneva.

Pinedo-Vasquez, M. & Padoch, C. 1993. Community and governmental experiences protecting biodiversity in the Lowland Peruvian Amazon. In *Perspectives on Biodiversity: Case Studies of Genetic Resource Conservation and Development*. Eds C.S. Potter, J.I. Cohen & D. Janczewski. pp. 199–211. AAAS (American Association for the Advancement of Science) Press, Washington, DC.

Playford, J., Bell, J.C. & Moran, G.F. 1993. A major disjunction in genetic diversity over the geographic range of *Acacia melanoxylon* R. Br. *Australian Journal of Botany*, 41, 355–68.

Plucknett, D.L., Smith, N.J.H., Williams, J.T. & Anishetty, N.M. 1987. *Gene Banks and the World's Food*. Princeton University Press, Princeton, NJ, USA.

Pollak, E. 1987. On the theory of partially inbreeding finite populations. I. Partial selfing. *Genetics*, 117, 353–60.

Porter, A.H. 1990. Testing nominal species boundaries using gene flow statistics: the taxonomy of two hybridizing Admiral Butterflies (Limenitis: Nymphalidae). *Systematic Zoology*, 39, 131–48.

Posey, D.A. 1985. Indigenous management of tropical forest ecosystems: the case of the Kayapó indians of the Brazilian Amazon. *Agroforestry Systems*, **3**, 139–58.

Posey, D.A. & Dutfield, G. 1996. *Beyond Intellectual Property: Towards Traditional Resource Rights for Indigenous Peoples and Local Communities*. IDRC, Canada.

Potts, B.M. & Wiltshire, R.J.E. 1997. Eucalypt genetics and genecology. In *Eucalypt Ecology: Individuals to Ecosystems*. Eds J. Williams and J. Woinarski. pp. 56–91. Cambridge University Press, Cambridge.

Powell, A.H. & Powell, G.V.N. 1987. Population dynamics of male euglossine bees in Amazonian forest fragments. *Biotropica*, 19, 176–9.

Powers, H.R. 1984. Control of fusiform rust of southern pines in the USA. *European Journal of Forest Pathology*, 14, 426–31.

Prabhu, R., Colfer, C.J.P., Venkateswarlu, P., Tan, L.C., Soekmadi, R. & Wollenberg, E. 1996. *Testing Criteria and Indicators for the Sustainable Management of Forests*. Phase I. Final Report. Center for International Forestry Research, Bogor, Indonesia.

Prescott-Allen, C. & Prescott-Allen, R. 1986. *The First Resource: Wild Species in the North American Economy*. Yale University Press, New Haven, CT, USA.

Pressey, R.L. & Logan, V.S. 1997. Inside looking out: findings of research on reserve selection relevant to }off-reserve' nature conservation. In *Conservation Outside Nature Reserves*. Eds P. Hale & D. Lamb. pp. 407–18. Centre for Conservation Biology, University of Queensland, Brisbane.

Pressey, R.L., Bedward, M. & Nicholls, A.O. 1990. Reserve selection in mallee lands. In *The Mallee Lands: A Conservation Perspective*. Eds J.C. Noble, P.J. Joss & G.K. Jones. pp. 167–78. CSIRO, Melbourne.

Preston, R.E. 1986. Pollen-ovule ratios in the Cruciferae. *American Journal of Botany*, 73, 1732–40.

Price, M.V. & Waser, N.M. 1979. Pollen dispersal and optimal outcrossing in *Delphinium nelsoni*. *Nature*, 277, 294–7.

Primack, R.B. & Kang, H. 1989. Measuring fitness and natural selection in wild plant populations. *Annual Review of Ecology and Systematics*, 20, 367–96.

Prober, S.M. & Brown, A.H.D. 1994. Conservation of the grassy white box woodlands—population genetics and fragmentation of *Eucalyptus albens*. *Conservation Biology*, 8, 1003–13.

Prus-Głowacki, W. & Godzik, St 1991. Changes induced by zinc smelter pollution in the genetic structure of pine (*Pinus sylvestris* L.) seedling populations. *Silvae Genetica*, 40, 184–8.

Prus-Głowacki, W. & Godzik, St 1995. Genetic structure of *Picea abies* trees tolerant and sensitive to industrial pollution. *Silvae Genetica*, 44, 62–5.

Prus-Głowacki, W. & Nowak-Bzowy, R. 1989. Demographic processes in *Pinus sylvestris* populations from regions under strong an weak anthropogenous pressure. *Silvae Genetica*, 38, 55–62.

Prus-Głowacki, W. & Nowak-Bzowy, R. 1992. Genetic structure of a naturally regenerating Scots pine population tolerant for high pollution near a zinc smelter. *Water, Air, and Soil Pollution*, 62, 249–59.

Prus-Głowacki, W. & Stephan, B.R. 1994. Genetic variation of *Pinus sylvestris* from Spain in relation to other European populations. *Silvae Genetica*, 43, 7–14.

Pushpakumara, D. 1997. The reproductive biology of *Artocarpus heterophyllus* Lam. D. Phil. Thesis, University of Oxford.

Quijada, A., Liston, A., Robinson, W. and Alvarez-Buylla, E. 1997. The ribosomal ITS region as a marker to detect hybridization in pines. *Molecular Ecology*, 6, 995–6.

Rachwał, L. & Oleksyn, J. 1987. Growth and development of black pine (*Pinus nigra* Arn.) and Norway spruce (*Picea abies* (L.) Karst.) in the Niepolomice forest provenance experiments. *Acta Agraria et Silvestria*, 24, 163–81.

Raddi, S.F., Stefanini, F.M., Camussi, A. & Giannini, R. 1994. Forest decline index and genetic variability in *Picea abies* (L.) Karst. *Forest Genetics*, 1, 33–40.

Rahim, A.N. & Harding D. 1992. Effects of selective logging methods on water yield and streamflow parameters in Peninsular Malaysia. *Journal of Tropical Forest Science*, 5 (2), 130–54.

Rajasekaran, B. & Warren, D.M. 1994. IK for socioeconomic development and biodiversity conservation: the Kolli Hills. *Indigenous Knowledge & Development Monitor*, 2, 13–17.

Ramesh, B.R., Menon, S. & Bawa, K.S. 1997. A vegetation based approach to biodiversity gap analysis in the Agastyamalai region, Western Ghats, India. *Ambio*, 26, 529–36.

Rand, D.M. & Harrison, R.G. 1989. Ecological genetics of a mosaic hybrid zone: mitochondrial, nuclear, and reproductive differentiation of crickets by soil type. *Evolution*, 43, 432–49.

Ranney, J.W., Broner, M.C. & Leverson, J.B. 1981. The importance of edge in the structure and dynamics of forest islands. In *Forest Island Dynamics in Man-dominated Landscapes*. Eds R.L. Burgess and D.M. Sharpe. pp. 67-96. Springer-Vortas, New York.

Rappaport, R.A. 1993. Distinguished lecture in general anthropology: the anthropology of trouble. *American Anthropologist*, 95, 295–303.

Rausher, M.D. & Fowler, N.L. 1979. Intersexual aggression and nectar defense in *Chauliognathus distiguendus* (Coleoptera, Cantharidae). *Biotropica*, 11, 96–100.

Read, D.J. 1968. Some aspects of the relationship between shade and fungal pathogenicity in an epidemic disease of pines. *New Phytologist*, 67, 39–48.

Redclift, M. & Woodgate, G. 1994. Sociology and the environment—discordant discourse. In *Social Theory and the Global Environment*. Eds M. Redclift & T. Benton. pp. 51–66. Routledge, London.

Rehfeldt, G.E. 1992. Early selection in *Pinus ponderosa*: compromises between growth potential and growth rhythm in developing breeding strategies. *Forest Science*, 38, 661–77.

Reich, P.B., Oleksyn, J. & Tjoelker, M.G. 1994. Relationship of aluminium and calcium to net CO_2 exchange among diverse Scots pine provenances under pollution stress in Poland. *Oecologia*, 97, 82–92.

Reid, W.V., Laird, S.A., Meyer, C.A., Gamez, R., Sittenfeld, A., Janzen, D.H., Gollin, M.A. & Juma, C. 1993. *Biodiversity Prospecting*. World Resources Institute, Washington, DC.

Renner, S.S. 1989. A survey of reproductive biology in Neotropical Melastomataceae and Memecylaceae. *Annals of the Missouri Botanical Gardens*, 76, 496–518.

Repetto, R. 1987. Economic incentives for sustainable production. *Annals of Regional Science*, 21, 44–59.

Repetto, R. 1988. *The Forest for the Trees? Government Policies and the Misuse of Forest Resources*. World Resources Institute, Washington, DC.

Repetto, R. & Gillis, M. (Eds). 1988. *Public Policies and the Misuse of Forest Resources*. Cambridge University Press, Cambridge.

Repetto, R., Dower, R.C., Jenkins, R. & Geoghegan, J. 1992. *Green Fees: How a Tax Shift can Work for the Environment and the Economy*. World Resources Institute, Washington, DC.

Reynolds, S. & El-Kassaby, Y.A. 1990. Parental balance in a Douglas-fir seed orchard: cone vs. seed production. *Silvae Genetica*, 39, 40–2.

Rhymer, J.M. & Simberloff, D. 1996. Extinction by hybridization and introgression. *Annual Review of Ecology and Systematics*, 27, 83–109.

Rhymer, J.M., Williams, M.J. & Braun, M.J. 1994. Mitochondrial analysis of gene flow between New Zealand mallards (*Anas platyrhynchos*) and grey ducks (*A. superciliosa*). *Auk*, 111, 970–8.

Rice, K.J., Richards, J.H. & Matzner, S.L. 1997. Patterns and process of adaptation in blue oak seedlings. USDA Forest Service Gen. Tech. Rep PSW-GTR-160. United States Department of Agriculture, Forest Service, Intermountain Research Station, Ogden, UT, USA.

Richards, A.J. 1986. *Plant Breeding Systems*. Allen & Unwin, London.

Richards, A.J. 1990. Studies in *Garcinia*, dioecious tropical forest trees: agamospermy. *Botanical Journal of the Linnean Society*, 103, 233–50.

Rieseberg, L.H. 1991. Hybridization in rare plants: insights from case studies in *Cercocarpus* and *Helianthus*. In *Genetics and Conservation of Rare Plants*. Eds D.A. Falk and K.E. Holsinger. pp. 171–81. Oxford University Press, New York.

Rieseberg, L.H. 1996. Rare trees. *Science*, 271, 16.

Rieseberg, L.H. & Gerber, D. 1995. Hybridization in the Catalina Island mountain mahogany (*Cercocarpus traskiae*): RAPD evidence. *Conservation Biology*, 9, 199–203.

Rieseberg, L.H. & Soltis, D.E. 1991. Phylogenetic consequences of cytoplasmic gene flow in plants. *Evolutionary Trends in Plants*, 5, 65–84.

Rieseberg, L.H. & Swensen, S.M. 1996. Conservation genetics of endangered island plants. In *Conservation Genetics: Case Histories from Nature*. Eds J.C. Avise and J.L. Hamrick. pp. 305–34. Chapman and Hall, New York.

Rieseberg, L.H. & Wendel, J.F. 1993. Introgression and its consequences in plants. In *Hybrid Zones and the Evolutionary Process*. Ed. R.G. Harrison. pp. 70–109. Oxford University Press, New York.

Rieseberg, L.H., Whitton, J. and Gardner, K. 1999. Hybrid zones and the genetic architecture of a barrier to gene flow between two sunflower species. *Genetics*, 152, 713–27.

Rieseberg, L.H., Zona, S., Aberbom, L. & Martin, T.D. 1989. Hybridization in the Island endemic, Catalina mahogany. *Conservation Biology*, 3, 52–8.

Rigg, J. 1991. Grass-roots development in rural Thailand: a lost cause? *World Development*, 19, 199–211.

Rigg, J. 1997. *Southeast Asia—The Human Landscape of Modernization and Development*. Routledge, London.

Riggs, L.A. 1990. Conserving genetic resources on-site in forest ecosystems. *Forest Ecology and Management*, 35, 45–68.

Rink, G., Carroll, E.R. & Kung, F.H. 1989. Estimation of *Juglans nigra* L. mating system parameters. *Forest Science*, 35, 623–7.

Ritland, K. 1983. Estimation of mating systems. In *Isozymes in Plant Genetics and Breeding*. Part A. Eds S.D. Tanksley & T.J. Orton. pp. 289–302. Elsevier Science Publishers bv, Amsterdam, Netherlands.

Ritland, K. 1984. The effective proportion of self-fertilization with consanguineous matings in inbred populations. *Genetics*, 106, 139–52.

Ritland, K. 1985. The genetic mating structure of subdivided populations I. Open-mating model. *Theoretical Population Biology*, 27, 51–74.

Ritland, K. 1986. Joint maximum likelihood estimation of genetic and mating structure using open-pollinated progenies. *Biometrics*, 42, 25–43.

Ritland, K. 1988. The genetic mating structure of subdivided populations. II. Correlated mating models. *Theoretical Population Biology*, 34, 320–46.

Ritland, K. 1989. Genetic differentiation, diversity and inbreeding in the mountain monkeyflower (*Mimulus caespitosus*) of the Washington Cascades. *Canadian Journal of Botany*, 67, 2017–24.

Ritland, K. 1990. A series of FORTRAN computer programs for estimating plant mating systems. *The Journal of Heredity*, 81, 235–7.

Ritland, K. & El-Kassaby, Y.A. 1985. The nature of inbreeding in a seed orchard of Douglas-fir as shown by an efficient multilocus model. *Theoretical and Applied Genetics*, 71, 375–84.

Ritland, K. & Ganders, F.R. 1985. Variation in the mating system of *Bidens menziesii* (Asteraceae) in relation to population substructure. *Heredity*, 55, 235–44.

Ritland, K. & Ganders, F.R. 1987. Covariation of selfing rates with parental gene fixation indices within populations of *Mimulus guttatus*. *Evolution*, 41, 760–71.

Ritland, K. & Jain, S. 1981. A model for the estimation of outcrossing rate and gene frequencies using n independent loci. *Heredity*, 47, 35–52.

Roberds, J.H. & Namkoong, G. 1989. Population selection to maximize value in an environmental gradient. *Theoretical and Applied Genetics*, 77, 128–35.

Roberds, J.H., Friedman, S.T. & El-Kassaby, Y.A. 1991. Effective number of pollen parents in clonal seed orchards. *Theoretical and Applied Genetics*, 82, 313–20.

Robinson, J.G. 1993. The limits to caring: sustainable living and the loss of biodiversity. *Conservation Biology*, 7, 20–28.

Rocha, O.J. & Lobo, J.A. 1996. Genetic variation and differentiation among five populations of the Guanacaste tree (*Enterolobium cyclocarpum* Jacq.) in Costa Rica. *International Journal of Plant Sciences*, 157, 234–9.

Rogers, D.L. & Ledig, F.T. (Eds) 1996. *The Status of Temperate North American Forest Genetic Resources*. Report No. 16, University of California Genetic Resources Conservation Program, Davis, CA, USA.

Rogstad, S.H., Nybom, H. & Schaal, B.A. 1991. The tetrapod 'DNA fingerprinting' M13 probe reveals genetic diversity and clonal growth in quaking aspen (*Populus tremuloides*, Salicaceae). *Plant Systematics and Evolution*, 175, 115–23.

Rogstad, S.H., Patton, J.C. & Schaal, B.A. 1988. M13 repeat probe detects DNA minisatellite-like sequences in gymnosperms and angiosperms. *Proceedings of the National Academy of Sciences, USA*, 85, 9176–8.

Roll-Hansen, F. 1989. *Phacidium infestans*: a literature review. *European Journal of Forest Pathology*, 19, 237–50.

Roose, M.L., Bradshaw, A.D. & Roberts, T.M. 1982. Evolution of resistance to gaseous air pollutants. In *Effects of Gaseous Air Pollution in Agriculture and Horticulture*. Eds M.H. Unsworth & D.P. Ormrod. pp. 379–409. Butterworth Scientific, London.

Roques, A., Kerjean, M. & Auclair, D. 1980. Effets de la pollution atmosphérique par le fluor et le dioxyde de soufre sur l'appareil reproducteur femelle de *Pinus sylvestris* en Forêt de Roumare (seine-Maritime, France). *Environmental Pollution (Series A)*, 21, 191–1201.

Ros-Tonen, M., Dijkman, W. & Lammerts van Bueren, E. 1995. *Commercial and Sustainable Extraction of Non-timber Forest Products*. The Tropenbos Foundation, Wageningen, Netherlands.

Rowe, R.E. 1995. The population biology of *Trochetiopsis*: a genus endemic to St. Helena. D. Phil. Thesis, University of Oxford, Oxford.

Rowe, R. & Cronk, Q. 1995. Applying molecular techniques to plant conservation: screening genes for survival. *Plant Talk*, 1, 18–19.

Rudin, D., Muona, O. & Yazdani, R. 1986. Comparison of the mating system of *Pinus sylverstris* in natural and seed orchards. *Hereditas*, 104, 15–19.

Saccheri, I., Kuussaari, M., Kankare, M., Vikman, P., Fortelius, W. & Hanski, I. 1998. Inbreeding and extinction in a butterfly metapopulation. *Nature*, 392, 491–4.

Sampson, J.F. 1998. Multiple paternity in *Eucalyptus rameliana* (Myrtaceae). *Heredity*, 81, 349–55.

Sampson, J.F., Coates, D.J. & Van Leeuwen, S.J. 1996. Mating system variation in animal-pollinated rare and endangered plant populations in Western Australia. pp. 187–95. In *Gondwanan Heritage: Past, Present and Future of the Western Australian Biota*. Eds S. Hopper, J. Chappill, M. Harvey, & A. George. Surrey Beatty & Sons, Chipping Norton, Australia.

Sampson, J.F., Collins, B.G. & Coates, D.J. 1994. Mixed mating in *Banksia brownii* Baxter ex R. Br. (Proteaceae). *Australian Journal of Botany*, 42, 103–11.

Sampson, J.F., Hopper, S.D. & James, S.H. 1988. Genetic diversity and the conservation of *Eucalyptus crucis* Maiden. *Australian Journal of Botany*, 36, 447–60.

Sandiford, M. 1998. A study of the reproductive biology of *Bombacopsis quinata* (Jacq.) Dugand. D.Phil. Thesis, University of Oxford, Oxford.

Saunders, D.A., Margules, C.R. & Hill, B. 1998. *Environment indicators for National State of the Environment Reporting—Biodiversity*. Environment Australia, Canberra. 68 p.

Saunier, R.E. & Meganck, R.A. 1995. *Conservation of Biodiversity and the New Regional Planning*. IUCN (World Conservation Union), Gland.

Savard, L., Li, P., Strauss, S.H., Chase, M.W., Michaud, M. & Bousquet, J. 1994. Chloroplast and nuclear gene sequences indicate late Pennsylvanian time for the last common ancestor of extant seed plants. *Proceedings of the National Academy of Sciences, USA*, 91, 5163–7.

Savolainen, O. 1996. Pines beyond the polar circle: adaptation to stress conditions. *Euphytica*, 92, 139–45.

Saxe, H. & Murali, N.S. 1989. Diagnostic parameters for selecting against novel spruce (*Picea abies*) decline: II. Response of photosynthesis and transpiration to acute NO_x exposure. *Physiologia Plantarum*, 76, 349–55.

Schaal, B.A. 1980. Measurement for gene flow in *Lupinus texensis*. *Nature*, 248, 450–1.

Schat, H., Vooijs, R. & Kuiper, E. 1996. Identical major gene loci for heavy metal tolerances that have independently evolved in different local populations and subspecies of *Silene vulgaris*. *Evolution*, 50, 1888–95.

Schierenbeck, K.A., Skupski, M., Lieberman, D. & Lieberman, M. 1997. Population structure and genetic diversity in four tropical tree species in Costa Rica. *Molecular Ecology*, 6, 137–44.

Schiller, G., Conkle, M.T., & Grunwald, L. 1983. Local differentiation among Mediterranean populations of Aleppo pine in their isozymes. *Silvae Genetica*, 35, 11–19.

Schlich, W. 1905. *Manual of Forestry. Vol. III. Forest Management*. 3rd edn, Bradbury, Agnew & Co., London.

Schmid, R. 1970. Notes on the reproductive biology of *Asterogyne martiana* (Palmae). Inflorescence and floral morphology phenology. *Pincipes*, 14, 3–9.

Schmidt–Adam, G., Young, A. & Murray, B.G.M. 2000. Low outcrossing rates and pollinator shift in New Zealand Pohutukawa (*Metrosideros excelsa*) (Myrtaceae). *American Journal of Botany* (in press).

Schmidtling, R.C. 1983. Genetic variation in fruitfulness in a loblolly pine (*Pinus taeda* L.) seed orchard. *Silvae Genetica*, 32, 76–80.

Schmidtling, R.C. 1994. Use of provenance tests to predict response to climatic change: loblolly pine and Norway spruce. *Tree Physiology*, 14, 805–17.

Schmidtling, R.C. 1997. Using provenance tests to predict response to climatic change. In *Ecological Issues and Environmental Impact Assessment*. Ed P.N. Cheremisinoff. Chapter 27. Gulf Publishing Co., Houston, TX, USA.

Schmidt, T. & Heslop-Harrison, J.S. 1998. Genomes, genes and junk: the large-scale organization of plant genomes. *Trends in Plant Science*, 3, 195–9.

Schmitt, D. & Perry, T.O. 1964. Self-sterility in sweetgum. *Forest Science*, 10, 302–5.

Schnabel, A. & Hamrick, J.L. 1995. Understanding the population genetic structure of *Gleditsia triacanthos* L.: the scale and pattern of pollen gene flow. *Evolution*, 49, 921–31.

Schnabel, A., Nason, J.D. & Hamrick, J.L. 1998. Understanding the population genetic structure of *Gleditsia triacanthos* L: seed dispersal and variation in female reproductive success. *Molecular Ecology*, 7, 819–32.

Schneiderbauer, A., Back, E. Sandermann, H. Jr. & Ernst, D. 1995. Ozone induction of extensin mRNA in Scots pine, Norway spruce and European beech. *New Phytologist*, 130, 225–30.

Schoen, D.J. 1982. The breeding system of *Gilia achilleifolia*: variation in floral charactersitics and outcrossing rate. *Evolution*, 36, 596–613.

Schoen, D.J. & Brown, A.H.D. 1991. Intraspecific variation in population gene diversity and effective population size correlates with mating system in plants. *Proceedings of the National Academy of Sciences, USA*, 88, 4494–7.

Schoen, D.J. & Stewart, S.C. 1986. Variation in male reproductive investment and reproductive success in white spruce. *Evolution*, 40, 1109–20.

Schoen, D.J. & Stewart, S.C. 1987. Variation in male fertilities and pairwise mating probabilities in *Picea glauca*. *Genetics*, 116, 141–52.

Schoen, D.J., Denti, D. & Stewart, S.C. 1986. Strobilus production in a clonal white spruce seed orchard: evidence for unbalanced mating. *Silvae Genetica*, 35, 201–5.

Scholz, F. & Bergmann, F. 1984. Selection pressure by air pollution as studied by isozyme-gene-systems in Norway spruce exposed to sulphur dioxide. *Silvae Genetica* 33, 238–41.

Scholz, F. & Reck, S. 1977. Effects of acids on forest trees as measured by titration *in vitro*, inheritance of buffering capacity in *Picea abies*. *Water, Air, and Soil Pollution*, 8, 41–5.

Scholz, F. & Venne, H. 1989. Structure and first results of a research program on ecological genetics of air pollution effects in Norway spruce. In *Genetic Effects of Air*

Pollutants in Forest Tree Populations. Eds F. Scholz, H.-R. Gregorius & D. Rudin. pp. 39–54. Springer, Berlin.

Scholz, F., Gregorius, H.-R. & Rudin, D. (Eds) 1989. *Genetic Aspects of Air Pollutants in Forest Tree Populations*. Springer Verlag, Berlin.

Scholz, F., Timmann, T. & Krusche, D. 1979. Untersuchungen zur Variation der Resistenz gegen HF-Begasung bei *Picea abies* Familien. pp. 249–58. *Tagungsbericht X. IUFRO S2.09, Ljubljana 1973*, Mitteilungen des Institutes für Forst- und Holzwirtschaft Ljublana, Jugoslawien.

Schröder, G., Brown, J.W.S. & Schröder, J. 1988. Molecular analysis of resveratrol synthase: cDNA, genomic clones and relationship with chalcone synthase. *European Journal of Biochemistry*, 172, 161–9.

Schubert, R, Bahnweg, G., Nechwatal, J., Jung, T., Cooke, D.E.L., Duncan, J.M., Müller-Starck, G., Langebartels, C., Sandermann Jr., H. & Oßwald, W. 1999. Detection and quantification of *Phytophthora* species which are associated with root-rot diseases in European deciduous forests by species-specific PCR. *European Journal of Forest Pathology*, 29, 169–88.

Schubert, R., Ernst, D., Sandermann, H. & Müller-Starck, G. 1997a. Gene discovery in Norway spruce based on cDNA sequencing. In *Proceedings of a Joint Meeting of IUFRO Working Parties 2.04-07 and 2.04-06*. Quebec, Canada.

Schubert, R., Fischer, R., Hain, R., Schreier, P.H., Bahnweg, G., Ernst, D. & Sandermann Jr. H. 1997b. An ozone-responsive region of the grapevine resveratrol synthase promoter differs from the basal pathogen-responsive sequence. *Plant Molecular Biology*, 34, 417–26.

Schubert, R., Müller-Starck, G., Sandermann, Jr, H., Ernst, D. & Häger, K.-P. 1997c. The molecular structure and evolutionary relationships of a 16.9 kDa heat shock protein from Norway spruce (*Picea abies* (L.) Karst.). *Forest Genetics*, 4, 131–8.

Schubert, R., Sperisen, Ch., Müller-Starck, G., La Scala, S., Ernst, D., Sandermann Jr, H. & Häger, K.-P. 1998. The cinnamyl alcohol dehydrogenase gene structure in *Picea abies* (L.) Karst.: genomic sequences, Southern hybridization, genetic analysis and phylogenetic relationships. *Trees: Structure and Function*, 12, 453–63.

Schultz, A. 1892. Beitrage zur morphologie und biologie der bluthen II. *Berichte der Deutschen Botanischen Gesellschaft*, 10, 395–409.

Scott, J.M., Csuti, B. & Ciacco, S. 1991. Gap analysis: assessing protection needs. In *Landscape Linkages and Biodiversity*. Ed. W.E. Hudson. Island Press, Washington, DC.

Seavey, S.R. & Bawa, K.S. 1986. Late-acting self-incompatibility. *Botanical Review*, 52, 196–217.

Sedgley, M. & Griffin, A.R. 1989. *Sexual Reproduction of Tree Crops*. Academic Press, London.

Sekar, C., Vinaya Rai, R.S. & Ramasany, C. 1996. Role of minor forest products in tribal economy of India: a case study. *Journal of Tropical Forest Science*, 8, 280–88.

Sekhran, N. 1996. Pursuing the 'D' in Integrated Conservation and Development Projects (ICADPs): Issues and Challenges. Network Paper 19b. Rural Development Forestry Network. Overseas Development Institute, London.

Shaw, C.R. & Prasad, R. 1970. Starch gel electrophoresis of enzymes—a compilation of recipes. *Biochemical Genetics*, 4, 297–320.

Shaw, D.V. & Allard, R.W. 1982. Estimation of outcrossing rates in Douglas-fir using isozyme markers. *Theoretical and Applied Genetics*, 62, 113–20.

Shea, K.L. 1987. Effects of population structure and cone production on outcrossing rates in Engelmann spruce and subalpine fir. *Evolution*, 41, 124–36.

Shea, S.R. 1979. Forest management and *Phytophthora cinnamomi* in Australia. In *Phytophthora and Forest Management in Australia*. Ed. K.M. Old. pp. 73–100. CSIRO, Melbourne.

Shear, C.L. & Stevens, N.E. 1913. The chestnut-blight parasite (*Endothia parasitica*) from China. *Science*, 38, 295–7.

Shearer, B.L. & Tippett, J.T. 1989. Jarrah dieback: the dynamics and management of *Phytophthora cinnamomi* in the jarrah (*Eucalyptus marginata*) forest of south-western Australia. Research Bulletin No. 3, Department of Conservation and Land Management, Western Australia.

Sheffield, V.C., Cox, D.R., Lerman, L.S. & Myers, R.M. 1989. Attachment of a 40-base pair G+C-rich sequence (GC-clamp) to genomic DNA fragments by the polymerase chain reaction results in improved detection of single base changes. *Proceedings of the National Academy of Sciences, USA*, 86, 232–6.

Shen, H.H., Rudin, D. & Lindgren, D. 1981. Study of the pollination pattern in a Scots pine seed orchard by means of isozyme analysis. *Silvae Genetica*, 30, 7–15.

Shriver, M.D., Jin, L., Chakraborty, R. & Boerwinkle, E. 1993. VNTR allele frequency distributions under the stepwise mutation model: a computer simulation approach. *Genetics*, 134, 983–93.

Silen, R. & Osterhaus, C. 1979. Reduction of genetic base by sizing of bulked Douglas-fir seed lots. *Tree Planter's Notes*, 30, 24–30.

Silen, R.R. & Keane, G. 1969. Cooling a Douglas-fir seed orchard to avoid pollen contamination. Res. Note PNW–101. USDA Forestry Service. Portland, OR, USA.

Sim, B.L. 1984. The genetic base of *Acacia mangium* Willd. in Sabah. In *Provenance and Genetic Improvement Strategies in Tropical Forest Trees*. Eds R.D. Barnes & G.L. Gibson. pp. 597–603. Mutare, Zimbabwe, April 1984. Commonwealth Forestry Institute, Oxford & Forest Research Centre, Harare.

Simons, A.J. 1996. Ecology and reproductive biology. In Gliricidia sepium: *Genetic Resources for Farmers*. Eds J.L. Stewart, G.E. Allison & A.J. Simons. pp. 19–31.

Tropical Forestry Papers No. 33. Oxford Forestry Institute, Oxford.

Sinclair, W.A. & Campana, R.J. 1978. Dutch elm disease. Perspectives after 60 years. *Search*, 5, 1–52.

Skole, D. & Tucker, C. 1993. Tropical deforestation and habitat fragmentation in the Amazon: satellite data from 1978–1988. *Science*, 260, 1905–10.

Skroch, P. & Nienhuis, J. 1995. Impact of scoring error and reproducibility of RAPD data on RAPD-based estimates of genetic distance. *Theoretical and Applied Genetics*, 91, 1086–91.

Slatkin, M. 1980. The distribution of mutant alleles in a subdivided population. *Genetics*, 95, 503–24.

Slatkin, M. 1985. Rare alleles as indicators of gene flow. *Evolution*, 39, 53–65.

Slatkin, M. & Barton, N.H. 1989. A comparison of three indirect methods for estimating average levels of gene flow. *Evolution*, 43, 1349–68.

Smalley, E.B. & Guries, R.P. 1993. Breeding elms for resistance to Dutch elm disease. *Annual Review of Phytopathology*, 31, 325–52.

Smith, C.C., Hamrick, J.L. & Kramer, C.L. 1998. The effects of stand density on frequency of filled seeds and fecundity in lodgepole pine (*Pinus contorta* Dougl.). *Canadian Journal Forest Research*, 8, 453–60.

Smith, D.B. & Adams, W.T. 1983. Measuring pollen contamination in clonal seed orchards with the aid of genetic markers. In *Proceedings of the 17th Southern Forest Tree Improvement Conference*, 6–9 June 1983. pp. 69–77. Athens, Georgia, copies available from The National Technical Information Service, Springfield, VA, USA.

Smith, D.N. & Devey, M.E. 1994. Occurrence and inheritance of microsatellites in *Pinus radiata*. *Genome*, 37, 977–83.

Smith, L.L. & Ferlito, K. 1997. Monterey pine forest: a forest at risk. *Fremontia*, 25, 3–4.

Smith, P.J., Pressey, R.L. & Smith, J.E. 1993. Birds of particular conservation concern in the Western Division of NSW. *Biological Conservation*, 69, 315–38.

Smouse, P. 1986. The fitness consequences of multiple-locus heterozygosity under the multiplicative overdominance and inbreeding depression models. *Evolution*, 40, 946–57.

Smouse, P.E. & Neel, J.V. 1977. Multivariate analysis of gametic disequilibrium in the Yanomama. *Genetics*, 85, 733–52.

Sorensen, F.C. 1971. Estimate of self-fertility in coastal Douglas fir from inbreeding studies. *Silvae Genetica*, 20, 115–20.

Sorensen, F.C. 1994. Genetic variation and seed transfer guidelines for ponderosa pine in Central Oregon. Research Paper, PNW-RP-472. United States Department of Agriculture, Forest Service, Pacific Northwest Research Station, Portland, OR, USA.

Sorensen, F.C. & Miles, R.S. 1974. Self-pollination effects on Douglas-fir and ponderosa pine seeds and seedlings. *Silvae Genetica*, 23, 135–8.

Sorensen, F.C. & White, T.L. 1988. Effect of natural inbreeding on variance structure in tests of wind-pollination Douglas-fir progenies. *Forest Science*, 34, 102–18.

Sork, V.L., Campbell, D., Dyer, R., Fernandez, J., Nason, J., Remy, P., Smouse, P. & Steinberg, E. 1998. *Proceedings from a Workshop on Gene Flow in Fragmented, Managed, and Continuous Populations*. Research Paper No. 3. National Center for Ecological Analysis and Synthesis, Santa Barbara, CA, USA. Available at http://www.nceas.ucsb.edu/nceas-web/projects/2057/nceas-paper3/

Sork, V.L., Sork, K.A. & Hochwender, C. 1993. Evidence for local adaptation in closely adjacent subpopulations of northern red oak (*Quercus rubra* L.) expressed as resistance to leaf herbivores. *American Naturalist*, 142, 928–36.

Sparvoli, F., Martin, C., Scienza, A., Gavazzi, G. & Tonelli, C. 1994. Cloning and molecular analysis of structural genes involved in flavanoid and stilbene biosynthesis in grape (*Vitis vinifera* L.). *Plant Molecular Biology*, 24, 743–55.

Squillace, A.E. 1976. Analysis of monoterpenes of conifers by gas-liquid chromatography. In *Modern Methods in Forest Genetics*. Ed. J.P. Miksche. pp. 120–57. Springer-Verlag, Berlin.

Stace, C. 1997. *New Flora of the British Isles*. Cambridge University Press, Cambridge.

Stacy, E.A., Hamrick, J.L., Nason, J.D., Hubbell, S.P., Foster, R.B. & Condit, R. 1996. Pollen dispersal in low density populations of three neotropical tree species. *American Naturalist*, 148, 275–98.

Steane, D.A., Byrne, M., Vaillancourt, R.E. & Potts, B.M. 1998. Chloroplast DNA polymorphism signals complex interspecific interactions in *Eucalyptus* (Myrtaceae). *Australian Systematic Botany*, 11, 25–40.

Stearns, S.C. 1977. The evolution of life history traits: a critique of the theory and a review of the data. *Annual Review of Ecology and Systematics*, 8, 145–71.

Stearns, S.C. 1980. A new view of life history evolution. *Access*, 35, 266–81.

Stebbins, G.L., Jr. 1950. *Variation and Evolution in Plants*. Columbia University Press, New York.

Stebbins, G.L., Jr. 1959. The role of hybridization in evolution. *Proceedings of the American Philosophical Society*, 103, 231–51.

Steinberg, E.K. & Jordan, C.E. 1997. Using molecular genetics to learn about the ecology of threatened species: the allure and illusion of measuring genetic structure in natural populations. In *Conservation Biology for the Coming Decade*. Eds P.L. Fiedler & P.M. Kareiva. pp. 440–60. Chapman and Hall, New York.

Steiner, K.C. & Davis, D.D. 1979. Variation among *Fraxinus* families in foliar response to ozone. *Canadian Journal of Forest Research*, 9, 106–9.

Steiner, K.C., McCormick, L.H. & Canaverda, D.S. 1980. Differential response of paper birch provenances to aluminum in solution culture. *Canadian Journal of Forest Research*, 10, 25–9.

Steinkellner, H., Fluch, S., Turetschek, E., Lexer, C., Streiff, R., Kremer, A., Burg, K. & Glossl, J. 1997. Identification and characterization of (GA/CT)$_n$-microsatellite loci from *Quercus petraea*. *Plant Molecular Biology*, 33, 1093–6.

Stephenson, S.L. 1986. Changes in a former chestnut-dominated forest after a half century of succession. *The American Midland Naturalist*, 116, 173–9.

Stern, K. & Roche, L. 1974. *Genetics of Forest Ecosystems*. Chapman and Hall, London.

Stettler, R.F. & Bradshaw, H.D. 1994. The choice of genetic material for mechanistic studies of adaptation in forest trees. *Tree Physiology*, 14, 781–96.

Stewart, S.C. 1994. Simultaneous estimation of pollen contamination and pollen fertilities of individual trees in conifer seed orchards using multilocus genetic data. *Theoretical and Applied Genetics*, 88, 593–6.

Stoehr, M.U. & El-Kassaby, Y.A. 1997. Levels of genetic diversity at different stages of the domestication cycle of interior spruce in British Columbia. *Theoretical and Applied Genetics*, 94, 83–90.

Storer, A.J., Gordon, T.R., Dallara, P.L. & Wood, D.L. 1994. Pitch canker kills pines, spreads to new species and regions. *California Agriculture*, 48, 9–13.

Stork, N.E., Boyle, T.J.B., Dale, V., Eeley, H., Finegan, B., Lawes, M., Manokaran, N., Prabhu, R., & Soberon, J. 1997. *Criteria and Indicators for Assessing the Sustainability of Forest Management: Conservation of Biodiversity*. Working Paper No. 17, Center for International Forestry Research, Bogor, Indonesia.

Strauss, S.H. & Tsai, C.-H. 1988. Ribosomal gene number variability in Douglas-fir. *Journal of Heredity*, 79, 453–8.

Strobeck, C. 1983. The use of coefficients of identity to study random drift. *Lecture Notes on Biomathematics*, 52, 1–13.

Stukely, M.J.C. & Crane, C.E. 1994. Genetically based resistance of *Eucalyptus marginata* to *Phytophthora cinnamomi*. *Phytopathology*, 84, 650–6.

Styles, B.T. 1972. The flower biology of the Meliaceae and its bearing on tree breeding. *Silvae Genetica*, 21, 175–82.

Surles, S.E., Arnold, J., Schnabel, A., Hamrick, J.L. & Bongarten, B.C. 1990. Genetic relatedness in open-pollinated families of two leguminous tree species, *Robinia pseudoacacia* L. and *Gleditsia triacanthos* L. *Theoretical and Applied Genetics*, 80, 49–56.

Swanson, T. & Barbier, E. 1992. *Economics for the Wild: Wildlife, Wildlands, Diversity and Development*. Earthscan Publications, London.

Szmidt, A.E. & Muona, O. 1985. Genetic effects of Scots pine (*Pinus sylvestris* L.) domestication. In *Population Genetics in Forestry*. Ed. H.R. Gregorius. pp. 242–52. *Lecture Notes in Biomathematics* 60, Springer Verlag, Berlin.

Tacconi, L. & Bennett, J. (Eds). 1997. *Protected Area Assessment and Establishment in Vanuatu: A Socioeconomic Approach*. Australian Centre for International Agricultural Research, Canberra.

Tanner, E.V.J. 1982. Species diversity and reproductive mechanisms in Jamaican trees. *Biological Journal of the Linnean Society*, 18, 263–78.

Tarasofsky, RG. 1995. *The international forests regime: legal and policy issues*. IUCN (World Conservation Union) and World Wide Fund for Nature-International, Gland.

Taroda, N. & Gibbs, P.E. 1982. Floral biology and breeding system of *Sterculia chicha* St. Hiz. (Sterculiaceae). *New Phytologist*, 90, 735–43.

Tasmanian Public Land Use Commission 1997. *Consultation Report: Mechanisms for Achieving Conservation Management on Private Forested Land*. Inquiry into lands to be reserved under the Tasmania–Commonwealth Regional Forest Agreement. PLUC, Hobart.

Taylor, G.E. Jr. 1994. Role of genotype in response of loblolly pine to tropospheric ozone: effects at the whole-tree, stand, and regional level. *Journal of Environmental Quality*, 23, 63–84.

Temesgen, B., Neale, D.B. & Harry, D.E. 1998. Sequence-tagged site markers for loblolly pine (*Pinus taeda* L.) revealed by denaturing gradient gel electrophoresis. Abstract, Plant and Animal Genome VI Conference, San Diego, 18–22 January 1998. (http://probe.nalusda.gov:8300/pag/6/review/446.html.).

Templeton, A.R. 1986. Coadaptation and outbreeding depression. In *Conservation Biology: The Science of Scarcity and Diversity*. Ed. M.E. Soule. pp. 105–15. Sinauer, Sunderland, MA, USA.

Templeton, A.R. 1995. Biodiversity at the molecular genetic level: experiences from disparate macroorganisms. In *Biodiversity. Measurement and Estimation*. Ed. D.L. Hawksworth. pp. 59–64. Chapman and Hall, London.

ten Kate, K. 1995. *Biopiracy or Green Petroleum? Expectations and Best Practice in Bioprospecting*. Overseas Development Administration, London.

Thackway, R. & Cresswell, I.D. 1995. *An Interim Biogeographic Regionalisation for Australia*. Australian Nature Conservation Agency, Canberra.

Thang, H.C. 1987. Forest management systems of tropical high forest, with special reference to Peninsular Malaysia. *Forest Ecology and Management* , 21, 3–20.

Thang, H.C. 1988. *Selective Management System: Concept and Practice (Peninsular Malaysia)*. Forestry Department Headquarters, Kuala Lumpur.

Thomson, K. 1996. Global initiatives in forest policy: context setting. In *Making Forest Policy Work 1996*. Ed. K.L. Harris. pp. 1–13. Oxford Forestry Institute, Oxford.

Thomson, J.D. & Plowright, R.C. 1980. Pollen carryover, nectar rewards, and pollinator behavior with special reference to *Diervilla lonicera*. *Oecologia*, 46, 68–74.

Tien, L.V. 1994. Vietnam. In *Non Wood Forest Products in Asia*. Eds P.B. Durst, W. Ulrich & M. Kashio. pp. 151–61. RAPA Publication 1994/28. Food and Agriculture Organization of the United Nations, Bangkok.

Timmis, R., Flewelling, J. & Talbertk, C. 1994. Frost injury prediction model for Douglas-fir seedlings in the Pacific Northwest. *Tree Physiology*, 14, 855–69.

Townsend, A.M. & Dochinger, L.S. 1974. Relationship of seed source and developmental stage to the ozone tolerance of *Acer rubrum* seedlings. *Atmospheric Environment*, 8, 957–64.

Tsumura, Y., Suyama, Y., Yoshimura K., Shirato, N. & Mukai, Y. 1997. Sequence-tagged-sites (STSs) of cDNA clones in *Cryptomeria japonica* and their evaluation as molecular markers in conifers. *Theoretical and Applied Genetics*, 94, 764–72.

Tulsieram, L.K., Glaubitz, J.C., Kiss G., Carlson, J.E. 1992. Single-tree genetic linkage mapping in conifers using haploid DNA from megagametophytes. *Bio/technology*, 10, 686–90.

Tzschacksch, O. 1982. Untersuchungen zur Erblichkeit der SO$_2$-Resistenz bei Kiefer (*Pinus silvestris* L.) und Douglasie (*Pseudotsuga menziesii* Mirb./Franco) mit Schlußfolgerungen für die Forstwirtschaft. *Beiträge für die Forstwirtschaft*, 16, 103–6.

Uma Shanker, Murali, K.S., Uma Shaanker, R., Ganeshaiah, K.N. & Bawa, K.S. 1996. Extraction of non-timber forest products in the forests of Biligiri Rangan Hills, India. 3. Productivity, extraction and prospects of sustainable harvest of amla *Phyllanthus emblica* (Euphorbiaceae). *Economic Botany*, 50, 270–9.

Ungerer, M.C., Baird, S., Pan, J. & Rieseberg, L.H. 2000. Rapid hybrid speciation in wild sunflowers. *Proceedings of the National Academy of Sciences, USA*, 95, 11757–62.

United Kingdom Parliament. House of Commons. *Wildlife and Countryside Act, Bill No. 168, 1980/1981 Session*. (Elizabeth II, 1981). London, HMSO, 1981. (Readex Microfiche).

United Nations Environment Programme 1994. *Convention on Biological Diversity*. Secretariat to the Convention on Biological Diversity, Geneva.

University of British Columbia & Universiti Pertanian Malaysia 1996. *Economic, Social and Political Issues in Certification of Forest Management*, University of British Columbia, Vancouver.

Vallejos, A. 1983. Enzyme activity staining. In *Isozymes in Plant Genetics and Breeding*. Eds S.D. Tanksley & T.J. Orton. pp. 469–516. Elsevier, Amsterdam.

Vanclay, J.K. 1992. Species richness and productive forest management. In *Wise Management of Tropical Forests*. Proceedings of the Oxford Conference on Tropical Forests 1992, Eds F.R. Miller & K.L. Adam. pp. 1–9. Oxford Forestry Institute, University of Oxford, Oxford.

Vancura, K. 1993. The growth of different Norway spruce [*Picea abies* (L.) Karst.] provenances in the exposed air

polluted region of the Kruöné Hory mountains. *Lesnictví- Forestry*, 39, 87–93.

Vane-Wright, R.I., Humphries, C.J. & Williams, P.H. 1995. What to protect? Systematics and the agony of choice. *Biological Conservation*, 55, 235–54.

Vannini, A. & Scarascia Mugnozza, G. 1991. Water stress: a predisposing factor in the pathogenesis of *Hypoxylon mediterraneum* on *Quercus cerris*. *European Journal of Forest Pathology*, 21, 193–201.

Van Noordwijk, M., van Schaik, C.P., de Foresta, H. & Tomich, T.P. 1996. Segregate or integrate nature and agriculture for biodiversity conservation. Proceedings of the Global Biodiversity Forum, 1–2 November, Jakarta, Indonesia, IUCN (World Conservation Union), Gland.

Van Raamsdonk, L.W.D. & Schouten, H.J. 1997. Gene flow and establishment of transgenes in natural plant populations. *Acta Botanica Neerlandica*, 46, 69–84.

van Treuren, R., Bijlsma, R., Ouberg, N.J. & van Delden, W. 1993. The effects of population size and plant density on outcrossing rates in locally endangered *Salvia pratensis*. *Evolution*, 47, 1094–1104.

Varela, M.C. & Eriksson, G. 1995. Multipurpose gene conservation in *Quercus suber* —a Portuguse example. *Silvae Genetica*, 44, 28–37.

Vayda, A.P. 1996. Improving the Livelihoods of Forest-dependent People: How and Where to Do Strategic Social Science Research. Center for International Forestry Research, Bogor, Indonesia.

Vendramin, G.G., Lelli, L., Rossi, P. & Morgante, M. 1996. A set of primers for the amplification of 20 chloroplast microsatellites in *Pinaceae*. *Molecular Ecology*, 5, 595–8.

Venne, H., Scholz, F. & Vornweg, A. 1989. Effects of air pollutants on reproductive processes of poplar (*Populus* spp.) and Scots pine (*Pinus sylvestris* L). In *Genetic Effects of Air Pollutants in Forest Tree Populations*. Eds F. Scholz, H.-R. Gregorius & D. Rudin. pp. 89–103. Springer, Berlin.

Verlaat, E. 1995. We don't want a road, it will bring development. *BOS NiEuWsLetter* No. 31, 14, 46–57.

Villani, F., Benedettelli, S., Paciucci, M., Cherubini, M. & Pigliucci, M. 1991. Genetic variation and differentiation between natural populations of chestnut (*Castanea sativa* Mill.) from Italy. In *Biochemical Markers in the Population Genetics of Forest Trees*. Eds S. Fineschi, M.E. Malvolti, F. Cannata & H.H. Hattemer. pp. 91–103. SPB Academic Publishing, The Hague.

Villar, M., Lefevre, F., Bradshaw Jr, H.D. & Tessier du Cros, E. 1996. Molecular genetics of rust resistance in poplars (*Melampsora larici-populina* Kleb/*Populus* sp.) by bulked segregant analysis in a 2x2 factorial mating design. *Genetics*, 143, 531–6.

Visser, M.B.H. 1992. Environmental ethics—a case for survival. In *In Harmony with Nature*. Eds S.K. Yap & S.W. Lee. pp. 534–44. Malayan Nature Society, Kuala Lumpur.

Vos, P., Hogers, R., Bleeker, M., Reijans, M., van de Lee, T., Hornes, M., Friters, A., Pot, J., Peleman, J., Kuiper, M. & Zabeau, M. 1995. AFLP: a new technique for DNA fingerprinting. *Nucleic Acids Research*, 23, 4407–14.

Vyse, A.H. & Rudd, J.D. 1974. Sowing rules for container nurseries. In *Proceedings of the North American Containerized Forest Tree Seedling Symposium*. Eds R.W. Tinus, W.I. Stein & W.E. Balmer. Great Plains Agricultural Council Publication. p. 68. Denver, CO, USA.

Wade, M.J. & McCauley, D.E. 1988. Extinction and recolonisation: their effects on the genetic differentiation of local populations. *Evolution*, 42, 995–1005.

Wagner, D.B. 1992. Nuclear, chloroplast, and mitochondrial DNA polymorphisms as biochemical markers in population genetic analyses of forest trees. *New Forests*, 6, 373–90.

Wagner, W.L., Herbst, D.R. & Sohmer, S.H. 1990. *Manual of the Flowering Plants of Hawaii*. University of Hawaii Press and Bishop Museum Press, Honolulu.

Wahlund, S. 1928. Zusammensetzung von Populationen und Korrelationserscheinungen von Standpunkt der Vererbungslehre aus betrachtet. *Hereditas*, 11, 65–106.

Wallace, B. 1968. *Topics in Population Genetics*. W.W. Norton, New York.

Wallace, R.D. 1984. New blight-resistant hybrid chestnut. *American Horticulturist*, 63, 4–5.

Walter, M.H., Grima-Pettenati, J., Grand, C., Boudet,A.M. & Lamb, C.J. 1988. Cinnamyl- alcohol dehydrogenase, a molecular marker specific for lignin synthesis: cDNA cloning and mRNA induction by fungal elicitor. *Proceedings of the National Academy of Sciences, USA*, 85, 5546–50.

Wang, D.G., Fan, J.B., Siao, C.J. et al. (27 others) 1998. Large-scale identification, mapping, and genotyping of single-nucleotide polymorphisms in the human genome. *Science*, 280, 1077–82.

Wang, X., Lindgren, D., Szmidt, A.E. & Yazdani, R. 1991. Pollen migration into a seed orchard of *Pinus sylvestris* L. and the methods of its estimation using allozyme markers. *Scandinavian Journal of Forest Research*, 6, 379–85.

Waser, N.M. & Price, M.V. 1985. Reciprocal transplant experiments with *Delphinium nelsonii* (Ranunculaceae): evidence for local adaptation. *American Journal of Botany*, 72, 1726–32.

Watano, Y., Imaszu, M. & Shimizu, T. 1995. Chloroplast DNA typing by PCR-SSCP in the *Pinus pumila-P. parviflora* var. *pentaphylla* complex (Pinaceae). *Journal of Plant Research*, 108, 493–9.

Watkins, L. & Levin, D.A. 1990. Outcrossing rates as related to plant density in *Phlox drummondii*. *Heredity*, 65, 81–9.

Wayne, R.K. & Jenks, S.M. 1991. Mitochondrial DNA analysis implying extensive hybridization of the endangered red wolf *Canis rufus*. *Nature*, 351, 565–8.

Webber, J.E. & Painter, R.A. 1996. *Douglas-fir Pollen Management Manual*. 2nd edn. Working Paper 02. Research Branch, Ministry of Forests, Victoria, BC, Canada.

Weber, J.L. & May, P.E. 1989. Abundant class of human DNA polymorphism which can be typed using the polymerase chain reaction. *American Journal of Human Genetics*, 44, 388–96.

Weinberg, W. 1908. Uber den Nachweis der Vererbung beim Menschen. Jahresh. Verein f. vaterl. Naturk. *Württemberg*, 64, 368–82.

Weinstein, L.H. 1977. Fluoride and plant life. *Journal of Occupational Medicine*, 19, 49–78.

Weir, R.J. 1977. *Genetic variation in loblolly pine (Pinus taeda) tolerance to ozone*. PhD thesis, North Carolina State University, Raleigh, NC, USA.

Weiss, E.A. 1997. *Essential Oil Crops*. CAB International, Wallingford, UK.

Wellburn, F.A.M, Lau, K.-K., Milling, P.M.K. & Wellburn, A.R. 1996. Drought and air pollution affect nitrogen cycling and free radical scavenging in *Pinus halepensis* (Mill.). *Journal of Experimental Botany*, 47, 1361–7.

Wells, M. 1992. Biodiversity conservation, affluence and poverty: mismatched costs and benefits and efforts to remedy them. *Ambio*, 21, 237–43.

Wells, M.P. 1994/95. *Biodiversity Conservation and Local Peoples' Development Aspirations: New Priorities for the 1990s*. Network Paper 18a. Rural Development Forestry Network. Overseas Development Institute, London.

Wells, M.P. 1995. Community-based forestry and biodiversity projects have promised more than they have delivered. Why is this and what can be done? In *Management of Tropical Forest: Towards an Integrated Perspective*. Ed. Ø. Sandbukt. pp. 269–86. Centre for Development and Environment, University of Oslo, Oslo.

Wells, M.P. & Brandon, K.E. 1993. The principles and practices of buffer zones and local participation in biodiversity conservation. *Ambio*, 22, 157–62.

Wells, O.O. & Wakeley, P.C. 1966. Geographic variation in survival, growth and fusiform rust infection of planted loblolly pine. *Forest Science Monograph*, 11, 1–40.

Wendel, J.F. & Weeden, N.F. 1989. Visualisation and interpretation of plant isozymes. In *Isozymes in Plant Biology*. Eds D.E. Soltis & P.S. Soltis. pp. 5–45. Chapman and Hall, London.

West, R.F. 1988. Tolling the chestnut. *American Forests*, 94, 10 and 76–7.

Weste, G. 1981. Vegetation changes as a result of sclerophyll shrubby woodland associated with invasion by *Phytophthora cinnamomi*. *Australian Journal of Botany*, 29, 261–76.

Western, D. 1984. Amboseli National Park: human values and the conservation of a savanna ecosystem. In *National Parks, Conservation, and Development: The Role of Protected Areas in Sustaining Society*.

Eds J.A. McNeely & K.R. Miller. pp. 93–100. Smithsonian Institution Press, Washington, DC.

Western, D. & Wright, R.M. 1994. The background to community-based conservation. In *Natural Connections: Perspectives in Community-based Conservation*. Eds D. Western, R.M. Wright & S.C. Strum. pp. 1–12. Island Press, Boulder, CO, USA.

Western, D., Wright, R.M. & Strum, S.C. (Eds) 1994. *Natural Connections: Perspectives in Community-based Conservation*. Island Press, Boulder, CO, USA.

Westman, W.E. 1977. How much are nature's services worth? *Science*, 197, 960–4.

Westoby, J.C. 1989. *Introduction to World Forestry*. Blackwell, Oxford.

Wheeler, N.C. & Bramlett, D.L. 1991. Flower stimulation treatments in a loblolly pine seed orchard. *Southern Journal of Applied Forestry*, 15, 44–50.

Wheeler, N.C. & Guries, R.P. 1982. Population structure, genetic diversity, and morphological variation in *Pinus contorta* Dougl. *Canadian Journal of Forest Research*, 12, 595–606.

Wheeler, N. & Jech, K. 1986. Pollen contamination in a mature Douglas-fir seed orchard. In *Proceedings, IUFRO Conference on Breeding Theory, Progeny Testing and Seed Orchards*. pp. 160–71, Williamsburg, VA, USA.

White, E.E. 1990. Chloroplast DNA in *Pinus monticola* 2. Surveys of within-species variability and detection of heteroplasmic individuals. *Theoretical and Applied Genetics*, 79, 251–5.

White, G.M. & Boshier, D.H. 2000. Fragmentation in Central American dry forests—genetic impacts on *Swietenia humilis*. In *Genetics, Demography and the Viability of Fragmented Populations*. Eds A.G. Young & G. Clarke. Cambridge University Press, Cambridge.

White, G. & Powell, W. 1997. Isolation and characterization of microsatellite loci in *Swietenia humilis* (Meliaciea)—an endangered tropical hardwood species. *Molecular Ecology*, 6, 851–60.

White, T.L. 1987. Drought tolerance of southwestern Oregon Douglas-fir. *Forest Science*, 33, 283–93.

Whitemore, A.T. & Schaal, B.A. 1991. Interspecific gene flow in sympatric oaks. *Proceedings of the National Academy of Sciences, USA*, 88, 2540–4.

Whitham, T.G. & Maschinski, J. 1996. Current hybrid policy and the importance of hybrid plants in conservation. In *Southwestern Rare and Endangered Plants: Proceedings of the Second Conference*. Eds J. Maschinski & D. Hammond. pp. 103–12. USDA Forest Service, Rocky Mountain Forest and Ranger Expt. Station, Ft Collins, CO, USA.

Whitham, T.G., Morrow, P.A. & Potts, B.M. 1991. Conservation of hybrid plants. *Science*, 254, 779–80.

Whitham, T.G., Morrow, P.A. & Potts, B.M. 1994. Plant hybrid zones as centers of biodiversity: the herbivore community of two endemic Tasmanian eucalypts. *Oecologia*, 97, 481–90.

Whitlock, M.C. 1992. Temporal fluctuations in demographic parameters and the genetic variance among populations. *Evolution*, 46, 608–15.

Whitlock, M.C. & Barton, N.H. 1997. The effective size of a subdivided population. *Genetics*, 146, 427–41.

Whitlock, M.C. & McCauley, D.E. 1990. Some population genetic consequences of colony formation and extinction: genetic correlations within founding groups. *Evolution*, 44, 1717–24.

Whitton, J., Wolf, D.E., Arias, D.M., Snow, A.A. & Rieseberg, L.H. 1997. The persistence of cultivar alleles in wild populations of sunflowers five generations after hybridization. *Theoretical and Applied Genetics*, 95, 33–40.

Wickneswari, R. & Norwati, M. 1993. Genetic diversity of natural populations of *Acacia auriculiformis*. *Australian Journal of Botany*, 41, 65–77.

Wickneswari, R., Lee, C.T., Norwati, M. & Boyle, T.J.B. 1997a. Immediate effects of logging on the genetic diversity of five tropical rainforest species in a ridge forest in Peninsular Malaysia. Paper presented at CIFOR Wrap-Up Workshop on Impact of Disturbance, Bangalore, India, August 1997. Center for International Forestry Research, Bogor, Indonesia.

Wickneswari, R., Lee, C.T., Norwati, M. & Boyle, T.J.B. 1997b. Effects of logging on the genetic diversity of six tropical rainforest species in a regenerated mixed dipterocarp lowland forest in Peninsular Malaysia. Paper presented at CIFOR Wrap-Up Workshop on Impact of Disturbance, Bangalore, India, August 1997. Center for International Forestry Research, Bogor, Indonesia.

Widen, B. 1993. Demographic and genetic effects on reproduction as related to population size in a rare, perennial herb, *Senecio integrifolius* (Asteraceae). *Biological Journal of the Linnean Society*, 50, 179–95.

Widen, B. & Andersson, S. 1993. Quantitative genetics of life-history and morphology in a rare plant, *Senecio integrifolius*. *Heredity*, 70, 503–14.

Wiens, D. 1984. Ovule survivorship, brood size, life history, breeding systems, and reproductive success in plants. *Oecologia*, 64, 47–53.

Wiens, J. 1997. Metapopulation dynamics and landscape ecology. In *Metapopulation Biology, Ecology, Genetics, and Evolution*. Eds I. Hanski & M.E. Gilpin. pp. 43–62. Academic Press, San Diego, CA, USA.

Wijewardana, D., Caswell, S.J. & Palmberg-Lerche, C. 1997. Criteria and indicators of sustainable forest management. In *Proceedings, XI World Forestry Congress*. Antalya, Turkey, October 1997, 6, 3–17.

Wildt, D.E., Bush, M., Goodrowe, K.L., Packer, C., Pusey, A.E., Brown, J.L., Joslin, P. & O'Brien, S.J. 1987. Reproductive and genetic consequences of founding isolated lion populations. *Nature*, 329, 328–31.

Wilkie, D.S. & Godoy, R.A. 1996. Trade, indigenous rain forest economies and biological diversity. In *Current Issues in Non-Timber Forest Products Research*. Eds M. Ruiz Pérez & J.E.M. Arnold. pp. 83–102.

Center for International Forestry Research, Bogor, Indonesia.

Willan, R.L. 1985. *A Guide to Forest Seed Handling, with Special Reference to the Tropics*. FAO Forestry Paper 20/2, FAO, Rome.

Williams, C.G. & Savolainen, O. 1996. Inbreeding depression in conifers: implications for breeding strategy. *Forest Science*, 42, 102–17.

Williams, C.G., Hamrick, J.L. & Lewis, P.O. 1995. Multiple-population versus hierarchical breeding populations: a comparison of genetic diversity levels. *Theoretical and Applied Genetics*, 90, 584–94.

Williams, J.G.K., Kubelik, A.R., Livak, K.J., Rafalski, J.A. & Tingey, S.V. 1990. DNA polymorphisms amplified by arbitrary primers are useful as genetic markers. *Nucleic Acids Research*, 18, 6531–5.

Williams, S.B. 1984. Protection of plant varieties and parts as intellectual property. *Science*, 225, 18–23.

Wilson, B.C. 1990. Gene-pool reserves of Douglas-fir. *Forest Ecology and Management*, 35, 121–30.

Wink, L. 1984. Chestnut blight report. *American Horticulturist*, 63, 6–7.

Winner, W.E. & Mooney, H.A. 1980a. Ecology of SO_2 resistance: II. Photosynthetic changes of shrubs in relation to SO_2 absorption and stomatal behavior. *Oecologia*, 44, 296–302.

Winner, W.E. & Mooney, H.A. 1980b. Ecology of SO_2 resistance: III. Metabolic changes of C3 and C4 *Atriplex* species due to SO_2 fumigation. *Oecologia*, 46, 49–54.

Winner, W.E., Cotter, I.S., Powers, H.R. & Skelly, J.M. 1987. Screening loblolly pine seedlings responses to SO_2 and O_3: analysis of families differing in resistance to fusiform rust disease. *Environmental Pollution*, 47, 205–20.

Wiselogel, A.E. 1986. Pollen contamination in a superior loblolly pine seed orchard. In *Proceedings of the Ninth North American Forest Biology Workshop*. Eds C.G. Tauer & T.C. Hennessey. pp. 274–8. Oklahoma State University, Stillwater, OK, USA.

Wiselogel, A.E., Bailey, J.K., Newton, R.J. & Fong, F. 1991. Growth of loblolly pine (*Pinus taeda* L.) seedlings to ozone fumigation. *Environmental Pollution*, 71, 43–56.

Wittig, R. 1993. General aspects of biomonitoring. In *Plants as Biomonitors*. Ed. B. Markert. pp. 3–27. Verlag VCH, Weinheim.

Wood, D. 1995. Conserved to death—are tropical forests being overprotected from people. *Land Use Policy*, 12, 115–35.

Woods, F.W. & Shanks, R.E. 1959. Natural replacement of chestnut by other species in the Great Smoky Mountains National Park. *Ecology*, 40, 349–61.

World Commission on Environment and Development. 1987. *Our Common Future*. Oxford University Press, Oxford.

World Commission on Protected Areas/IUCN 1997. *Protected Areas in the 21st Century: From Islands To Networks*. Proceedings of a WCPA/IUCN Meeting, Albany, Western Australia, 23–29 November 1997.

World Conservation Monitoring Centre 1998. http://www.wcmc.org.uk/forest (accessed November 1998).

World Conservation Union (IUCN)/World Wide Fund for Nature (WWF) 1998. *Protected Areas for a New Millennium*. Discussion Paper, World Wide Fund for Nature-International and IUCN (World Conservation Union), Gland.

World Wide Fund for Nature (WWF)/IUCN (World Conservation Union) 1996. *Forests for Life*. World Wide Fund for Nature-International and IUCN (World Conservation Union), Gland.

Worrall, J. 1983. Temperature-bud-burst relationships in amabilis and subalpine fir provenance tests replicated at different elevations. *Silvae Genetica*, 32, 203–9.

Wright, S. 1931. Evolution in Mendelian populations. *Genetics*, 16, 97–159.

Wright, S. 1938. Size of population and breeding structure in relation to evolution. *Science*, 87, 430–1.

Wright, S. 1940. The statistical consequences of Mendelian heredity in relation to speciation. In *The New Systematics*. Ed. J.S. Huxley. pp. 161–83. Oxford University Press, London.

Wright, S. 1942. Statistical genetics and evolution. *Bulletin of the American Mathematical Society*, 48, 223–46.

Wright, S. 1943. Isolation by distance. *Genetics*, 28, 114–38.

Wright, S. 1946. Isolation by distance under diverse systems of mating. *Genetics*, 31, 39–59.

Wright, S. 1951. The genetical stucture of populations. *Annals of Eugenics*, 15, 323–54.

Wright, S. 1965. the interpretation of population structure by F-statistics with special regard to systems of mating. *Evolution*, 19, 395–420.

Wright, S. 1978. *Evolution and the Genetics of Populations. Vol. 4. Variability Within and Among Natural Populations*. University of Chicago Press, Chicago.

Wyatt-Smith, J. 1963. *Manual of Malayan Silviculture for Inland Forests*. Volume 1. Malayan Forest Records No. 23. Forest Research Institute, Kepong, Malaysia.

Wyatt-Smith, J. & Foenander, E.C. 1962. Damage to regeneration as a result of logging. *Malaysian Forester*, XXV (1), 40–4.

Xie, C.Y. & Knowles, P. 1992. Associations between allozyme phenotypes and soil nutrients in a natural population of jack pine (*Pinus banksiana*). *Biochemical Systematics and Ecology*, 20, 179–85.

Xie, C.Y., Yeh, F.C., Dancik, B.P. & Strobeck, C. 1991. Joint estimation of immigration and mating system parameters in gymnosperms using the EM algorithm. *Theoretical and Applied Genetics*, 83, 137–40.

Yang, R.-C. & Yeh, F.C. 1992. Genetic consequences of *in situ* and *ex situ* conservation of forest trees. *Forestry Chronicle*, 68, 720–9.

Yang, R.-C., Yeh, F.C. & Yanchuk, A.D. 1996. A comparison of isozyme and quantitative genetic variation in *Pinus contorta* ssp. *latifolia* by Fst. *Genetics*, 142, 1045–52.

Yap, S.K. 1980. Phenological behaviour of some fruit tree species in a lowland dipterocarp forest of West Malaysia. In *Tropical Ecology and Development*. Proceedings of V International Symposium on Tropical Ecology. Ed. J.L. Furtado. pp. 161–7. International Society of Tropical Ecology, Kuala Lumpur.

Yap, S.K. & Chan, H.T. 1990. Phenological behaviour of some *Shorea* species in Peninsula Malaysia. In *Reproductive Ecology of Tropical Forest Plants*. Eds K.S. Bawa & M. Hadley. pp. 21–36. Man and the Biosphere Series, vol. 7. Unesco/Parthenon Publishing, Paris, Carnforth.

Yazdani, R. & Lindgren, D. 1991. Variation of pollen contamination in a Scots pine seed orchard. *Silvae Genetica*, 40, 243–6.

Yeh, F.C.H. 1989. Isozyme analysis for revealing population structure for use in breeding strategies. In *Breeding Tropical Trees—Population Structure and Genetic Improvement Strategy in Clonal and Seedling Forestry*. Proceedings of IUFRO Conference, Pattaya, Thailand, November 1988. Eds G.L. Gibson, A.R. Griffin & A.C. Matheson. pp. 119–31. Oxford Forestry Institute, Oxford, UK, & Winrock International, Arlington, VA, USA.

Yeh F.C. & Layton, C. 1979. The organization of genetic variability in central and marginal populations of lodgepole pine *Pinus contorta* spp. *latifolia*. *Canadian Journal of Genetics and Cytology*, 21, 487–503.

Yeh, F.C., Cheliak, W.M., Dancik, B.P., Illingworth, K., Trust, D.C. & Pryhitka, B.A. 1985. Population differentiation in lodgepole pine, *Pinus contorta* spp. *latifolia*: a discriminant analysis of allozyme variation. *Canadian Journal of Genetics and Cytology*, 27, 210–18.

Yeh, F.C., Shi, J.S, Yang, R.C., Hong, J.S. & Ye, Z.H. 1994. Genetic diversity and multilocus associations in *Cunninghamia lanceolata* (Lamb.) Hook from People's Republic of China. *Theoretical and Applied Genetics*, 88, 465–71.

Young, A.G. & Brown, A.H.D. 1999. Paternal bottlenecks in fragmented populations of the endangered grassland daisy *Rutidosis leptorrhynchoides*. *Genetical Research*, 73, 111–17.

Young, A. & Mitchell, N. 1993. Microclimate and vegetation edge effects in a fragmented podocarp-broadleaf forest in New Zealand. *Biological Conservation*, 67, 63–72.

Young, A., Boyle, T. & Brown, T. 1996. The population genetic consequences of habitat fragmentation for plants. *Trends in Ecology and Evolution*, 11, 413–18.

Young, A.G., Brown, A.H.D. & Zich, F.C. 1999. Genetic structure of fragmented populations of the endangered grassland daisy *Rutidosis leptorrhynchoides*. *Conservation Biology*, 13, 256–65.

Young, A.G., Merriam, H.G. & Warwick, S.I. 1993a. The effects of forest fragmentation on genetic variation in *Acer saccharum* Marsh. (sugar maple) populations. *Heredity*, 71, 277–89.

Young, A.G., Warwick, S.I. & Merriam, H.G. 1993b. Genetic variation and structure at three spatial scales for *Acer saccharum* (sugar maple) in Canada and the implications for conservation. *Canadian Journal of Forest Research*, 23, 2568–78.

Yu, Y.G., Buss, G.R. & Saghai-Maroof, M.A. 1996. Isolation of a superfamily of candidate disease-resistance genes in soybean based on a conserved nucleotide-binding site. *Proceedings of the National Academy of Sciences, USA*, 93, 11751–6.

Zapata, T.R. & Arroyo, M.T.K. 1978. Plant reproductive ecology of a secondary deciduous tropical forest in Venezuela. *Biotropica*, 10, 221–30.

Zeevart, J.A.D. & Creelman, R.A. 1988. Metabolism and physiology of abscisic acid. *Annual Review of Plant Physiology and Plant Molecular Biology*, 39, 439–73.

Ziehe, M. & Hattemer, H.H. 1989. Genetische Variation und Züchtung von Waldbäumen. *Allgemeine Forst- und Jagdzeitung*, 159, 88–92.

Ziehe, M., Hattemer, H.H., Müller-Starck, R. & Müller-Starck, G. 1999. Genetic structures as indicators for adaptation and adaptational potentials. In *Forest Genetics and Sustainability*. Ed. C. Mátyás. pp. 75–89. Kluwer Academic Publishers, Dordrecht.

Zinser, C., Ernst, D. & Sandermann Jr., H. 1998. Induction of stilbene synthase and cinnamyl alcohol dehydrogenase mRNAs in Scots pine (*Pinus sylvestris* L.) seedlings. *Planta*, 204, 169–76.

Zobel, B.J. 1982. Developing fusiform-resistant trees in the south-eastern United States. In *Resistance to Disease and Pests in Forest Trees*. Eds H.M. Heybroek, B.R. Stephan & K. von Weissenberg. pp. 417–26. Centre for Agricultural Publishing and Documentation, Wageningen, Netherlands.

Zobel, B.J. & Talbert, J.T. 1984. *Applied Forest Tree Improvement*. John Wiley and Sons, New York.

Zobel, B.J. & van Buijtenen, J.P. 1989. *Wood Variation: Its Cause and Control*. Springer-Verlag, New York.

Zube, E.H. & Busch, M.L. 1990. Park–people relationships: an international review. *Landscape and Urban Planning*, 19, 117–31.

Zulkifli, Y. & Anhar, S. 1994. Effects of selective logging methods on suspended solids concentration and turbidity level in streamwater. *Journal of Tropical Forest Science*, 7, 199–219.